Microfluidics and Bio-MEMS

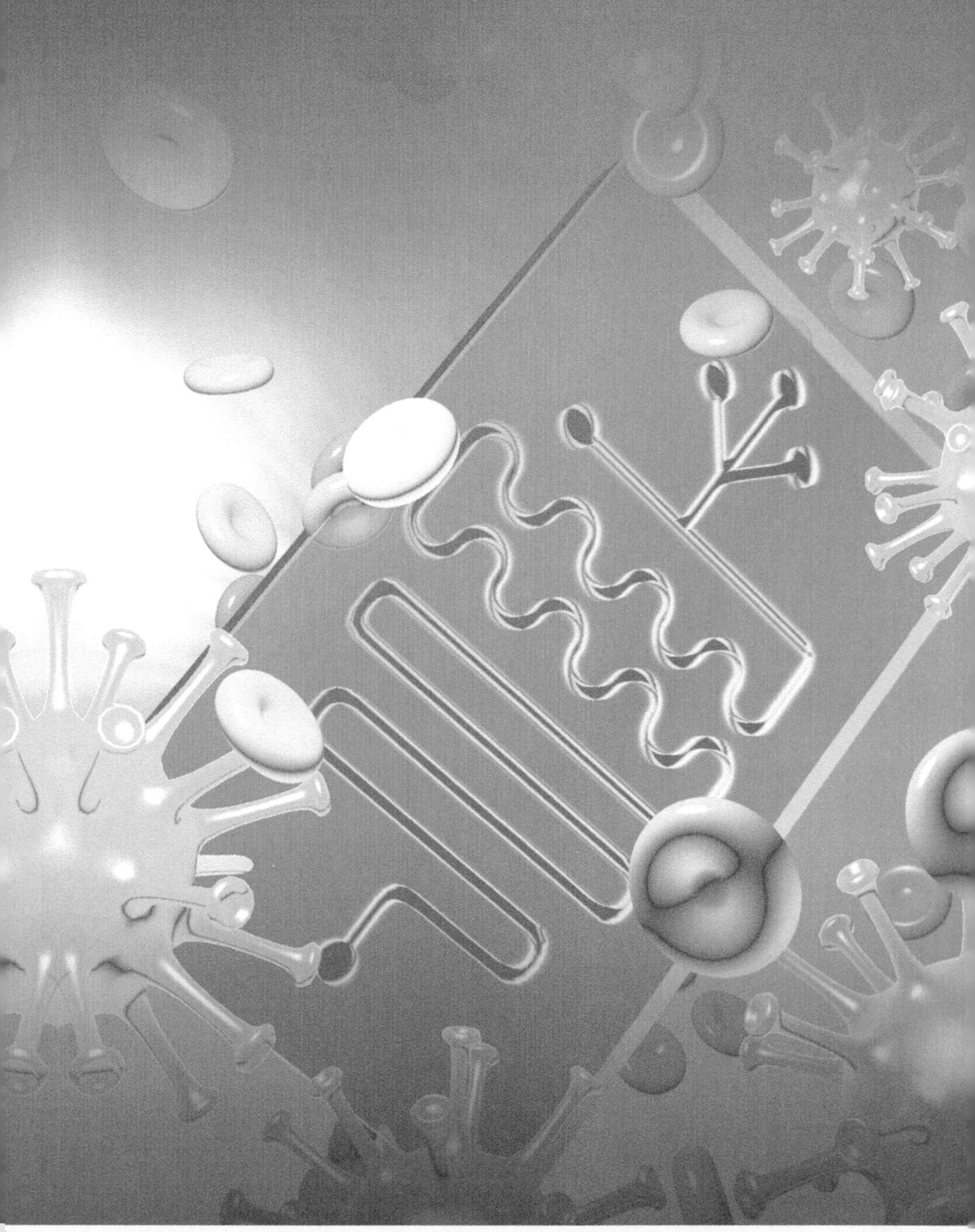

Microfluidics and Bio-MEMS

Devices and Applications

edited by

Tuhin S. Santra

JENNY STANFORD
PUBLISHING

Published by

Jenny Stanford Publishing Pte. Ltd.
Level 34, Centennial Tower
3 Temasek Avenue
Singapore 039190

Email: editorial@jennystanford.com
Web: www.jennystanford.com

British Library Cataloguing-in-Publication Data
A catalogue record for this book is available from the British Library.

Microfluidics and Bio-MEMS: Devices and Applications

ISBN 978-981-4800-85-3 (Hardcover)
ISBN 978-1-003-01493-5 (eBook)

Contents

Preface xv

1 Microfluidic Technologies for Cell Manipulation, Therapeutics, and Analysis **1**
Amogh Kumar, Pallavi Shinde, Loganathan Mohan, Pallab Sinha Mahapatra, and Tuhin S. Santra
1.1 Introduction 2
1.2 Microfluidic Cell Capture Techniques 3
1.2.1 Microdroplet-Based Cell Trapping 4
1.2.2 Cell Trapping through Microarray Devices 9
1.2.3 Cell Trapping through Hydrodynamic Systems and Microvortices 11
1.2.4 Miscellaneous Techniques for Cell Trapping 13
1.3 Microfluidic Single-Cell Therapy and Analysis 17
1.3.1 Electroporation 17
1.3.2 Mechanoporation 21
1.3.3 Optoporation 22
1.4 Microfluidic Cell Diagnosis and Analysis 24
1.4.1 Cell Diagnosis 24
1.4.2 Cell Analysis 28
1.4.2.1 Droplet-based analytical techniques 28
1.4.2.2 Microfluidic devices for massively parallel cell analysis 32
1.4.2.3 Miscellaneous analysis employing microfluidic devices 35
1.5 Future Prospects 38
1.6 Conclusions 39

2 Optical Manipulation of Cells **49**
Srabani Kar, Pallavi Shinde, Moeto Nagai, and Tuhin S. Santra
2.1 Introduction 50
2.2 Photodynamic and Electrokinetic Phenomena 51
2.2.1 Dielectrophoresis 51
2.2.2 AC Electro-osmosis 53
2.2.3 Electrothermal Effect 54
2.3 Optical Trapping 56
2.3.1 Working Principle 56
2.3.2 Configurations of Optical Traps 59
2.3.2.1 Single-beam optical trap 60
2.3.2.2 Dual-beam optical trap 60
2.3.2.3 Multiple optical traps 61
2.3.3 Applications of Optical Trapping in Biology 63
2.3.3.1 Cell manipulation 63
2.3.3.2 Studies of cell-to-cell interactions 65
2.3.3.3 Robot-tweezer manipulation system 66
2.3.3.4 Manipulation of subcellular organelles 66
2.4 Optoelectronic Tweezers 67
2.4.1 Device Design, Working Principle, and Developments 68
2.4.2 Applications of an Optoelectronic Tweezer 73
2.4.2.1 Cell lysis 73
2.4.2.2 Cell trapping, alignment, and patterning 74
2.4.2.3 Electroporation 74
2.4.2.4 Cell selection, identification, and separation 75
2.4.2.5 Cellular intrinsic properties 76
2.5 Rapid Electrokinetic Patterning 77
2.5.1 Setup and Working Principle 78
2.5.2 Biological Applications of REP 81
2.5.2.1 Patterning, translating, and sorting 81
2.6 Conclusions 83

3 Micro-Robots/Microswimmers for Biomedical Applications **95**
Naveen Kumar Agrawal, Pallab Sinha Mahapatra, and Tuhin S. Santra
3.1 Introduction 96
3.2 Propulsion Mechanism 98
3.2.1 Magnetic Propulsion 99
3.2.2 Bubble Propulsion 105
3.2.3 Biological Propulsion 109
3.2.4 Self-Thermophoresis 113
3.3 Materials and Fabrication Techniques 114
3.3.1 Tubular Micro-Robots 115
3.3.1.1 Rolled-up technology 116
3.3.1.2 Template synthesis 118
3.3.2 Helical Micro-Robots 119
3.3.3 Flexible Tail Micro-Robots 122
3.3.4 Janus Micro-Robots 124
3.4 Biomedical Applications 125
3.4.1 Delivery 127
3.4.2 Surgery 130
3.4.3 Sensing and Diagnosis 132
3.5 Discussion and Future Scope 136

4 Microfluidics in Neuroscience **149**
Pallavi Gupta, Nandhini Balasubramaniam, Kiran Kaladharan, Fan-Gang Tseng, Moeto Nagai, Hwan-You Chang, and Tuhin S. Santra
4.1 Introduction 150
4.2 Traditional Microfluidic Devices 152
4.3 Current Approaches 153
4.3.1 Compartmentalized Microfluidics 154
4.3.2 Single-Cell Monitoring 156
4.3.3 Co-culturing 162
4.3.3.1 Types of co-culture systems 162
4.3.3.2 Blood–brain barrier 163
4.3.3.3 Co-culturing of neural cells 164
4.3.4 Integrated Microfluidic/Microelectrode Array 166
4.3.5 Hydrogel Gradients 169

4.4 Applications 171
4.4.1 Neuron Differentiation and Polarity 172
4.4.1.1 Axon guidance 173
4.4.1.2 Patterned substrates 173
4.4.2 Biochemical Gradients 175
4.4.3 Electrophysiological Recordings 177
4.4.4 Dendritic Signaling and Synapse Formation 180
4.4.5 Developmental Study at Cell Population/Tissue/Organ-on-a-Chip (Brain-on-a-Chip) Level 184
4.4.6 Neurodegenerative Studies 188
4.5 Future Prospects 194
4.6 Summary 196

5 Vascularized Microfluidic Organ on a Chip and Its Applications 213
Qiyue Sun, Jianghua Pei, Qinyu Li, and Xiaolin Wang
5.1 Introduction 214
5.2 In vitro Vascularization Strategies 215
5.2.1 EC Lining–Based Methods 216
5.2.1.1 Microneedle-based removable method 217
5.2.1.2 Micropatterned planar hydrogel slab bonding method 218
5.2.1.3 Dissolvable material–based sacrificial micromolding method 218
5.2.1.4 EC lining inside a PDMS-based microfluidic channel 219
5.2.2 Vasculogenesis- and Angiogenesis-Based Methods 221
5.2.2.1 Vasculogenesis 221
5.2.2.2 Angiogenesis 222
5.2.2.3 Hybrid methods 223
5.3 Vascular-Inducing Factors 226
5.3.1 Biomechanical Factors 226
5.3.2 Extracellular (or Diffusible) Signaling Molecules 228
5.3.3 Cell Source and Cell–Cell Interaction 230
5.4 Selective Vascular Barrier 230

5.5 Application of Engineered Microvascular Networks to Cancer Biology 234
5.5.1 Tumor Angiogenesis 235
5.5.2 Tumor Intravasation 237
5.5.3 Tumor Extravasation 237
5.5.4 Tumor Microenvironment 238
5.5.5 Application of Vascularized Tumor on a Chip 239
5.5.5.1 Anticancer drug screening 239
5.5.5.2 Different tumor therapies 240
5.6 Conclusions and Future Perspectives 241

6 DNA Gene Microarray Biochip and Applications 255
Jian-Chiun Liou
6.1 Introduction 256
6.2 Combining Nanotechnology's Biochips 259
6.3 Simulation of the Injection Performance of a Single-Channel Injection Chamber 260
6.4 Experiment and Results 266
6.5 Conclusions 271

7 Microneedles: Current Trends and Applications 275
Hima Manoj, Pallavi Gupta, Loganathan Mohan, Moeto Nagai, Syrpailyne Wankhar, and Tuhin S. Santra
7.1 Introduction 276
7.2 History of Microneedles 278
7.3 Mechanism of Drug Delivery via Microneedles 279
7.4 Types of Microneedles 283
7.4.1 Solid Microneedles 283
7.4.1.1 Solid, durable microneedles 283
7.4.1.2 Solid, degradable microneedles 283
7.4.2 Hollow Microneedles 284
7.4.3 Polymer Microneedles 287
7.4.3.1 Dissolving microneedles 288
7.4.3.2 Biodegradable polymers 291
7.4.3.3 Swellable polymers 293
7.5 Microneedle Material and Its Properties 295
7.5.1 Silicon 295
7.5.2 Metal 295

7.5.3 Ceramic 295
7.5.4 Silica Glass 297
7.5.5 Carbohydrate 297
7.5.6 Polymer 297
7.6 Microneedle Patch Design Parameters and Evaluation 298
7.6.1 Microneedle and Patch Geometry 298
7.6.2 Drug and Vaccine Stability in Microneedles 299
7.6.3 Mechanical Strength of Microneedles 300
7.6.4 Characterization of Penetration 300
7.6.5 In vivo Animal Model Studies 301
7.7 Applications of Microneedles 301
7.7.1 Delivery to the Skin 301
7.7.2 Cosmetology 302
7.7.3 Microneedles for Ocular Drug Delivery 304
7.7.4 Insulin Delivery 309
7.7.5 Oral and Gastrointestinal Drug Delivery 312
7.7.6 Protein and Vaccine Delivery 316
7.7.7 Miscellaneous Applications 319
7.8 Drawbacks of Microneedles 321
7.9 Future Scope 323
7.10 Summary 324

8 Microfluidic Electroporation and Applications 343
Koyel Dey, Srabani Kar, Pallavi Shinde, Loganathan Mohan, Saumendra Kumar Bajpai, and Tuhin S. Santra
8.1 Introduction 344
8.2 Brief Overview of Electroporation 345
8.3 Recent Advancement in Electroporation 348
8.3.1 Gold Nanoparticle–Enhanced Electroporation 349
8.3.2 Carbon Nanotube–Based Electroporation 351
8.4 Recent Advancements in Single-Cell Electroporation 351
8.4.1 Single-Cell Trapping and Electroporation 352
8.4.1.1 Mechanical trap 353
8.4.1.2 Optical tweezer 353
8.4.1.3 Droplets microfluidics 354
8.4.1.4 Magnetic trap 355
8.4.2 Micro-/Nanochannel-Based Single-Cell Electroporation 357

8.4.3 Magnetoelectroporation 363
8.4.4 Optoelectronic Tweezer for Electroporation 365
8.4.5 Waveguide-Based Electroporation 368
8.4.6 Microcavity-Based Electroporation 370
8.4.7 Electrofusion 371
8.5 Applications 375
8.5.1 Imaging Based on Electrical Properties 375
8.5.2 Cell Reprogramming 377
8.5.3 Tissue Engineering and Regenerative Medicine 378
8.5.4 Intracellular Recording 378
8.5.5 Cancer Studies 379
8.5.6 Gene Transfer and DNA Vaccination 380
8.6 Limitations and Future Directions 380
8.7 Conclusions 381

9 Electrical Cell Lysis on Microfluidic Devices 401
Bashir I. Morshed
9.1 Cell Membrane and Electrical Lysis 402
9.2 Mechanism of Pore Formation and Electrical Lysis 405
9.2.1 Theories 405
9.2.2 Transmembrane Potential and Pore Formation 406
9.2.3 Transition States and Dynamics 408
9.3 Microfluidic Devices for On-Chip Electrical Lysis 412
9.3.1 On-Chip Electrical Lysis 412
9.3.2 Microchannel and Electrode Geometry 413
9.4 Electrical Lysis Parameters and Considerations 417
9.4.1 PEF Parameters 417
9.4.2 Critical Pore Formation Time 418
9.4.3 Excitation Type 418
9.4.4 Fluidic Medium 419
9.4.5 Cell Line 419
9.5 EL-Associated Phenomena 420
9.5.1 Secondary Effects 420
9.5.2 Cellular Spin Resonance 420
9.5.3 Electrofusion 420
9.5.4 Electroinsertion 421
9.5.5 Electroactivation 421

9.6 On-Chip Electrical Lysis Applications 421
9.6.1 Advantages of On-Chip Electrical Lysis 421
9.6.2 Bioanalysis Device 422
9.6.3 Biofouling Treatment 422
9.6.4 Single-Cell Analysis 422
9.6.5 Cancer and Tumor Treatment 423
9.7 Summary 423

10 Microfluidics-Based Metallic Nanoparticle Synthesis and Applications **429**
Kavitha Illath, Ashwin Kumar Narasimahan, Moeto Nagai, Syrpailyne Wankhar, and Tuhin S. Santra
10.1 Introduction 430
10.1.1 Nanoparticle Formation 432
10.1.2 Metallic Nanomaterials 434
10.2 Microfluidics 435
10.2.1 Single-Phase Microfluidics 436
10.2.2 Multiphase Microfluidics 437
10.3 Microfluidic Devices for Metallic Nanoparticle Synthesis 438
10.3.1 Continuous-Flow Microfluidics 439
10.3.2 Segmented-Flow Microfluidics 439
10.3.3 Mixing Strategies in Microfluidics 440
10.3.4 Controlling Parameters in Metallic Nanoparticle Synthesis 446
10.3.5 Synthesis of Gold Nanoparticles 461
10.3.6 Synthesis of Silver Nanoparticles 467
10.3.7 Synthesis of Other Metallic Nanoparticles for Biomedical Applications 472
10.4 Biomedical Applications 475
10.4.1 Bioimaging 475
10.4.2 Biosensing 481
10.4.3 Photothermal Therapy 487
10.5 Future Prospects 490
10.6 Conclusions 491

11 Microfluidic Particle Separation and Biomedical Applications 503
Ida Laila Ahmad and Mohd Ridzuan Ahmad
11.1 Introduction 504
11.1.1 Theory and Mechanism 506
11.1.2 Flow Resistance 508
11.2 Materials and Methods 509
11.2.1 Flow Streamlines Simulation 511
11.2.2 Microparticle Separation Simulation 512
11.2.3 Microparticle Separation Technique 513
11.3 Results and Discussions 518
11.3.1 Separation of Polystyrene Microbeads 518
11.3.2 Separation of Polydisperse Samples 520
11.4 Conclusions 524

Index 531

Preface

Microelectromechanical systems (MEMS) are defined as miniaturization of the mechanical as well as electromechanical components (devices and structures) at the microscale level. The past two decades have seen rapid development of MEMS technology, with the integration of chemical engineering, biomedical engineering, chemistry, and life sciences, generating a new branch bio-microelectromechanical systems (bio-MEMS). Bio-MEMS, or lab-on-chip devices or micro–total analysis systems (μTAS), have proven to be powerful and inevitable tools for cellular analysis in a complex micro-/nanofluidic environment with their strengths of minimum sample consumption and contamination. These devices can potentially manipulate and detect biosamples, reagents, or biomolecules in a micro-/nanoscale environment and precisely perform cellular analysis because of their real-time operations, range of designing, easy fluid control, monitoring, and programmatic switching. The devices are not only useful for optical, mechanical, electrical, and biochemical characterization at the cellular and subcellular levels but also capable of cell trapping, manipulation, isolation, separation, and lysis. Even the cellular structure and function can be analyzed in detail with the aid of induced flow fields, such as shear stress, by employing micro-/nanofluidic devices.

This book compiles details of cutting-edge research, including the techniques of cell trapping, manipulation, separation, and analysis, utilizing advanced microfluidic and bio-MEMS devices. It encompasses several aspects of micro-/nanofluidic devices, including neuronal cell activity in the microfluidic compartment, vascularized microfluidic organs-on-a-chip and their implementation, different physical techniques for drug delivery, microarrays for gene patterning, micro-/nanoparticle preparation and size-based separation in a micro-/nanofluidic environment, and their feasible biomedical applications. The chapters describe the recent advancements in single-cell analysis (SCA), including single-cell manipulation, separation, lysis, omics analysis, therapeutics, and diagnostics. These SCA techniques are more beneficial compared to cellular bulk analysis. SCA facilitates an in-depth understanding of several fundamental characteristics at cellular and subcellular levels, such as cellular heterogeneity characteristics and molecular dynamics; a precise biological context, such as functional mutation and copy number variation of cells; cell–cell interactions; and cellular responses to the environment, and all of this knowledge is critical for therapeutic and diagnostic purposes.

This book comprises 11 chapters broadly covering several aspects of microfluidics and bio-MEMS-based devices for cellular analysis. Each chapter, written by the experts, provides in-depth analysis, potential problems, limitations, and future prospects. Chapter 1 discusses cell manipulation, therapy, and their feasibility for clinical applications using various microfluidic devices. It mainly focuses on high-throughput cell capture or trapping and therapeutic analysis at the single-cell level and discusses the advantages of this method compared to conventional techniques. Chapter 2 demonstrates the optical manipulation of cells, including advanced opto-electrokinetic technology for biological cell manipulation with their different sizes, shapes, and properties. A different optophysical mechanism has been demonstrated, including the light control electric field, optothermal effect, optical radiative force, and light-driven vortices. Chapter 3 discusses different types of micro-robots or microswimmers and microswimmer-based devices for biomedical and health care applications, with their existing challenges and future prospects. Chapter 4 demonstrates neuronal cell activity in a microfluidic environment. Microfluidic devices can provide a microenvironment with fluid control, which is an essential requirement for neuroscience research. The re-creation of neurodegenerative disorders and neural stem cell (NSC) behavior using microfluidic and bio-MEMS platforms with their advantages and limitations has been discussed in this chapter, and their processes can be used as a basis to construct brain-on-a-chip devices. Chapter 5 elaborates on different strategies of on-chip vascularization to create organ-specific models encompassing healthy as well as pathological vascularized microtissues by using microfluidic technologies. The chapter describes how the integration of tumor cells/spheroids into engineered microvasculature can model vascularized tumors on a chip, which is potentially beneficial for studying cancer metastasis or anticancer drug screening. Chapter 6 discuss the formation of DNA gene microarray biochips that can help us analyze DNA sequences, gene expression profiles, and diseases and perform embryogenesis studies; search for tissue-specific genes; and analyze toxicity of drugs. Chapter 7 discusses the advanced transdermal drug delivery system using microneedles. This chapter emphasizes in detail on the microneedle history, types, fabrications, and working mechanism for drug delivery. It also highlights therapeutic agent delivery using various shapes, sizes, and composition-based microneedles. Chapter 8 focuses on recent cutting-edge technologies for bulk and single-cell electroporation, with their advantages, limitations, potential clinical applications, and future prospects. In continuation, electrical cell lysis using microfluidic and bio-MEMS devices is demonstrated in Chapter 9. The chapter briefly discusses recent on-chip electrical lysis, its related theory and dynamics, and applications for cellular analysis. Chapter 10 demonstrates different metallic nanoparticle synthesis using microfluidic devices and potential biomedical applications such as therapeutics and diagnostics, sensing and imaging, etc. The advantages, limitations, and future challenges in nanoparticle preparation using microfluidic devices are also discussed. Finally, Chapter 11 emphasizes on microfluidic particle separation and its potential biomedical applications, such as for point-of-care diagnosis and μTAS.

This book will be interesting to the readers, especially undergraduate and graduate students, and it will be useful for scientists performing academic and industrial research.

Tuhin S. Santra
2021

Chapter 1

Microfluidic Technologies for Cell Manipulation, Therapeutics, and Analysis

Amogh Kumar,[a] Pallavi Shinde,[a] Loganathan Mohan,[a] Pallab Sinha Mahapatra,[b] and Tuhin S. Santra[a]

[a] *Department of Engineering Design, Alumni Ave, Indian Institute of Technology, Chennai, Tamil Nadu 600036, India*
[b] *Department of Mechanical Engineering, Indian Institute of Technology, Chennai, Tamil Nadu 600036, India*
tuhin@iitm.ac.in, santra.tuhin@gmail.com

Biological cell studies commonly involve thousands or millions of cells together, especially in the case of clinical applications. Toward this end, conventional techniques used for research, diagnostic, and therapeutic purposes in cellular biology fall short due to their low throughput and inability to provide information about individual cells. In recent years, microfluidic technologies have come to the frontiers of research due to their seemingly vast applications and adaptability to various requirements for conducting comprehensive cell studies on chip. On the other hand, microfluidic-based single-cell analysis has emerged over the past two decades owing to high throughput combined with the ability it provides to study

Microfluidics and Bio-MEMS: Devices and Applications
Edited by Tuhin S. Santra

ISBN 978-981-4800-85-3 (Hardcover), 978-1-003-01493-5 (eBook)
www.jennystanford.com

individual cells precisely. This chapter discusses in detail some of the prominent microfluidic strategies employed for cellular manipulation, analysis, and treatment and their feasibility in clinical applications.

1.1 Introduction

In the last three decades, the advances in microfabrication techniques have enabled us to develop several microsystem devices with various applications. The application of microsystems in biological cell studies is a growing research field that has shown several advantages over conventional methods [1–3]. The development of microsystem technology integrated with chemistry, chemical engineering, and life sciences in the form of biomicroelectromechanical systems or lab-on-a-chip (LOC) devices provides a powerful tool for cellular analysis in a micro- or nanofluidic environment. To understand cellular functions and their interactions precisely, a conventional or bulk approach involving millions of cells together can provide average information, but it cannot reveal the cellular heterogeneity characteristics and molecular dynamics [4–8]. On the other hand, micro- or nanofluidic devices have the ability to analyze cellular function precisely at the single-cell level. Micro- and nanofluidic devices can not only perform cellular manipulation, separation, and isolation but also help in cellular therapeutics and diagnostics [6–8].

In their natural environment, cells exist in heterogeneous states, with extreme variations in intracellular components, such as DNA, RNA, and proteins [9, 10]. This has been the main inspiration behind the development of micro- and nanofluidic devices for performing single-cell analysis, with such systems termed as single-cell technologies (SCTs) [4, 11]. SCTs can provide highly accurate information about individual cells and enable cell manipulation and therapeutics in a highly efficient manner [12, 13]. Obtaining extensive knowledge about the behavior of cells and various cellular and molecular pathways is the first step toward understanding the mechanisms through which diseases originate or progress as well as discovering pathways to treat those diseases. SCTs offer the possibility to conduct dynamic studies on individual cells and analyze cellular compartments [5, 14].

Recently, more advanced microfluidic techniques have emerged that allow one to study cellular functions and mechanisms precisely. The major advantage of microfluidic techniques over other conventional methods is that they provide high-throughput devices for the manipulation and analysis of cells. Microfluidic devices are fabricated using well-established microfabrication techniques, such as photolithography, soft lithography using polydimethylsiloxane (PDMS), and wet etching [15, 16]. The devices require picoliter to microliter sample volumes for performing their desired functions and enable simultaneous analysis of many cells. This, in turn, shows great promise toward adapting these devices for clinical diagnostic, therapeutic, and regenerative medicine purposes in the future [7].

This chapter discusses several microfluidic techniques currently employed for studying cellular biology. Many high-throughput microfluidic devices have been used for single-cell trapping using electro-osmotic flow manipulation, microvortex manipulation, etc. Then various microfluidic-based physical techniques, such as electroporation and mechanoporation, are discussed for single-cell therapeutic and diagnostic purposes. Finally, the prospects and applications of microfluidic technologies in cellular biology have been discussed.

1.2 Microfluidic Cell Capture Techniques

Cell capture or trapping is the first step for cellular analysis or many other applications, such as treatment, culture, and monitoring. Cells often need to be trapped or moved to a specific location on a chip for further incubation, analysis, or treatment. Many microfluidic devices have been developed to trap cells with high efficiency [17]. These devices usually do not employ any external force fields, such as optical, electric, or magnetic fields. Other conventional techniques for cell trapping, such as microscopy or patch clamp, suffer disadvantages such as high skill requirement for operation. In comparison to this, the most difficult step in employing microfluidic devices for cell trapping is the complex

Table 1.1 Microfluidic techniques for cell trapping

S. no.	Microfluidic cell trapping options	Key features	References
1	Microdroplets	Provides high throughput Allows further analysis after cell capture	[19, 20]
2	Microarray devices	Provides high throughput Provides the possibility for incubation on the same chip after trapping Allows the selection and retrieval of specific cells from the microenvironment without losing any information	[21–23]
3	Microvortices	Involves relatively simple device fabrication and operation, unlike other methods Allows cell sorting by size	[24, 25]
4	Dielectrophoresis and electro-osmotic flow	Allows easy patterning of single cells on a substrate	[26, 27]

design and fabrication process [18]. Microfluidic devices generally allow integration with other techniques for analysis or monitoring of micro- to picoliter samples. Both droplet microfluidics and flow channels containing hydrodynamic traps are used for single-cell trapping. Recent advancements in microfabrication techniques have enabled the design of several novel microfluidic devices for single-cell trapping, a few of which are discussed in this section (Table 1.1).

1.2.1 Microdroplet-Based Cell Trapping

Fan et al. [19] introduced a microfluidic device that employed a change in flow resistance due to trapping of droplets in a channel as the driving principle for separating single cells on a chip. The device consisted of two dispensing modules: one dispensed oil, which acted as a continuous phase, and the other module dispensed the cell sample, which acted as the dispersed phase. The two inlets met at a T-junction, as shown in Fig. 1.1. Due to the shear force

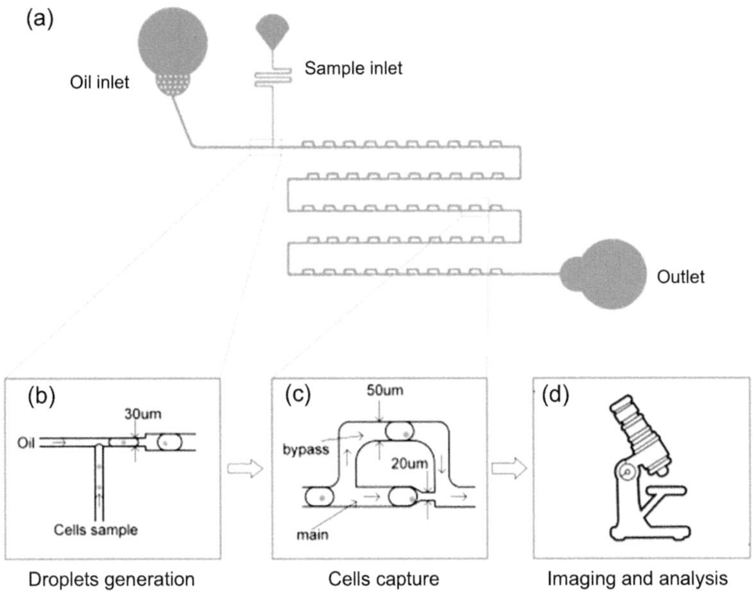

Figure 1.1 (a) Overall schematic of a microfluidic device. (b) The T-junction where microdroplets are formed. (c) Schematic showing the main and bypass channels. (d) Imaging and analysis following cell capture. Reprinted from Ref. [19].

created by the flow rate of the continuous phase, microdroplets of the sample would break off and flow along the main channel., thus forming the dispersed phase. The main channel consisted of several snake-like bypass channels and was directly connected with them. The diameter of the main channel was equal to that of the bypass channel. Without any droplet trapping, the flow rate in the straight channels was higher than that in the bypass channels due to the variation in length.

However, once a droplet was trapped in the straight channel due to the decrease in the radius of the channel after a certain length, the flow resistance of the straight channel became much higher than that of the bypass channel. As a result, the oil and the sample droplets started flowing in the bypass channel. Thus, a single droplet containing the cell was trapped at each junction in the main

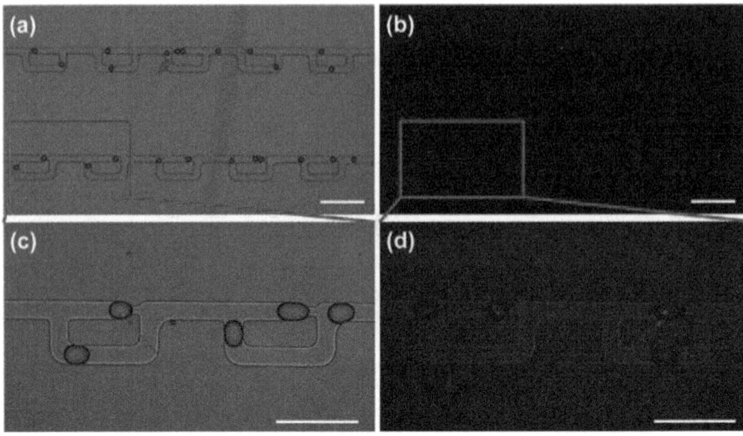

Figure 1.2 (a, c) Bright-field images of cells in the device. (b, d) Fluorescence images of cells trapped in the device. Reprinted from Ref. [19].

channel corresponding to a bypass channel. The diameters of the microdroplets were larger than the widths of the narrow parts of the main channel to ensure the droplets would not flow through the channel and to prevent the continuous phase from passing through it.

The concentration of cells in the sample and the size of the microdroplets were adjusted accordingly to obtain the maximum number of droplets encapsulating single cells. Human liver hepatocellular carcinoma (HepG2) cells were used for experimental studies. The efficiency of trapping the droplets containing single cells reached up to 60%. Following cell capture, fluorescence studies were conducted to demonstrate the device efficiency in the single-cell analysis (Fig. 1.2). Thus, with further optimization, this device could show great promise in biological applications.

Sauzade and Brouzes [20] introduced a similar device for single-cell trapping using droplet microfluidics. They focused on the optimization of single-cell encapsulation in droplets. The cell trapping technique was similar to the one employed by Fan et al. [19]. However, instead of the microdroplets, the cells acted as the plug blocks in the main channel. Encapsulation of the cell occurred after the cell plugged the main channel. The authors identified two

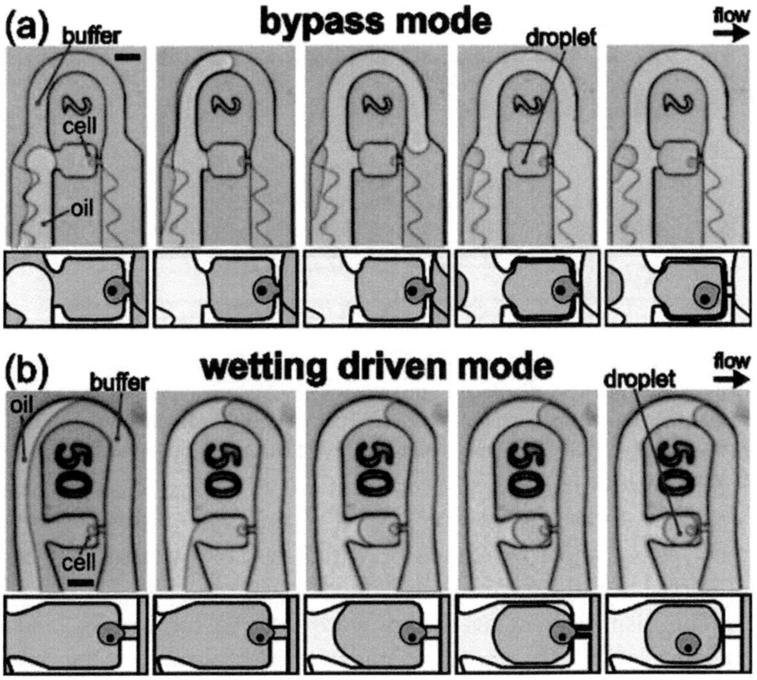

Figure 1.3 The different encapsulation modules; (a) bypass mode and (b) wetting-driven mode. Republished with permission of The Royal Society of Chemistry, from Ref. [20], copyright (2017); permission conveyed through Copyright Clearance Center, Inc.

modes of encapsulation: bypass mode and wetting-driven mode (Fig. 1.3). The bypass mode involved the oil phase filling the entire bypass channel while simultaneously covering both exits of the main channels. Thus, a microdroplet of the sample containing a single cell was formed in the bypass channel. In the wetting-driven mode, the surface of the device had great affinity toward wetting by oil and thus the oil covered the surface in the channels. Wetting of the entire inner surface of the main channel left behind an encapsulated sample droplet containing a cell. However, the wetting-driven mode needed excellent oil wettability on the inner surface of the channels and a slow oil flow rate to ensure surface wetting occurred before complete displacement of the sample volume from

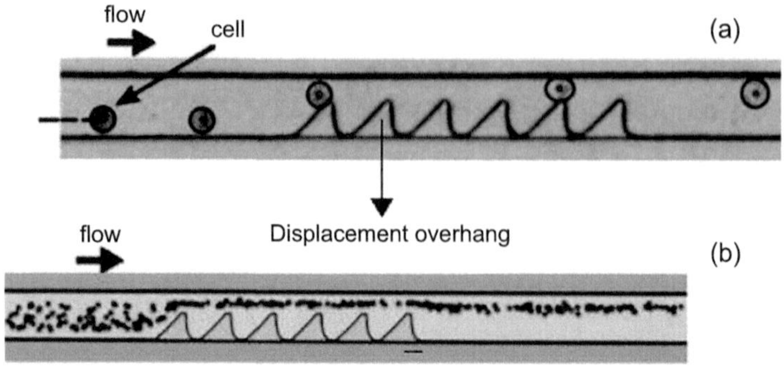

Figure 1.4 Time-lapse pictures of (a) a single cell and (b) multiple cells showing their displaced trajectories in the presence of overhangs. Republished with permission of The Royal Society of Chemistry, from Ref. [20], copyright (2017); permission conveyed through Copyright Clearance Center, Inc.

the bypass channel. Due to these complexities, the bypass mode was preferred over the wetting-driven mode.

The displacement overhangs were used in the bypass mode to drive cells into the streamlines, which were flowing into the main channel (Fig. 1.4). These overhangs were fabricated following simulations in MATLAB showing their ability to control the streamlines in which cells were made to flow. A498 cancer cells were used for the trapping experiments. It was observed that there was a 93.8% probability of cell capture at the first trapping site. If trapping was not successful, then the cell was trapped at the second trapping site in series in 5.6% of the cases. Overall, single-cell trapping efficiency greater than 96% was observed in the proposed device. The encapsulation efficiency of the trapped single cells using bypass encapsulation mode was found to be 78% for live cells and 86% for fixed cells. The lower failure rate could be attributed to the presence of insignificant deformable cells that could pass through the trapping site. Using this device, the encapsulation efficiency was found to be 58% for HeLa cells owing to their variation in size from 10 to 26 μm.

Following encapsulation, the encapsulated single cells were retrieved by reversing the flow of the oil. The reverse flow was at a higher rate than the initial flow rate to displace the droplets from

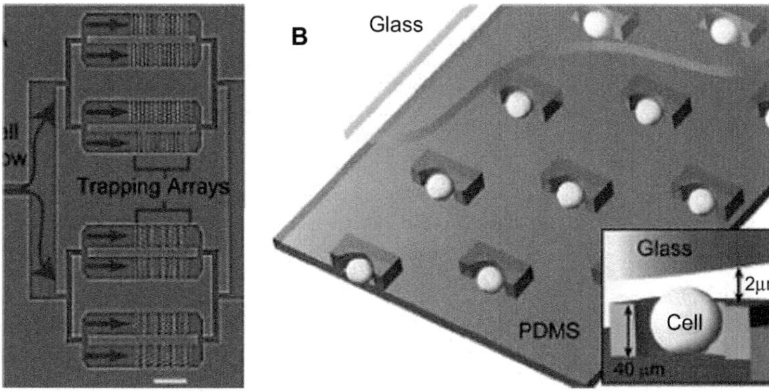

Figure 1.5 (a) Schematic of the fabricated chip; (b) visualization of the U-shaped traps on the device along with captured cells. Republished with permission of The Royal Society of Chemistry, from Ref. [21], copyright (2006); permission conveyed through Copyright Clearance Center, Inc.

the trapping sites. Thus, the platform was able to efficiently trap, encapsulate, and retrieve single cells using droplet microfluidics.

1.2.2 Cell Trapping through Microarray Devices

Apart from droplet microfluidics, cell trapping is achieved by using physical barriers to capture cells flowing in a stream. While devices employed with these techniques often require complex fabrication, their operation is simpler than techniques employed in droplet microfluidics. Many microarray devices have been designed for high-throughput single-cell trapping. These arrays often consist of blockage of different sizes and shapes or use of microwells to trap cells from the flowing sample.

Di Carlo et al. [21] recently developed a microarray chip for single-cell trapping and culture. The device had an array of U-shaped PDMS structures, which were used for hydrodynamic trapping of cells flowing through the chip (Fig. 1.5). Higher trapping efficiency was observed when there was asymmetry in the ordering of the array. The cells showed adhesion in the trapping region after capture. In this device, the trapping efficiency of HeLa cells was 70% and up to 85% of the captured cells were retained in the traps due to adhesion.

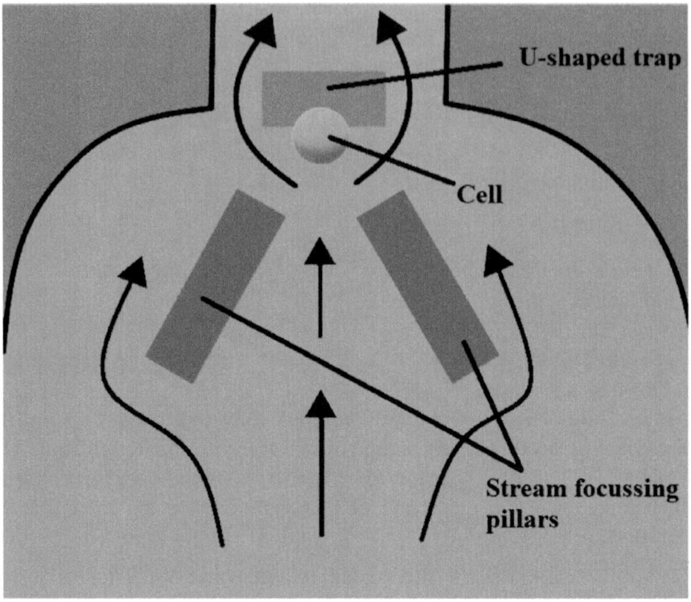

Figure 1.6 Schematic representation of the device showing cell capture and stream flow lines [22].

Luo et al. [22] proposed a microarray device for cell trapping and subsequent cell culture on the same chip. Each trapping site on this chip contained two-stream focusing arms and a U-shaped structure to capture the cell. There was a microwell and a sieve at the bottom of the U-shape structure to enable cell culture (Fig. 1.6). The device contained 512 trapping sites for high-throughput cell capture. To test the efficiency of the device, polystyrene beads were used followed by MC 3T3 cells. The performance of the device with and without stream focusing arms was compared. For the polystyrene beads, the capture efficiency was 29.2% in the absence of the stream and 81.3% in the presence of the stream. In the case of the MC 3T3 cells, the capture efficiency was 44.8% and 82.2% in the absence and presence of the stream, respectively. Following capture, the cells were incubated for at least 6 days. Gradually cell colony formation was observed on the trapping sites, thus confirming the entire functionality of the device.

Guan et al. [23] described a microfluidic device that performed cell trapping using a very narrow channel alongside two bypass channels. They identified the key parameters that needed to be optimized to achieve maximum trapping efficiency, such as the size of the cells, the density of the cells in the flow sample, the velocity of the flow, the number of trap sites, and the width of the narrow channel. When trap sizes are too large, the smaller cells would pass through them without getting captured. A higher cell density in the sample could cause choking of the channels. A high flow rate or flow pressure could cause cell deformation and squeezing through the trap sites. Thus, these parameters of the device needed to be considered for cell trapping.

1.2.3 Cell Trapping through Hydrodynamic Systems and Microvortices

Single-cell trapping can also be performed by employing hydrodynamic techniques such as microvortices to capture cells in them. Khojah et al. [24] introduced a device employing microvortices for single-cell capture by using cavities fabricated in the main channel (Fig. 1.7). They observed that the size of the cells captured depends on the Reynolds number (Re) of the stream flow. Three major phases

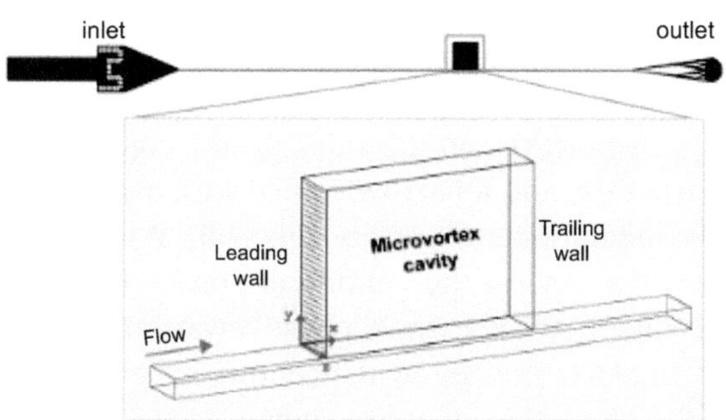

Figure 1.7 Schematic representation of the cavity of a microvortex for cell trapping showing the cavity, the leading wall, the trailing wall, and the direction of flow. Republished with permission of The Royal Society of Chemistry, from Ref. [24], copyright (2017); permission conveyed through Copyright Clearance Center, Inc.

of interest were identified: Phase I (Re 100–175), where clusters of cells were captured in the microvortex formed in the cavity; Phase II (Re 175–225), where large cells were captured; and Phase III (Re 225–300), where small cells were captured. Trapping efficiency was tested using different aspect ratios of cavities. Under the same flow rate, it was observed that particles obtained equilibrium in different orbits as per their size. If the orbit of the particles was completely within the cavity, they would end up trapped, whereas if their orbit went beyond the cavity, the particles would eventually escape it.

Initially, polystyrene beads of different sizes were used to verify the particle capture efficiency. It was observed that the core of the microvortex formed in the cavity shifted from being in proximity of the leading wall to being in proximity of the trailing wall as the Re of the flow was increased. Additionally, in Phase I, larger beads were observed in proximity to the vortex core, while the smaller beads revolved in orbits farther away. Compared to this, in Phase III, smaller beads were observed to move toward the core of the vortex and larger beads moved toward the walls and eventually exited the cavity. Further testing with MCF-7 and MDA-MB-231 cells showed that Phase I was ideal for capture of MCF-7 cells and Phase III for MDA-MB-231 cells. Thus, with precise tuning, the device was able to capture different types of single cells as well as cell clusters.

Kamalakshakurup and Lee [25] designed a chip that used microvortices for cell trapping and then the cells would be dispensed in microdroplets from the chip. The trapping structure of the chip consists of an entry gap to allow the dispersed medium containing cells to flow, two orifices on the sides for the continuous phase to enter the structure, and a narrow orifice with a variable diameter opposite the medium's entry orifice (Fig. 1.8). When the dispersed phase enters the device, the continuous phase flowing from the side orifices diverts the dispersed phase toward the narrow orifice. The coshearing force created by the continuous phase on each side results in the formation of two symmetric microvortices at the phase flow–focusing junction. These vortices are directed toward the narrow orifice at the phase boundaries and reverse-oriented at the center of the junction. The particles in the dispersed phase keep flowing in these vortices until the width of the narrow orifice is increased to an extent that allows them to escape. The width of the

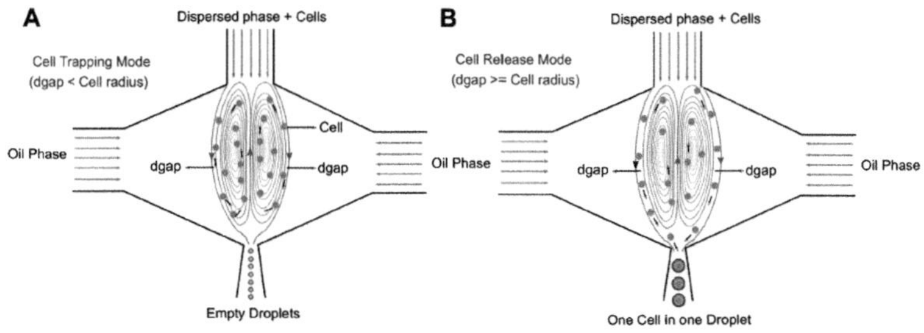

Figure 1.8 Schematic representation of the device. (a) The device in trapping mode; the orifice is too small for cell encapsulation and escape. (b) The device in release mode; the size of the orifice becomes comparable to the size of the entrapped cells. Republished with permission of The Royal Society of Chemistry, from Ref. [25], copyright (2017); permission conveyed through Copyright Clearance Center, Inc.

narrow orifice being larger than the cells implies that any droplet dispersed through the orifice would not be able to encapsulate the cells. Thus, if the orifice size is comparable to the radius of the cell, each cell flowing near the orifice would be able to enter in a droplet and exit the vortices.

This device can also be used to separate single cells of different sizes. By gradually increasing the size of the narrow orifice, the cells could be sorted by size and dispersed into droplets. The authors first used polystyrene beads to verify the trapping and encapsulation of cells in droplets upon increasing the size of the orifice. Following this, single-cell encapsulation efficiency greater than 40% was observed by passing the medium containing K-562 cells through the device.

1.2.4 Miscellaneous Techniques for Cell Trapping

There are several other microfluidics techniques used for cell trapping and manipulation that employ forces such as the electric field and the magnetic field in addition to flow manipulation to trap cells. Toriello et al. [26] used surface functionalization of cells to improve their capture efficiency on activated gold electrodes (Fig. 1.9). K1 and M1WT3 Chinese hamster ovary (CHO) cells were used for the experiments. These cells were functionalized

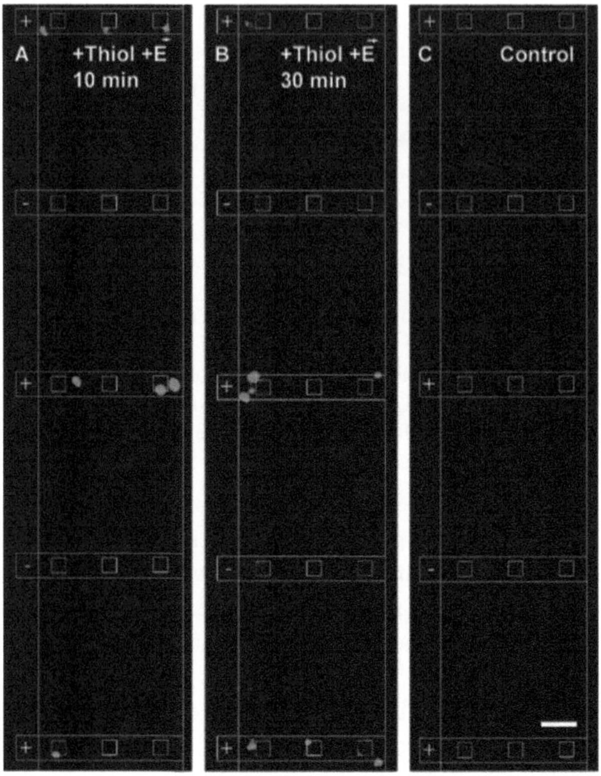

Figure 1.9 Images of the device after experiments, showing cells trapped on the gold electrode's surface. Adhesion of cells with a thiol end and in the presence of an electric field for experiment duration of (a) 10 min and (b) 30 min and (c) cells captured in control experiment. Reprinted with permission from Ref. [26]. Copyright (2005) American Chemical Society.

with the RGD peptide CCRRGDWLC, which contained a thiol end (cysteine). Cell labeling was performed by using cell-permeant blue and green dyes. The microfluidic chip contained gold electrodes at the bottom of a flow channel. The electric field was used to electrophoretically drive the negatively charged cells toward the surface of the electrode. The cell surface functionalization with a thiol end helped to strengthen the adhesion of the cell surface to the gold electrodes after contact was established. Thus, cells were captured. The trapping of single or multiple cells on each electrode was controlled by limiting the duration of the experiment. For a run of 10 min, around 63% of the electrodes contained single-cell adhered to each of them. Cell clumping was observed with a run time of 30 min. In the absence of functionalization with the peptide and an electric field, almost no cells were observed to adhere to the gold electrodes. Functionalization of cells instead of the substrates also enabled further use of the gold electrodes in sensing applications.

Fan et al. [27] used a novel sandwiched microfluidic chip for single-cell trapping with an efficiency of 83.7%. The chip consisted of two layers of indium tin oxide electrodes separated by a membrane channel for sample flow. There was a substrate above the top electrode layer and below the bottom electrode layer. Microwells were fabricated on the bottom substrate, and alignment was performed in such a way that there was one microwell at the center of the intersection of the electrodes. These microwells could only accommodate a single cell but were made shallow to prevent obstruction of the trapped cells and permit cell release after the cells were trapping. The electric field was set up using the electrodes with signals such that the field was strongest at the microwells, thus driving the cells to the microwells via dielectrophoresis. Upon application of a negative charge on top electrodes and a positive charge on bottom electrodes, single cells were attracted to the wells by the dielectrophoretic force. Further on, this device has the ability for cell patterning by switching off the field in certain parts of the electrode array. When this occurred, the cells previously trapped in microwells corresponding to the parts where the field is switched off could be washed away by fluid flow (Fig. 1.10).

The device can form diverse patterns using this technique, and it has a total of 300 trapping sites with 83.7% single-cell trapping efficiency; 9% of the traps showed multiple-cell and 7.3% of the traps showed no-cell microwells. The author used cDFA-SE cells to test the cell patterning efficiency and 97L-GFP cells for single-cell trapping efficiency. Various K-shaped and triangular patterns were successfully obtained by trapping single cells using this device.

Flow cytometry is the most widely used technique for sorting cells on the basis of various physical properties, such as size and granularity [28]. This sorting takes place by the detection of

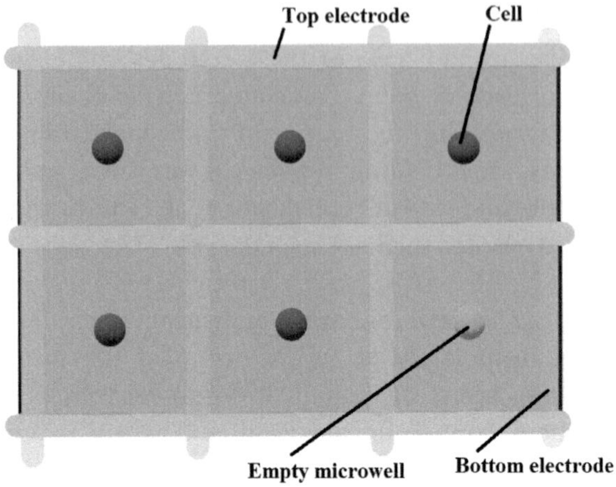

Figure 1.10 (a) Schematic diagram of a fabricated device showing the electrodes and the microwells. An empty microwell is obtained by turning off the corresponding electrodes. Redrawn from Ref. [27].

the fluorescence emitted by the cells, caused by the presence of intracellular or extracellular antibodies and biomarkers in the cells. The biggest advantage of flow cytometry is its high throughput analysis, being able to analyze up to 100,000 cells per second [29]. A flow cytometry device consists of a fluidics component and an optics component. Fluids are required to transport cells across various stages in flow cytometry. Different flow rates can be adopted for different purposes, such as fast flow being used for quantitative analysis whereas slow flow rates being used for complex analysis of intracellular DNA content, etc. [30]. A huge variety of laser setups are used in flow cytometry as per requirements, such as argon laser being used to excite synthetic dyes (such as fluorescein isothiocyanate) and ultraviolet (UV) lasers being used to excite fluorochromes in the 630 nm range. Electrostatic sorting is used to separate cells on the basis of their features for further analysis. Cells are first dispensed in droplets in a stream. These droplets pass through laser beams and get charged by an electrode. Following this, according to their charge, the droplets get deflected toward a negatively or a positively charged plate. Thus, the droplets and cells get separated [28].

1.3 Microfluidic Single-Cell Therapy and Analysis

It is crucial to study the cellular biological functions and response of cells after foreign molecules are delivered inside them, such as different drugs, foreign DNA, and viruses. However, most of these foreign cargos cannot be delivered into a cell due to the selective permeability of the cell membrane [4, 6, 8, 31]. Thus, it is important to develop some techniques that enable the delivery of different drugs/biomolecules into cells without causing any adverse effects on the cells. Many techniques have been developed for cellular transfection, such as liposome-mediated delivery [32], viral vectors [33], electroporation [34–36], optoporation [37–41], microinjection [4, 41], and mechanoporation [41–43]. However, each technique has some drawbacks as well as some advantages. Manual microinjection, one of the oldest cell transfection techniques, has greater than 99% efficiency when performed by a highly skilled operator [44]. Viral vectors are currently used in over 90% of clinical trials because viruses have their own natural pathways for entering cells [33]. However, viral vectors follow cell-specific delivery and can cause an immunogenic response from the host cells and the process of packaging them with cargo changes with any change in the cell type or the cargo [45]. Meanwhile, physical techniques, such as electroporation, optoporation, and mechanoporation, rely on disrupting the cell membrane and creating transient pores on it through which the cargo enters the cell [4]. This section discusses a few microfluidic devices and techniques that have been employed toward achieving cellular therapy and analysis (Table 1.2).

1.3.1 Electroporation

Electroporation refers to the formation of transient membrane pores on the cell membrane by the application of a high external voltage. Due to the application of this external voltage, the electric field strength generated inside and outside of the cell membrane is completely different and this potential difference is called transmembrane potential. When transmembrane potential exceeds the cell membrane threshold voltage, hydrophilic membrane pores can be formed from initially created hydrophobic pores and thus

Table 1.2 Microfluidic techniques for cell therapeutics

S. no.	Physical therapeutics technique	Features	References
1	Electroporation	High transfection efficiency High throughput Low cell viability after treatment Cell toxicity from anions and cations prevented by separation of buffer and cell suspension flow	[34, 35, 46]
2	Mechanoporation	High transfection efficiency Cargo delivered directly into the cell cytosol High cell viability and throughput Intranuclear delivery of cargo of different sizes difficult	[43, 47]
3	Optoporation	Few optoporation techniques involving microfluidics considered Possibility of delivering a large-sized cargo High cell transfection efficiency and high cell viability observed with devices	[39, 40, 48]

biomolecules can be gently delivered into cells through them [34, 35, 49, 50]. This technique mainly requires electrodes and cells in a medium for experiments. Many microfluidic devices have been developed to enhance electroporation efficiency. The first account of electroporation using microfluidic devices was presented by Lundqvist et al. [51], where they used carbon electrodes in the proximity of cells to cause deformation of the cell membrane. Electroporation provides many advantages for cellular transfection, such as different molecular delivery into cells with respect of their size, delivery into different cell types, and high transfection efficiency. However, this method suffers from low cell viability after an electroporation experiment due to an electrolysis effect [34, 35]. In the last two decades, several microfluidic devices have been developed to overcome the limitations of conventional electroporation [34, 35, 52]. Santra et al. [53] designed a chip for localized single-cell micro-/nanoelectroporation. This enabled pore formation in a small area of the cell membrane and delivered molecules gently into the cells. In their design, electrodes were passivated with a dielectric layer to limit the electrolysis effect and other adverse effects that arise from conventional electroporation. This device reported up to 98% cell viability after treatment [54–61].

Wei et al. [62] introduced a microfluidic device that focused cells into a concentrated laminar flow before electroporation. This prevented direct contact of cells with the electrodes, thus avoiding many harmful effects usually observed in electroporation, such as heating, hydrolysis, and pH change of the buffer containing cells. Their chip had three segments: (i) a concentration of cells and molecules in the sample along with one sheath buffer layer above and one below in a triple-layer laminar, (ii) one segment containing electrodes for electroporation, and (iii) one segment for the collection of the cells after electroporation treatment (Fig. 1.11). The presence of a sheath buffer layer above and one below the cell concentration can protect cells from the electrodes during the electroporation process. After experiment, the electroporation efficiency of this device was compared with conventional flow electroporation efficiency. Up to 60% transfection efficiency and 90% cell viability was observed after electroporation using this device in comparison with conventional flow electroporation (35% transfection efficiency and 40% cell viability). Thus, employing triple-layer laminar flow with the sample enveloped by the buffer on both sides helped to increase the transfection efficiency as well as cell viability after electroporation.

Huang et al. [63] proposed a similar microfluidic device to protect cells from the harmful effects of being in direct contact with the electrodes. The device used the principle of the formation of a Dean vortex [64, 65] to manipulate an inflow of cells and electrolytic buffer such that the cell sample was covered by the buffer on both sides during electroporation. Following this, due to the Dean vortex flow, the buffer and the cell sample would segregate into separate halves in the flow channel, thus allowing for the simple collection of treated cells. The device used a long microfluidic channel containing three major parts: an initial curved

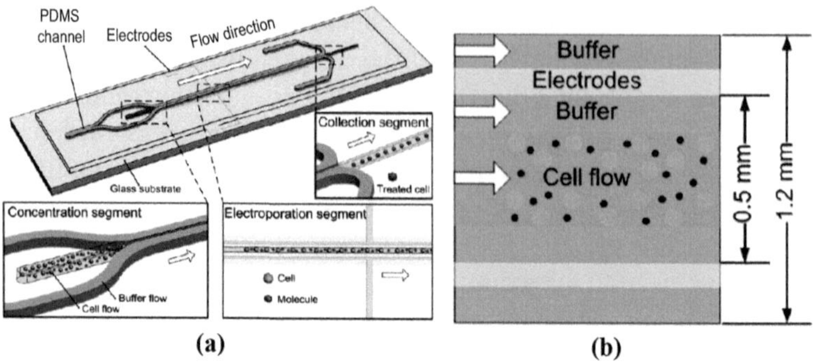

Figure 1.11 (a) Schematic diagram of the device along with its three important segments. (b) Diagram showing the sample flow containing cells and molecules with a buffer layer above and one below. Reprinted with permission from Ref. [62]. Copyright (2011) American Chemical Society.

channel to drive the cell sample into the middle of the buffer layer, a straight segment for electroporation, and finally another curved channel for separating the cell sample from the buffer (Fig. 1.12). The Dean number (De) is representative of the characteristics of the Dean vortex flow in a channel and is directly proportional to the Re of the flow and inversely proportional to the radius of the curved channel. Thus, for a low Re, it was observed that the Dean vortex was symmetric and stable. A low value of De (De $\approx$1) does not significantly impact the flow of fluids in the curved channels. For De $\approx$10, the transverse force acting on the fluid in a curved channel is strong enough to transport it from the inner wall to the outer wall of the channel [66].

The cell suspension and the buffer flow into the microfluidic channel at the same rate through different inlets. The formation of the Dean vortex in the first curved channel pushes the cell sample into the middle level of the channel, which is enveloped by the buffer above and below. This configuration is maintained throughout the flow in the straight channel. The cells also rotate in the channel during electroporation due to the flow, thus increasing the effective cell membrane area exposed to the electric field. Following this, in the second curved channel, a second Dean vortex is formed that pushes the cell sample suspension to one side of the channel and recombines the two buffer layers on the other side (Fig. 1.12).

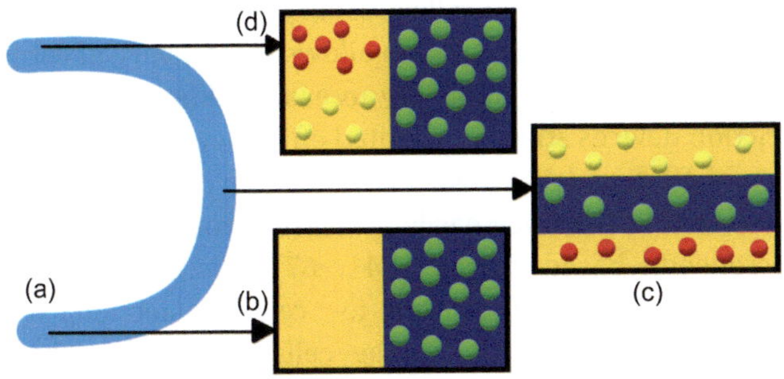

Figure 1.12 Representation of the composition of flow in cross sections at different parts of the microchannel [63]. (a) Representation of the curved channel used, (b) before entering the electroporation region, (c) in the electroporation region, and (d) near collection outlets. The red spheres represent cations, the yellow spheres represent anions, and the green spheres represent the cells. The cell suspension is denoted by the blue fluid, and the buffer solution is denoted by the orange fluid [63].

This leads to faster neutralization of anions and cations formed in these layers during the electroporation process and thus reduces cytotoxicity. Finally, suspension cells are collected from the cell outlet and the buffer from the waste outlet.

Huang et al. optimized the cell collection and cell transfection efficiency by varying the flow rate in the microfluidic channel. They identified the highest cell collection efficiency of 96% with a flow rate of 2.2 mL/min at De = 12.63. Upon testing the device for the delivery of plasmid DNA into HeLa, HL-60, and HUVECs it was found that the device provided higher cell viability (over 20% increase) and transfection rate (over 25% increase) compared to electroporation in a straight microfluidic channel. Thus, this microfluidic device was able to increase the cell viability and transfection efficiency of flow electroporation.

1.3.2 Mechanoporation

Mechanoporation refers to the formation of transient pores on the cell membrane through the application of mechanical shear forces or pressure directly on the membrane. These forces can be applied through different mechanical processes, such as microinjection, nanoneedle impinging, and cells passing through microconstriction devices [41, 43]. Mechanoporation allows the delivery of cargo into a variety of cells due to its ability to easily adapt to the cell type. Constriction devices still show limited efficiency in the intracellular delivery of DNA and RNA, but cytosolic delivery of cells that are hard to transfect has been easily achieved through microinjection- or microneedle-based delivery [41, 67]. Microinjection is one of the earliest techniques used for cell transfection, whereas a microneedle is used to impinge cells and deliver the cargo through the needle itself [44]. This technique offers a very limited throughput coupled with very high transfection efficiency. Due to this, several improvements and designs have been made to obtain high-throughput microinjection systems. These new systems often employ multiple microneedles in parallel for high-throughput cargo delivery or microfluidic flow systems for the automation of the injection process [68, 69].

Sharei et al. [47] developed a chip that forced cells to flow through microconstriction channels and deliver molecules into cells (Fig. 1.13). The device contained parallel microchannels, and each contained one or more microconstrictions. When cells passed through these constrictions, the shear force caused on the cell membrane due to the squeezing resulted in the formation of transient nanopores on the membrane, and thus molecules surrounding fluid could diffuse into the cell cytosol. Sharei et al. [47] identified key parameters affecting the performance of the device, such as the number of constrictions in a channel, the flow rate, and the size of the microconstrictions. To prove the functionality of their device across various cell types, they successfully performed the delivery of dextran of different sizes into primary dendritic cells, primary fibroblasts, embryonic stem cells, and blood-derived immune cells. Szeto et al. [67] further used the cell squeeze platform to prove the device potential for developing cellular vaccines.

1.3.3 Optoporation

Optoporation is a technique that involves the formation of transient membrane pores by using a continuous laser or laser pulses. One of the major optoporation techniques involves metal nanoparticles

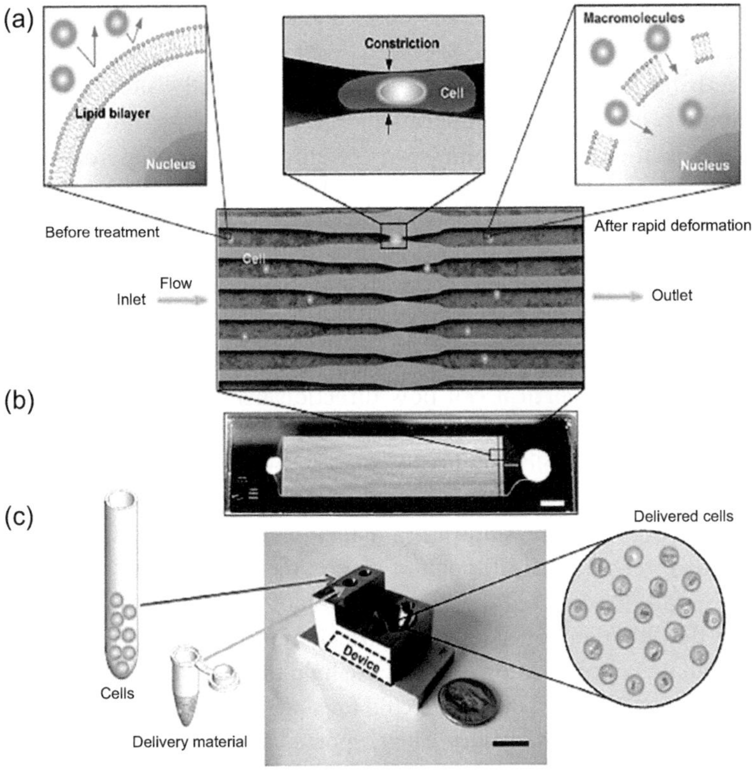

Figure 1.13 (a) Representation of pore formation on the cell membrane after cells are passed through the microconstriction channel. (b) Image of the fabricated device. (c) Representation of the fluids given to the input port of the device and delivered cells collected from the output port. Reproduced from Ref. [47] with permission of PNAS.

(MNPs) dispersed in a medium containing adhered cells and allowed to sediment on them. Following this, a continuous or pulsed laser is incident directly on the sample. This causes localized surface plasmon resonance of the MNPs, which results in the heating up of MNPs and the induction of plasmonic micro- or nanobubbles. These micro-/nanobubbles can rapidly grow, coalesce, and collapse to induce explosion, resulting in very strong fluid flow at MNP and cell membrane interfaces. Thus, the cellular plasma membrane can disrupt and create membrane pores and deliver cargo from outside to inside of the cells [37, 38, 40].

Using lasers to directly form pores on a cell's membrane is another method of optoporation. Although only a few devices have been developed to improve the efficiency of optoporation by employing microfluidic techniques, there seems to be scope for future applications.

Breunig et al. [48] developed a technique to achieve high-throughput optoporation with high cell viability by employing microfluidics with a femtosecond near-infrared laser. They used a standard cell flow cytometry system, upon which a laser was scanned along vertical cell flow direction using scanning mirrors on a multiphoton microscope. Experiments were performed with CHO cells, and cell viability was observed above 96% after 30 min of treatment. Transfection was confirmed by green fluorescent protein (GFP) expression in the daughter cells of the cells treated by this technique. Although a very low percentage (<2%) of cells was transfected by this technique, it shows promise for the development of a microfluidics platform for continuous and automated optoporation of cells with high throughput.

Wu et al. [39] introduced a chip for massively parallel optoporation using the BLAST device. They reported a throughput of 100,000 cells per min with their device. The device used laser energy pulses to heat up thin titanium films present on SiO_2 pillars on the chip. Cells adhered to these pillars before treatment. When the titanium films heat up, they form cavitation bubbles that grow and collapse in nanoseconds. This rapid growth and explosion causes the formation of pores on the plasma membranes of the cells adhered to the pillars. Cell viability greater than 90% was observed with optimized parameters in experiments conducted using HeLa cells. The average delivery efficiency was about 58% for the device.

1.4 Microfluidic Cell Diagnosis and Analysis

1.4.1 Cell Diagnosis

Microfluidic devices are most commonly used for diagnosis and screening due to their high throughput and accuracy. Rare cells, like circulating tumor cells (CTCs), are hard to detect by other conventional screening techniques. A nanovelcro microchip was used for rare-cell assays, and it has shown tremendous improvement in specificity and sensitivity for CTC screening from different forms of cancer, such as lung cancer and prostate cancer. Flow cytometry, such as fluorescence-activated flow cytometry or magnetically activated cell sorters, is a widely used method for high-throughput cell separation. The major disadvantage of flow cytometry is the requirement for cells to be removed from their environment for passing through the device. This can result in the loss of crucial information, such as the loss of extracellular proteins present in the environment of the cells or loss of information about different generations of cells. Toward this end, microfluidic chips have been developed to enable diagnostics of cells without affecting their environment.

Chen et al. [70] fabricated a device that enabled the identification of gene expression signatures while preserving cell lineage data (Fig. 1.14). Their device consisted of a composite film of carbon nanotubes (CNTs) and PDMS layer upon which cells were cultured. Any desired cell on this substrate was then irradiated with a laser pulse for 7 ns, causing heating of the CNTs, resulting in the generation of high stress on the PDMS layer. This stress applied

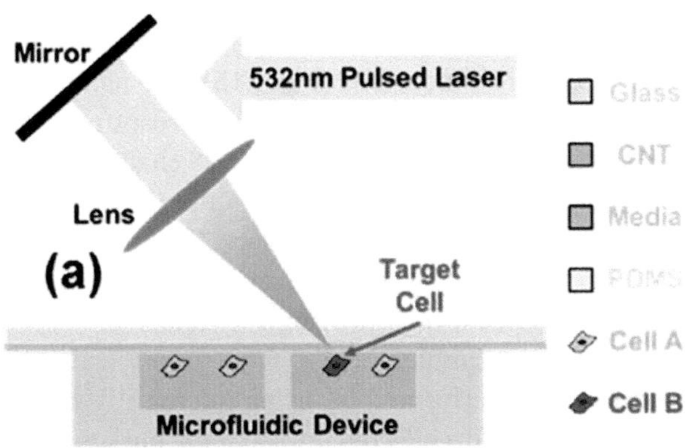

Figure 1.14 Schematic diagram of the device fabricated by Chen et al. Reprinted with permission from Ref. [70]. Copyright (2017) American Chemical Society.

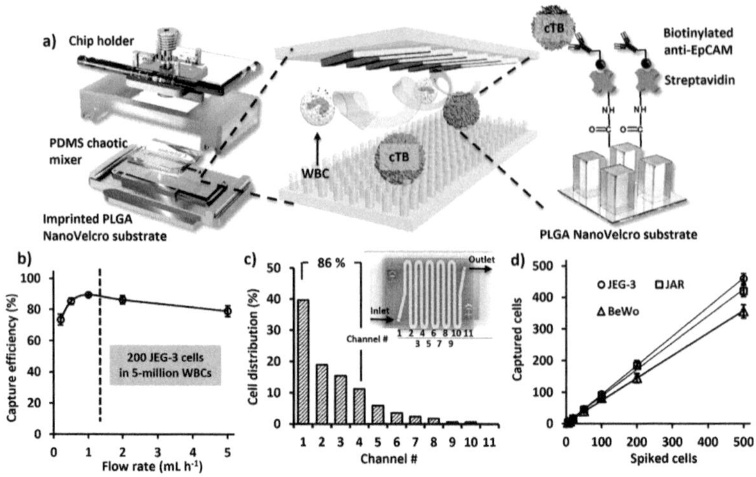

Figure 1.15 Representation of the device showing the functionalized nanopillars in the inset image. Reprinted with permission from Ref. [71]. Copyright (2017) American Chemical Society.

to the PDMS layer was enough to detach the desired cell from its surroundings and deliver it into a microfluidic well, from where it could be collected for further analysis. Thus, cells of specific families or colonies could be extracted with this device.

Hou et al. [71] developed a nanovelcro chip to detect circulating fetal nucleated cells (CFNCs). The device consisted of an array of nanopillars of specific dimensions (Fig. 1.15). The nanopillars were functionalized with the Anti-Ep Cam antibody to capture CFNCs from blood. The substrate was held against a PDMS chaotic mixer having lanes for the blood to flow through. When the CFNCs in the blood encountered the functionalized nanopillars, they were captured by the antibody. Capture of CFNCs was observed within the first three lanes of the mixer itself, thus proving the ability of the device to capture rare cells in a sample.

Diagnosis is also performed by using microarray chips, where cells are separated and collected in separate outlets. Song et al. [72] proposed a microfluidic chip for the capture and activation of CD4+ T-cells with major histocompatibility complex class II monomers (Fig. 1.16). The chip consists of an array of subnanoliter

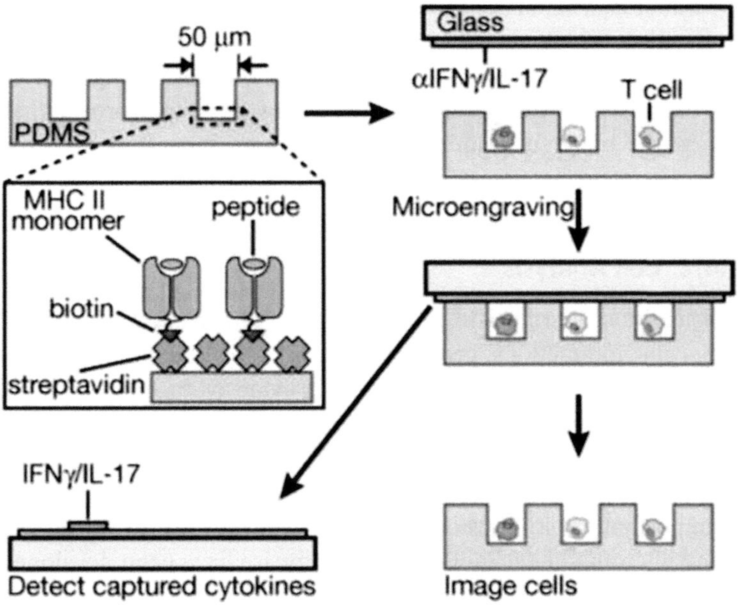

Figure 1.16 Schematic diagram of the entire process of studying activation of CD4+ T-cells using a microarray chip. Reprinted with permission from Ref. [72]. Copyright (2010) American Chemical Society.

wells on which cells are loaded after functionalization. Following incubation of the cells, to allow activation and proliferation, a glass slide containing cytokines secreted is separated and incubated for 2 h. Finally, imaging is used to calculate the average number of cells per well and the number of cytokines captured is studied by labeling.

Detection and isolation of CTCs for further analysis or enumeration is a crucial part of cancer diagnosis. Cancer cells are distributed throughout the human body during the metastatic growth of tumors. Detection of these cells can enable early diagnosis of cancers and enable effective treatment. In the past decades, microfluidic technologies have shown great advantages over conventional techniques in detecting CTCs in blood [73]. Zhou et al. [74] developed a multiflow microfluidic channel for the isolation of CTCs with greater than 87% purity. The buffers flow from the buffer inlet and the sample from the sample inlet in such a way that a three-flow stream is formed in the main channel, with the buffer flowing in the middle to collect the large target CTCs due to size-dependent inertial forces. The total flow rate and the sample flow rate were the main parameters to obtain CTCs of high purity. There was a trifurcation region at the end of the main microchannel, where the buffer containing CTCs would get separated from the other two streams.

1.4.2 Cell Analysis

Cell analysis is crucial in determining cell-to-cell variation, their molecular interactions, cell growth stages, etc. The bulk measurement of millions of cells together can provide average information; however, it cannot unlock the mystery of cellular heterogeneity characteristics or molecular dynamics in certain cell population levels. For accurate measurement and characterization of cells, it is important to study them at the single-cell level [4, 6, 7, 41]. Many microfluidic technologies and software have been developed recently to enable such studies.

1.4.2.1 Droplet-based analytical techniques

Droplet microfluidics is frequently employed for single-cell analysis due to its high specificity and potentiality to isolate single cells. Segaliny et al. [75] developed a microfluidic chip to study the activation of single TCR T-cells upon the detection of specific tumor cells (Fig. 1.17). Their chip contained two inlets, one for the inflow of the sample containing the target cells and another for the inflow of specific and nonspecific TCR T-cells. Upon activation of the T-cells, expression of eGFP would be triggered, thus allowing for further sorting and downstream analysis. Each droplet of the device consisted of a target cell and a T-cell isolated from the rest of the sample. After the droplets were formed, they were delivered to a site containing an inverted floating droplet array (iFDA). This array consists of microwells of specific dimensions such that each could accommodate only one droplet. After filling of the majority of the microwells, the iFDA chip was disconnected from the rest of the device for the incubation of the cells. Specific droplets from the chip were expelled by creating a heat-induced cavitation bubble through

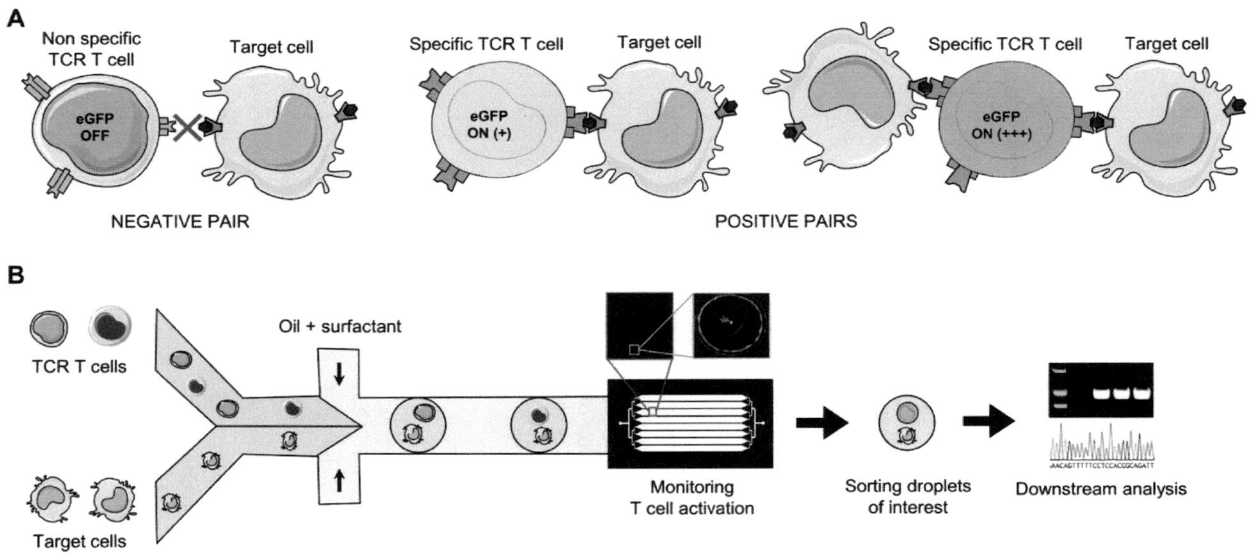

Figure 1.17 Schematic representation of the process for T-cell activation and analysis using droplet-based analysis. Republished with permission of The Royal Society of Chemistry, from Ref. [75], copyright (2018); permission conveyed through Copyright Clearance Center, Inc.

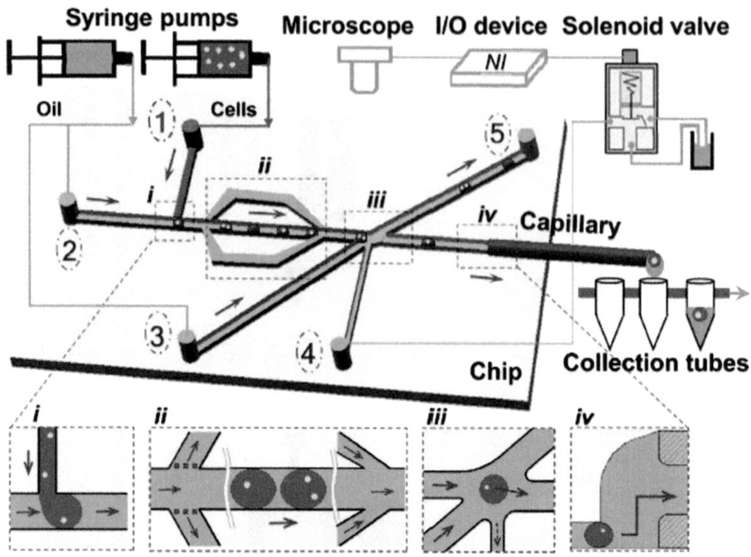

Figure 1.18 Visualization of the device fabricated for single-cell isolation in droplets and integration with gene analyses, showing the components and four major sections of the device: (i) T-junction for cell encapsulation in a droplet, (ii) deceleration of droplets and oil, (iii) sorting of single-cell droplets, and (iv) collection of sorted droplets in tubes. Reprinted from Ref. [76] under a Creative Commons License Attribution 4.0 International (CC BY 4.0).

nanopulsed UV-laser irradiation at those locations. Carrier oil was injected into the iFDA chip to transport the ejected droplets for further analysis. Greater than 90% efficiency was achieved using this method to separate and release the droplets from the chip, following which the droplets were recovered in a polymerase chain reaction (PCR) tube. The major drawback of this device was its low throughput and requirement of around 15 min for the separation and analysis of each droplet.

Zhang et al. [76] developed a device for the separation of droplets encapsulating single cells from droplets containing multiple cells by manipulating the flow acceleration and velocity (Fig. 1.18). The device was fabricated with four sections: one for the encapsulation of cells in droplets at the T junction, one for the deceleration of the flowing droplets by using a branched channel structure, one containing a solenoid valve structure [77] for sorting of single-cell droplets, and final one for the collection of single-cell droplets via a microtube for further analysis. The branched structure consisted of two microchannels branching out symmetrically from the main channel. Due to a shunting effect, the flow of oil through the main channel was decelerated. However, no droplets entered the branches due to their hydrodynamic resistance. The flow of each droplet in the main channel of the branched structure was carefully observed under a microscope. When a droplet containing multiple cells or no cells passed through the channel, the oil flow would drive it into the waste flow channel. However, when a single cell droplet passed through the channel, a digital switch would be activated to provide a suction force on the droplet and redirect it toward the collection channel. The time this selection process took was 1 s per droplet. Greater than 90% efficiency in the isolation of single-cell droplets was observed using this technique. Finally, Zhang et al. achieved successful integration of this device with gene-specific analyses and performed quantitative polymerase chain reaction (qPCR) assays to find the amplification efficiency.

Lan et al. [78] introduced a single-cell genome sequencing technique using droplet microfluidics. They enabled sequencing with picoliter volume droplets and provided throughput of greater than 50,000 cells per run. Single-cell genome sequencing involves the labeling of all DNA fragments originating from a genome with a barcode (identifier) unique to that cell. It is required to isolate and fragment cells before barcoding them. This was achieved by combining a stream containing cells with a flow of molten agarose, resulting in the encapsulation of each cell in agarose microgels. The microgels were cooled, solidified, and transferred to an aqueous phase for lysis and genome purification (Fig. 1.19). Cell lysis was performed by the incubation of the droplets with lytic enzymes, which dissolved the cell membranes. Following this, purification was performed by using detergents and proteases to break down the proteins and lipids. Thus, highly purified DNA was obtained in the droplets. Fragmentation of these DNA was performed by re-encapsulating them. Finally, the droplets were barcoded by using a microfluidic device to merge each DNA-containing droplet with a barcode droplet and PCR reagents.

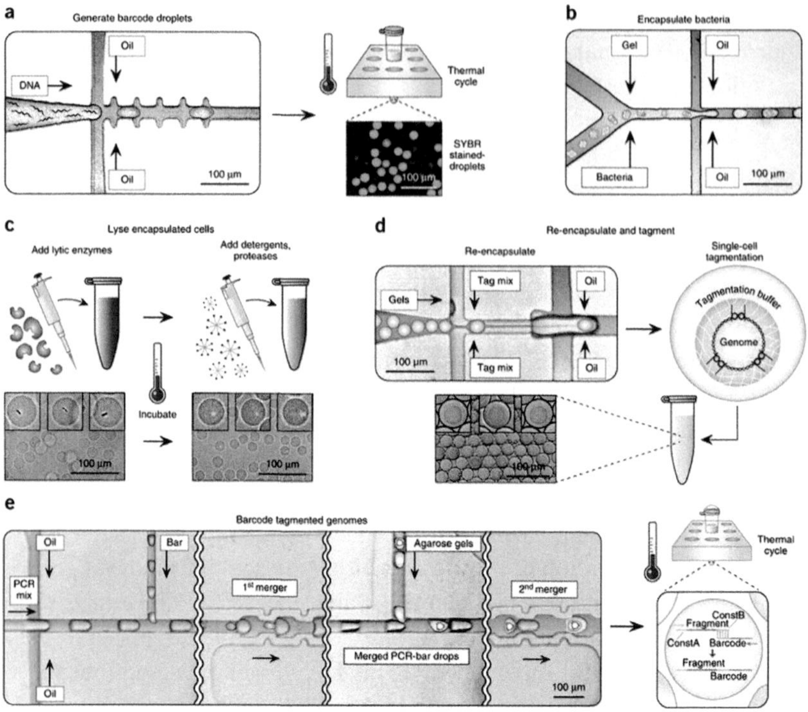

Figure 1.19 Workflow for single-cell genome sequencing using droplet microfluidics. (a) Generation of the barcode droplets. (b) Encapsulation of the bacterial cells with agarose to form the droplets. (c) Purification of the droplets to obtain pure DNA. (d) Re-encapsulation of the droplets for the fragmentation of the DNA. Reprinted with permission from Ref. [78]. Copyright (2017) Springer Nature.

1.4.2.2 Microfluidic devices for massively parallel cell analysis

Aside from droplet microfluidics, there are other microfluidic techniques used for cell isolation and analysis. Many of them aim for high-throughput parallel analysis using microfluidic array devices. Such devices offer a variety of applications in the study of cell biology, biomarkers, etc. Jimenez-Valdes et al. [79] introduced a microarray device containing microwells for high-throughput parallel isolation and analysis of cells. The device contained 74,240 microwells, and each microwell could accommodate up to four cells. The authors optimized various parameters, such as flow rate in the chambers, the concentration of cells, and the settling time of the cells. While the sample flowed through the chambers, the trapping efficiency was observed to be very low and most of the cells were carried away and did not get deposited into the wells. Thus, the authors allowed continuous flow until the sample containing cells filled the chambers and then the flow was stopped to allow the cells to settle at the bottom of the microwells. This process was repeated for many cycles until the percentage of the microwells occupied by cells reached a steady value. It was observed that the occupancy of microwells was faster with a number of cycles in the case of suspensions where cell concentration is greater though the final steady occupancy for different concentrations remained comparable. Following capture, a cell viability assay (calcein-AM/EthD-I) was used to demonstrate the ability of the device in assisting high-throughput cell analysis and check it for any loss of cell viability. After 3.5 h, cell viability of around 80% was observed in this device. Around 80% of the microwells were filled with cells, and up to 36% of the microwells contained a single cell. Furthermore, trapping rates of cells of various sizes were tested with this device. Studies with THP-1 cells showed that most of the microwells were filled with single cells, whereas studies with B16F10 cells (having a larger diameter than THP-1 cells) showed that very few wells were filled with cells.

Finally, the device was used to study neutrophil extracellular trap (NET) induction [80], which has been associated with autoimmune diseases [81] and chronic inflammation [82]. NET induction is caused by the activation of neutrophils using agonists such as phorbol myristate acetate (PMA), which causes disintegration of the nuclear envelope and rupture of the cell membrane. Fluorescence imaging was used to study the effect of the treatment of neutrophils with PMA. The device enabled temporal study of NETosis of the neutrophils and showed that the activation time was proportional to the concentration of the agonist in the environment. Thus, the device proved its potentiality in enabling parallel isolation and temporal analysis of cells.

Le Gac et al. [83] introduced a device for single-cell trapping, integrated with several functions, such as cell membrane permeabilization, cell staining, lysis, and cell content retrieval on the chip

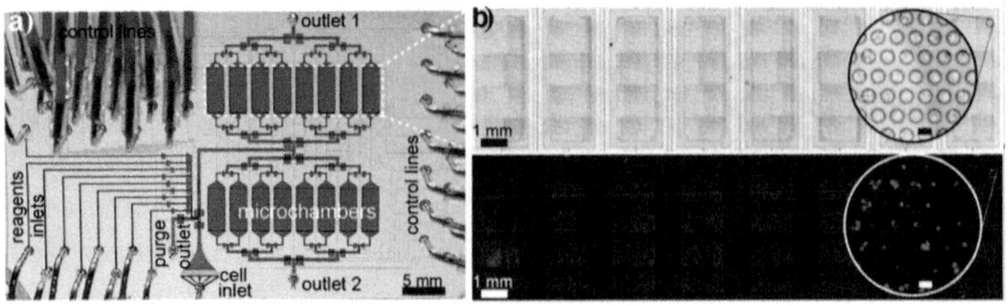

Figure 1.20 (a) Representation of a fabricated device showing the microchambers and connecting ports. (b) Fluorescence images of neutrophils having their nuclei stained after isolation in the microwells. Reprinted with permission from Ref. [79]. Copyright (2017) American Chemical Society.

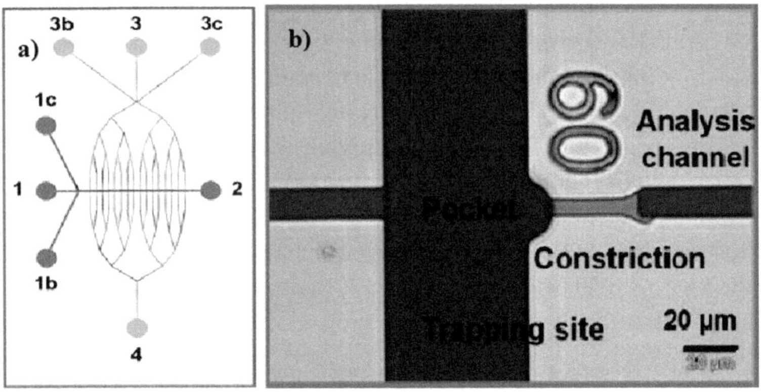

Figure 1.21 (a) Representation of the device fabricated by Le Gac et al., showing the main channel (1–2) and the multiple analysis channels connected to negative pressure reservoirs. (b) A close view of the trapping site where cells are captured in the device. Reprinted with permission from Ref. [83]. Copyright (2017) Springer Science Business Media, LLC.

(Fig. 1.21). The device consists of a main channel having 16 or 32 half-circular pockets for cell trapping. These pockets are directly connected to analysis channels, as shown in Fig. 1.20, which in turn are connected to a common reservoir. The size of the analysis channels is much smaller than the cell size to prevent cells from flowing into them. Upon application of negative pressure in the common reservoir connected to analysis channels, the cells will flow and get diverted toward the pockets and get trapped in them.

Following this, various cell studies could be performed on the same chip. Cell staining was successfully performed on a population of heterogenous cells captured by the device, in order to distinguish them. EpCAM-positive MCF-7 and EpCAM-negative P3X63Ag8 cells were trapped in the device, and labeling was performed by passing a solution containing Hoechst and PE-Anti-EpCAM dyes through the device. Under fluorescence microscopy, MCF-7 exhibited both blue and red staining, suggesting the intake of both dyes, while P3X63Ag8 cells showed only blue staining, suggesting the intake of Hoechst dye alone. Cell permeabilization was also performed on the same chip by passing a mild (digitonin) or strong detergent (LiDS) through the main channel, which induced transient and permanent permeabilization of the cell membrane, respectively. Cell membrane permeabilization by electroporation was also successfully performed on the same chip. Electrodes were inserted in reservoirs 3 and 4 to generate an intense electric field of a magnitude greater than 1 kV/cm across the trapping site of the cell. This voltage difference created membrane deformation, resulting in the formation of membrane pores on the trapped cells, causing electroporation. This was confirmed by the observation of cytosolic leakage of previously loaded calcein from the cells after electroporation. For retrieval of intracellular contents, cell lysis using detergents was performed by applying low negative pressure at the common reservoir, which drives the cellular contents into the analysis channels. In the case of irreversible electroporation, the electric field can produce an electro-osmotic flow in the side channels connected to the reservoirs containing electrodes. As a result, the cell components were driven out into these reservoirs and collected. Thus, the device showed the potential for a variety of cellular analysis.

1.4.2.3 Miscellaneous analysis employing microfluidic devices

Additionally, microfluidic technologies are widely used to analyze single-cell intracellular components, such as DNA, RNA, proteins, and amino acids. Analysis of these components on a single-cell level is important due to the extreme variation in the composition of proteins within a population of cells. Similarly, due to frequent

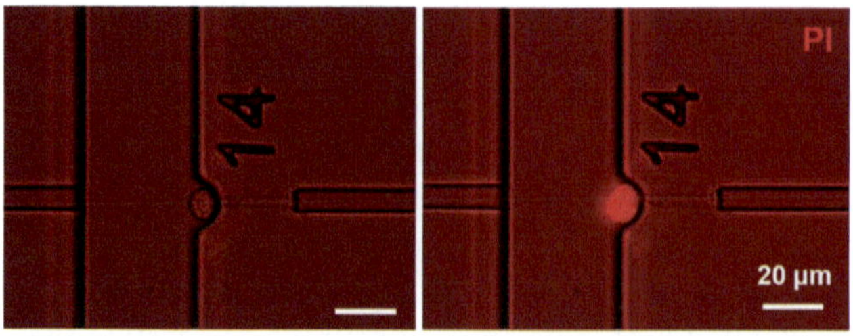

Figure 1.22 Images were taken at 0 s (left) and 240 s (right) after the reversible breakdown of cell membrane using digitonin and PI dye for staining with the device. Bright-red fluorescence and swelling of the cell are observed in the image on the right, confirming membrane breakdown. Reprinted with permission from Ref. [83]. Copyright (2017) Springer Science Business Media, LLC.

changes of RNA expression, their composition and abundance can vary per cell even in homogenous environments [9]. Microfluidic technologies enable the analysis of cells without lysis, such as enabling impedance studies of cells to identify changes in protein expression, DNA content, and nature of the membrane. Jang and Wang [84] fabricated a chip containing micropillars for cell trapping (Fig. 1.23). In this device, the electrodes were deposited at these trapping sites to enable impedance studies. They were successfully able to capture single HeLa cells between the micropillars and study impedance at frequencies between 1 to 100 kHz. They observed a decrease in the impedance of the cell upon the increase of frequency, which could be explained by the capacitive nature of the cell (Fig. 1.22). Major advantages of impedance studies are the ability to study cells without labeling them and the adaptability to flow cytometric approaches [85, 86]. Similarly, as highlighted in this section, microfluidic devices also enable fluorescence-based studies to a great extent owing to the presence of well-defined flow channels and easy access to cells on the chip. Most of these chips are made by using PDMS and allow the flow of oxygen, which enables cell culture on closed chips after trapping [87].

Recently, there has been a rising trend of microfluidic devices being used for bioanalysis applications. Many new microfluidic devices have been developed to perform PCR with high throughput and without loss in efficiency [88, 89]. PCR involves the generation

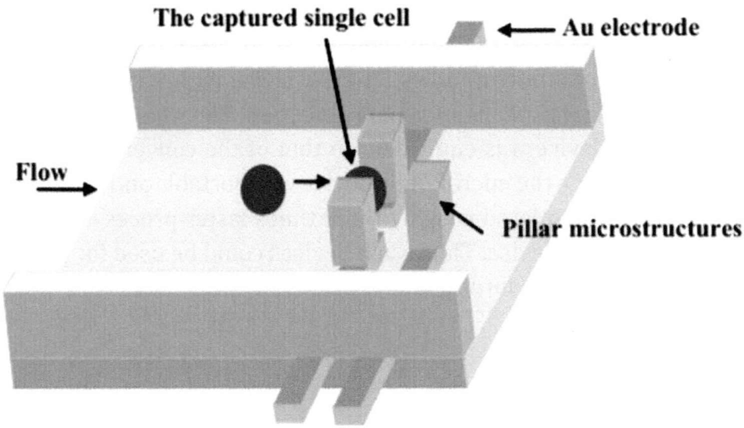

Figure 1.23 A 3D view of the device by Jang and Wang for cell trapping and impedance studies. Reprinted with permission from Ref. [84]. Copyright (2007) Springer Science Business Media, LLC.

of millions of copies of a target DNA from a single copy by a few cycles of treatment. In PCR, a double-stranded DNA is first denatured into two single strands. Following this, primers get attached to the single strands and a polymerase gets attached to each such complex. On increasing the temperature, the polymerase is activated and synthesizes a complementary double-stranded DNA from the free nucleotides already present in the solution. Thus, millions of double-stranded DNA can be obtained by repeating this entire process several times [90]. Furthermore, fluorescent dyes can be added to the reaction mixture to enable real-time monitoring of the progress and determine initial copies of DNA present in the sample. This process is called quantitative real-time PCR (qRT-PCR) [88]. Microfluidic devices also enable the isolation of cells in compartments to perform PCR. Sanchez-Freire et al. [91] presented a microfluidic device for performing qRT-PCR with high throughput using a biomark dynamic array-real-time PCR chip to study the development of cells. A microfluidic dynamic array containing 96 wells connected through integrated fluidic circuits was used for the amplification of the specific target DNA. The array enabled qPCR detection and quantification of specific DNA in each microwell and distribution of the synthesized cDNA across all microwells such that each rarest target gene would contain enough cDNA for detection. Wulff-Burchfield et al. [92] compared the performance of a microfluidics RT-PCR device with a conventional device for *Mycoplasma pneumoniae* detection. The sensitivity of the microfluidic device was equivalent to that of the conventional PCR device. Moreover, the microfluidic device was portable and relatively inexpensive and offered roughly three times faster processing than the conventional device. Thus, such devices could be used for point-of-care testing in the future.

1.5 Future Prospects

Many conventional methods for cell manipulation, such as optical tweezers [93], micropipette arrays [94, 95], and automated micropipettes [96], suffer from low throughput. Compared to these techniques, microfluidic devices offer much higher throughput while retaining similar efficiency. Due to their potential for fabrication in batch mode using well-established microfabrication techniques, the devices can be expected to translate to commercial products in future. There has been a huge advancement of microfluidic techniques described for cell manipulation and adaptability for different biological studies [7]. This suggests the possibility of further interest toward clinical applications. These devices also allow for the study of specific cells in their microenvironments without loss of any relevant information, such as extracellular proteins or intracellular composition, while conventional techniques require the removal of cells from their environments and their resuspension in a fresh solution before analysis [9]. In recent decades, the development of lab-on-a-chip (LOC) devices has specifically enabled the study of cells in their microenvironments to a remarkable extent [97].

Microfluidic technologies have contributed toward the diagnosis and study of several cellular phenomena such as tumor initiation [98], detection of rare CTCs [99], and progressive studies on neural diseases such as Parkinson's disease [100]. To enable the widespread use of microfluidic technologies, the gap between the technology available and the establishment of science must be bridged to fabricate robust devices for clinical or research applications. Achieving this would enable the study of the vast applications of microfluidic systems in cellular biology in greater depths than ever before.

1.6 Conclusions

This chapter discusses the recent developments in microfluidic technologies for cellular manipulation, therapy, and analysis. Different microfluidic-based cell capture techniques are demonstrated using microdroplets, microarray, and microvortices, with their advantages and disadvantages. Using these techniques, more than 96%, 85%, and 40% of cells, respectively, were successfully captured and analyzed. Intracellular delivery and analysis are performed using different microfluidic-based physical approaches, such as electroporation, photoporation, and mechanoporation devices. Among them, electroporation is the most efficient for cellular therapy. However, there is lower cell viability due to electrolysis. On the other hand, photoporation and mechanoporation are upcoming techniques that can reduce the toxicity effect. However, they are only able to deliver molecules in specific cell types. More research is needed in the future to improve these techniques. Cellular diagnostics and analysis are described using different PDMS-based devices, nanovelcro chips, as well as microfluidic-based PCR techniques. Most of these devices are inexpensive and reproducible with a faster processing time compared to any conventional approaches, and in the future, these devices may be used for point-of-care diagnostic or clinical trial purposes.

Acknowledgments

The authors greatly appreciate the financial support from the Science and Engineering Research Board (SERB) under grant number ECR/2016/001945, Department of Science and Technology (DST), Government of India, and the Wellcome Trust/DBT India Alliance Fellowship under grant number IA/E/16/1/503062. We acknowledge all authors and publishers who provided the copyright permissions.

References

1. M. Hamon, J. W. Hong (2013). New tools and new biology: recent miniaturized systems for molecular and cellular biology. *Mol Cells*, **36**(6):485–506.
2. A. Akhtar, E. Fuchs, T. Mitchison, R. J. Shaw, et al. (2011). A decade of molecular cell biology: achievements and challenges. *Nat Rev Mol Cell Biol*, **12**(10):669.
3. J. R. Heath, A. Ribas, P. S. Mischel (2016). Single-cell analysis tools for drug discovery and development. *Nat Rev Drug Discov*, **15**(3):204.
4. F.-G. Tseng, T. S. Santra (2016). Essentials of single-cell analysis: concepts, applications and future prospects. In *Essentials of Single-Cell Analysis: Concepts, Applications and Future Prospects*. Springer, pp. 85–129.
5. F.-G. Tseng, T. S. Santra (2016). *Single Cell Analysis in Biotechnology and System Biology*. MDPI AG Basel, Switz.
6. T. Santra, F. Tseng (2014). Micro/nanofluidic devices for single cell analysis. *Micromachines*, **5**(2):154–157.
7. P. Shinde, L. Mohan, A. Kumar, K. Dey, A. Maddi, A. Patananan, et al. (2018). Current trends of microfluidic single-cell technologies. *Int J Mol Sci*, **19**(10):3143.
8. T. S. Santra, F.-G. Tseng (2021). *Handbook of Single Cell Technologies*. Springer Verlag, Singapore.
9. T.-C. Chao, A. Ros (2008). Microfluidic single-cell analysis of intracellular compounds. *J R Soc Interface*, **5**(Suppl 2):S139–S150.
10. B. Bruijns, A. van Asten, R. Tiggelaar, H. Gardeniers (2016). Microfluidic devices for forensic DNA analysis: a review. *Biosensors*, **6**(3):41.
11. K. D. Korthauer, L.-F. Chu, M. A. Newton, Y. Li, J. Thomson, R. Stewart, et al. (2016). A statistical approach for identifying differential distributions in single-cell RNA-seq experiments. *Genome Biol*, **17**(1): 222.
12. S.-B. Liang, L.-W. Fu (2017). Application of single-cell technology in cancer research. *Biotechnol Adv*, **35**(4):443–449.
13. T. H. Schulte, R. L. Bardell, B. H. Weigl (2002). Microfluidic technologies in clinical diagnostics. *Clin Chim Acta*, **321**(1–2):1–10.

14. F. S. O. Fritzsch, C. Dusny, O. Frick, A. Schmid (2012). Single-cell analysis in biotechnology, systems biology, and biocatalysis. *Annu Rev Chem Biomol Eng*, **3**:129–155.
15. V. I. Vullev, J. Wan, V. Heinrich, P. Landsman, P. E. Bower, B. Xia, et al. (2006). Nonlithographic fabrication of microfluidic devices. *J Am Chem Soc*, **128**(50):16062–16072.
16. B. Gale, A. Jafek, C. Lambert, B. Goenner, H. Moghimifam, U. Nze, et al. (2018). A review of current methods in microfluidic device fabrication and future commercialization prospects. *Inventions*, **3**(3):60.
17. H. Wu, J. Zhu, Y. Huang, D. Wu, J. Sun (2018). Microfluidic-based single-cell study: current status and future perspective. *Molecules [Internet]*, **23**(9):2347. Available from: http://www.ncbi.nlm.nih.gov/pubmed/30217082
18. V. Narayanamurthy, S. Nagarajan, A. Y. Firus Khan, F. Samsuri, T. M. Sridhar (2017). Microfluidic hydrodynamic trapping for single cell analysis: mechanisms, methods and applications. *Anal Methods [Internet]*, **9**(25):3751–3772. Available from: http://xlink.rsc.org/?DOI=C7AY00656J
19. W. Fan, M. Qiao, Y. Jin, H. Zhou, Y. Ge, Q. Jin, et al. (2019). High efficiency single-cell capture based on microfluidics for single cell analysis. *J Micromech Microeng [Internet]*. **29**(3):035004. Available from: http://stacks.iop.org/0960-1317/29/i=3/a=035004?key=crossref.1adc9381eea4fab4a9cb1dee7bff25ad
20. M. Sauzade, E. Brouzes (2017). Deterministic trapping, encapsulation and retrieval of single-cells. *Lab Chip [Internet]*, **17**(13):2186–2192. Available from: http://www.ncbi.nlm.nih.gov/pubmed/28585962
21. D. Di Carlo, L. Y. Wu, L. P. Lee (2006). Dynamic single cell culture array. *Lab Chip [Internet]*, **6**(11):1445. Available from: http://xlink.rsc.org/?DOI=b605937f
22. T. Luo, J. Hou, Y. T. Chow, D. Sun (2016). A microarray platform for high-throughput single-cell capture and culture. In *2016 IEEE 10th International Conference on Nano/Molecular Medicine and Engineering (NANOMED)*, pp. 69–72.
23. A. Guan, A. Shenoy, R. Smith, Z. Li (2015). Streamline based design guideline for deterministic microfluidic hydrodynamic single cell traps. *Biomicrofluidics [Internet]*, **9**(2):024103. Available from: http://aip.scitation.org/doi/10.1063/1.4914469
24. R. Khojah, R. Stoutamore, D. Di Carlo (2017). Size-tunable microvortex capture of rare cells. *Lab Chip*, **17**(15):2542–2549.

25. G. Kamalakshakurup, A. P. Lee (2017). High-efficiency single cell encapsulation and size selective capture of cells in picoliter droplets based on hydrodynamic micro-vortices. *Lab Chip [Internet]*, **17**(24):4324–4333. Available from: http://xlink.rsc.org/?DOI=C7LC00972K
26. N. M. Toriello, E. S. Douglas, R. A. Mathies (2005). Microfluidic device for electric field-driven single-cell capture and activation. *Anal Chem*, **77**(21):6935–6941.
27. L. Fan, T. Luo, D. Sun (2018. High-throughput single cell trapping and patterning using a sandwiched microfluidic chip. In *IEEE International Conference on Robotics and Biomimetics (ROBIO)*, pp. 1508–1513.
28. A. Adan, G. Alizada, Y. Kiraz, Y. Baran, A. Nalbant (2017). Flow cytometry: basic principles and applications. *Crit Rev Biotechnol*, **37**(2):163–176.
29. Y. Han, Y. Gu, A. C. Zhang, Y.-H. Lo (2016). Imaging technologies for flow cytometry. *Lab Chip*, **16**(24):4639–4647.
30. M. J. Wilkerson (2012). Principles and applications of flow cytometry and cell sorting in companion animal medicine. *Vet Clin Small Anim Pract*, **42**(1):53–71.
31. D. Cross, J. K. Burmester (2006). Gene therapy for cancer treatment: past, present and future. *Clin Med Res*, **4**(3):218–227.
32. J. Zabner (1997). Cationic lipids used in gene transfer. *Adv Drug Deliv Rev*, **27**(1):17–28.
33. X. Zhang, W. T. Godbey (2006). Viral vectors for gene delivery in tissue engineering. *Adv Drug Deliv Rev*, **58**(4):515–534.
34. S. Kar, M. Loganathan, K. Dey, P. Shinde, H.-Y. Chang, M. Nagai, et al. (2018). Single-cell electroporation: current trends, applications and future prospects. *J Micromech Microeng*, **28**(12):123002.
35. T. S. Santra, F. G. Tseng (2013). Recent trends on micro/nanofluidic single cell electroporation. *Micromachines*, **4**(3):333–356.
36. T. S. Santra, P.-C. Wang, F. G. Tseng (2013). Electroporation based drug delivery and its applications. In *Advances in Micro/Nano Electromechanical Systems and Fabrication Technologies*. InTech.
37. T. S. Santra, T.-H. Wu, E. P. Chiou (2016). Photothermal microfluidics. In *Optical MEMS for Chemical Analysis and Biomedicine*. Institution of Engineering and Technology, pp. 289–323.
38. Y.-C. Wu, T. Santra, T.-H. Wu, D. L. Clemens, B.-Y. Lee, X. Wen, et al. (2016). Photothermal nanoblades for delivery of large-sized cargo into mammalian cells at high throughput. In *IEEE 16th International Conference on Nanotechnology (IEEE-NANO)*, pp. 47–50.

39. Y.-C. Wu, T. H. Wu, D. L. Clemens, B. Y. Lee, X. Wen, M. A. Horwitz, et al. (2015). Massively parallel delivery of large cargo into mammalian cells with light pulses. *Nat Methods*, **12**(5):439–444.
40. T. S. Santra, F.-G. Tseng (2016). Pulse laser activated photoporation for high efficient intracellular delivery using nano-corrugated mushroom shape gold nanoparticles. In *20th International Conference on Miniaturized Systems for Chemistry and Life Sciences, MicroTAS 2016*, pp. 1097–1098.
41. T. S. Santra, F.-G. Tseng (2020). *Handbook of Single Cell Technology + eReference*. Springer, Singapore.
42. A. Sharei, J. Zoldan, A. Adamo, W. Y. Sim, N. Cho, E. Jackson, et al. (2013). A vector-free microfluidic platform for intracellular delivery. *Proc Natl Acad Sci*, **110**(6):2082–2087.
43. A. Kumar, L. Mohan, P. Shinde, H. Chang, M. Nagai, T. S. Santra (2018). Mechanoporation: toward single cell approaches. In *Handbook of Single Cell Technologies*. Springer, Singapore.
44. C. Ratledge, B. Kristiansen (2006). *Basic Biotechnology*. Cambridge University Press.
45. R. Waehler, S. J. Russell, D. T. Curiel (2007). Engineering targeted viral vectors for gene therapy. *Nat Rev Genet*, **8**(8):573.
46. T. S. Santra, M. A. Teitell, P. E. Chiou (2018). Device for massively parallel high throughput single cell electroporation and uses thereof. United States Patent Application 20180066222.
47. A. Sharei, J. Zoldan, A. Adamo, W. Y. Sim, N. Cho, E. Jackson, et al. (2013). A vector-free microfluidic platform for intracellular delivery. *Proc Natl Acad Sci*, **110**(6):2082–2087.
48. H. G. Breunig, A. Uchugonova, A. Batista, K. König (2014). High-throughput continuous flow femtosecond laser-assisted cell optoporation and transfection. *Microsc Res Tech*, **77**(12):974–979.
49. T. S. Santra, F.-G. Tseng (2016). Electroporation for single-cell analysis. In *Essentials of Single-Cell Analysis*. Springer, pp. 55–83.
50. F.-G. Tseng, T. S. Santra (2014). Nano-electrode based chip. Google Patents.
51. J. A. Lundqvist, F. Sahlin, M. A. I. Åberg, A. Strömberg, P. S. Eriksson, O. Orwar (1998). Altering the biochemical state of individual cultured cells and organelles with ultramicroelectrodes. *Proc Natl Acad Sci*, **95**(18):10356–10360.
52. T. Geng, C. Lu (2013). Microfluidic electroporation for cellular analysis and delivery. *Lab Chip*, **13**(19):3803–3821.

53. T. S. Santra, P.-C. Wang, H.-Y. Chang, F.-G. Tseng (2013). Tuning nano electric field to affect restrictive membrane area on localized single cell nano-electroporation. *Appl Phys Lett*, **103**(23):233701.
54. T. S. Santra, C.-W. Chen, H.-Y. Chang, F.-G. Tseng (2016). Dielectric passivation layer as a substratum on localized single-cell electroporation. *RSC Adv*, **6**(13):10979–10986.
55. T. S. Santra, H.-Y. Chang, P.-C. Wang, F.-G. Tseng (2014). Impact of pulse duration on localized single-cell nano-electroporation. *Analyst*, **139**(23):6249–6258.
56. S.-C. Chen, T. S. Santra, C.-J. Chang, T.-J. Chen, P.-C. Wang, F.-G. Tseng (2012). Delivery of molecules into cells using localized single cell electroporation on ITO micro-electrode based transparent chip. *Biomed Microdevices*, **14**(5):811–817.
57. T. S. Santra, S. Kar, J. Borana, P.-C. Wang, F.-G. Tseng (2014). Nanolocalized single-cell-membrane nanoelectroporation: for higher efficiency with high cell viability. *IEEE Nanotechnol Mag*, **8**(1):30–34.
58. T. S. Santra, J. Borana, P.-C. Wang, F.-G. Tseng (2014). Nanoelectroporation and controllable intracellular delivery into localized single cell with high transfection and cell viability. In *IEEE 27th International Conference on Micro Electro Mechanical Systems (MEMS)*, pp. 865–868.
59. T. Santra, P. Tseng (2013). Nano-intensified electric field for multi-localized single cell electroporation. In *17th Int Conf on Miniatnrized Systems for Chemistry and Life Sciences* (Freiburg, Germany, 27–31 October 2013), pp. 1036–1038.
60. T. S. Santra, C. Chiu, N. Agarwal, A. Ganatathi, P.-C. Wang, F.-G. Tseng (2013). Nanofocused electric field for localized single cell nanoelectroporation with membrane reversibility. In *The 8th Annual IEEE International Conference on Nano/Micro Engineered and Molecular Systems*, pp. 961–964.
61. T. Santra, S. Chen, C. Chang, T. Chen, P.-C. Wan, F.-G. Tseng (2012). Electrically mediated gene delivery and their diffusion mechanism on localized single cell using ito microelectrode based transparent chip \$μ\$TAS. In *16th Int Conf Miniaturized Syst Chem Life Sci*, pp. 962–964.
62. Z. Wei, D. Zhao, X. Li, M. Wu, W. Wang, H. Huang, et al. (2011). A laminar flow electroporation system for efficient DNA and siRNA delivery. *Anal Chem*, **83**(15):5881–5887.
63. D. Huang, D. Zhao, J. Li, Y. Wu, W. Zhou, W. Wang, et al. (2017). High cell viability microfluidic electroporation in a curved channel. *Sens Actuators, B*, **250**:703–711.

64. W. R. Dean XVI (1927). Note on the motion of fluid in a curved pipe. *London, Edinburgh, Dublin Philos Mag J Sci*, **4**(20):208–223.

65. W. R. Dean LXXII (1928). The stream-line motion of fluid in a curved pipe (second paper). *London, Edinburgh, Dublin Philos Mag J Sci*, **5**(30):673–695.

66. D. R. Gossett, D. Di Carlo (2009). Particle focusing mechanisms in curving confined flows. *Anal Chem*, **81**(20):8459–8465.

67. G. Lee Szeto, D. Van Egeren, H. Worku, A. Sharei, B. Alejandro, C. Park, et al. (2015). Microfluidic squeezing for intracellular antigen loading in polyclonal B-cells as cellular vaccines. *Sci Rep*, **5**:10276.

68. D. Matsumoto, R. R. Sathuluri, Y. Kato, Y. R. Silberberg, R. Kawamura, F. Iwata, et al. (2015). Oscillating high-aspect-ratio monolithic silicon nanoneedle array enables efficient delivery of functional biomacromolecules into living cells. *Sci Rep*, **5**:15325.

69. D. Matsumoto, A. Yamagishi, M. Saito, R. R. Sathuluri, Y. R. Silberberg, F. Iwata, et al. (2016). Mechanoporation of living cells for delivery of macromolecules using nanoneedle array. *J Biosci Bioeng*, **122**(6):748–752.

70. Y.-C. Chen, H. W. Baac, K.-T. Lee, S. Fouladdel, K. Teichert, J. G. Ok, et al. (2017). Selective photomechanical detachment and retrieval of divided sister cells from enclosed microfluidics for downstream analyses. *ACS Nano*, **11**(5):4660–4668.

71. S. Hou, J.-F. Chen, M. Song, Y. Zhu, Y. J. Jan, S. H. Chen, et al. (2017). Imprinted NanoVelcro microchips for isolation and characterization of circulating fetal trophoblasts: toward noninvasive prenatal diagnostics. *ACS Nano*, **11**(8):8167–8177.

72. Q. Song, Q. Han, E. M. Bradshaw, S. C. Kent, K. Raddassi, B. Nilsson, et al. (2009). On-chip activation and subsequent detection of individual antigen-specific T cells. *Anal Chem*, **82**(2):473–477.

73. H. Cho, J. Kim, H. Song, K. Y. Sohn, M. Jeon, K.-H. Han (2018). Microfluidic technologies for circulating tumor cell isolation. *Analyst*, **143**(13):2936–2970.

74. J. Zhou, A. Kulasinghe, A. Bogseth, K. O'Byrne, C. Punyadeera, I. Papautsky (2019). Isolation of circulating tumor cells in non-small-cell-lung-cancer patients using a multi-flow microfluidic channel. *Microsyst Nanoeng*, **5**(1):8.

75. A. I. Segaliny, G. Li, L. Kong, C. Ren, X. Chen, J. K. Wang, et al. (2018). Functional TCR T cell screening using single-cell droplet microfluidics. *Lab Chip*, **18**(24):3733–3749.

76. Q. Zhang, T. Wang, Q. Zhou, P. Zhang, Y. Gong, H. Gou, et al. (2017). Development of a facile droplet-based single-cell isolation platform for cultivation and genomic analysis in microorganisms. *Sci Rep*, **7**:41192.
77. Q. Zhang, P. Zhang, Y. Su, C. Mou, T. Zhou, M. Yang, et al. (2014). On-demand control of microfluidic flow via capillary-tuned solenoid microvalve suction. *Lab Chip*, **14**(24):4599–4603.
78. F. Lan, B. Demaree, N. Ahmed, A. R. Abate (2017). Single-cell genome sequencing at ultra-high-throughput with microfluidic droplet barcoding. *Nat Biotechnol*, **35**(7):640.
79. R. J. Jimenez-Valdes, R. Rodriguez-Moncayo, D. F. Cedillo-Alcantar, J. L. Garcia-Cordero (2017). Massive parallel analysis of single cells in an integrated microfluidic platform. *Anal Chem*, **89**(10):5210–5220.
80. V. Brinkmann, A. Zychlinsky (2012). Neutrophil extracellular traps: is immunity the second function of chromatin? *J Cell Biol*, **198**(5):773–783.
81. C. Carmona-Rivera, M. J. Kaplan (2013). Low-density granulocytes: a distinct class of neutrophils in systemic autoimmunity. In *Seminars in Immunopathology*, pp. 455–463.
82. E. Hacbarth, A. Kajdacsy-Balla (1986). Low density neutrophils in patients with systemic lupus erythematosus, rheumatoid arthritis, and acute rheumatic fever. *Arthritis Rheum Off J Am Coll Rheumatol*, **29**(11):1334–1342.
83. S. Le Gac (2017). Microfluidic platform for parallel single cell analysis for diagnostic applications. In *Microchip Diagnostics*. Springer, pp. 187–209.
84. L.-S. Jang, M.-H. Wang (2007). Microfluidic device for cell capture and impedance measurement. *Biomed Microdevices*, **9**(5):737–743.
85. S. Gawad, K. Cheung, U. Seger, A. Bertsch, P. Renaud (2004). Dielectric spectroscopy in a micromachined flow cytometer: theoretical and practical considerations. *Lab Chip*, **4**(3):241–251.
86. K. Cheung, S. Gawad, P. Renaud (2005). Impedance spectroscopy flow cytometry: on-chip label-free cell differentiation. *Cytom Part A*, **65**(2):124–132.
87. D. A. Markov, E. M. Lillie, S. P. Garbett, L. J. McCawley (2014). Variation in diffusion of gases through PDMS due to plasma surface treatment and storage conditions. *Biomed Microdevices*, **16**(1):91–96.
88. C. D. Ahrberg, A. Manz, B. G. Chung (2016). Polymerase chain reaction in microfluidic devices. *Lab Chip*, **16**(20):3866–3884.

89. S. L. Crane, J. van Dorst, G. C. Hose, C. K. King, B. C. Ferrari (2018). Microfluidic qPCR enables high throughput quantification of microbial functional genes but requires strict curation of primers. *Front Environ Sci*, **6**:145.
90. L. Garibyan, N. Avashia (2013). Research techniques made simple: polymerase chain reaction (PCR). *J Invest Dermatol*, **133**(3):e6.
91. V. Sanchez-Freire, A. D. Ebert, T. Kalisky, S. R. Quake, J. C. Wu (2012). Microfluidic single-cell real-time PCR for comparative analysis of gene expression patterns. *Nat Protoc*, **7**(5):829.
92. E. Wulff-Burchfield, W. A. Schell, A. E. Eckhardt, M. G. Pollack, Z. Hua, J. L. Rouse, et al. (2010). Microfluidic platform versus conventional real-time polymerase chain reaction for the detection of Mycoplasma pneumoniae in respiratory specimens. *Diagn Microbiol Infect Dis*, **67**(1):22–29.
93. H. Zhang, K.-K. Liu (2008). Optical tweezers for single cells. *J R Soc interface*, **5**(24):671–690.
94. L. M. Lee, A. P. Liu (2014). The application of micropipette aspiration in molecular mechanics of single cells. *J Nanotechnol Eng Med*, **5**(4):40902.
95. M. Nagai, K. Oohara, K. Kato, T. Kawashima, T. Shibata (2015). Development and characterization of hollow microprobe array as a potential tool for versatile and massively parallel manipulation of single cells. *Biomed Microdevices*, **17**(2):41.
96. E. Shojaei-Baghini, Y. Zheng, Y. Sun (2013). Automated micropipette aspiration of single cells. *Ann Biomed Eng*, **41**(6):1208–1216.
97. S. Gupta, K. Ramesh, S. Ahmed, V. Kakkar (2016). Lab-on-chip technology: a review on design trends and future scope in biomedical applications. *Int J Bio-Sci Bio-Technol*, **8**:311–322.
98. Y.-C. Chen, B. Humphries, R. Brien, A. E. Gibbons, Y.-T. Chen, T. Qyli, et al. (2018). Functional isolation of tumor-initiating cells using microfluidic-based migration identifies phosphatidylserine decarboxylase as a key regulator. *Sci Rep*, **8**(1):244.
99. T. A. Burinaru, M. Avram, A. Avram, C. Mărculescu, B. Ţîncu, V. Ţucureanu, et al. (2018). Detection of circulating tumor cells using microfluidics. *ACS Comb Sci*, **20**(3):107–126.
100. J. T. S. Fernandes, O. Chutna, V. Chu, J. P. Conde, T. F. Outeiro (2016). A novel microfluidic cell co-culture platform for the study of the molecular mechanisms of Parkinson's disease and other synucleinopathies. *Front Neurosci*, **10**:511.

Chapter 2

Optical Manipulation of Cells

Srabani Kar,[a,b] Pallavi Shinde,[b] Moeto Nagai,[c] and Tuhin S. Santra[b]

[a] *Department of Electrical Engineering, University of Cambridge, 9 JJ Thomson Ave, Cambridge CB3 0FA, UK*
[b] *Department of Engineering Design, Alumni Ave, Indian Institute of Technology, Chennai, Tamil Nadu 600036, India*
[c] *Department of Mechanical Engineering, Toyohashi University of Technology, Toyohashi, Aichi, Japan*
tuhin@iitm.ac.in, santra.tuhin@gmail.com

Unveiling cellular characteristics is appealing for biomedical research and diagnostics. Recent advances in optoelectrokinetic technology have enabled the manipulation of biological cells of different sizes, shapes, and properties and thus opened up a versatile era in biological science. Optical manipulation techniques provide a powerful and versatile platform to manipulate cells into arbitrary configurations with microscale precision. In this chapter, we have summarized the progress in optical manipulation techniques of biological cells using microfluidic devices and their related applications. Various approaches based on different optophysical mechanisms have been discussed, including direct optical radiative forces, light-controlled electric fields, opto- or photothermal effects, and laser-driven vortices. With flexible and versatile capabilities, optical manipulation techniques have proven to be a promising and

Microfluidics and Bio-MEMS: Devices and Applications
Edited by Tuhin S. Santra

ISBN 978-981-4800-85-3 (Hardcover), 978-1-003-01493-5 (eBook)
www.jennystanford.com

powerful tool that can be applied in numerous fields of biology, medicine, and pharmacy.

2.1 Introduction

Understanding of intra- and extracellular behavior in response to an external stimuli is the basic step toward next-generation cell manipulation and analysis [1–4]. The enhanced demand for the precise manipulation of cells to unveil detailed cell-specific characteristics has enforced the advancement of micro- and nanomanipulation techniques [1, 2, 5, 6]. Several advanced cell manipulation techniques, such as optical tweezers [7–12], dielectrophoresis (DEP) [13–15], electrophoresis [16–18], acoustophoresis [19, 20], and hydrodynamic forces, have been demonstrated so far in the literature. The base of these techniques is to establish an interaction with the optophysical properties of the cells by applying external stimuli. Among these approaches, cell biological research has tremendously benefited through the use of optical manipulation techniques. Manipulation and sorting of cells with a high recovery rate are of particular importance toward a wide range of physiological applications. In this context, optical manipulation becomes an extremely popular tool because it is non-invasive, contact free, and easily programmable, with existing microscopic imaging techniques. This chapter describes the versatile applications of optical manipulations techniques in the context of cell biological research.

Over the past few decades, optical techniques that have been employed for cell manipulation can be categorized into two streams: (i) the use of direct radiation pressure from incident light, called optical trapping (OT), and (ii) integration of laser light and electric field in the same microfluidic device for creating optoelectrokinetic phenomena that serve the desired purposes. The latter one includes optoelectronic trapping (OET) and rapid electrokinetic patterning (REP) techniques.

In this chapter, we have described multiple ways of utilizing optical energy through advanced optoelectronic devices to manipulate biological objects dynamically in a non-invasive manner. We have summarized the discovery and evolution of different aspects of optical manipulation techniques toward improving the level of precision and serving multiple purposes. While an optical tweezer has already been established as a potential tool for cell manipulation, OET and REP are still being explored in the field of biomedical research.

The chapter is sectioned according to the aspects of optical manipulation techniques. The next section will discuss a few common interesting and important photodynamic and electrokinetic phenomena that occur during OET and REP. These phenomena include DEP, AC electro-osmosis (ACEO), and electrothermal flow (ETF), playing the key role in the device operation of OET and REP. The following sections will describe the OT, optoelectronic tweezer, and REP techniques and their relevant applications.

2.2 Photodynamic and Electrokinetic Phenomena

2.2.1 Dielectrophoresis

DEP was first introduced and described by Herbert A. Pohl in 1951 [21]. It became one of the eminent tools in biological research in the late 1980s [22]. In principle, a dielectric particle experiences a force called DEP force when exposed to an external non-uniform electric field. This electrostatic phenomenon is called DEP. The time-averaged DEP force acting on a spherical dielectric particle is given by [22]

$$\mathbf{F}_{\mathrm{DEP}} = 2\pi\varepsilon_0\varepsilon_{\mathrm{m}}R^3\mathrm{CM}\left(\nabla|E|^2\right), \tag{2.1}$$

where ε_0 is the permittivity of free space, ε_{m} is the relative permittivity of the surrounding medium, R is the radius of the spherical particle, CM is the Clausius Mossotti factor related to the effective polarizability of the particle, E is the root mean square amplitude of the electric field, and ∇ represents the gradient operator. The above equation gives us the following information:

- In a uniform electric field, $\nabla E = 0$. Therefore, the DEP force is zero for a charged neutral particle.

- Larger particles experience a greater force when other parameters are kept fixed.
- The DEP force is independent of the time derivative of E; therefore, it can be generated using either a direct current (DC) or an alternating current (AC) electric field.
- The DEP force depends on the surrounding environment and also on the electrode geometry and their arrangements because the spatial distribution of the electric field depends on the arrangements and geometry of the electrodes.

Now coming to the CM factor, at frequencies below 50 kHz [22],

$$\mathrm{CM} \approx \varepsilon_{\mathrm{m}} \left(\frac{\sigma_{\mathrm{p}} - \sigma_{\mathrm{m}}}{\sigma_{\mathrm{p}} + 2\sigma_{\mathrm{m}}} \right), \tag{2.2}$$

and at frequencies above 50 MHz [22],

$$\mathrm{CM} \approx \varepsilon_{\mathrm{m}} \left(\frac{\varepsilon_{\mathrm{p}} - \varepsilon_{\mathrm{m}}}{\varepsilon + 2\varepsilon_{\mathrm{m}}} \right). \tag{2.3}$$

According to the above expressions of CM, the amplitude and polarity of the DEP force depend on the conductive properties of the particle and the surrounding medium at low frequencies. On the other hand, at high frequencies, it depends on the permittivity of the particle and the surrounding medium. At intermediate frequencies, both conductivity and permittivity determine the effective DEP force and its polarity.

The effective permittivity and hence CM factor can be either positive or negative, depending on the particle or the surrounding environment. Therefore, according to Eq. 2.1, the effective dielectrophoretic force can act either in the direction of increasing field strength or in the opposite direction. While the particle is attracted toward high electric field intensity, the motion is ascribed to be positive dielectrophoresis (p-DEP) and the opposite movement, that is, when the particle is repelled from the high-intensity region, is called negative dielectrophoresis (n-DEP). These two phenomena are depicted in Fig. 2.1. The non-uniform electric field creates unequal Coulomb forces acting on the two poles of the induced dipole (Fig. 2.1b,c). When the particle permittivity is greater than that of the surrounding medium, the net force acting on the particle is toward intense fields, resulting in p-DEP (Fig. 2.1b). On the other

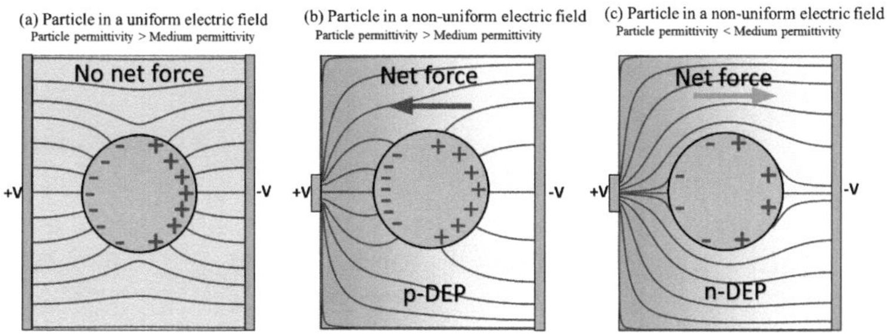

Figure 2.1 (a) Schematic illustration of the interaction of a dielectric spherical particle in a homogeneous electric field distribution when particle permittivity is greater than the medium permittivity, showing that the net force on the particle is zero. The particle is in an inhomogeneous electric field with the medium having (b) lesser and (c) greater permittivity than that of the particle. The field amplitudes are high on the left, decreasing toward the right [23, 24].

hand, when the medium permittivity is higher than that of the particle, it experiences forces in the opposite direction, that is, n-DEP occurs (Fig. 2.1c).

2.2.2 AC Electro-osmosis

When a non-uniform AC electric field is applied through coplanar microelectrode structures in microfluidic devices, it causes the flow of an aqueous electrolyte solution, called ACEO. Due to the application of an AC electric field, charges get accumulated at the electrode-electrolyte interface, creating an electrical double layer, as shown in Fig. 2.2a. The access-induced charge at the electric double layer (EDL) reacts to the tangential component of the electric field, resulting in an electro-osmotic flow over the surface of electrodes. The time-averaged velocity of the electro-osmotic flow is expressed by the Helmholtz–Smoluchowski equation [25, 26]

$$\langle u_{\text{slip}} \rangle = -\frac{\varepsilon_{\text{m}}}{2\eta} \text{Re}\left[\tilde{\Psi}_{\text{D}} \tilde{E}_{\text{t}}^{*}\right], \tag{2.4}$$

where η is the viscosity of the medium, $\tilde{\Psi}_{\text{D}}$ is the complex potential drop across the EDL, $\tilde{E}_{\text{t}}^{*}$ is the complex conjugate of the tangential electric field, and Re[A] refers to the real part of A. Typically for coplanar electrode arrangements, the maximum velocity of the fluid

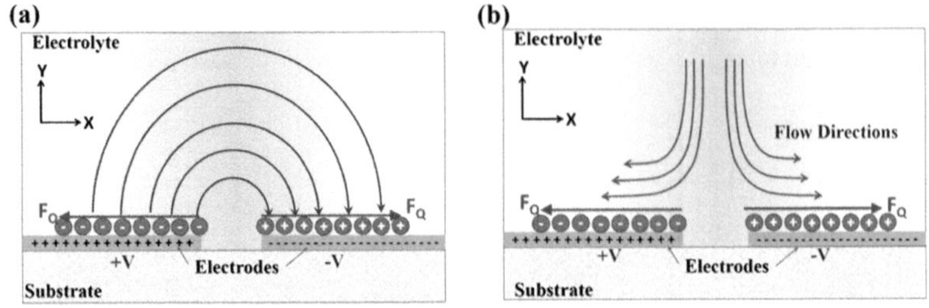

Figure 2.2 A schematic illustration of AC electro-osmosis for coplanar electrode arrangements separated by a narrow gap. (a) The induced electric double layer of the charge and the electric field distribution. The tangential component of the electric field in the x direction results in a force $\mathbf{F}_Q$ on the charges at the electrode surfaces. (b) Schematic illustration of fluid flow because of the force on charges [26].

is observed at the edge, which reduces rapidly with an increasing distance from the edge [24]. As a result, a circulation of flow is generated, with its center close to the electrode edge. This restricts the flow toward the center of the gap between two electrodes, resulting in two counter-rotating fluid vortices circulating above the electrode, as shown in Fig. 2.2b [24]. When a microparticle is added to such a flow, the hydrodynamic force will act on it and move it from the high-velocity region near the electrode edge. When the frequency of the applied electric field is very high (above 100 kHz), ion migration cannot follow the rapid change of electric field polarity and, therefore, the electro-osmotic flow will be suppressed. Thereafter, only DEP and particle weight will control the net force acting on the particle.

2.2.3 Electrothermal Effect

While a microfluidic environment is irradiated with laser light, Joule heating is one of the physical phenomena that lead to a local temperature gradient within a photo-illuminated area (Fig. 2.3). Apart from laser illumination, an electric current that is flowing within the device can also induce such a temperature gradient due to the resistive nature of the materials. The temperature variation in the fluid causes a local variation in the conductivity and the permittivity, resulting in a local inhomogeneity that creates a net

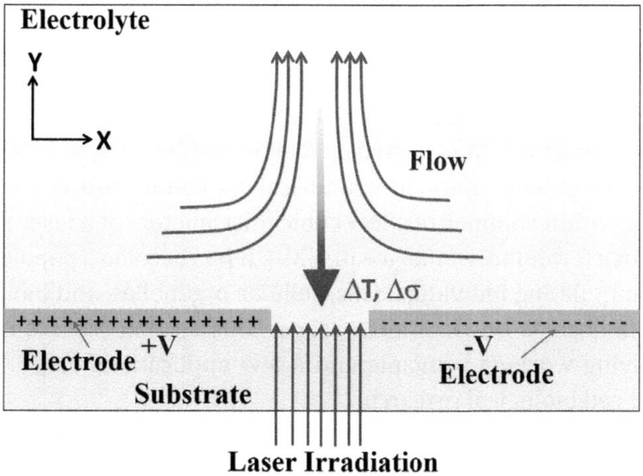

Figure 2.3 Schematic illustration of AC electro-osmotic flow due to Joule heating resulting in a temperature gradient followed by a conductivity and permittivity gradient in the electrolyte solution [28].

force acting on the fluid. The time-averaged effective force per unit volume is given by [27]

$$\langle \mathbf{F}_e \rangle = \frac{1}{2}\mathrm{Re}\left[\frac{\sigma\varepsilon\,(\alpha-\beta)}{\sigma+i\omega\varepsilon}\left(\nabla T.\mathbf{E}\right)\mathbf{E}^* - \frac{1}{2}\sigma\varepsilon\,|\mathbf{E}|^2\,\nabla T\right], \tag{2.5}$$

where Re[...] denotes the real part of the whole expression inside the bracket, **E** is local electric field, **E*** is the complex conjugate of **E**, ω is the frequency of the AC electric field, σ is the conductivity and ε is the permittivity of the fluid, $\alpha = 1/\varepsilon(\delta\varepsilon/\delta T)$, and $\beta = 1/\sigma(\delta\sigma/\delta T)$. Equation 2.5 suggests that temperature gradient plays an important role in electrothermal flow. The first term represents the Coulomb force, and the second term represents the dielectric force. With the increase in Joule heating, the temperature gradient increases and that enhances the electrothermal flow. In contrast, the increase of AC frequency of the applied voltage diminishes the effect as ω approaches the charge carrier relaxation rate of σ/ε. A significant decrease in ETF occurs when the frequency goes above a certain critical limit.

2.3 Optical Trapping

An optical tweezer is a contact-free, non-invasive optical manipulation technique developed in 1970 [29]. The use of optical traps in biology was first demonstrated by Ashkin and Dziedzic in 1987 [30]. They successfully confined cylindrical individual tobacco mosaic viruses within volumes of a few cubic micrometers at a laser power of about a few hundred megawatts [30]. It has become a popular tool for manipulating individual cells, cellular organelles, and biological macromolecules under a microscope. This section focuses on the underlying working principle and a few applications of OT in the field of cell biological research.

2.3.1 Working Principle

An optical trap is created by tightly focused laser light by using an objective lens of a high numerical aperture (NA). When light impinges on dielectric particles some part gets absorbed, some gets reflected, and the rest is scattered in various directions, resulting in a net momentum transfer. The net transfer of momentum of the incident light gives rise to the trapped force [31]. This force can be attractive or repulsive to a high-intensity area depending on the dielectric properties of the particle and the surrounding environments. The trapped force can be divided into two components: (i) scattering force, in the direction of light propagation, and (ii) gradient force, in the direction of the spatial light gradient [31]. In the case of isotropic scattering, the net force is canceled in all but the forward direction and an effective scattering cross section can be obtained. However, if there is a gradient intensity profile impinging on the particle, it experiences a gradient force in the direction, as the name suggests, of the gradient intensity profile. In an optical trap, the inhomogeneous electric field at the focus induces dipoles; the interactions between these dipoles and the inhomogeneous electric field give rise to the gradient trapping force. The gradient force depends on both the optical intensity gradient at the focus and the polarizability of the dielectric particle.

The assessments of the underlying physics of an optical trap have been described by Neumann et al. [32]. It relies on two factors, the wavelength of the incident light and the size of the trapped object. Considering a spherical trapped object of radius a, there are three cases:

(I) The size of the trapped object is much bigger than the wavelength (λ) of the incident light, that is, $a \gg \lambda$; here the conditions for Mie scattering are satisfied.
(II) The size of the trapped object is much less than the incident wavelength of the light, that is, $a \ll \lambda$; here the conditions for Raleigh scattering are satisfied.
(III) The size of the trapped object is comparable to the wavelength, that is, $a \sim \lambda$.

Case I: The condition $a \gg \lambda$ covers most of the cases of whole-cell trapping. Here, the trapping force is calculated following ray optics. Consider a ray of light is tracing a spherical dielectric sphere with a refractive index more than that of the surrounding environment, as shown in Fig. 2.4. Some part of the light is reflected, some part gets absorbed, and the rest is refracted. For simplicity of understanding, only the refracted part is considered here. The ray experiences a change of momentum at the point of incidence and at the point of leaving the object. Thus there is a net change of momentum that induces a net force on the object in a direction opposite to the momentum change, as per Newton's third law of motion. For ease of understanding, this force is divided into two components, one in the direction of light propagation, called scattering force, and the other in a direction perpendicular to the light propagation, that is, across the gradient of light intensity, named gradient force. At this point, we consider two light rays, as the parts of the whole Gaussian beam, incident on two equivalent points (angles of the incident are the same) of the object. The intensities of the rays are different because of the Gaussian intensity profile of the incident beam. The change in the net momentum of these rays is shown by dashed vectors ($\Delta \mathbf{p}_1$ and $\Delta \mathbf{p}_2$ in Fig. 2.4), and the corresponding net forces experienced by the object are $\mathbf{F}_1$ and $\mathbf{F}_2$, respectively. Thus all the equivalent points have to be considered to obtain the net scattering force and gradient force. When a laser beam impinges on a dielectric object, the net scattering force moves it toward a high-intensity focus in the direction of light propagation for the object having a

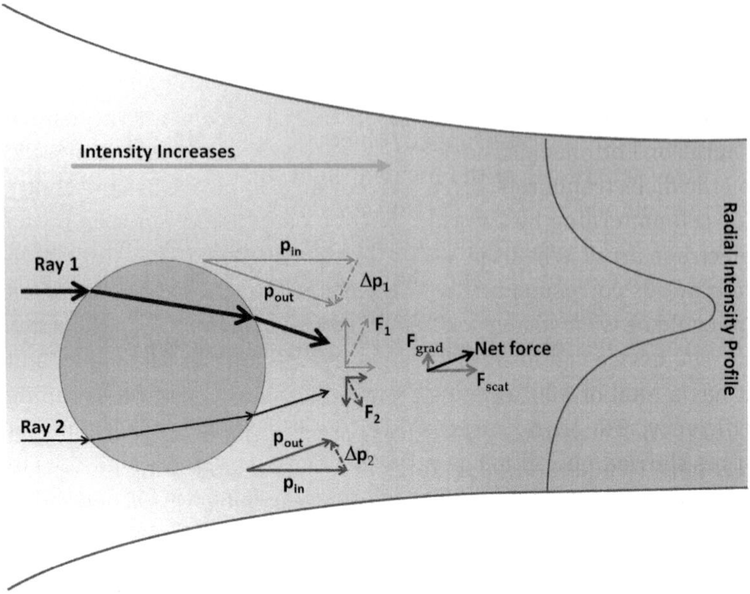

Figure 2.4 Schematic illustration of OT following ray optics. A Gaussian beam of light, with intensity increasing from left to right, illuminates a spherical particle with a refractive index greater than the medium's index. $\mathbf{F}_1$ and $\mathbf{F}_2$ are the forces arising due to momentum transfer corresponding to Ray 1 and Ray 2, respectively. The net force acting on the particle is decomposed into scattering ($\mathbf{F}_{\text{scat}}$) and gradient ($\mathbf{F}_{\text{grad}}$) force components [33].

greater refractive index than that of the surrounding medium. In contrast, the object experiences a repelling force, that is, a force in the opposite direction (away from the high intensity) for a low-refractive-index dielectric object. On the other hand, the gradient force moves the object toward high intensity for a high-refractive-index object or away from high intensity for a low-refractive-index object in the direction of the intensity gradient perpendicular to the laser propagation.

Case II: The condition $a \ll \lambda$ is valid for trapping of cell organelles. In such cases, the object can be treated as a point dipole. The scattering force arises because of the absorption and reradiation of light by the dipole. When light with a wavelength λ is incident on a sphere of radius a, the force is given by [32]

$$\mathbf{F}_{\text{scat}} = \frac{I_0 \sigma n_{\text{m}}}{\mathbf{c}} \tag{2.6}$$

and

$$\sigma = \frac{128\pi^5 a^6}{3\lambda^4} \left(\frac{m^2 - 1}{m^2 + 2} \right)^2, \tag{2.7}$$

where I_0 is the intensity of the incident light, σ is the scattering cross section of the sphere, n_{m} is the index of refraction of the medium, **c** is the velocity of light in vacuum, and m is the ratio of the index of refraction of the particle and the surrounding medium ($n_{\text{p}}/n_{\text{m}}$). From Eq. 2.6, the scattering force is proportional to the intensity of light and points toward the direction of the incident light. On the other hand, the interaction of the induced dipole with the inhomogeneous field results in a time-averaged value of gradient force and is given by [32]

$$\mathbf{F}_{\text{grad}} = \frac{2\pi\alpha}{\mathbf{c} n_{\text{m}}^2} \nabla I_0 \tag{2.8}$$

and

$$\alpha = n_{\text{m}}^2 a^3 \left(\frac{m^2 - 1}{m^2 + 2} \right) \tag{2.9}$$

is the polarizability of the sphere. The gradient force is proportional to the value of the intensity gradient and is directed toward the gradient direction for $m > 1$.

Case III: The particle size is comparable to the laser wavelength ($a \sim \lambda$). In such a case, the above two approaches are not applicable. More complete electromagnetic theories are required for describing the net force. Since we mainly focus on the manipulation of whole biological cells, we are not discussing the underlying theoretical description of this case in this chapter.

2.3.2 Configurations of Optical Traps

Traditional OT uses a freely propagating laser beam with TEM_{00} mode of a Gaussian intensity profile and a high-numerical-aperture objective lens. This lens is required to obtain a tight focus, which can act as a trapping center. Such experimental setups are often bulky and expensive and require a transparent substrate. In cellular analysis, these limitations often restrict versatile applications of OT, such as measuring cellular forces associated with cell proliferation and differentiation. For fulfilling versatile purposes different kinds of OT, such as non-Gaussian laser beam traps, dual beams, and multiple traps, have been demonstrated in the literature. In the following sections, we will discuss different configurations of OT that use free propagating optical beams as well as optical fibers for trapping and manipulating biological cells.

2.3.2.1 Single-beam optical trap

A single-beam optical trap uses a single laser beam. The simplest OT uses a single laser beam with a zero-order fundamental mode Gaussian intensity profile in the radial direction. However, this approach fails to trap particles more than a few micrometers in multiple locations of 3D space apart from the axial direction because the beam strongly diverges after the focal plane and the trapped object distorts the beam profile. Thus the performance is limited by depth, which allows single object trapping once at a time. To address this problem, a laser beam with a Bessel intensity profile has been proposed. Compared to the Gaussian beam, the Bessel beam contains concentric rings of light field patterns of decreasing intensity with a diffraction-less and self-healing nature, which can efficiently trap multiple spatially separated objects once at a time [34]. This approach has been successfully implemented to trap and manipulate multiple yeast cells in 3D traps [35]. Therefore, the use of different beam intensity profiles has a significant contribution to the development of OT techniques. Similarly, Laguerre–Gaussian beams with LG_{01} mode were used to trap and manipulate spermatozoa with reduced photodamage of the cells, without compromising the trapping efficiency, compared to TEM_{00} mode of the Gaussian beam [36].

2.3.2.2 Dual-beam optical trap

Two counterpropagating or unidirectional propagating optical beams are used to create dual-beam optical traps (DBOTs) [37].

For unidirectional configuration, the beams have also been used to produce an interference pattern, where the intensity distribution gives rise to a gradient optical force that can be used to trap rod-shaped particles [38]. The counterpropagating configuration, also known as an optical stretcher, has been applied to trap and deform biological cells. Light powers in the range of a few hundreds of milliwatts in each beam can result in surface forces up to hundreds of piconewtons [39]. A schematic illustration of a counterpropagating DBOT is shown in Fig. 2.5a–c. Compared to free laser beams, fibers are widely used for creating DBOTs. When two identical rays of light pass through the object from the opposite sides, the net scattering force acting on the object is $\mathbf{F}_1-\mathbf{F}_2$. For perfect collinear alignment of these beams (Fig. 2.5a), the two forces cancel each other, making the net force zero. But the force on the surface is additive and equals $(\mathbf{F}_1 + \mathbf{F}_2)/2$, which can stretch the objects along the beam axis [37]. Furthermore, by adjusting the transverse offset distance d between two fibers, the net scattering and gradient forces can be tuned to induce spin and orbital rotation of cells, as shown in Fig. 2.5b,c [40]. In general, when these configurations are integrated with microfluidic devices, it turns out to be a potential platform for the characterization of cells by their viscoelastic signature, without any mechanical contact. These cell-specific characteristics can be utilized for marker-free sorting of stem cells from heterogeneous populations and cytological diagnosis of cancer for therapeutic approaches in regenerative medicine [41]. Since these fiber-based configurations do not use any focused beam, it reduces the possibility of photodamage or any kind of two-photon-induced damage to the trapped cell.

2.3.2.3 Multiple optical traps

The integration of an OT system with microfluidic technologies has provided in-depth insights at the cellular and subcellular levels. The current state of biomedical research stresses more on the parallel manipulation of thousands of cells with precise control over individual cells. In this context, multiple optical traps is one of the pathways to accomplish such purposes. The vertical cavity surface emitting laser arrays have been exploited to create

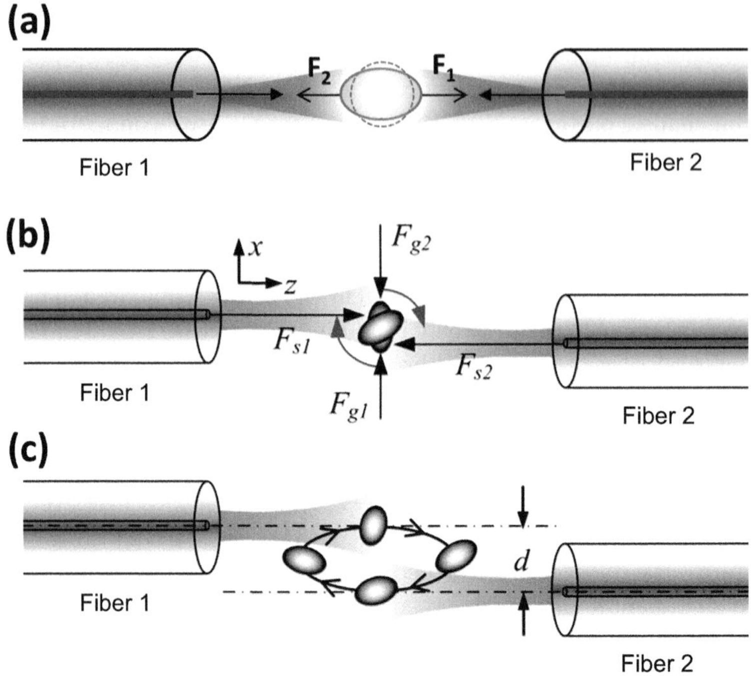

Figure 2.5 Configurations for dual-beam optical traps. (a) DBOT for counterpropagating collinear alignments of a fiber laser. The figure has been redrawn from Refs. [37, 42]. (b, c) The spin (b) orbital (c) rotation of cells in the dual-beam fiber-optic trap with transverse offset distance d. $\mathbf{F}_{s1}$ and $\mathbf{F}_{s2}$ denote the scattering forces, and $\mathbf{F}_{g1}$ and $\mathbf{F}_{g2}$ denote the gradient forces from the corresponding fibers. (b, c) Reprinted from Ref. [40] under a CC BY license (https://creativecommons.org/licenses/by/3.0/).

arrays of individually addressable optical traps to capture multiple biological cells in parallel [43, 44]. These traps are capable of manipulating objects in 3D without any need of individual diode or gas lasers. However, the relatively low output power of this technique restricts its application in larger cell capture or trapping cells of lower refractive indices [43]. Another approach is the use of generalized phase contrast in combination with a spatial light modulator to generate dynamic high-contrast light patterns in the far field [45]. Also, acousto-optic-deflector-based time-sharing-basis OT has been demonstrated [46]. In addition, a holographic optical tweezer generated from advanced and real-time SPM technology offers complete flexibility for the creation of arbitrary optical landscapes and desired light patterns for manipulating multiple cells independently [47].

2.3.3 Applications of Optical Trapping in Biology

The significant capability of the OT technique is to exert a driving force in the range of a few tens of femtonewtons to hundreds of piconewtons, which conveniently overlaps with the range of forces experienced by biological molecules in their native environments. This makes optical tweezers an ideal tool for versatile applications in biological science. Optical tweezers can be easily integrated with an optical microscope merely by focusing a laser beam through the microscope objective. The radiative force allows any dielectric microparticles to be trapped near the beam focus in a stable manner due to the interaction between the laser light and the particle itself, as discussed in the previous paragraphs. In the following section, we will discuss how these techniques have been applied for fulfilling several biological and medical research purposes, including cell and subcellular organelle manipulation, cell sorting, and nucleic acid and molecular motor analysis.

2.3.3.1 Cell manipulation

In 1987, the first experiment of OT in the field of biological science was demonstrated by Ashkin and Dziedzic [30]. They used a single-beam gradient trap for trapping and manipulating individual tobacco mosaic viruses, live motile bacteria, and the less motile *Escherichia coli* bacteria. All of those experiments were performed using a 524 nm green argon laser [30]. Along with the success of trapping, one key concern raised was about the associated temperature rise, which can lead to significant optical damage to the trapped cell. To reduce photoinduced damage, Ashkin et al. proposed optical traps by using a near-infrared (NIR) laser of 1064 nm wavelength in their next study [48]. They showed a way to pluck out long filaments of the cytoplasm for conducting internal cell surgery to manipulate "local rheological properties of essentially unperturbed living cells." They successfully manipulated the locations of relatively large organelles. such as chloroplasts and nuclei, which was, indeed, a remarkable step in unveiling the mechanical properties of the cytoplasm [48]. The above observations also suggest that the laser wavelength is an important parameter in biological experiments to acquire high cell viability after manipulation, which is indeed a benchmark of experimental success. Thereafter, several thorough investigations were carried out in the visible and infrared (IR) wavelength range, leading to the conclusion that in practice, the wavelength range should be cell specific, and in particular, the IR range is more admissible to avoid the possible heating effect and optical damage [49–51]. This conclusion was reinforced by cell cloning experiments performed on Chinese hamster ovary (CHO) cells using optical traps of various wavelengths [52]. Wavelength dependence of clonal growth was recorded showing the maximum clonability at 950–990 nm and the least clonability at 740–760 nm and 900 nm [52]. In addition, at a wavelength of 1064 nm, a strongly reduced clonability was observed, depending upon the power density and the exposure time [52]. However, experiments performed on liposomes and CHO cells revealed non-negligible sample heating and a temperature rise of several degrees centigrade using IR optical tweezers, as recorded by the spatially resolved fluorescence detection technique [53]. The increase in the sample temperature also depends on the nature of the cell and scales linearly with applied optical power [53]. The most common wavelength used for optical cell trapping experiments is 1064 nm. Ohta et al. developed a strategy to trap cells flowing in a microfluidic environment [54]. They used a telescopic pair of conical lenses to generate a ring-like beam, rather than a conventional single Gaussian beam, in combination with a high-NA objective to create a focused evanescent field near a surface, which can build a stable trap for micron-sized particles. Erythrocytes under flow were examined, showing successful trapping of the cells horizontally against the surface. This approach was particularly favorable for monitoring the cell membrane [54]. The sorting of specific target cells is of particular importance for regenerative medicine and tissue engineering applications [55, 56]. The optical tweezer integrated with microfluidic technologies offers feasible cell sorting methodologies in continuous flow using a microscopic environment. In addition to cell trapping, there are several experimental demonstrations illustrating the capability of optical tweezers in controlled the sorting of targeted biological cells, such as yeast cells [57, 58], human embryonic stem cells [59], and bovine blood cells [60].

OT was also applied to perform laser-induced cell fusion [61]. Steubing et al. demonstrated an OT (1064 nm) setup that allows the execution of the selective fusion of two single cells without any need for chemical or electrical treatment [61]. Another application was to analyze cell fusion on different combinations of sample cells, such as human embryonic stem cells and Jurkat cells, in the microwell array-based microfluidic chip, which allows high selectivity and controllability at the single-cell level [62]. The combined application of optical tweezers and optical scissors facilitated cell pairing and fusion with high controllability. Further biophysical assessments on the fused cells showed that the fusion efficiency of the homotypical pairs is higher than that of the heterogenic pairs [62]. Furthermore, a noncontact micro-operation was employed to trap and manipulate red blood cells within subdermal capillaries in living mice, which showed excellent success in clearing out a blocked microvessel [63]. This success expands the application of optical tweezers to studying cell dynamics in live animals beyond in vitro measurements [63].

2.3.3.2 Studies of cell-to-cell interactions

In-depth investigation of cell-to-cell and cell-to-extracellular-matrix interactions is of particular importance for understanding cellular functions such as cell adhesion, migration, differentiation, proliferation, and apoptosis, which can promote new ideas for advanced drug discovery. An optical tweezer in combination with fluorescent microscopy has been applied to study the adhesion properties of leukemia cells on bone marrow stromal cells. This assessment revealed that leukemia cancer cells can migrate on the stromal cell layer depending on the adhesion state [64].

2.3.3.3 Robot-tweezer manipulation system

Micromanipulation of cells and biomolecules by optical tweezers is now a well-established technique. Consistent technical improvements have made this technique readily accessible, and thus it is becoming a standard platform for the study of biological systems. One of the major drawbacks of OT in most of the laboratories arises from the inescapable issues of the manual intervention of the devices that slow down the speed of manipulation and results in low throughput. In this context, robotic automation can tremendously help through the development of robotic-user interfacing techniques. To maximize the operational efficiency for obtaining feasible and advanced high-throughput manipulation systems, automation can be applied specifically for readjusting traps and gripper positions, sketching optimal trajectories for cell transportation to desired locations in the assays, and controlling trap intensities [65].

2.3.3.4 Manipulation of subcellular organelles

Apart from whole-cell manipulation, OT techniques for subcellular organelles have already seen tremendous progress. In terms of spatial and temporal resolution, these techniques are capable of measuring angstrom-scale conformational changes of subcellular organelles. Here we have briefly mentioned a few applications of optical tweezers for exploring static and dynamic characteristics of these subcellular components. There are several experiments measuring the flexural rigidity of microtubules, which are important components of cytoskeletal structures. Microtubules, in combination with actin and intermediate filaments, impart the static as well as the dynamic framework of cell shape and resist various internal/external forces to maintain cell structure [66]. They also support motor proteins to generate the force required for cell movement and shape change [66]. Molecular motors are fascinating biological proteins associated with the cytoskeleton and are responsible for intracellular movement by converting chemical energy into mechanical work [67, 68]. There are three types of cytoplasmic motors: myosins, which move along actin filaments, and dyneins and kinesins, which use microtubules as tracks. In-depth information about myosins, kinesins, and processive DNA enzymes can be explored by using optical tweezers. To conduct such experiments, one or more dielectric microspheres are entangled with the molecular motors. These microspheres are trapped by optical tweezers for handling and manipulating molecular motors for performing experiments of interests. Those techniques have been extensively used to study motor protein properties such as speed, spin, force, torque, and processivity in detail [69, 70]. Thus, optical tweezers allow the mapping of the dynamics of molecular motors at ultrahigh spatial and temporal resolutions. The ability to access the discrete steps of motors at angstrom size provides important insights into their operation. The development of an ultrastable OT system with angstrom-scale resolution facilitates monitoring transcriptional elongation by single molecules in RNA polymerase (RNAP) [71–73]. The studies revealed that RNAP moves stepwise through sequence-dependent pause sites with a DNA template by a single base pair per nucleotide addition to the nascent RNA during transcription [74]. Assessments have been made to describe how the pause sites reduce the forward transcription rate of nearly all RNAP molecules and the nascent RNA structures act as modulators, which either enhance or attenuate pausing, depending on the sequence context [75, 76]. Thus OT techniques explore intracellular activities that are extremely beneficial for advanced biological research and applications.

With the tremendous advancements in optical tweezer–related techniques, optical energy has also been applied to induce optoelectrokinetic phenomena, which opens up two more efficient techniques of cell manipulation, OET and REP. In the following sections, we will discuss these techniques and their applications in the field of biological research.

2.4 Optoelectronic Tweezers

Optoelectronic tweezers were introduced to the scientific community in 2005 [77]. They are capable of achieving parallel manipulation of thousands of cells at individually addressable single-cell resolution. An optoelectronic tweezer uses the spatially varying non-uniform electric field generated by optically induced virtual electrodes on a photoconductive substrate. This light-induced electric field leads to ACEO, DEP, and ETF. These electro-optokinetic forces collectively act to trap and manipulate dielectric particles suspended in the medium. The virtual electrodes can be moved to control real-time manipulation of the particles. Compared to an optical tweezer, an optoelectronic tweezer requires very low light intensity (~100,000 times lower [77]). In addition, the use of an incoherent optical source is sufficient to control the DEP forces over a large area, which facilitates massively parallel cell manipulation. Thus a new platform was established with more flexibility for cell manipulation [77]. Furthermore, optoelectronic tweezer devices do not require photolithography in the fabrication process. Therefore, the devices are less expensive, making them highly attractive for disposable applications. On the basis of DEP forces, there are two operational modes for optoelectronic tweezer–based optical manipulation—positive optoelectronic tweezer, which helps attract particles toward the laser-illuminated area, and negative optoelectronic tweezer, which repels the particles away from the illuminated spot. The transformation of a positive optoelectronic tweezer into a negative optoelectronic tweezer or vice versa is possible by tuning the frequency of the applied AC electric field and the dielectric properties of the surrounding medium.

2.4.1 Device Design, Working Principle, and Developments

The schematic configuration of the device is shown in Fig. 2.6. The simplest optoelectronic tweezer device consists of two optically transparent substrates, such as glass, coated with a few hundreds of nanometers thick indium tin oxide (ITO) coating. The substrate at the bottom is coated with a few tens of nanometers thick heavily doped hydrogenated amorphous silicon (a-Si:H) on top of the ITO layer followed by a few micrometers thick undoped a-Si:H and a few nanometers thick nitride layer, as shown in the enlarged picture in Fig. 2.6. This undoped amorphous silicon layer acts as a photoconductive layer to generate a patterned virtual electrode,

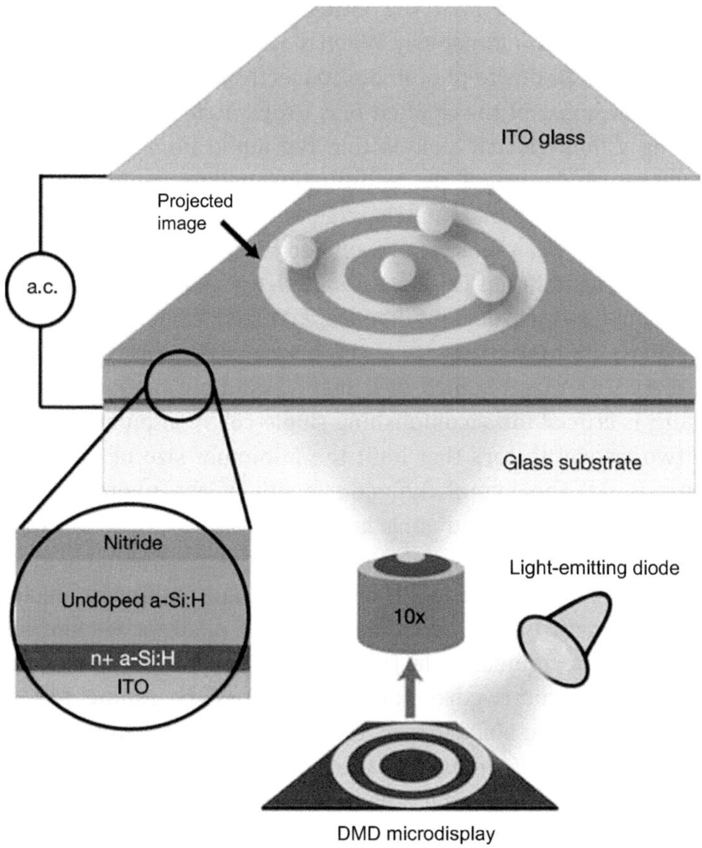

Figure 2.6 The schematic configuration of an optoelectronic tweezer device and the experimental setup. Reprinted by permission from Springer Nature Customer Service Centre GmbH: Springer Nature, *Nature*, Ref. [77], copyright (2005).

while the nitride layer acts as a surface passivation layer. The topside of the device is coated with a few tens of millimeters thick SU-8 layer or a double-sided tape as the spacer, which is patterned to define the channel geometry. Then two substrates are compiled to allow the cell-containing solution to be sandwiched between the ITO layers. The bottom ITO layer serves as a virtual electrode while the device is illuminated by an incident light. To manipulate cells, a vertical AC electric field is applied between the two ITO electrodes and the device at the bottom is illuminated by a projected patterned light simultaneously. When the a-Si layer is illuminated, its resistance drops due to photoinduced electron-hole pair generation. As a result, most of the applied bias drops across the liquid layer, creating a high electric field within the liquid immediately above the illuminated area of the a-Si. In the unilluminated areas, the resistance of the a-Si is much larger, which causes a significant potential drop from most of the applied bias. As a result, the liquid just above the unilluminated areas experiences low electric fields. In this manner, electric field gradients are created, which drive the DEP force in the optoelectronic tweezer device in response to the light patterns. The maintenance of a high-resolution virtual electrode pattern is crucial for accomplishing single-cell manipulation. There are two critical factors that limit the minimum size of the virtual electrode: (i) the lateral diffusion length of the photogenerated carriers in the photoconductive layer and (ii) the optical diffraction of the objective lens. It is reported that a large number of electronic defect states in undoped a-Si:H result in a short ambipolar electron diffusion length ($\sim$100 nm). Therefore, the effective resolution is eventually determined by the optical diffraction limit. Furthermore, the optoelectronic tweezer force (Eq. 2.1) is the gradient of the square of the local electric field. As a consequence, a well-confined force space is created within the area of the individual virtual electrode. So this technique operates more precisely at the single-cell resolution. The frequency of the AC electric field can be tuned to control the AC electrokinetic flow for performing the selective manipulation of particles of various sizes and dielectric properties.

The operation of a conventional optoelectronic tweezer device is limited to the low-conductivity cell suspending media. This arises from the limited photoconductivity of amorphous silicon. In the context of biological applications, typical cell culture media or physiological buffers have significantly higher conductivities ($\sim$1.4 S/m). Therefore, during optoelectronic tweezer experiments, the salts in the cell culture media solutions were replaced by osmotically equivalent amounts of non-electrolytes, which are nonphysiological. Hence cells lose their normal functions, such as proliferation and growth, resulting in short-term sustainment of cell

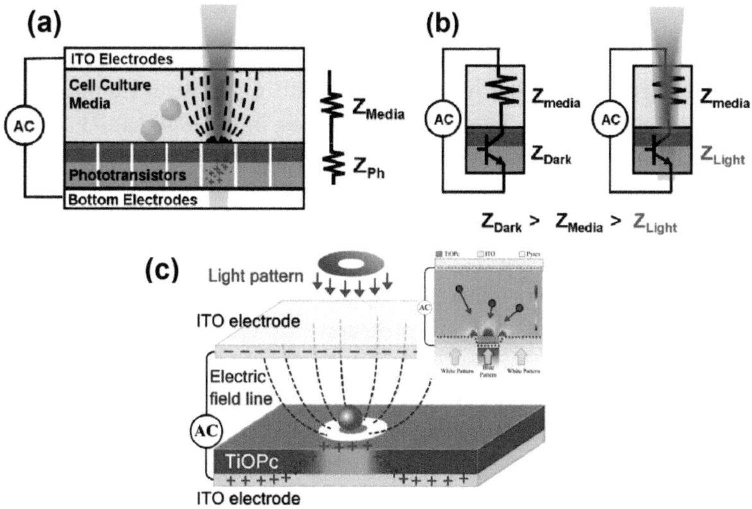

Figure 2.7 (a) The schematic of a phototransistor-based optoelectronic tweezer device and (b) its effective circuit representation. Republished with permission of The Royal Society of Chemistry, from Ref. [78], copyright (2010); permission conveyed through Copyright Clearance Center, Inc. (c) TiOPc-based device where an a-Si:H photoconductive layer is replaced by a Y-TiOPc photoconductive layer. Reprinted with permission from Ref. [80] © The Optical Society.

viability. To address this issue, phototransistor-based optoelectronic tweezers (Ph– optoelectronic tweezers) have been developed, with a few hundred times higher photoconductivity than that of amorphous silicon [78, 79]. A typical NPN phototransistor consists of a highly n-doped emitter, a moderately p-doped base, and a lightly n-doped collector region. The schematic device and its working principle are depicted in Figs. 2.7a and 2.7b. To operate a Ph–optoelectronic tweezer, a bipolar junction phototransistor must exhibit a high photoconductance (10 times higher than the liquid conductivity) while the dark conductivity should be 10 times lower. In the dark, there is no base current and the phototransistors are in the off state, resulting in a low dark conductance. As a result, most of the voltage drops across the transistors. Under illumination, the absorbed photons generate an effective base current in the base-collector reverse-biased junction, which is further amplified by the phototransistor gain. Thus the phototransistors are in the on state, resulting in a high photoconductance, which transfers most of the voltage drop across the liquid media. Thus a significant amount of non-uniform electric field is generated to induce DEP in the suspended particles or cells. Therefore, the most important parameters of a phototransistor are photoconductivity and dark conductivity, which can be tuned by varying the doping profile. Apart from Ph–optoelectronic tweezers, methods such as organic photoconductive materials were also proposed to replace the amorphous silicon. Yang et al. demonstrated such an optoelectronic tweezer device with Y-TiOPc as the photoconductive layer [80]. Figure 2.7c shows a TiOPc-based device where an a-Si:H photoconductive layer is replaced by a Y-TiOPc photoconductive layer. This TiOPc-based optoelectronic tweezer (Ti–optoelectronic tweezer) device was used to deliver a light-driven negative DEP force to the cells to pattern them in a specific manner [80].

It has been realized that an optoelectronic tweezer is a powerful tool for the manipulation of non-adhesive particles such as polystyrene beads [81–83]. In contrast, it is more challenging to manipulate adherent mammalian cells because of their nonspecific adhesion to optoelectronic tweezer device surfaces. This nonspecific adhesion of cells is a well-known event that most of the biomedical devices experience [84, 85], and an optoelectronic tweezer is not an exception. The native oxide that is present on the surface of an optoelectronic tweezer device acts as a charged hydrophilic surface [86]. The mammalian cells tend to adhere to this surface with a force on the order of a few nanonewtons [86]. Thus, to prevent cell adhesion to the substrate surfaces in the optoelectronic tweezer chamber, the optoelectronic tweezer chamber surface has to be cleaned thoroughly before adding the cells [87]. However, this is not a permanent solution for multiple uses of the same device. Another remedy is the proper passivation of the surface of the optoelectronic tweezer chamber. The passivation layers must be sufficiently thin, uniform, and nonconductive to avoid unwanted voltage drop followed by an undesired change of the field distribution, which can affect the performance of optoelectronic tweezer devices. In 2009, Lau et al. proposed two types of antifouling coatings that can fulfill these requirements: a bovine serum albumin coating and a poly(ethylene glycol) coating [86].

2.4.2 Applications of an Optoelectronic Tweezer

The effectively reconfigurable virtual electrode patterns, generated by optically projected images in an optoelectronic tweezer microfluidic chip, exclude the used of fixed electrodes that facilitate dynamic, parallel, and programmable micro-/nanoscaled manipulation of dielectric particles. As a consequence, this technique has been extensively exploited in a series of applications in biological fields as a potential non-invasive and nondestructive platform. There are several lab demonstrations of optoelectronic tweezers for the trapping and manipulation of several biological cells, such as human B cells [77], protozoan cells [88], HeLa cells [78, 86, 89–92], murine red and white blood cells [93–95], human sperm cells [96], fibroblasts cells and oral cancer cells [97, 98], human prostate cancer cells [99], human liver cells [80], porcine oocytes [100], and fibrosarcoma cells [101]. The applications of this technique in cell biological research have been briefly described in the following paragraphs.

2.4.2.1 Cell lysis

Cell lysis is a crucial phenomenon to extract protein and nucleic acid for a variety of biological research fields. There are several successful techniques for cell lysis, such as chemical [102, 103], enzyme-induced [104, 105], optical [106, 107], electrical [108–110], thermal [111, 112], acoustic [113, 114], mechanical [115], and electrochemical [116, 117] approaches. Among these techniques, the superiority of the optoelectronic tweezer technique is its ability to lyse the membrane of a specific single cell within a group of cells without damaging the cell nucleus. Red blood cells, fibroblast cells, and oral cancer cells were lysed by using an optoelectronic tweezer [97, 98, 118]. It was reported that a bigger illumination spot diameter and a higher illumination power density can cause a higher cell lysis rate due to the generation of more electron-hole pairs in amorphous silicon. High power results in a higher potential drop across the liquid layer [97, 98].

2.4.2.2 Cell trapping, alignment, and patterning

Since the strength of a light-driven DEP force depends on the incident light intensity, an optoelectronic tweezer is capable of exerting variable forces at the same instance by using a gray-scale optical image. This technique was demonstrated by Choi et al. [88] on highly motile protozoan cells to align them in a vertical direction. The behaviors of the aligned cells depend on the brightness of different positions, due to the variations in the strength of the local electric field acting on each cell [88]. Optoelectronic tweezer devices were also used to pattern white blood cells [95] and the delicate human hepatocellular liver carcinoma cell line (HepG2 cells) [80]. The real-time image patterning of HepG2 cells is shown in Fig. 2.8.

2.4.2.3 Electroporation

In conventional electroporation, a high external voltage is applied in a cuvette containing millions of cells together, resulting in the generation of a strong electric field (a few kilovolts per centimeter) in the vicinity of the cells. The potential drop varies throughout the cell due to variation in its dielectric properties, and it is completely different from the outside to the inside of the cell membrane. This potential difference across the cell membrane is called transmembrane potential. When the values of the transmembrane potential exceed the cell membrane threshold value, hydrophilic transient membrane pores are created that allow the delivery of molecules from the outside to the inside of the cell [2, 119–128]. An optoelectronic tweezer device was also demonstrated to perform electroporation of a HeLa cell [89]. It was mainly a two-step process: cell manipulation and cell electroporation. In the first step, a low electric field (intensity $\sim$0.2 kV/cm) was applied to create a 2×2 cell array through photoinduced DEP. In the next step, a sufficiently high electroporation bias (1.5 kV/cm) was applied, which can create

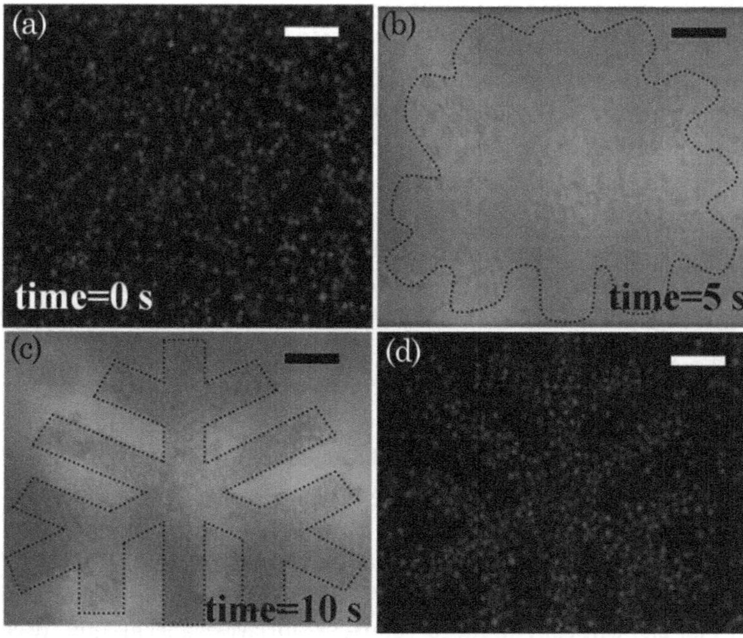

Figure 2.8 The real-time image recording of HepG2 cell patterning by using the light-driven optoelectronic n-DEP manipulation. (a–c) light-induced optoelectronic n-DEP occurs when the AC potential at 30 kHz is applied to the Ti–optoelectronic tweezer chip. Initially, HepG2 cells are randomly distributed and then trapped within the contours of the unilluminated region shown in (a) and (b), respectively. When the applied frequency is increased to megahertz, the n-DEP turns into light-induced p-DEP, which is an attractive force for cells. (d) Cell pattern by prelabeling HepG2 cells. Reprinted with permission from Ref. [80] © The Optical Society.

permeable transient pores at the surface of the cell membrane, resulting in electroporation followed by PI dye uptake [89].

2.4.2.4 Cell selection, identification, and separation

The selection of viable sperms from among nonmotile human sperms is crucial for reproductive technologies. It was reported that viable sperms were attracted to optoelectronic tweezer fields and can be induced to move at an average maximum velocity of 8.8 ± 4.2 μm s^{-1}, while nonviable sperms were repelled away from optoelectronic tweezer fields and were induced to move at an average maximum velocity of -0.8 ± 1 μm s^{-1} [96]. Thus, an optoelectronic tweezer was successfully applied to identify viable nonmotile sperms by comparing their responses to the applied non-uniform electric field [96]. In the field of cancer research, an optoelectronic tweezer was also shown to be a potential platform. To analyze the drug-resistant gene expression of cancer patients, highly pure circulating tumor cells (CTCs) are essential, which, indeed, allows the selection of an appropriate therapeutic regimen for achieving personalized chemotherapy. Higher order of CTC purity helps to achieve higher analytical sensitivity. An optoelectronic tweezer was applied to isolate CTCs from the mixture of leukocytes in a cell sample, with cell purity as high as 100% under the guidance of fluorescent microscopic observation [99, 129]. Another example of cell selection and storage has been demonstrated by Zhang et al. [83] on MCF-7 breast cancer cells and ARPE-19 human retinal pigment epithelial cells. The processes of cell selection and storage by using optoelectronic tweezer operation are depicted in Fig. 2.9. Optically induced cellular translation and self-rotation responses in an optoelectronic tweezer device for various applied voltages, frequencies, and electrode patterns can be used as a unique identification tool of cells [95, 130–132]. Therefore, these distinct responses can be used as label-free biomarkers for various clinical applications. This technique has also been demonstrated to separate Raji cells from red blood cells in a label-free process [131].

2.4.2.5 Cellular intrinsic properties

The intrinsic properties of cells, such as volume, density, mass, membrane capacitance, and conductance, are the key intrinsic information seamlessly correlated with the cellular morphology and dielectric properties. In-depth knowledge of these parameters is crucial for clinical applications. Measuring these parameters at the single-cell level in an accurate, selective, and repeatable manner is a challenging task in cell biology. An optically induced dielectrophoresis (ODEP) platform was applied to measure the mass and density of yeast cells, human leukemic cells (HL-60), and mouse lymphocytic leukemia cells (L1210) [133, 134]. The

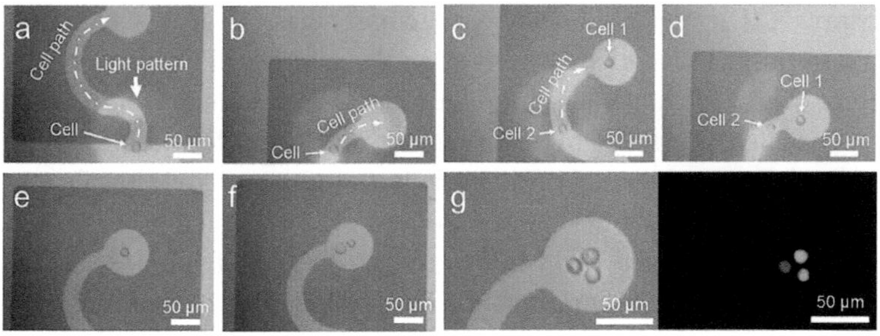

Figure 2.9 (a–g) Steps in the selection and movement of MCF-7 breast cancer cells in optoelectronic tweezer doughnut traps followed by their transfer and storage in a p-optoelectronic tweezer trap. Reproduced from Ref. [83], copyright (2018) by permission of John Wiley and Sons.

assessment was made on the basis of the particle's settlement in a fluidic media with different viscosity and density values in an ODEP device. The results showed good agreement with the results of the existing method of the suspended microchannel resonator [133, 134]. Furthermore, the cell membrane capacitance and conductance of Raji cells, MCF-7 cells, HEK293 cells, and K562 cells were successfully determined using an ODEP microfluidic chip by monitoring the translational behaviors of cells under various applied frequencies and determining the crossover frequencies [135]. Thus an optoelectronic tweezer turns out to be a powerful tool for determining the intrinsic properties of individual cells in a non-invasive manner. It has proven to be a versatile and strongly potential cell manipulation technique in the field of advanced biological research and its multidisciplinary applications.

2.5 Rapid Electrokinetic Patterning

REP is a complex opto-electrokinetic process driven by the simultaneous application of laser illumination and a uniform AC electric field. It was first proposed by Williams et al. in 2008 [136]. The difference between REP and OT lies in the underlying principle of utilizing optical energy in two completely different landscapes. For OT, light is directly used to create radiation pressure on a particle. In contrast, REP involves highly focused patterns of laser light to create a non-uniform local heating effect in the microfluidic environment. Here, in the presence of illumination, a high-frequency alternating electric field is applied to the drive vortex flow inside microchannels of microfluidic devices [136]. The velocity of the microvortex flow is decided by the laser illumination intensity, the dielectric properties of the medium, and the voltage and frequency of the applied AC signal [136, 137]. The microvortex can be exploited for multipurpose activities of cell manipulation in advance biological research such as sorting, trapping, translating and patterning of biological particles and the fabrication of bacterial biofilms, etc. The underlying working principle of REP is described next.

2.5.1 Setup and Working Principle

The schematic REP chip and the arrangement of the experimental setup are illustrated in Figs. 2.10a and 2.10b, respectively. The REP chip consists of two conductive ITO or metal-coated transparent glass plates separated in parallel by a spacer with a thickness of a few tenths to hundreds of micrometers. Typically, a commercial double-sided adhesive tape, patterned with a razor or laser cutter, is used as the spacer. The laser-irradiated spot is applied on the bottom plate. The coated glass substrates serve as electrodes for applying the AC voltage. Instead of an ITO coating, more conductive metals, such as copper, gold, and chrome-coated electrodes, can be used, which can absorb sufficient illumination to induce the ETF. Typically, an NIR (1064 nm) laser is used for optical illumination.

Laser illumination induces a non-uniform thermal gradient in the medium, which in turn causes local variation in the fluid permittivity and conductivity in the microfluidic system. As a result, a bulk ETF is generated around the laser-illuminated area. In addition, a convective flow is also created, which is observed to be negligible in these systems. When the voltage is applied between the electrodes, ACEO is induced because the local non-uniform AC electric field acts on the ions of the fluidic medium. Although the applied electric field is supposed to be uniform at the space between the two electrodes, the local electric fields adjacent to the particles are distorted by the particle itself, resulting in non-uniform fields that can induce

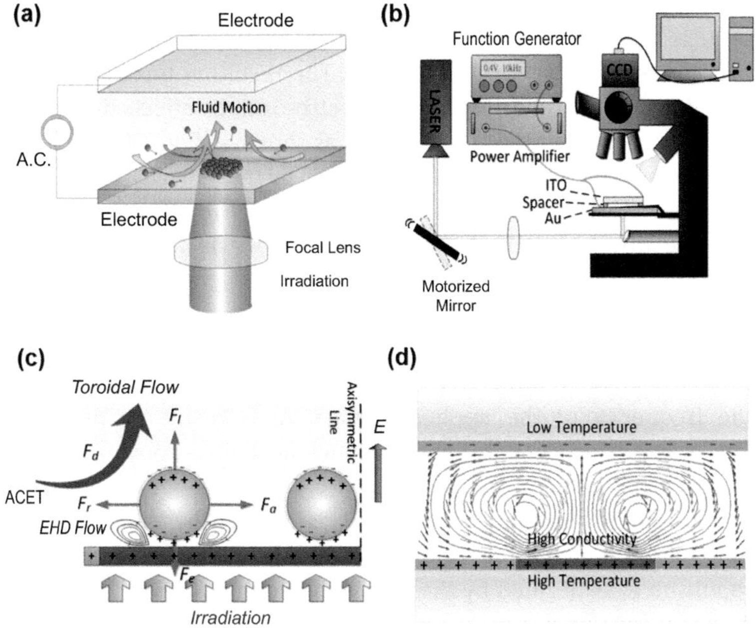

Figure 2.10 (a) Illustration of REP operation. (b) Schematic of the experimental setup. (c) The schematic representation of an induced dipole and the effective forces acting on the particle. (d) The toroidal vortex flow from a global viewpoint. Republished with permission of The Royal Society of Chemistry, from Ref. [138], copyright (2014); permission conveyed through Copyright Clearance Center, Inc.

DEP and ACEO. The ACEO results in a unidirectional flow and is effective in the frequency range of 10 Hz to 100 kHz. The details of these phenomena are described after the introduction section of this chapter. The combination of these mechanisms acts collectively in microfluidic systems to induce an effective electrohydrodynamic (EHD) force [139]. EHD forces create liquid motion that drags micro-objects along streamlines, and these forces can be applied to fulfill several purposes, such as trapping, patterning, and particle separation. For simplicity of understanding, the effective forces that determine the overall optoelectrokinetic behavior of the particles can be divided into four components, as illustrated in Fig. 2.10c [138]. Here, $\mathbf{F}_l$ represents a lift force, which moves the particles away from the bottom electrode. $\mathbf{F}_a$ denotes the force for particle-particle attraction. $\mathbf{F}_r$ represents particle-particle repulsion, and $\mathbf{F}_e$ denotes particle-electrode adhesion. The resulting toroidal vortex flow due to electrothermal and AC electro-osmotic effects drives the fluid toward the center of irradiation, as shown in Fig. 2.10d.

The effective lift force $\mathbf{F}_l$ is determined by the bulk ETF [140]. The particle-electrode-and particle-wall-induced hydrodynamic force and the disruption of the bulk flow in particle aggregates may also contribute to $\mathbf{F}_l$ [138]. If the particle aggregation becomes highly compact, it causes instability within the aggregation. Then the electric-field-induced dipole-dipole interaction may also contribute to the effective lift force. ETF decreases exponentially on increasing the frequency of the applied voltage. As a result, $\mathbf{F}_l$ peaks at low frequencies ($f < f_L$) and becomes negligible above a critical frequency ($f > f_H$) [140].

The particle-particle attraction $\mathbf{F}_a$ is decided by the local EHD flow and dipole-dipole interaction [138]. The particle-particle repulsion $\mathbf{F}_r$ is decided by the dipole-dipole interaction [138, 141]. The induced dipoles result in attraction between particles when particles are aligned parallel to the electric field but turn into repulsion when the particles are perpendicular to the electric field, regardless of the frequency [138, 141]. At low frequencies ($f < f_L$), the toroidal ETF drives the particles to the irradiated spot and then builds a compact cluster through particle-particle attraction $\mathbf{F}_a$ [138]. The trapped particles stay near the bottom electrode surface since the particle-electrode attraction $\mathbf{F}_e$ decays rapidly with distance. They float slightly above the bottom electrode surface due to the lift force $\mathbf{F}_l$. The particle-electrode interaction arises due to the EDL induced by ACEO, gravitational force, and DEP [138, 142]. The EDL-induced $\mathbf{F}_e$ peaks appear at low frequencies but disperse when the frequency is above the critical value f_L ($f > f_L$). This happens because the ions become incapable of migrating as fast as the polarized electric field [138]. While EDL polarization relaxes above f_L, the Maxwell–Wagner (MW)-induced dielectrophoretic forces start to dominate. In the transitional phase from EDL polarization to MW polarization ($f_L < f < f_M$), $\mathbf{F}_a$, $\mathbf{F}_r$, and $\mathbf{F}_e$ drop significantly [138]. Since the ETF-induced force $\mathbf{F}_l$ decreases more slowly than the MW-induced particle-electrode adhesion $\mathbf{F}_e$, the particles cannot aggregate strongly on the electrode surface [138]. As a result, there is no particle aggregation throughout this frequency range. When the frequency increases in-between f_M and f_H ($f_M < f < f_H$), particle reaggregation can be observed again because the ETF constantly reduces, leading $\mathbf{F}_r$ and $\mathbf{F}_l$ to drop below $\mathbf{F}_e$ [138]. However, this reaggregation is loose and weak compared to the trapping at low frequencies, due to the reduced local EHD-induced attraction ($\mathbf{F}_a$) and weak ETF [138]. Furthermore, gravity-induced particle sedimentation is significantly observed due to a lack of a strong drag force to lift the particles [138]. Eventually, the optoelectrokinetic trapping is terminated for the frequency above f_H because of the extinction of the ETF.

2.5.2 Biological Applications of REP

Most of the experiments of REP are conducted on particles such as nanoparticles, nanobeads, and carbon nanotubes. This technique is yet to be explored in the field of advanced cell biological research. There are few demonstrations of biological samples. In the following section, we discuss the application of REP in the aspect of cell biological research.

2.5.2.1 Patterning, translating, and sorting

Kwon et al. demonstrated the manipulations of motile *Shewanella oneidensis* MR-1 bacteria with the REP technique [143]. They applied uniform AC electric signals with a frequency 18.69 kHz, a peak-to-peak voltage of 17.8 V, and laser (1064 nm) power of around 20 mW. Significant aggregation of the bacteria was observed due to the REP process, which was explained stepwise as follows: When an AC electric field was applied, the EHD flow attracted them to an electrode surface. There, they experienced reorientation caused by the frequency-dependent torque due to their geometric shape (prolate ellipsoids). This torque aligned the principal axis of the bacterial body parallel to the applied electric field lines. At the same time, they experienced the repulsive forces due to induced dipole moments. The competition between these repulsive forces and attractive forces resulted in an irregular configuration

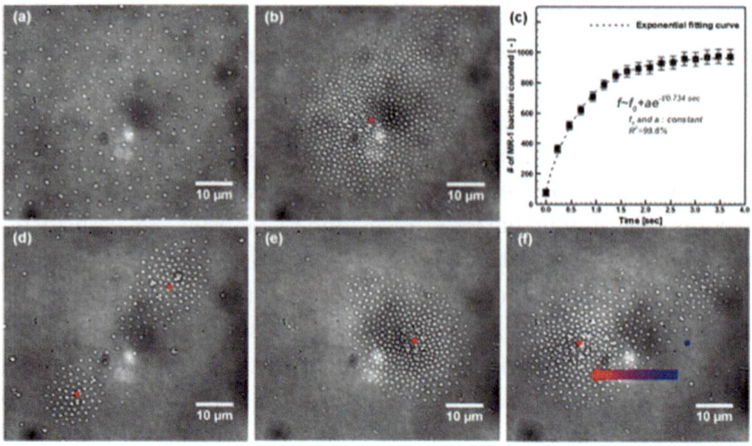

Figure 2.11 Patterning of S. *Oneidensis* MR-1 bacteria with the REP technique. (a) Irregular configuration of bacteria due to competition between the attractive forces due to induced dipole moment and repulsive force due to EHD forces. (b) Aggregation of the bacteria by the REP process at the laser-illuminated area. The red dot represents the position of the focused laser beam. (c) The plot of the number of aggregated bacteria with time after application of the optoelectrokinetic technique. (d) Patterning the bacteria with two laser illumination spots. (e–f) Translation of the assembly by moving the laser beam. The blue dot in (f) represents the initial location and the actual size of the beam before the movement, and the arrow represents the direction of the laser movement. Republished with permission of The Royal Society of Chemistry, from Ref. [143], copyright (2012); permission conveyed through Copyright Clearance Center, Inc.

of bacteria on the surface of the electrode (Fig. 2.11a). Upon laser illumination, temperature-dependent gradients of permittivity and conductivity in the medium were induced, resulting in an electrothermal vortex. The toroidal microvortex continuously drove the upright MR-1 bacteria toward the illuminated area on the electrode, resulting in a bacterial aggregation at the non-uniform temperature region (Fig. 2.11b). Around 700 MR-1 bacteria were observed to aggregate on the electrode surface in a time period of 0.73 s, and the process was saturated when all of the bacteria within the electrothermal vortex had been collected (Fig. 2.11c). Furthermore, two focal spots were used to examine the patterning capability of REP (Fig. 2.11d). Experiments were conducted to translate the assembly by changing the position of laser spots (Fig. 2.11e,f). All of these experiments showed a negligible impact of bacterial death, revealing the biocompatibility of the REP process. Therefore, REP can be exploited potentially in cell biological research.

2.6 Conclusions

Since the first experiments on optical manipulation of biological cells in the 1970s, it has emerged as a powerful and versatile tool to explore the mysteries of cellular behavior. With progress in optical instrumentations, several strategies have been discovered to implement them in manipulation techniques. Optical tweezers, with their unique control capabilities of manipulation, exhibit intrinsic properties of cells even at the subcellular level. Furthermore, several clever contrivances have been applied for the indirect manipulation of single molecules to uncover their static and dynamic properties through force measurements in the range of piconewtons. This is indeed a strength of OT compared to other methods, such as the microelectromechanical approach, and allows measurements down to nanonewtons. Therefore, OT provides finer manipulation opportunities with its high spatial (å) and temporal (μS) resolution capabilities for exploring cell biological characteristics. The current state of the art is to improve the speed of these systems through a robotic controlled programming interface, which not only speeds up the measurements but also improves the accuracy for rapid biological characterizations. Instead of imparting direct optical radiation pressure, OET and REP invoke indirect involvement of optical energy by exploiting their postexposure light-matter interactions. For instance, OET allows the precise control of a large number of cells for parallel manipulation, which can be useful for statistical studies of the heterogeneous behavior of cells. REP is still an emerging technique that involves a diverse level of complex physical processes. These techniques are easy to implement with lab-on-a-chip or microfluidic devices. In combination with advanced microscopy techniques, they have become powerful tools to unveil intracellular and extracellular behavior of cells. Thus optical manipulation has extensive applications in the field of biology, medicine, and pharmacy and offers versatile new opportunities to explore specific targeted therapy and diagnostics toward the treatment of human diseases.

Acknowledgments

The authors greatly appreciate the financial support from the Science and Engineering Research Board (SERB) under grant number ECR/2016/001945, Department of Science and Technology (DST), Government of India, and the Wellcome Trust/DBT India Alliance Fellowship under grant number IA/E/16/1/503062. We acknowledge all authors and publishers who provided the copyright permissions.

References

1. T. S. Santra, F.-G. Tseng (2021). *Handbook of Single Cell Technology*. Springer Nature.
2. F.-G. Tseng, T. S. Santra (2016). *Essentials of Single-Cell Analysis*. Springer-Verlag, Berlin Heidelberg, Germany.
3. F.-G. Tseng, T. S. Santra (2016). *Single Cell Analysis in Biotechnology and System Biology*. MDPI AG Basel, Switzerland.
4. F.-G. Tseng, T. S. Santra (2015). *Micro/Nano Fluidic Devices for Single Cell Analysis*. MDPI AG Basel, Switzerland.
5. P. Shinde, et al. (2018). Current trends of microfluidic single-cell technologies. *Int J Mol Sci*, **19**(10):3143.
6. T. S. Santra, T.-H. Wu, P. Y. Chiou (2016). Photothermal microfluidics. In *Optical MEMS for Chemical Analysis and Biomedicine*, H. Jiang, ed. Institution of Engineering and Technology.
7. J. Li, L. Lin, Y. Inoue, Y. Zheng (2018). Opto-thermophoretic tweezers and assembly. *J Micro Nano-Manuf*, **6**(4):040801.
8. A. Keloth, O. Anderson, D. Risbridger, L. Paterson (2018). Single cell isolation using optical tweezers. *Micromachines*, **9**:434.
9. P. Grigaravičius, K. O. Greulich, S. Monajembashi (2009). Laser microbeams and optical tweezers in ageing research. *ChemPhysChem*, **10**(1):79–85.

10. J. Van Mameren (2002). Single molecule mechanics of biopolymers: an optical tweezers study.
11. M. Uchida, M. Sato-Maeda, H. Tashiro (1995). Micromanipulation: whole-cell manipulation by optical trapping. *Curr Biol*, **5**(4):380–382.
12. Y. Z. Yoon, P. Cicuta (2010). Optical trapping of colloidal particles and cells by focused evanescent fields using conical lenses. *Opt Express*, **18**(7):7076.
13. H. Li (2002). Dielectrophoretic separation and manipulation of live and heat-treated cells of Listeria on microfabricated devices with interdigitated electrodes. *Sens Actuators, B*, **86**(2–3):215–221.
14. H. A. Pohl, J. S. Crane (1971). Dielectrophoresis of cells. *Biophys J*, **11**(9):711–727.
15. R. Pethig, A. Menachery, S. Pells, P. De Sousa (2010). Dielectrophoresis: a review of applications for stem cell research. *J Biomed Biotechnol*, **2010**:1–7.
16. L. Kremser, D. Blaas, E. Kenndler (2004). Capillary electrophoresis of biological particles: viruses, bacteria, and eukaryotic cells. *Electrophoresis*, **25**(14):2282–2291.
17. J. N. Mehrishi, J. Bauer (2002). Electrophoresis of cells and the biological relevance of surface charge. *Electrophoresis*, **23**(13):1984.
18. A. J. Dickinson, P. M. Armistead, N. L. Allbritton (2013). Automated capillary electrophoresis system for fast single-cell analysis. *Anal Chem*, **85**(9):4797–4804.
19. J. D. Adams, C. L. Ebbesen, R. Barnkob, A. H. J. Yang, H. T. Soh, H. Bruus (2012). High-throughput, temperature-controlled microchannel acoustophoresis device made with rapid prototyping. *J Micromech Microeng*, **22**(7):075017.
20. P. Augustsson, C. Magnusson, M. Nordin, H. Lilja, T. Laurell (2012). Microfluidic, label-free enrichment of prostate cancer cells in blood based on acoustophoresis. *Anal Chem*, **84**(18):7954–7962.
21. H. A. Pohl (1951). The motion and precipitation of suspensoids in divergent electric fields. *J Appl Phys*, **22**(7):869–871.
22. R. Pethig (2010). Review article—Dielectrophoresis: status of the theory, technology, and applications. *Biomicrofluidics*, **4**(2):022811.
23. J. Voldman (2006). Electrical forces for microscale cell manipulation. *Annu Rev Biomed Eng*, **8**(1):425–454.
24. D. Lee, B. Hwang, B. Kim (2016). The potential of a dielectrophoresis activated cell sorter (DACS) as a next generation cell sorter. *Micro Nano Syst Lett*, **4**(1):2.

25. M. Walid Rezanoor, P. Dutta (2016). Combined AC electroosmosis and dielectrophoresis for controlled rotation of microparticles. *Appl Phys Rev*, **10**(2):024101.
26. N. G. Green, A. Ramos, A. González, H. Morgan, A. Castellanos (2002). Fluid flow induced by nonuniform AC electric fields in electrolytes on microelectrodes. III. Observation of streamlines and numerical simulation. *Phys Rev E Stat Nonlin Soft Matter Phys*, **66**(2 Pt 2):026305.
27. A. Salari, M. Thompson (2018). Recent advances in AC electrokinetic sample enrichment techniques for biosensor development. *Sens Actuators, B*, **255**:3601–3615.
28. W. Liu, Y. Ren, Y. Tao, B. Yao, Y. Li (2018). Simulation analysis of rectifying microfluidic mixing with field-effect-tunable electrothermal induced flow. *Electrophoresis*, **39**(5–6):779–793.
29. A. Ashkin (1970). Acceleration and trapping of particles by radiation pressure. *Phys Rev Lett*, **24**(4):156–159.
30. A. Ashkin, J. Dziedzic (1987). Optical trapping and manipulation of viruses and bacteria. *Science (80-)*, **235**(4795):1517–1520.
31. A. Ashkin (1992). Forces of a single-beam gradient laser trap on a dielectric sphere in the ray optics regime. *Biophys J*, **61**(2):569–582.
32. K. C. Neuman, S. M. Block (2004). Optical trapping. *Rev Sci Instrum*, **75**(9):2787–2809.
33. A. Ashkin (1997). Optical trapping and manipulation of neutral particles using lasers. *Proc Natl Acad Sci USA*, **94**:4853–4860.
34. V. Garcés-Chávez, D. McGloin, H. Melville, W. Sibbett, K. Dholakia (2002). Simultaneous micromanipulation in multiple planes using a self-reconstructing light beam. *Nature*, **419**(6903):145–147.
35. Y. Zhang, et al. (2017). Multiple particles 3-D trap based on all-fiber bessel optical probe. *J Lightwave Technol*, **35**(18):3849–3853.
36. R. Dasgupta, S. Ahlawat, R. S. Verma, S. Shukla, P. K. Gupta (2010). Optical trapping of spermatozoa using Laguerre-Gaussian laser modes. *J Biomed Opt*, **15**(6):065010.
37. H. Zhang, K.-K. Liu (2008). Optical tweezers for single cells. *J R Soc Interface*, **5**(24):671–690.
38. A. E. Chiou, W. Wang, G. J. Sonek, J. Hong, M. Berns (1997). Interferometric optical tweezers. *Opt Commun*, **133**(1–6):7–10.
39. J. Guck, R. Ananthakrishnan, H. Mahmood, T. J. Moon, C. C. Cunningham, J. Käs (2001). The optical stretcher: a novel laser tool to micromanipulate cells. *Biophys J*, **81**(2):767–784.

40. X. Chen, G. Xiao, X. Han, W. Xiong, H. Luo, B. Yao (2017). Observation of spin and orbital rotation of red blood cell in dual-beam fibre-optic trap with transverse offset. *J Opt*, **19**(5):055612.
41. B. Lincoln, et al. (2007). Reconfigurable microfluidic integration of a dual-beam laser trap with biomedical applications. *Biomed Microdevices*, **9**(5):703–710.
42. A. Constable, J. Kim, J. Mervis, F. Zarinetchi, M. Prentiss (1993). Demonstration of a fiber-optical light-force trap. *Opt Lett*, **18**(21):1867.
43. A. L. Birkbeck, R. A. Flynn, M. Ozkan, D. Song, M. Gross, S. C. Esener (2003). VCSEL arrays as micromanipulators in chip-based biosystems. *Biomed Microdevices*, **5**(1):47–54.
44. B. Shao, S. Zlatanovic, M. Ozkan, A. L. Birkbeck, S. C. Esener (2006). Manipulation of microspheres and biological cells with multiple agile VCSEL traps. *Sens Actuators, B*, **113**(2):866–874.
45. J.-M. Fournier, F. Merenda, J. Rohner, P. Jacquot, R. P. Salathé (2008). Comparison between various types of multiple optical tweezers. *Proc SPIE*, **7038**:70381G.
46. M. Righini, et al. (2018). Full molecular trajectories of RNA polymerase at single base-pair resolution. *Proc Natl Acad Sci*, **115**(6):201719906.
47. G. Dwivedi, A. Gupta, M. Shukla, S. Kanaujia, S. Yede, J. T. Andrews (2014). Trapping sub-micron size particles in holographic optical tweezers. *J Phys Conf Ser*, **534**(1):012059.
48. A. Ashkin, J. M. Dziedzic (1989). Internal cell manipulation using infrared laser traps. *Proc Natl Acad Sci*, **86**(20):7914–7918.
49. K. C. Neuman, E. H. Chadd, G. F. Liou, K. Bergman, S. M. Block (1999). Characterization of photodamage to Escherichia coli in optical traps. *Biophys J*, **77**(5):2856–2863.
50. Z. Pilát, J. Ježek, M. Šerý, M. Trtílek, L. Nedbal, P. Zemánek (2013). Optical trapping of microalgae at 735–1064 nm: photodamage assessment. *J Photochem Photobiol, B*, **121**:27–31.
51. K. König, Y. Tadir, P. Patrizio, M. W. Berns, B. J. Tromberg (1996). Effects of ultraviolet exposure and near infrared laser tweezers on human spermatozoa. *Hum Reprod*, **11**(10):2162–2164.
52. H. Liang, et al. (1996). Wavelength dependence of cell cloning efficiency after optical trapping. *Biophys J*, **70**(3):1529–1533.
53. Y. Liu, D. K. Cheng, G. J. Sonek, M. W. Berns, C. F. Chapman, B. J. Tromberg (1995). Evidence for localized cell heating induced by infrared optical tweezers. *Biophys J*, 68(5):2137–2144.

54. T. A. Ohta, et al. (2010). Optoelectronic tweezers for the manipulation of cells, microparticles, and nanoparticles. In *Recent Optical and Photonic Technologies*, InTech, pp. 18–388.

55. Y. Oshima, A. Suzuki, K. Kawashimo, M. Ishikawa, N. Ohkohchi, H. Taniguchi (2007). Isolation of mouse pancreatic ductal progenitor cells expressing CD133 and c-met by flow cytometric cell sorting. *Gastroenterology*, **132**(2):720–732.

56. L. Qiaozhen, et al. (2009). Development of different analysis platforms with LC-MS for pharmacokinetic studies of protein drugs. *Anal Chem*, **81**(21):8715–8723.

57. J. A. Lundqvist, F. Sahlin, M. A. Aberg, A. Strömberg, P. S. Eriksson, O. Orwar (1998). Altering the biochemical state of individual cultured cells and organelles with ultramicroelectrodes. *Proc Natl Acad Sci USA*, **95**(18):10356–10360.

58. X. Wang, Z. Wang, D. Sun (2010). Cell sorting with combined optical tweezers and microfluidic chip technologies. In *2010 11th International Conference on Control Automation Robotics & Vision*, pp. 201–206.

59. X. Wang, et al. (2011). Enhanced cell sorting and manipulation with combined optical tweezer and microfluidic chip technologies. *Lab Chip*, **11**(21):3656.

60. R. Applegate, J. Squier, T. Vestad, J. Oakey, D. Marr (2004). Optical trapping, manipulation, and sorting of cells and colloids in microfluidic systems with diode laser bars. *Opt Express*, **12**(19):4390–4398.

61. R. W. Steubing, S. Cheng, W. H. Wright, Y. Numajiri, M. W. Berns (1991). Laser induced cell fusion in combination with optical tweezers: the laser cell fusion trap. *Cytometry*, **12**(6):505–510.

62. X. Wang, S. Chen, Y. T. Chow, C. W. Kong, R. A. Li, D. Sun (2013). A microengineered cell fusion approach with combined optical tweezers and microwell array technologies. *RSC Adv*, **3**(45):23589–23595.

63. M.-C. Zhong, X.-B. Wei, J.-H. Zhou, Z.-Q. Wang, Y.-M. Li (2013). Trapping red blood cells in living animals using optical tweezers. *Nat Commun*, **4**(1):1768.

64. X. Gou, R. Wang, S. S. Y. Lam, J. Hou, A. Y. H. Leung, D. Sun (2015). Cell adhesion manipulation through single cell assembly for characterization of initial cell-to-cell interaction. *Biomed Eng Online*, **14**(1):114.

65. A. G. Banerjee, S. Chowdhury, W. Losert, S. K. Gupta (2011). Survey on indirect optical manipulation of cells, nucleic acids, and motor proteins. *J Biomed Opt*, **16**(5):051302.

66. M. Welch (2015). *Biology: The Dynamic Science*, 4th ed.
67. M. Schliwa, G. Woehlke (2003). Molecular motors. *Nature*, **422**:759.
68. B. Alberts, A. Johnson, J. Lewis, M. Raff, K. Roberts, P. Walter (2002). *Molecular Biology of the Cell*, 4th ed. Garland Science, New York.
69. J. A. Spudich, S. E. Rice, R. S. Rock, T. J. Purcell, H. M. Warrick (2011). Optical traps to study properties of molecular motors. *Cold Spring Harb Protoc*, **2011**(11):pdb.top066662-pdb.top066662.
70. J. A. Spudich, S. E. Rice, R. S. Rock, T. J. Purcell, H. M. Warrick (2011). The optical trapping dumbbell assay for nonprocessive motors or motors that turn around filaments. *Cold Spring Harb Protoc*, **2011**(11):1372–1374.
71. J. Zhou, K. S. Ha, A. La Porta, R. Landick, S. M. Block (2011). Applied force provides insight into transcriptional pausing and its modulation by transcription factor NusA. *Mol Cell*, **44**(4):635–646.
72. K. M. Herbert, et al. (2006). Sequence-resolved detection of pausing by single RNA polymerase molecules. *Cell*, **125**(6):1083–1094.
73. K. C. Neuman, E. A. Abbondanzieri, R. Landick, J. Gelles, S. M. Block (2003). Ubiquitous transcriptional pausing is independent of RNA polymerase backtracking. *Cell*, **115**(4):437–447.
74. E. A. Abbondanzieri, W. J. Greenleaf, J. W. Shaevitz, R. Landick, S. M. Block (2005). Direct observation of base-pair stepping by RNA polymerase. *Nature*, **438**(7067):460–465.
75. R. Gabizon, A. Lee, H. Vahedian-Movahed, R. H. Ebright, C. J. Bustamante (2018). Pause sequences facilitate entry into long-lived paused states by reducing RNA polymerase transcription rates. *Nat Commun*, **9**(1):2930.
76. B. Zamft, L. Bintu, T. Ishibashi, C. Bustamante (2012). Nascent RNA structure modulates the transcriptional dynamics of RNA polymerases. *Proc Natl Acad Sci*, **109**(23):8948–8953.
77. P. Y. Chiou, A. T. Ohta, M. C. Wu (2005). Massively parallel manipulation of single cells and microparticles using optical images. *Nature*, **436**(7049):370–372.
78. H. Hsu, A. T. Ohta, P.-Y. Chiou, A. Jamshidi, S. L. Neale, M. C. Wu (2010). Phototransistor-based optoelectronic tweezers for dynamic cell manipulation in cell culture media. *Lab Chip*, **10**(2):165–172.
79. Y. Yang, Y. Mao, K.-S. Shin, C. O. Chui, P.-Y. Chiou (2016). Self-locking optoelectronic tweezers for single-cell and microparticle manipulation across a large area in high conductivity media. *Sci Rep*, **6**(1):22630.
80. S.-M. Yang, T.-M. Yu, H.-P. Huang, M.-Y. Ku, L. Hsu, C.-H. Liu (2010). Dynamic manipulation and patterning of microparticles and cells

by using TiOPc-based optoelectronic dielectrophoresis. *Opt Lett*, **35**(12):1959.

81. M. J. Zheng, Y. L. Qu, Y. Z. Zhang, Z. L. Dong (2013). Optically induced dielectrophoresis based automatic assembly of micro/nano-devices. *Integr Ferroelectr*, **145**(1):24–31.
82. H.-H. Chen, W.-T. Tien, H.-H. Lo, C.-C. Lin, J.-C. Chen (2013). Optically-induced dielectrophoretic technology for particles manipulation and separation. In *2013 8th International Microsystems, Packaging, Assembly and Circuits Technology Conference (IMPACT)*, pp. 258–259.
83. S. Zhang, et al. (2018). Patterned optoelectronic tweezers: a new scheme for selecting, moving, and storing dielectric particles and cells. *Small*, **14**(45):1803342.
84. J. Yin, S. Luan (2016). Opportunities and challenges for the development of polymer-based biomaterials and medical devices. *Regen Biomater*, **3**(2):129–135.
85. S. Halldorsson, E. Lucumi, R. Gómez-Sjöberg, R. M. T. Fleming (2015). Advantages and challenges of microfluidic cell culture in polydimethylsiloxane devices. *Biosens Bioelectron*, **63**:218–231.
86. A. N. K. Lau, et al. (2009). Antifouling coatings for optoelectronic tweezers. *Lab Chip*, **9**(20):2952.
87. A. Zarowna-Dabrowska, et al. (2010). Miniaturised optoelectronic tweezers controlled by GaN micro-pixel light emitting diode arrays. In *2010 IEEE Photinic Society's 23rd Annual Meeting*, pp. 102–103.
88. W. Choi, S.-W. Nam, H. Hwang, S. Park, J.-K. Park (2008). Programmable manipulation of motile cells in optoelectronic tweezers using a grayscale image. *Appl Phys Lett*, **93**(14):143901.
89. J. K. Valley, S. Neale, H.-Y. Hsu, A. T. Ohta, A. Jamshidi, M. C. Wu (2009). Parallel single-cell light-induced electroporation and dielectrophoretic manipulation. *Lab Chip*, **9**(12):1714.
90. S. L. Neale, A. T. Ohta, H.-Y. Hsu, J. K. Valley, A. Jamshidi, M. C. Wu (2009). Trap profiles of projector based optoelectronic tweezers (OET) with HeLa cells. *Opt Express*, **17**(7):5232.
91. J. K. Valley, S. NingPei, A. Jamshidi, H.-Y. Hsu, M. C. Wu (2011). A unified platform for optoelectrowetting and optoelectronic tweezers. *Lab Chip*, **11**(7):1292.
92. K.-W. Huang, T.-W. Su, A. Ozcan, P.-Y. Chiou (2013). Optoelectronic tweezers integrated with lensfree holographic microscopy for wide-field interactive cell and particle manipulation on a chip. *Lab Chip*, **13**(12):2278.

93. H. Hwang, Y.-J. Choi, W. Choi, S.-H. Kim, J. Jang, J.-K. Park (2008). Interactive manipulation of blood cells using a lens-integrated liquid crystal display based optoelectronic tweezers system. *Electrophoresis,* **29**(6):1203–1212.
94. S. L. Neale, N. Mody, C. Selman, J. M. Cooper (2012). Optoelectronic tweezers for the measurement of the relative stiffness of erythrocytes. *Opt Trapp Opt Micromanipulation IX,* **8458**:845827.
95. L.-H. Chau, et al. (2013). Self-rotation of cells in an irrotational AC E-field in an opto-electrokinetics chip. *PLoS One,* **8**(1):e51577.
96. A. T. Ohta, et al. (2010). Motile and non-motile sperm diagnostic manipulation using optoelectronic tweezers. *Lab Chip,* **10**(23):3213.
97. Y.-H. Lin, G.-B. Lee (2009). An optically induced cell lysis device using dielectrophoresis. *Appl Phys Lett,* **94**(3):033901.
98. Y.-H. Lin, G.-B. Lee (2010). An integrated cell counting and continuous cell lysis device using an optically induced electric field. *Sens Actuators, B,* **145**(2):854–860.
99. T.-K. Chiu, et al. (2016). Application of optically-induced-dielectrophoresis in microfluidic system for purification of circulating tumour cells for gene expression analysis- cancer cell line model. *Sci Rep,* **6**(1):32851.
100. H. Hwang, D.-H. Lee, W. Choi, J.-K. Park (2009). Enhanced discrimination of normal oocytes using optically induced pulling-up dielectrophoretic force. *Biomicrofluidics,* **3**(1):014103.
101. M. Grad, A. W. Bigelow, G. Garty, D. Attinger, D. J. Brenner (2013). Optofluidic cell manipulation for a biological microbeam. *Rev Sci Instrum,* **84**(1):014301.
102. K. Tamura, T. Aotsuka (1988). Rapid isolation method of animal mitochondrial DNA by the alkaline lysis procedure. *Biochem Genet,* **26**(11–12):815–819.
103. H. C. Bimboim, J. Doly (1979). A rapid alkaline extraction procedure for screening recombinant plasmid DNA. *Nucleic Acids Res,* **7**(6):1513–1523.
104. B. A. Andrews, J. A. Asenjo (1987). Enzymatic lysis and disruption of microbial cells. *Trends Biotechnol,* **5**(10):273–277.
105. O. Salazar, J. A. Asenjo (2007). Enzymatic lysis of microbial cells. *Biotechnol Lett,* **29**(7):985–994.
106. H. Li, C. E. Sims, H. Y. Wu, N. L. Allbritton (2001). Spatial control of cellular measurements with the laser micropipet. *Anal Chem,* **73**(19):4625–4631.

107. P. A. Quinto-Su, H.-H. Lai, H. H. Yoon, C. E. Sims, N. L. Allbritton, V. Venugopalan (2008). Examination of laser microbeam cell lysis in a PDMS microfluidic channel using time-resolved imaging. *Lab Chip*, **8**(3):408.
108. S. K. Ameri, P. K. Singh, M. R. Dokmeci, A. Khademhosseini, Q. Xu, S. R. Sonkusale (2014). All electronic approach for high-throughput cell trapping and lysis with electrical impedance monitoring. *Biosens Bioelectron*, **54**:462–467.
109. F. Jiang, J. Chen, J. Yu (2016). Design and application of a microfluidic cell lysis microelectrode chip. *Instrum Sci Technol*, **44**(2):223–232.
110. N. de Lange, T. M. Tran, A. R. Abate (2016). Electrical lysis of cells for detergent-free droplet assays. *Biomicrofluidics*, **10**(2):024114.
111. K. Zhu, H. Jin, Z. He, Q. Zhu, B. Wang (2006). A continuous method for the large-scale extraction of plasmid DNA by modified boiling lysis. *Nat Protoc*, **1**(6):3088–3093.
112. B. H. Johnson, M. H. Hecht (1994). Recombinant proteins can be isolated from E. coli cells by repeated cycles of freezing and thawing. *Nat Biotechnol*, **12**(12):1357–1360.
113. J. M. Greenly, J. W. Tester (2015). Ultrasonic cavitation for disruption of microalgae. *Bioresour Technol*, **184**:276–279.
114. D. Taller, et al. (2015). On-chip surface acoustic wave lysis and ion-exchange nanomembrane detection of exosomal RNA for pancreatic cancer study and diagnosis. *Lab Chip*, **15**(7):1656–1666.
115. C. R. Engler, C. W. Robinson (1981). Disruption of Candida utilis cells in high pressure flow devices. *Biotechnol Bioeng*, **23**(4):765–780.
116. J. D. Besant, J. Das, E. H. Sargent, S. O. Kelley (2013). Proximal bacterial lysis and detection in nanoliter wells using electrochemistry. *ACS Nano*, **7**(9):8183–8189.
117. C. M. Gabardo, A. M. Kwong, L. Soleymani (2015). Rapidly prototyped multi-scale electrodes to minimize the voltage requirements for bacterial cell lysis. *Analyst*, **140**(5):1599–1608.
118. C. Kremer, C. Witte, S. L. Neale, J. Reboud, M. P. Barrett, J. M. Cooper (2014). Shape-dependent optoelectronic cell lysis. *Angew Chem Int Ed*, **53**(3):842–846.
119. T. S. Santra, P.-C. Wang, F.-G. Tseng (2013). Electroporation based drug delivery and its applications. In *Advance Micro-Nano Electromechanical Systems and Fabrication Technology*, InTech, EU, pp. 61–98.
120. T. S. Santra, M. Teitell, P.-Y. Chiou (2016). Device for massively parallel high throughput single cell electroporation and uses thereof. US15/673,255.

121. S. Kar, et al. (2018). Single-cell electroporation: current trends, applications and future prospects. *J Micromech Microeng*, **28**(12).

122. T. S. Santra, C.-W. Chen, H.-Y. Chang, F.-G. Tseng (2016). Dielectric passivation layer as a substratum on localized single-cell electroporation. *RSC Adv*, **6**(13):10979–10986.

123. T. S. Santra, H.-Y. Chang, P.-C. Wang, F.-G. Tseng (2014). Impact of pulse duration on localized single-cell nano-electroporation. *Analyst*, **139**(23):6249–6258.

124. T. S. Santra, F.-G. Tseng (2013). Recent trends on micro/nanofluidic single cell electroporation. *J Micromach*, **4**:333–356.

125. T. S. Santra, P.-C. Wang, H.-Y. Chang, F.-G. Tseng (2013). Tuning nano electric field to affect restrictive membrane area on localized single cell nano-electroporation. *Appl Phys Lett*, **103**(23):233701.

126. T. S. Santra, C.-W. Lee, S. Kar, J. Borana, P.-C. Wang, F.-G. Tseng (2014). Nanolocalized single cell membrane nanoelectroporation. In *9th IEEE International Conference on Nano/Micro Engineered and Molecular Systems, IEEE-NEMS 2014*.

127. S. C. Chen, T. S. Santra, C. J. Chang, T. J. Chen, P. C. Wang, F. G. Tseng (2012). Delivery of molecules into cells using localized single cell electroporation on ITO micro-electrode based transparent chip. *Biomed Microdevices*, **14**(5):811–817.

128. F.-G. Tseng, T. S. Santra (2013). Nano-electrode based Chip. U.S. Patent Pending, Application No 13/753,036.

129. C. Liao, et al. (2018). An optically induced dielectrophoresis (ODEP)-based microfluidic system for the isolation of high-purity CD45neg/EpCAMneg cells from the blood samples of cancer patients—demonstration and initial exploration of the clinical significance of these cells. *Micromachines*, **9**(11):563.

130. W. Liang, K. Zhang, X. Yang, L. Liu, H. Yu, W. Zhang (2015). Distinctive translational and self-rotational motion of lymphoma cells in an optically induced non-rotational alternating current electric field. *Biomicrofluidics*, **9**(1):014121.

131. W. Liang, et al. (2014). Rapid and label-free separation of Burkitt's Lymphoma cells from red blood cells by optically-induced electrokinetics. *PLoS One*, **9**(3):e90827.

132. Y. Zhao, D. Jia, X. Sha, G. Zhang, W. Li (2018). Determination of the three-dimensional rate of cancer cell rotation in an optically-induced electrokinetics chip using an optical flow algorithm. *Micromachines*, **9**(3):118.

133. Y. Zhao, H. S. S. Lai, G. Zhang, G.-B. Lee, W. J. Li (2015). Measurement of single leukemia cell's density and mass using optically induced electric field in a microfluidics chip. *Biomicrofluidics*, **9**(2):022406.
134. Y. Zhao, H. S. S. Lai, G. Zhang, G.-B. Lee, W. J. Li (2014). Rapid determination of cell mass and density using digitally controlled electric field in a microfluidic chip. *Lab Chip*, **14**(22):4426–4434.
135. W. Liang, Y. Zhao, L. Liu, Y. Wang, W. J. Li, G.-B. Lee (2017). Determination of cell membrane capacitance and conductance via optically induced electrokinetics. *Biophys J*, **113**(7):1531–1539.
136. S. J. Williams, A. Kumar, S. T. Wereley (2008). Electrokinetic patterning of colloidal particles with optical landscapes. *Lab Chip*, **8**(11):1879.
137. M. Nakano, S. Katsura, G. G. Touchard, K. Takashima, A. Mizuno (2007). Development of an optoelectrostatic micropump using a focused laser beam in a high-frequency electric field. *IEEE Trans Ind Appl*, **43**(1):232–237.
138. K.-C. Wang, A. Kumar, S. J. Williams, N. G. Green, K. C. Kim, H.-S. Chuang (2014). An optoelectrokinetic technique for programmable particle manipulation and bead-based biosignal enhancement. *Lab Chip*, **14**(20):3958–3967.
139. C. Vaillier, T. Honegger, F. Kermarrec, X. Gidrol, D. Peyrade (2014). Comprehensive analysis of human cells motion under an irrotational AC electric field in an electro-microfluidic chip. *PLoS One*, **9**(4):e95231.
140. A. Kumar, J.-S. Kwon, S. J. Williams, N. G. Green, N. K. Yip, S. T. Wereley (2010). Optically modulated electrokinetic manipulation and concentration of colloidal particles near an electrode surface. *Langmuir*, **26**(7):5262–5272.
141. K. Heatley (2016). Elctric-field assembly of a binary mixture of spherical colloids. Mines Theses & Dissertations.
142. S. J. Williams, A. Kumar, N. G. Green, S. T. Wereley (2010). Optically induced electrokinetic concentration and sorting of colloids. *J Micromech Microeng*, **20**(1):015022.
143. J.-S. Kwon, S. P. Ravindranath, A. Kumar, J. Irudayaraj, S. T. Wereley (2012). Opto-electrokinetic manipulation for high-performance on-chip bioassays. *Lab Chip*, **12**(23):4955.

Chapter 3

Micro-Robots/Microswimmers for Biomedical Applications

Naveen Kumar Agrawal,[a] Pallab Sinha Mahapatra,[a] and Tuhin S. Santra[b]

[a] *Department of Mechanical Engineering, Indian Institute of Technology Madras, Chennai 600036, India*

[b] *Department of Engineering Design, Indian Institute of Technology Madras, Chennai 600036, India*

pallab@iitm.ac.in

Microfluidic platforms, lab-on-a-chip devices, and microswimmers of different kinds attract researchers for various biomedical applications. Microswimmers propel themselves in the medium with their self-propelled forces. Microswimmers are now slowly changing many aspects of biomedical and health care sectors. Remotely controlled artificial micro-robots, as well as living microswimmers, have the potential of targeted drug delivery (TDD), microsystem steering, chemical mixing, etc. Biodegradable protein-based rockets are used for efficient drug encapsulation and targeted release. Researchers have created artificial Janus micro-robots that can be controlled externally from a microscale device, for cancer cell treatment. Sometimes external fields like electrical fields or ultrasounds are also used to release drugs into targeted areas. Here, we provide a comprehensive review of different types of microswimmers

Microfluidics and Bio-MEMS: Devices and Applications
Edited by Tuhin S. Santra

ISBN 978-981-4800-85-3 (Hardcover), 978-1-003-01493-5 (eBook)
www.jennystanford.com

and microswimmer-based devices for biomedical and health care applications. Finally, we discuss the existing challenges and future potential of micro-robots and microswimmers.

3.1 Introduction

The advancement in the technology of microactuation, microcontrolling, microfabrication, and microsensors opened new horizons in the development of untethered, miniaturized robots. These miniaturized robots have vast potential applications in various fields like biomedicine [1–16], environmental remediation [17–21], industry [22–24], food [25], etc. Research in the sophistication of micro-robots, efficient propulsion mechanisms, and precision steering methods has increased in the past few years. The small-scale robots are envisaged to perform tasks that are difficult or impossible, in an environment, and out of human reach. The advancement in electronics and communication control technology advanced the untethered control of miniaturized robots and scaled down the large-scale movement to the submillimeter scale. The miniaturized untethered robots have a large potential of performing complex tasks at the micro- and cellular length scales, specifically in the field of biomedicine, for targeted therapeutic delivery [4, 6, 10, 14, 26], medical diagnosis [7, 27], sensing [1, 9, 13, 92], minimally invasive surgery [2, 12, 15, 29], and detoxification [30]. A significant revolution in the biomedical field occurred in 2001 with the Food and Drug Administration (FDA) approval of capsule endoscopy. These endoscopy capsules are capable of real-time image/video capturing and wireless data transmission. These types of capsules are easily available in the market and are used to diagnose gastrointestinal (GI) tract problems [7].

Mobile, untethered, miniature robots can be classified depending upon the size and the operational method as [31]:

(1) Based upon the size of the robot:

  (a) Micro-robot: Size of the mobile, miniaturized robot ranges from 1 μm to 1 mm. The dynamics of these robots is dominated by microscale forces like surface tension and viscous forces.

(b) Millirobot: Size of the mobile, miniaturized, robot ranges from 1 mm to the palm size. The dynamics at the millimeter sale is governed by macroscale forces like buoyancy and inertial forces.

(2) Based upon operation:

  (a) On-board approach: All the necessary components required for the operation of a robot, like an actuator, sensor, battery, etc., are mounted on the robot itself. The robot can be controlled autonomously or with the use of a remote.

  (b) Off-board approach: The mobile robot does not have all the components aboard. A few components are outside of the working domain of the robot. These types of robots are sensed, operated, controlled, and powered by using external components.

The mobile robots of microscale size and operated with the on-board approach are best suited for biomedical applications. However, at the microscale level, the necessary components of a robot are not readily available. Therefore, the micro-robots with the off-board approach are prominent in the biomedical and other fields. At the microscale, the locomotion of a robot/swimmer is challenged by the low Reynolds number ($Re < 1$) flow and Brownian motion in a fluid. The propulsion dynamics at the microscale is far different than that in the macro-/large scale, where the $Re \gg 1$ condition prevails. The Scallop theorem [32] sets the condition for net nonzero displacement at the low Re condition. A variety of propulsion mechanisms (detailed in Section 3.2) have been conceptualized and are being explored in the research, like magnetic propulsion [6, 10, 33–35], bubble propulsion [4, 26, 36, 37], biological propulsion [38–41], self-thermophoresis [42–44], etc. The advantage of an untethered micro-robot is the accessibility to closed and confined regions, which are difficult or not possible to reach with conventional tethered robots. The primary objective of this chapter is to discuss miniaturized, mobile, untethered robots, which are used or conceptualized to have potential use in the biomedical field. The chapter is divided into multiple subsections. The different types of propulsion mechanisms are discussed in Section 3.2. The existing fabrication techniques are discussed in Section 3.3. The applications of different types of micro-robots and microswimmers are discussed in Section 3.4. Lastly, the future scope of micro-robots is discussed in Section 3.5.

3.2 Propulsion Mechanism

The propulsion mechanism of a robot in a fluid is far different in the microscale than that in its macroscale counterpart. In the microscale, the fluid dynamics is dominated by forces like viscous force, surface tension, and drag, whereas the forces dominating the dynamics at the macroscale, like inertial force and buoyancy, are negligible in microscale dynamics. According to Purcell's scallop theorem [32], in a Newtonian fluid a microswimmer can have nonzero displacement by performing nonreciprocal motion only. Also, the microswimmer has to have more than one degree of freedom to produce nonzero displacement. With only one degree of freedom, the microswimmer undergoes reciprocal motion only, therefore producing a net zero displacement [32]. However, most of the biological fluids are non-Newtonian, and thus, the scallop theorem is no longer valid [45].

The motion of a micro-robot/microswimmer is prominent mainly in a low-Reynolds-number (Re $\ll$ 1) regime. In the low-Re regime, the fluid dynamics is governed by the Stokes equations as follows:

$$\nabla \cdot u = 0 \tag{3.1}$$

$$-\nabla p + \mu \nabla^2 u = 0. \tag{3.2}$$

Here p is the pressure, μ is the dynamic viscosity, and $\vec{u}$ is the velocity of the fluid. The drag force experienced by a spherical body immersed in an infinite extent of fluid is linearly dependent on the velocity. From the Stokes law of resistance in a low-Reynolds-number (Re $<$ 1) condition the drag force is given as

$$F_{\text{drag}} = 3\pi \mu d U. \tag{3.3}$$

Here F_{drag} is the magnitude of the drag force exerted by the surrounding fluid on the spherical body, μ is the dynamic viscosity of the fluid, d is the diameter of the spherical body, and U is the speed of the body. The drag force is exerted along the direction opposite to the direction of the velocity of the body. The force balance on a self-propelling body (microswimmer/micro-robot) can be given by

$$F_{net} = F_{resistance} + F_{propulsion}. \tag{3.4}$$

Here F_{net} is the net force acting on a microswimmer body, $F_{resistance}$ is the sum of F_{drag} and other forces opposing the propulsion of a body, and $F_{propulsion}$ includes all the forces contributing in the propulsion of the body. At equilibrium $F_{net} = 0$. Therefore for producing a net displacement of a microswimmer in a fluid, the propulsion force ($F_{propulsion}$) should be greater than (or at least equal to) the resistive force ($F_{resistance}$) experienced by the microswimmer.

The propulsive force on a micro-robot can be provided either by external power sources like magnetic and electric fields, and optical and thermal gradients or by onboard mechanisms like bacterial swimming, chemical reaction, bubble generation, etc. Some of these mechanisms are shown in Fig. 3.1.

3.2.1 Magnetic Propulsion

There are studies advocating the successful propulsion and steering of micro-robots exploiting external magnetic fields [6, 10, 27, 46]. The reasons behind the popularity of magnetic fields include wireless transmission of high forces and torques with all six degrees of freedom. A magnetic field can penetrate almost any environment, and its use facilitates fuel-free propulsion of untethered micro-robots. Figure 3.1a shows a schematic view of a magnetic micro-robot. The micro-robots embedded with magnetically responsive materials (like Ni, Co, NdFeB, Fe) can be used to exploit external magnetic fields for navigation and orientation. An external magnetic field can be generated by using a magnetic resonance imaging (MRI) system [47], a coil system [48–50], or permanent magnets [6, 46]. The micro-robot can be oriented by application of a magnetic torque and can be propelled by applying a magnetic force. The magnetic

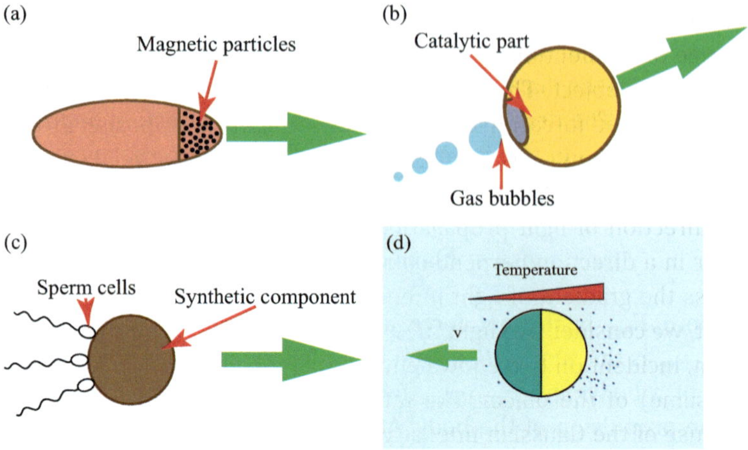

Figure 3.1 Schematic view of different types of propulsion mechanisms: (a) Magnetic propulsion, (b) bubble propulsion, (c) biological propulsion, and (d) self-thermophoresis. The green arrow shows the direction of micro-robot propulsion.

force acting on a micro-robot is given by [51]

$$\vec{F}_{\text{magnetic}} = \mu_0 V(\vec{M} \cdot \nabla)\vec{H}, \tag{3.5}$$

and the magnetic torque is given by [51]

$$\vec{T}_{\text{magnetic}} = \mu_0 V(\vec{M} \times \vec{H}), \tag{3.6}$$

where V is the volume (m^3) of the micro-robot, $\mu_0 = 4\pi \times 10^{-7} T \cdot m/A$ is permeability of free space, $\vec{M}$ is the magnetization (A/m) of the micro-robot, and $\vec{H}$ is the magnetic field (A/m). Both Eqs. 3.5 and 3.6 are based on the assumptions that the size of a micro-robot is very small compared to the spatial changes of the magnetic field H and the magnetic field is uniform across the micro-robot. From Eq. 3.5, the applied magnetic force can be altered by altering the magnetic field gradient for a given micro-robot. From Eq. 3.6, it is clear that the applied magnetic torque can be varied by varying the angle between the magnetization of the micro-robot and the magnetic field or by varying the magnetic field strength for a given micro-robot. We can broadly classify the different types of magnetic propulsion into three categories: (a) magnetic gradient pulling, (b) helical propulsion, and (c) planar wave propulsion. The helical

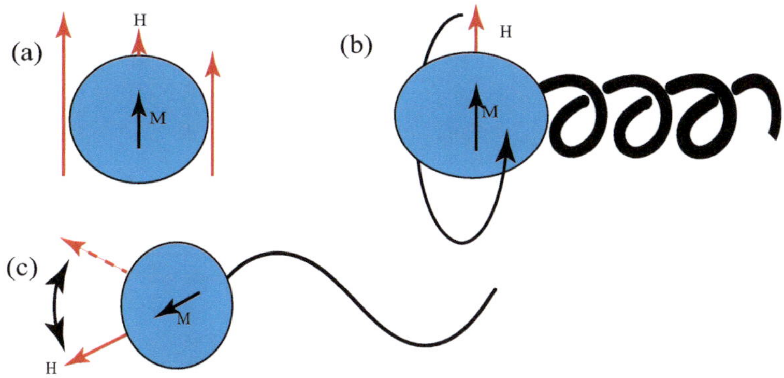

Figure 3.2 Different types of magnetic propulsion. Red arrows show the direction of the applied external magnetic field ($\vec{H}$). (a) Magnetic gradient pulling. The magnetic field strength is increasing while moving from right to left. (b) Helical propulsion under the application of a rotational magnetic field. (c) Planar wave propulsion under the application of an oscillating magnetic field. All the micro-robots are moving from right to left.

propulsion and planar propulsion of micro-robots are inspired by the biological microswimmers like *Escherichia coli* and spermatozoa, respectively. Figure 3.2 shows a schematic view of these three types of magnetic propulsion methods.

Magnetic gradient pulling: A micro-robot with magnetization $\vec{M}$ in a magnetic field $\vec{H}$ can be propelled by a magnetic force applied on it due to the magnetic gradient, as given in Eq. 3.5. A schematic view of a magnetic-gradient-pulled micro-robot is shown in Fig. 3.2a. Mathieu et al. [47] demonstrated the first use of an MRI system for gradient pulling of a ferromagnetic spherical body. Carpi et al. [46] demonstrated translation, rotation, and roto-translation of the wireless endoscope capsule (WEC) within tubular structures coated with bovine tissues. The hand-operated, permanent magnets NdFeB were used to generate external magnetic fields for the capsule locomotion. In another work, Carpi et al. [27] demonstrated the magnetic steering of the WEC in the main regions of the GI tract: esophagus, stomach, small bowel, and colon of a domestic pig. They used the static field of the already available Stereotaxis instrument for the magnetic maneuvering of the WEC. As the

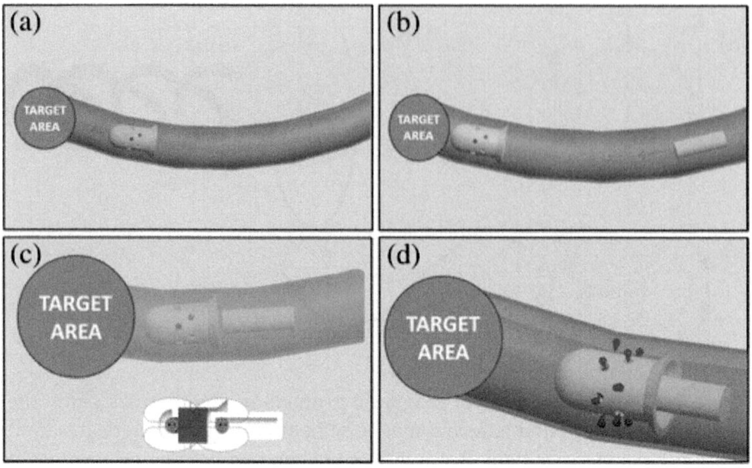

Figure 3.3 The propulsion, docking, and release of the drug at the target by the two-component micro-robot. The micro-robot components are magnetically pulled. (a) Individual navigation of the drug-loaded carrier to the target. (b) Then the piston was individually navigated to the docking site, while maintaining the carrier at the target. (c) The piston was brought to the critical docking distance. At this docking site, the magnetic force between the components dominated the external magnetic field force. (d) Docking occurs, causing the release of the drug. Reprinted by permission from Springer Nature Customer Service Centre GmbH: Springer Nature, *Biomedical Microdevices*, Ref. [6], copyright (2015).

commercially available WEC does not have any motion-controlling mechanism, they glued a magnetic-rubber shell to control the WEC magnetically. Iacovacci et al. [6] demonstrated individual steering of components (carrier and piston) of a near neutrally buoyant, millimeter-sized robot, exploiting the distinct magnetic property of the two components, as shown in Fig. 3.3.

Permanent magnets were used to generate a magnetic field of 5.2 mT and a magnetic field gradient of 0.05 T/m for the individual propulsion of the components to the target. An additional 50 mT magnetic field was used for holding the carrier at the target, while the piston was moved to the docking site. At the docking site, the magnetic force between the two components caused docking of the piston in the carrier, and thus compression of hydrogel released the embedded drug through the holes in the carrier.

Controlled navigation of a swarm of micro-robots is also possible by using magnetic gradient pulling. For individual controlling of each micro-robot in a swarm, they can be made heterogeneous. The heterogeneity can be included in the swarm through variation in geometrical design or the magnetic property of the micro-robots. The heterogeneity causes a different magnetic torque on each micro-robot in a rotating magnetic field. The first demonstration of independent 3D control of multiple micro-robots was done by Diller et al. [52]. The micro-robots were distinct in design, resulting in individual control of the micro-robots by application of different magnetic forces in the same magnetic field.

Helical wave propulsion: Bacteria (like *E. coli*) propel by performing corkscrew-like motion by rotation of the flagellum, producing helical waves. The tactic of the bacteria can also be mimicked for the propulsion of a micro-robot in a low-Reynolds-number regime. Honda et al. [53] demonstrated the first use of a centimeter-sized, magnetically propelled, spiral robot. Micro-robots with a helical propeller are also called artificial bacterial flagella (ABF). These ABF convert rotational motion into translational motion, as shown in Fig. 3.2b. These ABF have at least one of the components magnetized, either the head [50, 53] or the helical tail [48, 49]. The micro-robot component is magnetized in a way such that its magnetic moment is perpendicular to the longitudinal axis of the micro-robot. A uniform, rotating magnetic field is applied in a plane perpendicular to the axis of the helical propeller, which employs a torque, as given in Eq. 3.6, and rotates the micro-robot along its longitudinal axis. With the rotation of this micro-robot, helical waves are generated and propagated along the tail, in turn, and the micro-robot propels by converting rotational motion into linear motion. The main control parameter for controlling the propulsion of the micro-robot is the frequency of the rotating magnetic field. The swimming velocity of the micro-robot increases linearly with the frequency up to a step-out frequency, beyond which the swimming velocity decreases drastically [10, 51]. Therefore, for optimum performance, the helical micro-robots should operate at the step-out frequency, for which the applied magnetic torque on the micro-robot is maximum. Steering of the micro-robot can be done by rotating the axis of the external

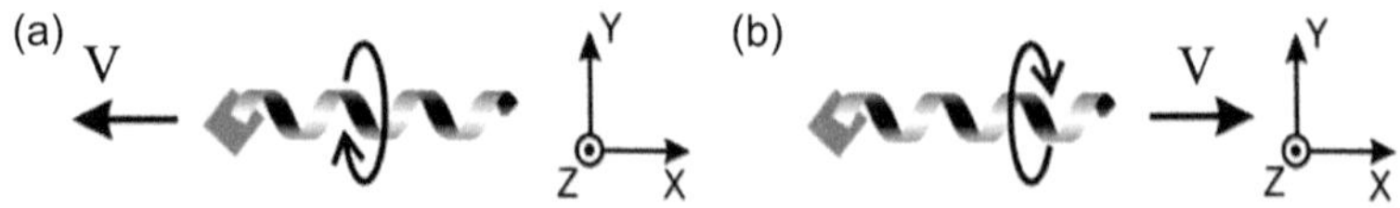

Figure 3.4 Forward and backward movement of a helical micro-robot. The rotational magnetic field is applied perpendicular to the *x* axis. The micro-robot is moving toward the left (a). When the direction of magnetic field rotation is reversed, the micro-robot started moving toward the right, without actual turning of the micro-robot. Reprinted from Ref. [50], with the permission of AIP Publishing.

magnetic field so that the magnetic dipole of the micro-robot tries to align itself along the external magnetic field. One of the merits of the helical micro-robot is that it can be made to propel in the backward direction just by reversing the direction of the applied rotating magnetic field and without actually turning the micro-robot, as shown in the schematic in Fig. 3.4. Helical propulsion is suitable for precise navigation of even smaller micro-robots of lengths 1–2 μm [48]. Tung et al. [54] demonstrated propulsion of a rectangular micro-robot called RodBot on a surface immersed in a solution of deionized (DI) water and poly(ethylene glycol) (PEG). They reported the formation of a vertex parallel to and above the RodBot, which can be used to entrap and transport an object.

Planar wave propulsion: Magnetic micro-robots can also be propelled by mimicking the propulsion mechanism of spermatozoa. The spermatozoa produce planar waves along their tails to propel in a liquid medium. Mimicking the same kind of mechanism, a micro-robot can also be propelled by producing planar waves along its thin elastic tail. These micro-robots have a magnetic head attached to an elastic tail, as shown in Fig. 3.2c. The head provides the magnetic dipole, and the tail produces the propulsive force by producing planar waves. A uniform magnetic field oscillating in time can be employed to propel the micro-robot. The uniform magnetic fields are oriented toward the direction of motion with a periodic component orthogonal to the direction of propulsion [33]. The oscillating magnetic field applies torque, as given in the Eq. 3.6, on the magnetic head of the micro-robot to align it along the magnetic field $\vec{H}$, thus oscillating the micro-robot. This oscillation generates a local bending moment along the tail and causes the elastic tail to produce planar waves, similar to spermatozoa. The tail surface governs the ratio of the perpendicular viscous coefficient ($\zeta_{\perp}$) to the tangential viscous coefficient ($\zeta_{\parallel}$) [55], which, in turn, depicts the flow of the micro-robot in the direction of wave propagation or along the direction opposite to it. Khalil et al. [34] demonstrated propulsion of a magneto sperm having an elliptic head (patterned with $Co_{80}Ni_{20}$ for magnetic steering) and a flexible tail, using the external weak magnetic field (less than 5 mT). El-Etriby et al. [33] demonstrated the micromanipulation of two nonmagnetic payloads (100 μm diameter each), moving with an average speed of 0.16 body length/s and 0.035 body length/s during collision-free locomotion and manipulation, respectively. Dreyfus et al. [55] fabricated a micro-robot with the flexible magnetic flagellum. Their micro-robot propelled toward the free end, unlike the other micro-robots discussed before, which propelled toward their heads. Khalil et al. [35] reported a nonlinear increase in the drag force with an increase in the frequency of the oscillating magnetic field and swimming speed. For improved functionality, these micro-robots can also be made with multiple flagella. The speed of the micro-robot increases linearly with an increase in the number of flagella [56].

3.2.2 Bubble Propulsion

Similar to the propulsion of rockets by exhaust gases, robots at the microscale can also be propelled by ejected gas bubbles, as shown in Fig. 3.1b. Gas bubbles can be produced during a chemical reaction in a fuel medium. These gas bubbles after ejection propel the micro-robots in the opposite direction. Bubble-propelled micro-robots can perform autonomous motion by locally converting chemical energy, from the surrounding, into kinetic energy. These micro-robots are embedded with a metallic catalyst like gold (Au), copper (Cu), or platinum (Pt) or a reactive metal like zinc (Zn), magnesium (Mg), or aluminum (Al), which decomposes the fuels like H_2O_2, acid, and water into gases like oxygen and hydrogen. Propulsion reactions

Table 3.1 Chemical reactions of different fuel and material combinations for bubble propulsion

Fuel	Material	Reaction
H_2O_2	Pt	$2\,H_2O_2 \longrightarrow 2\,H_2O + O_2$
Acid	Zn	$2\,H^+(aq) + Zn(s) \longrightarrow Zn^{2+}(aq) + H_2(g)$
Acid	Mg	$Mg(s) + 2\,H^+(aq) \longrightarrow Mg^{2+}(aq) + H_2(g)$
Water	Al	$2\,Al(s) + 6\,H_2O(l) \longrightarrow 2\,Al(OH)_3(aq) + 3\,H_2(g)$
Water	Mg	$Mg + 2\,H_2O \longrightarrow Mg(OH)_2 + H_2(g)$

involving different combinations of catalyst/reactant material and fuel are shown in Table 3.1.

The speed of catalytic micro-robots ranges from 100 to 1000 μm/s [26], which is very high compared to other micro-robots. The speed of these micro-robots increases with the increase in the frequency of gas bubble generation and ejection due to the increase in the concentration of the fuel [57]. However, the speed decreases with the decrease in the size of ejected bubble [58]. The micro-robot speed is almost equal to the product of ejected bubble radius and frequency of bubble ejection [57, 59]. The presence of a surfactant also affects the speed of a bubble-propelled micro-robot. Addition of a surfactant reduces the surface tension and bubble radius but increases the frequency of bubble generation. Manesh et al. [60] reported a decrease in the speed of bubble-propelled, tubular micro-robots by addition of Triton X-100 surfactant to H_2O_2 fuel, whereas in contrast, Gao et al. [57] reported an increase in speed by adding sodium cholate surfactant to H_2O_2 fuel. The effect of physiological temperature on speed is also reported [61]. Performance of micro-robots in the presence and absence of salt in the propulsion medium is reported [60]. Manesh et al. [60] observed bubble generation and micro-robot speed in salt-free and salt reach (NaCl, up to 1 M) media and reported a small (1%–3%) decrease in the speed of the micro-robot in the latter condition. Two key requirements in a bubble-propelled micro-robot to cause bubble ejection along a specific direction are [26] that (1) the micro-robot structure needs to be anisotropic and (2) the catalyst/reactant should be located only at a portion of and not on the overall micro-robot. Pt and H_2O_2 are two of the widely explored catalyst and fuel combinations [26, 38, 59,

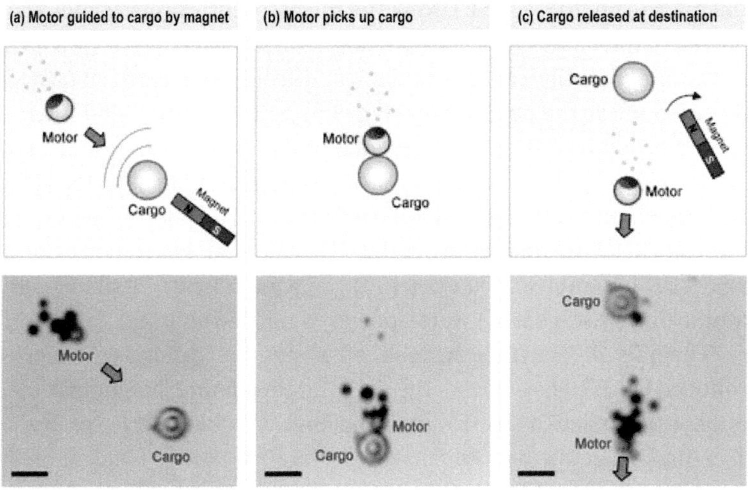

Figure 3.5 Pt/Fe NP-embedded soft catalytic micro-robot. The top panel shows the schematic view, and the bottom panel shows the experiment images. (a) Pick-up: The micro-robot is propelled to the cargo location under the influence of an external magnet. With a head-on collision, the cargo attaches to the micro-robot. (b) Transport: The adhered cargo is transported to the target location. (c) Drop-off: At the target location, the cargo is dropped-off with a flick in the external magnet. Reprinted with permission from Ref. [26]. Copyright (2016) American Chemical Society.

61, 62] in the development of bubble-propelled micro-robots. Lu et al. [26] demonstrated the autonomous, linear, and circular motion of a soft, catalytic (Pt) micro-robot in H_2O_2 solution. Their micro-robot was embedded with Pt/Fe nanoparticles (NPs). The Pt shell was exploited for bubble generation and the Fe core for magnetic navigation control using a neodymium magnet with a field strength of 0.57 T, as shown in Fig. 3.5.

The speed observed was 300 μm/s in 8 wt% H_2O_2 solution. As shown in Fig. 3.5, their micro-robot was able to successfully pick up, transport, and drop off a larger, soft, inert cargo (without NPs) made with the same material. With a particular polarity of the external magnet, the micro-robot was moved to the cargo location (Fig. 3.5a). Then the micro-robot and cargo were made to have a head-on collision, and due to its soft nature, the cargo adhered to the micro-robot (Fig. 3.5b). Now the micro-robot–cargo system was moved to the targeted location by magnetic steering. There, by a flick in the magnet, the cargo was detached from the micro-robot and dropped off at the destination (Fig. 3.5c). Bubble-propelled micro-robots based on other combinations of catalyst/reactant and fuel, Mg + acid [4, 37], Zn + acid [36], Al + water, and Mg + water [30], have also been explored. A Mg-based soft micro-robot propelled in the gastric acidic fluid is tested by de Avila et al. [4] to perform in vivo TDD in a mouse. Wu et al. [30] reported a first of its kind cell-mimicking (RBC-coated) water-powered micro-robot.

The use of enzymes instead of a metallic catalyst in micro-robots [19, 63] has paved the way for designing biocompatible, bubble-propelled micro-robots. Enzymes are biological catalysts. The first effective use of enzymes (catalase) as a catalyst in a bubble-propelled, tubular micro-robot is reported by Sanchez et al. [63]. Enzymes are useful in finding alternative fuels for high-speed propulsion along with cutting down the threshold fuel concentration required for the propulsion of bubble-propelled micro-robots. Wu et al. [64] showed the catalase could propel the micro-robot even at a low concentration of 0.1% of H_2O_2 fuel. Catalase is among the most efficient enzymes. Each catalase molecule can decompose millions of H_2O_2 molecules in a second [63]. The micro-robots with a biocatalyst (catalase) are reported to be more efficient than their metallic catalyst counterparts [63]. Sanchez et al. [63] demonstrated a hybrid micro-robot functionalized with 3-mercaptopropionic acid (3-MPA) propelling in H_2O_2 solution with an average speed of 226.1 ± 21.1 μm/s. They also observed the rotation of the micro-robot due to the bubble formed at the mouth of the tubular micro-robot. This bubble formation at the micro-robot mouth can be used for autonomous steering of the tubular micro-robot.

Bubble-propelled micro-robots can be controlled by using light [65], ultrasound [66], temperature [67], etc., as shown in Fig. 3.6. Solovev et al. [65] demonstrated the control on propulsion of a Ti/Cr/Pt catalytic micro-robot by using white light. The illumination of white light causes a local decrease in the H_2O_2 fuel and surfactant concentration, thus decreasing or switching off the bubble generation and thus the propulsion of the micro-robot. By using these external stimuli, the bubble-propelled micro-robot can

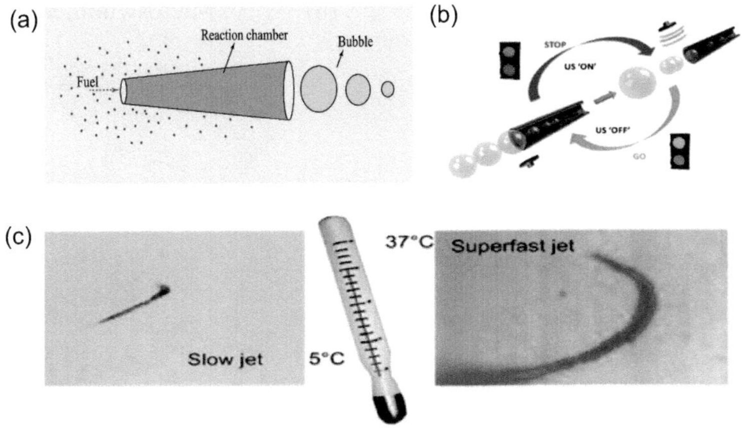

Figure 3.6 (a) Schematic view of a bubble-propelled micro-robot in tubular shape. Showing the inlet of fuel from the front opening, then bubble production in the reaction chamber, and ejection of the bubbles from the rear opening. (b) Effect of ultrasound on micro-robot propulsion. Reprinted with permission from Ref. [66]. Copyright (2014) American Chemical Society. (c) Effect of temperature on the acceleration of the micro-robot. Reprinted with permission from Ref. [67]. Copyright (2011) American Chemical Society.

be switched on/off as well. Use of the UV spectrum is an effective method but harmful to living entities. There is also a chance that by exposure of the UV spectrum, the micro-robot material may dissolve. More studies are required on the control of bubble-propelled micro-robots.

3.2.3 Biological Propulsion

Biological microswimmers have evolved over millions of years and developed excellent swimming capability in a low-Reynolds-number environment. These biological microswimmers propel themselves by using flagella (Fig. 3.7a) and cilia (Fig. 3.7b) extended from their body and navigate by using senses and taxes. Taxis is the motive response of a microorganism toward the gradient of a stimulus. Biological propulsion of a micro-robot is achieved by the aid of biomicroswimmers like bacteria, algae, and contractile cells integrated with the micro-robots. These micro-robots are termed

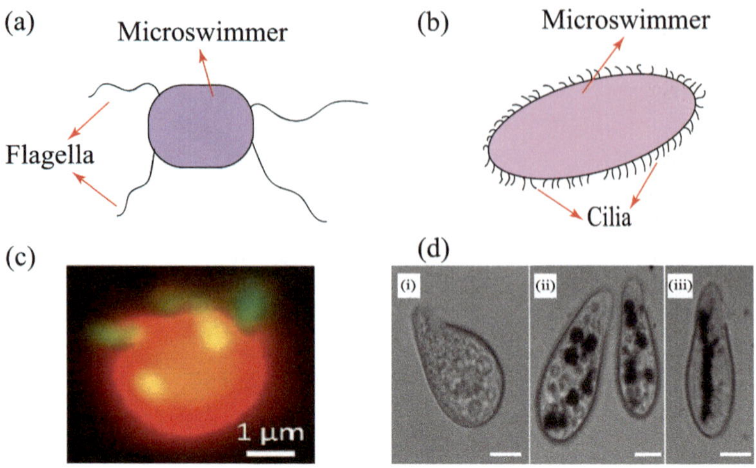

Figure 3.7 Schematic view of biomicroswimmers. Showing flagella (a) and cilia (b). (c) Fluorescent image of a biohybrid micro-robot. Multiple bacteria (yellow-green) are attached to the spherical polystyrene bead (red). Reprinted from Ref. [68] under a Creative Commons license (CC BY 4.0). (d) Artificial magnetization of *Tetrahymena pyriformis* cell. (i) Normal *T. pyriformis* cell. (ii) *T. pyriformis* cells with internalized iron oxide particles (magnetite) by oral ingestion. (iii) After magnetization, the magnetite NPs arranged in a line. Scale bar: 10 μm. Reprinted from Ref. [69], with the permission of AIP Publishing.

"biohybrid micro-robots." A biohybrid micro-robot propelled by attached bacteria is shown in Fig. 3.7c. The use of biomicroswimmers to propel a micro-robot is a very promising actuation method. It eliminates the need of an onboard synthetic propulsion mechanism and power source, thus sophisticating the micro-robots for biomedical use. The bioswimmer component of a biohybrid micro-robot can also function as a biosensor and navigation controller. In the biohybrid micro-robots, the biomicroswimmers can be integrated to the micro-robot by free or fixed attachment. The freely attached biomicroswimmers can be replaced at the end of their life span, whereas this flexibility is not available if the biomicroswimmers are fixed on the micro-robots as the whole micro-robot has to be changed after the fixed biomicroswimmers are dead. *E. coli* [40], *Serratia marcescens* [41, 68, 70–72], *Salmonella typhimurium*, *Vibrio alginolyticus*, *Bacillus subtilis*,

T. pyriformis [69], strains of magnetotactic bacteria (MTB) MC-1 [39] and AMB-1 [38, 73], muscle cells (myocytes), and algae like *Chlamydomonas reinhardtii* [74] and others can be used for the synthesis of a biohybrid micro-robot. The propulsion of a biohybrid micro-robot is limited by the stochastic random walk nature of the motion of biomicroswimmers [70]. Therefore navigation control is required and can be achieved by using external fields and taxes like chemotaxis [41, 70], aerotaxis [39], magnetotaxis [38, 73], phototaxis [71, 72, 74], pH-taxis [68], thermotaxis, galvanotaxis, etc.

Zhuang et al. [68] demonstrated pH-taxis of biohybrid micro-robots (a polystyrene [PS] bead attached with multiple *S. marcescence*). The micro-robot is shown in Fig. 3.7c. They reported the propulsion behavior of micro-robots in three different types of pH gradients: gradient 1 (pH: 6.0–7.6), gradient 2 (pH: 3.8–5.4), and gradient 3 (pH: 8.2–9.8). For the gradient 1 system, they found a bidirectional pH-taxis of micro-robots, that is, away from both strongly acidic and alkaline regions. A dense band of micro-robots was observed at optimal pH of slightly more than 7.0. A unidirectional movement was observed for the gradient 2 and gradient 3 systems, that is, away from strong acidic and alkaline regions, respectively. They also reported the persistent orientation and higher speed of the micro-robots while moving toward the favorable pH region.

Weibel et al. [74] demonstrated steering of a PS bead actuated by *C. reinhardtii*, using LED light with wavelengths $\lambda \approx 500$ nm. At the high intensity of the light, cells showed negative phototaxis (movement away from the light source) and positive phototaxis (movement toward the light source) at low intensity of light. They also demonstrated the release of microscale cargo using UV irradiation ($\lambda = 365$ nm). Steager et al. [72] demonstrated the exposure of UV light on a micro-robot propelled by *S. marcescens* reduces the rotational rate of the micro-robot. Other taxis like chemotaxis and magnetotaxis are discussed below.

Chemotaxis: Chemotactic micro-robots can be steered using a chemoattractant or chemorepellent. Micro-robots move toward increasing concentration of a chemical (positive chemotaxis) in the presence of a chemoattractant and move toward decreasing concentration of the chemical (negative chemotaxis) in the presence of a chemorepellent. Chemoattractants like L-threonine [70], L-aspartate [75], L-glutamate, and glycine are favorable for attracting flagellated bacteria [70]. The first study of chemotaxis steering of *S. marcescens* propelling a biohybrid micro-robot was done by Kim et al. [70]. They investigated the chemotactic effects on the micro-robots near the wall. Their micro-robots exhibited positive chemotaxis with an increase of 64.2% in propulsion speed as compared to the speed in the absence of any stimulus. The diffusion of the injected chemical plays an important factor in chemotaxis. If the chemical diffuses fast, then the concentration gradient cannot be maintained for a long time and long-lasting steering is not possible. The chemical should be administered near to the micro-robot; otherwise, the concentration might be lower than that detectable by the bioactuator attached to the micro-robot.

***Magnetotaxis*:** The magnetic field used for navigation control of a synthetic micro-robot is described in Section 3.2.1. It can also be exploited for controlling biohybrid micro-robots. The advantage of the magnetic field is that it can penetrate almost all environments without any physical contact. Magnetotactic bacteria (MTB) possess a chain of magnetosomes, which causes them to orient along the magnetic field lines as weak as the earth's magnetic field and propel by the force of flagellar rotation. Strains of MTB are used as actuators in biohybrid micro-robots [38]. Yu et al. [73] demonstrated the controlled steering of the MTB AMB-1 for delivery of magnetic microparticles as cargoes. They showed the bacterium made a U-turn by reversing the direction of the magnetic field. The diameter of the U-turn trajectory was found to decrease with an increase in the magnetic field strength. The biomicroswimmers can also be magnetized artificially to exploit an external magnetic field for navigation control of biohybrid micro-robots [69]. Kim et al. [69] prepared artificially magnetized microswimmers. In their experiment, magnetite NPs of size 50 nm were orally ingested by *T. pyriformis* cells. A normal *T. pyriformis* cell is shown in Fig. 3.7d-i. After the magnetite NPs were ingested orally by the cells (Fig. 3.7d-ii), the magnetite NPs were magnetized with the help of an external magnetic field (Fig. 3.7d-iii). Magnetized cells could be steered using external magnetic fields. Another method of magnetic steering control is using the magnetic property of the synthetic part of the micro-robot. Carlsen et al. [76] demonstrated the magnetic steering of a superparamagnetic bead actuated by *S. marcescens* bacteria.

Bioswimmers propel by producing a range of propulsive forces. Propulsive forces ranging from 0.5 pN for *E. coli* and *S. marcescens* to 4 pN for *Magnetospirillum marine coccus* (MC-1) have been observed [77]. Therefore the propulsive force required for a micro-robot decides the minimum number of biomicroswimmers to be attached on it. For controlled navigation, the maximum torque exerted by the external field should be greater than the propulsive torque produced by the cells in order to control the orientation of the micro-robot. Therefore, the maximum external torque that can be applied on the micro-robot for navigation control limits the maximum number of biomicroswimmers in a micro-robot. The positions of the biomicroswimmers on a micro-robot may also affect the net propulsive force and the speed of a biohybrid micro-robot.

3.2.4 Self-Thermophoresis

Propulsion of a micro-/nanoparticle along a temperature gradient in a fluidic environment is known as thermophoresis. Self-thermophoresis is different from the usual thermophoresis in a sense that in self-thermophoresis, the local temperature gradient is induced by the particle itself under an external stimulus. Micro-robots can also be propelled by self-thermophoresis in a fluidic environment, as shown in Fig. 3.1d. A unidirectional movement of the micro-robot can be obtained by generating an asymmetric temperature gradient across it. The asymmetric temperature gradient can be created by using a magnetic field [42], ultrasonic waves, light irradiation [43], or heat released during a catalytic reaction [44]. The propulsion mechanism by self-thermophoresis lies in the imbalance of momentum transfer between the faces of the micro-robot and the surrounding. As shown in Fig. 3.1d, due to the temperature gradient across the Janus micro-robot in a fluid, the collision between the right (yellow) face of the micro-robot and the surrounding molecules (blue particles) is stronger as compared to the collisions of molecules at the other (left) face, which causes a net thermophoretic force and propels the micro-robot toward the left (away from the yellow face).

Jiang et al. [43] measured the temperature distribution and the thermal slip flow field around a Janus microparticle. They propelled Au–silica Janus microparticle by self-thermophoresis using a defused laser beam. Their Janus micro-robot moved toward the colder region (positive Soret coefficient) in water. They observed that the Janus micro-robot moved toward the hotter region (negative Soret coefficient) after addition of surfactant Triton X-100 (0.05 wt%) to the solution, which shows the change in the thermophoretic property of the micro-robot by absorption of the surfactant on the micro-robot surface. They also showed rotation of a two Au–silica microparticle system (diameter 1 μm each). The twin micro-particle started rotating after a threshold laser power, and the rotational speed was found to be increased linearly with the laser power.

The effect of Janus micro-robot size on the propulsion was reported by Wu et al. [14]. Wu et al. found that a large-sized micro-robot produces more heat than a smaller one, but a smaller micro-robot moves faster under the same laser power. The reason behind such a phenomenon is the stronger temperature gradient across the smaller micro-robot than that across the larger micro-robot. Also, at higher laser power, the thermophoretic force is higher, resulting in a higher velocity of the micro-robot [78]. The other phoretic effects like self-electrophoresis [18, 79, 80], self-diffusiophoresis [81, 82] are also explored for the self-propulsion mechanism of a micro-robot in a fluidic environment. The advantage of self-phoretic propulsion is that the state of motion (on/off) can be controlled by switching the external stimuli source on/off.

3.3 Materials and Fabrication Techniques

The material and fabrication technique depends upon the type of micro-robot and its application environment. Synthesis of a microscale robot is a major challenge, especially for biomedical applications. For secure application in biological mediums, the micro-robot material needs to be biocompatible. For example, polydimethylsiloxane (PDMS) [6, 46], hydrogels [6], polycarbonate (PC), IP-L [83], SU-8 polymer [54, 83], chitosan biopolymer [26, 61], and sodium alginate (ALG) [61] are widely used in the fabrication of micro-robots for potential biomedical applications. Commercial PDMS, Dow Corning Sylgard-184 (Midland, MI, USA), is biocompatible and has good stiffness for use in bio-micro-robot fabrication [46]. Hydrogels are 3D networks comprising chemically or physically cross-linked hydrophilic polymer chains [6]. A hydrogel can be cast into almost any shape, size, or form and can absorb thousands of times its dry weight of water. Hydrogels, based on agar and gelatin, are biocompatible and used for drug embedding [6]. Their compressive strength can be varied by varying the composition of agar and gelatin [6]. For fabricating magnetically actuated or navigated micro-robots, magnetically active materials are used to make the micro-robots responsive to the applied external magnetic fields. For example, $SmCo_5$ [53], Co [48, 54], Ni [49, 54], Fe [59], etc., are used to fabricate magnetically actuated/controlled micro-robots.

In the case of biomedical applications, micro-robots have to be employed in difficult-to-reach and very delicate regions (like the central nervous system, pancreas, heart, etc.) of the living body. Therefore, specific fabrication methods are required to control the dimensions of the micro-robots at the submillimeter scale precisely. Some widely used fabrication techniques for different types of micro-robots are discussed below.

3.3.1 Tubular Micro-Robots

The shape of a tubular micro-robot is required to be anisotropic for the propulsion in a low-Reynolds-number environment, as stated earlier in Section 3.2.2. The conical shape of a tubular micro-robots has been reported with excellent performance for directional-specific propulsion [60]. The propulsion speed and steering of these micro-robots also depend upon the diameter of opening and length of the micro-robot [84]. The outer surface of the micro-robot is generally made up of nonreactive, stable material and should reduce viscous drag for efficient propulsion. The outer layer of tubular micro-robots can also be functionalized to perform specific tasks, as discussed in Section 3.4.3. For bubble-propelled tubular micro-robots, the catalytic/reactive layer cannot be exposed to the working fluid and fuel directly [58] but is rather used as the inner layer of the tubular micro-robot. Thus, the inner space serves as a reaction chamber and gas-collecting cavity [59] for the controlled generation and ejection of the bubble. The gas bubble moves toward the wider opening due to the Laplace pressure difference and gets ejected from the wider opening of the tube. Micro-robots having a rocket-like tubular shape are fabricated widely either by rolled-up technology or by template synthesis. The schematic view of the two techniques is shown in Fig. 3.8 and discussed below.

3.3.1.1 Rolled-up technology

Rolled-up technology is a strain-engineered fabrication method. It utilizes the strain gradient of the different deposited layers for fabrication of a microtube. The schematic view of the technique is shown in Fig. 3.8a. It consists of a substrate, a sacrificial layer, and then deposited layers of the materials, to be fabricated as a microtube. The sacrificial layer can be made up of a polymer (e.g., photoresist). Over the sacrificial layer, the prestressed material layers are sequentially deposited by any of the deposition methods, for example, dip coating [85], e-beam deposition [59], atomic layer deposition (ALD) [86], etc. The outermost layer of a micro-robot is deposited at the bottom (the first layer) over the sacrificial layer, and the innermost layer of the micro-robot is deposited at the top (the last layer) over the sacrificial layer. In between these two layers, other functional layers can be deposited, for example, a layer of Fe [59] for magnetic steering of the micro-robot. The material layers are removed from the substrate by etching the sacrificial layer, which also rolls up the layers in a microtubular structure. Selective etching of the sacrificial layer is required to avoid dissolution of the micro-robot material itself. The diameter of the micro-robot depends upon the thickness of the deposited layers and the built-in strain gradient [86]. The thickness of the layers can be controlled by controlling the rate of deposition, whereas the strain state of the deposited layers can be controlled by [86]:

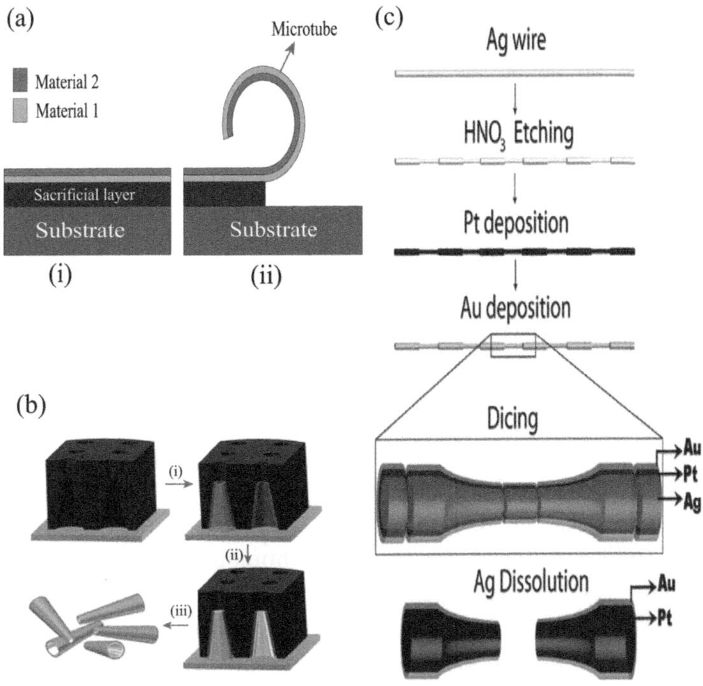

Figure 3.8 (a) Fabrication of a tubular micro-robot by rolled-up technology. (b) Fabrication of PANI/Pt microtubes by template synthesis. Reprinted with permission from Ref. [57]. Copyright (2011) American Chemical Society. (c) Fabrication of a Au/Pt micro-robot by the simplified template method. Reprinted with permission from Ref. [60]. Copyright (2010) American Chemical Society.

(1) Difference in the thermal expansion: The thermal expansion of a layer depends upon the thermal expansion coefficient (TEC). Therefore, by controlling the difference in the TECs of the deposited layers and substrate, the strain state of the layers can be controlled. This effect can also be enhanced by using different temperatures of the substrate for each layer deposition.
(2) Deposition rate: Variation in the deposition rate may lead to the formation of a different grain size, which exerts different stress levels in the deposited layers. Thus, the strain state can also be controlled by controlling the deposition rate of the layers.
(3) Stress development during the deposition itself.

An extensive exploration of this fabrication technique was done by Mei et al. [86]. Their group verified this fabrication technique by using various materials like metals, oxide layers, and different depositing techniques like e-beam evaporator, plasma immersion ion implantation and deposition (PIII&D), and ALD. They also used different photoresist sacrificial layers like ARP-3510, polydimethylsiloxane (PMMA), PDMS, and hydrogel. For removing the sacrificial layer, they used acetone, ethanol, or isopropanol. A new mechanism of rolling by selective swelling of the bottom layer of the deposited polymer or composite bilayer film was demonstrated by Luchnikov et al. [85]. In their method, the layers bent due to unequal changes in the specific volume of the layers. They fabricated micro- and nanosized tubes with a bilayer film of PS and poly(4-vinyl pyridine) (P4VP). Sanchez et al. [63] demonstrated a novel approach of fabricating high-efficient biocatalytic tubular micro-robots. Xi et al. [15] fabricated sharp-edged microdrillers with *Ti/Fe/Cr* material by rolled-up technology. Nearly any type of inorganic layer can be deposited and rolled up to make tubular micro-robots. The process consists of complex photolithographic processes and also requires a clean room facility, which is a factor in the increased cost of the method.

3.3.1.2 Template synthesis

Template synthesis involves the deposition of materials inside a template. A tubular micro-robot can be fabricated by the layer-by-layer deposition of materials inside the pores of a template. After that, the tubular micro-robot can be obtained by dissolution of the template. The diameter of the tube is determined by the diameter of the pores in the template membranes. Gao et al. [57] fabricated polyaniline/Pt bilayer micro-robots in double-conical micropores of cyclopore PC membranes. The fabrication procedure is shown in Fig. 3.8b. The fabrication steps followed by them were as follows: (i) First the deposited aniline monomers were allowed to polymerize on the inner wall of the membrane, forming a polyaniline (PANI) film. (ii) Using the galvanostatic method a Pt layer was plated inside the PANI tube. (iii) The bilayer tube was obtained by the dissolution of the template. Their micro-robot exhibited a maximum speed of more than 3 mm/s (i.e., more than 375 body length/s) corresponding to 10% H_2O_2 and 1.6% sodium cholate solution. As compared to the smooth surface of a rolled-up micro-robot, template electrodeposition produces a rough inner catalytic surface, which increases the activity of the catalyst. For the same reason the required minimum concentration of the H_2O_2 fuel is reported as 0.2% for a templated micro-robot [57] as compared to 1.5% for a rolled-up micro-robot [63]. Wu et al. [61] fabricated a tubular micro-robot using biodegradable chitosan (CHI) and ALG. CHI and ALG were alternately absorbed into porous PC membranes, and Pt NPs were subsequently assembled inside the CHI/ALG nanotubes. The micro-robots were obtained by dissolving the PC templates in CH_2Cl_2. Manesh et al. [60] used simplified template synthesis to fabricate concave, conical Au/Pt micro-robots using Ag wire templates, as shown in Fig. 3.8c. The Ag wire was etched at evenly spaced segments (for conical shape) using 30% HNO_3 for 30 s. Pt was galvanostatically deposited over the etched Ag wire, and Au was electrodeposited over Pt. Then these coated wires were diced at the desired location. The Au/Pt micro-robots were obtained by dissolving the Ag wire template in 25% (v/v) HNO_3 for 5 min.

As compared to the rolled-up technology, the cost of clean room requirement can be eliminated by using the template synthesis method. By eliminating strain engineering (as in the rolled-up method), the template synthesis method also simplifies the geometric consideration of the micro-robot to the choice of porous template and control on the thickness of the deposited layers.

3.3.2 Helical Micro-Robots

Helical micro-robots propel by generating helical waves, as discussed in Section 3.2.1. A helical micro-robot consists of a helical tail and a head attached to it. These micro-robots are also called ABF. Under the influence of an external rotating magnetic field, the helical tail of the micro-robot produces helical waves, which causes propulsion of the micro-robot. The helical micro-robot can be fabricated by template synthesis [49], self-scrolling technology [50], and glancing angle deposition (GLAD) [87].

The template synthesis method discussed for the fabrication of tubular micro-robots in Section 3.3.1.2 can also be used to fabricate helical micro-robots. It involves electrodeposition of different materials into cylindrical micro-pores of the template membranes, followed by dissolution of the template. It is a fast method by which thousands of micro-robots can be prepared in a few hours. The diameter of the helix can be controlled by selecting the proper pore size of the template membrane. The length of the helical tail can be controlled by charge density transported through electrodeposition [49]. The tailoring of helical pitch can be done through controlling the composition of plating solution [49]. PC membranes are commercially available (110605 and 110607, Whatman, NJ, USA), with the pore diameter ranging from 0.015 μm to 12 μm, and can be used as a template to fabricate the helical tail of a micro-robot [49]. Anodized aluminum oxide (AAO) available in the market (6809-6022, Whatman, Maidstone, UK) can also be used as a template [49, 88]. The process flow of template synthesis of a helical micro-robot is schematically shown in Fig. 3.9a.

Zhang et al. [50] fabricated helical micro-robots by the self-scrolling technique. They fabricated two types of InGaAs/GaAs semiconductor bilayer and InGaAs/GaAs/Cr hybrid semiconductor–metal trilayer tails attached with a soft magnetic head made up of Cr/Ni/Au. The trilayer tail was fabricated as shown in Fig. 3.9b-i–vi, and a field emission scanning electron microscopy (FESEM) image of the untethered helical micro-robot is shown in Fig. 3.9b-vii.

The InGaAs/GaAs bilayer was epigrown by molecular beam epitaxy over an AlGaAs sacrificial layer on a GaAs (001) wafer substrate. Then a Cr layer was deposited on the bilayer by e-beam evaporation. The InGaAs/GaAs/Cr trilayer was patterned to a ribbon-like mesa for making a helical tail by reactive ion etching. Then the soft magnetic head was prepared by e-beam evaporation of Cr/Ni/Au thin layers and the lift-off process. Finally, the AlGaAs sacrificial layer was selectively etched, by which the ribbons self-organized to form ABF. Then the ABF were untethered from the substrate by micromanipulation: cut, pick, and release.

Another method of fabricating a helical micro-robot is GLAD [87]. In this method, the substrate is highly tilted (i.e., away from the perpendicular to the stream of atoms) and materials are deposited obliquely over the substrate, inside a vacuum chamber, as shown in

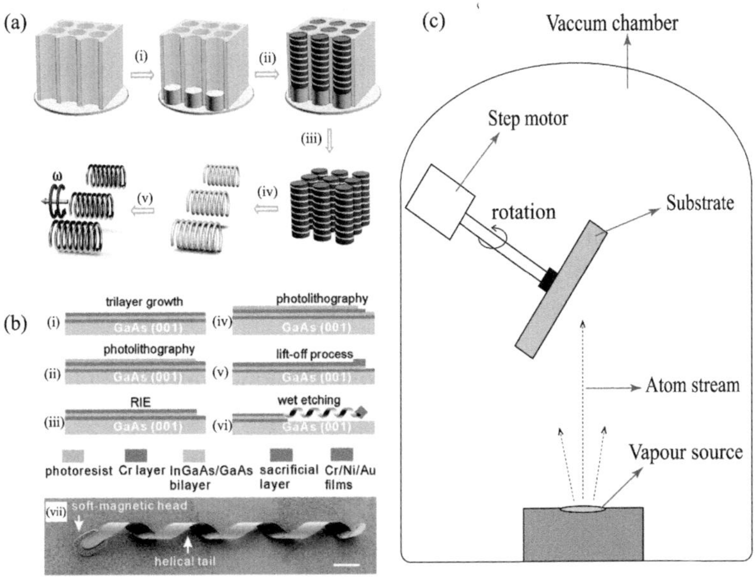

Figure 3.9 (a) Template synthesis of Pd helical micro-robot in an AAO template. (i) Deposition of gold. (ii) Deposition of Pd/Cu. (iii) Dissolution of the AAO template. (iv) Etching of copper from Pd/Cu micro-rods. (v) Magnetic nickel layer deposition on Pd micro-rods to make magnetically propelled, helical micro-robots. Republished with permission from Ref. [49]; permission conveyed through Copyright Clearance Center, Inc. (b) Fabrication steps of a trilayer helical micro-robot by the self-scrolling method (i–vi). FESEM image of an untethered helical tail with a soft magnetic head (vii). Reprinted from Ref. [50], with the permission of AIP Publishing. (c) Schematic view of the GLAD method for fabrication of helical micro-robots. The substrate is rotating by a step motor, and atoms are deposited obliquely on the substrate.

the schematic view of the method in Fig. 3.9c. The atoms condense on the substrate, and continuous deposition of atoms results in a columnar structure oriented toward the vapor source. Ghosh and Fischer [48] fabricated one of the smallest artificial micro-robots of width 0.2–0.3 μm and length 1–2 μm with SiO_2 using the GLAD technique. The helical tail can be grown by slow continuous rotation (~0.07 rpm) of the substrate during the material deposition [48].

Some of the materials used for making helical tails are Ni-Ti [10], Cu [53], SiO_2 [48], and Pd [49, 88].

3.3.3 Flexible Tail Micro-Robots

Flexible tail micro-robots propel by generating planar waves, as discussed earlier in Section 3.2.1. The main structural feature of a flexible tail micro-robot is its ultrathin, flexible tail, as shown in the schematic of Fig. 3.10a-iii,iv. The flexible tail is generally attached to a magnetic head, which can be made by using magnetically active materials, as discussed in Section 3.3, to provide a magnetic dipole moment. The flexible tail produces planar waves under the influence of an oscillating magnetic field, which results in micro-robot propulsion.

Electrospinning is the fabrication technique widely used for the fabrication of flexible tail micro-robots [33–35, 89]. This method is capable of fabricating ultrafine fibers with a diameter in the range of micrometer to nanometer. In the electrospinning process, a polymer solution is pumped out from a syringe pump at a controlled rate and deposited on a collector. A strong electric potential difference generally 1–30 kV is created between the needle of the syringe and the collector, which draws the solution to the collector [90]. The solution solidifies while traveling to the collector and gets deposited on the collector. The main components of the electrospinning apparatus are shown in Fig. 3.10a-i. The morphology and diameter of the electrospun fiber depend on various parameters [90] like the type of polymer, concentration, viscosity of the solution, electrical conductivity, surface tension of the solvent, distance between collector and needle, the strength of the electric field, etc. The formation of beads occurs due to Rayleigh instability [90]. After the solidification, the collected beaded fibers are cut and extracted by using tweezers under high magnification [89]. Khalil et al. [35] fabricated micro-robotic sperm from PS dissolved in dimethylformamide and mixed with iron oxide, as shown in Fig. 3.10a.

The iron oxide NPs were contained in the beads of fibers, forming the magnetic head of the micro-robots.

Dreyfus et al. [55] fabricated a flexible tail micro-robot made up of superparamagnetic particles, as shown in the schematic of Fig. 3.10b. In their study, the superparamagnetic particles were coated with streptavidin (red cross). The particles were attached by

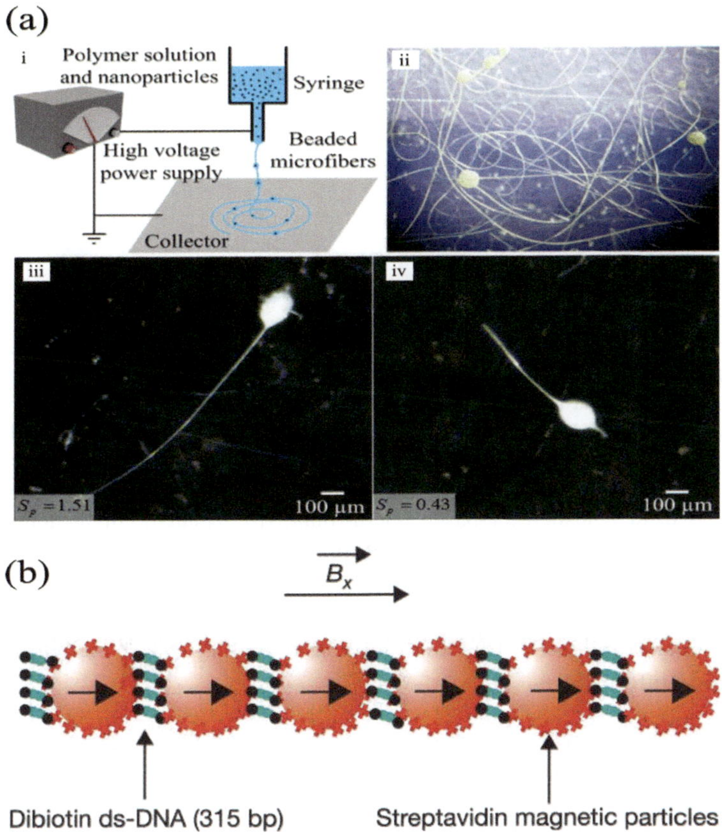

Figure 3.10 (a) Fabrication of magnetic microsperm by the electrospinning method using polystyrene dissolved in dimethylformamide and mixed with iron oxide. (i) Use of a syringe pump for fabrication of a flexible tail. (ii) Collecting the micro-robots formed by the electrospinning method, the iron oxide NPs are contained in the heads of the micro-robots. (iii, iv) Individual micro-robots with different tail lengths. Reprinted from Ref. [35], with the permission of AIP Publishing. (b) Artificial magnetic flagellum. Superparamagnetic particles coated with streptavidin (red cross) and connected with DNA. Reprinted by permission from Springer Nature Customer Service Centre GmbH: Springer Nature, *Nature*, Ref. [55], copyright (2005).

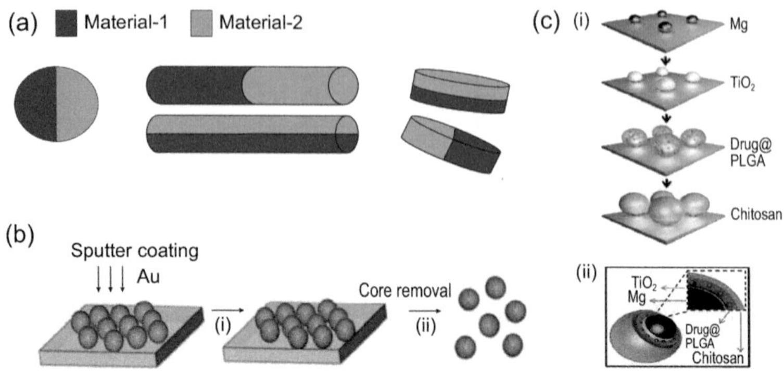

Figure 3.11 (a) Janus micro-robots with different shapes. (b) Template fabrication of Au-functionalized $(PSS/PAH)_5$ Janus micro-robots. Reprinted by permission from Springer Nature Customer Service Centre GmbH: Springer Nature, *Nano Research*, Ref. [14], copyright (2016). (c) Mg-based micro-robot for in vivo drug delivery. (i) Schematic image of preparation steps of drug-loaded Mg-based micro-robot. (ii) Schematic dissection of the Mg-based micro-robot. Reprinted from Ref. [4] under the terms of the Creative Commons CC BY license.

double-stranded DNA having biotin at each end, via streptavidin–biotin interaction.

3.3.4 Janus Micro-Robots

Micro-robots having two regions with a distinct physical/chemical property at the surface are known as Janus micro-robots. Figure 3.11a shows some common shapes of Janus micro-robots. Different groups fabricated various types of Janus micro-robots using different techniques. Wu et al. [14] fabricated Au-functionalized poly(styrenesulfonate) and poly(allylamine hydrochloride) (PSS/PAH) Janus micro-robots, demonstrating self-thermophoretic propulsion. They used the template-assisted, layer-by-layer deposition method similar to that discussed in Section 3.3.1.2. In their study, five bilayers of PSS/PAH were deposited on spherical silica templates via electrostatic interaction. A Au layer was deposited by sputtering over the half portion of the PSS/PAH particles (Fig. 3.11b-i). The Au-functionalized $(PSS/PAH)_5$ micro-robots were obtained by dissolution of the silica template in hydrofluoric acid (Fig. 3.11b-ii). The micro-robots can be coated with biological entities to protect them inside a biological environment. For example, a coat of RBC membranes on a Janus micro-robot prevents protein fouling on the micro-robot [30]. In this way, the effective working life of a micro-robot can be increased in an in vivo environment. De Avila et al. [4] fabricated Mg-based Janus microparticles for propulsion in an acidic fluid. The fabrication steps are depicted in Fig. 3.11c.

Other examples of micro-robot fabrication are as follows: Iacovacci et al. [6] used 3D-printed molds to caste a bicomponent micro-robot made with biocompatible PDMS material. Wehner et al. [62] fabricated a novel soft octobot (Fig. 3.12a) using molding and embedded 3D (EMB3D) printing. Ceylan et al. [90] 3D-printed a double-helical, magnetically propelled, biodegradabale micro-robot for theranostic cargo delivery applications (Fig. 3.12b).

Lu et al. [26] used a simple microfluidic tubing approach for synthesizing a soft micro-robot, as shown in Fig. 3.13.

In their fabrication process cross-linker glutaraldehyde (GA) was used for cross-linking of chitosan to form soft capsules. These soft capsules encapsulated NPs made up of an Fe core and a Pt shell. Fe provides magnetic responsiveness, and Pt provides catalytic property. During cross-linking these NPs were collected at the bottom part of the micro-robot, forming patches, by using neodymium magnets. This approach can be followed to synthesize other soft micro-robots as well.

3.4 Biomedical Applications

The advancement in micro-robot fabrication and their biocompatibility advanced the reach of biomedical science into the previously difficult-to-reach, intricate parts of the human body. The sophistication of the robotic system to the microscale have great potential to develop various applications down to the cellular scale, like targeted delivery [4, 6, 14, 26, 83, 92–95], active diagnosis [96–98], minimally invasive surgery [2, 12, 15, 99], biodetoxification [18, 21, 30], tissue engineering [100, 101], cell manipulation [102], and isolation of biological targets [1, 11, 28]. The bare micro-robot

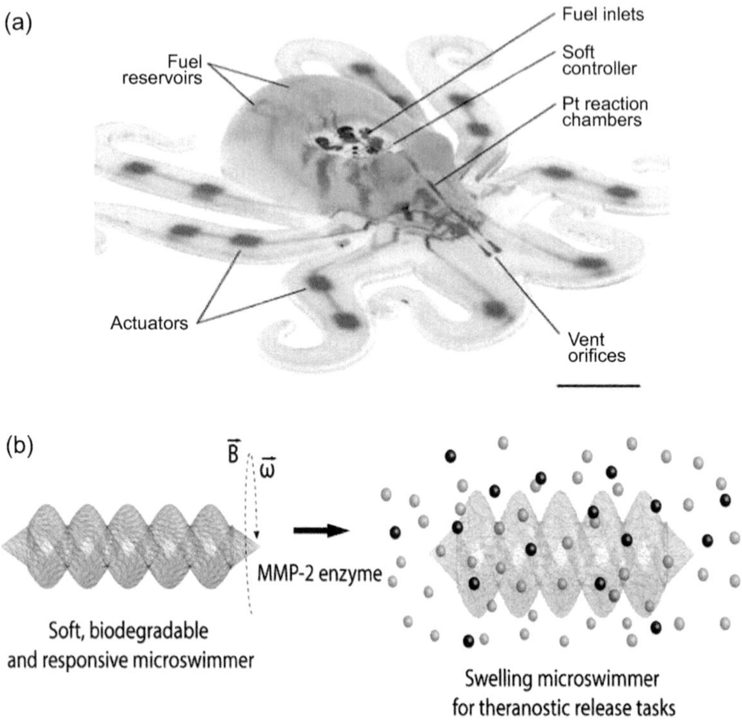

Figure 3.12 (a) Fully soft, autonomous robot, octobot controlled via embedded micro-fluidic soft controllers and propelled by the bubble propulsion mechanism. Reprinted by permission from Springer Nature Customer Service Centre GmbH: Springer Nature, *Nature*, Ref. [62], copyright (2016). (b) 3D-printed biodegradable, double-helical micro-robot. The micro-robot propels by application of a rotating magnetic field (left) and releases the embedded cargo in the presence of matrix metalloproteinase-2 (right). Reprinted with permission from Ref. [91]. Copyright (2019) American Chemical Society. <https://pubs.acs.org/doi/abs/10.1021/acsnano.8b09233>. Further permissions related to the material excerpted should be directed to the ACS.

just with a propulsion mechanism cannot be employed for these applications readily. It needs some mountings to execute the desired medical applications, like onboard sensors, drug loading, delivery mechanisms, surgical tools, computing and feedback components, etc. Some of the biomedical applications are discussed here. We are

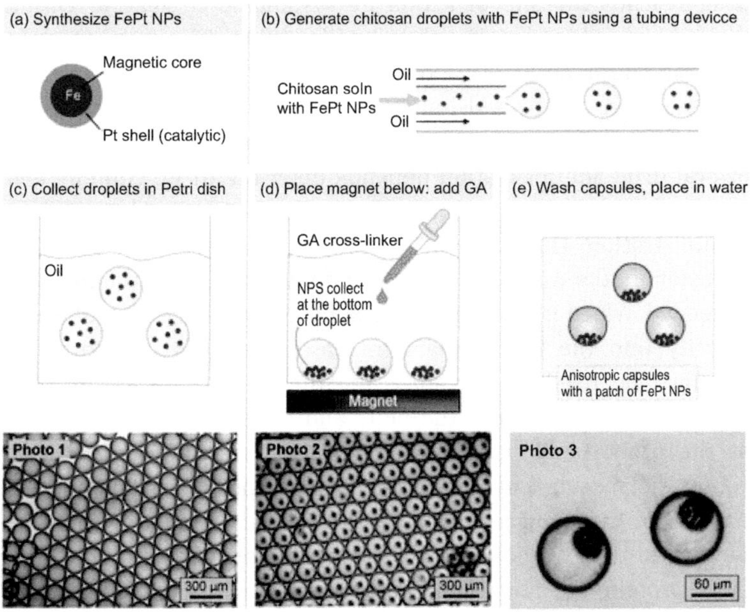

Figure 3.13 Synthesis of bubble-propelled, soft micro-robots using the microfluidic tubing approach. (a) The NPs had a magnetic (Fe) core and a catalytic (Pt) shell. (b) The NPs were dispersed in the chitosan solution, and the solution was used as the inner disperse phase in a coflow tubing device, with the continuous outer phase being oil. This microfluidic approach resulted in the formation of droplets embedded with NPs. (c) The droplets were collected in a Petri dish. Photo 1 shows the micrograph of the droplets. (d) Then a neodymium magnet was used to collect the NPs at the bottom of the droplets, as shown in Photo 2. Meanwhile, cross-linker GA converted the droplets into patchy capsules. (e) Capsules collected in water. Photo 3 shows individual capsules with embedded NPs. Reprinted with permission from Ref. [26]. Copyright (2016) American Chemical Society.

classifying these biomedical applications into three categories of delivery, surgery, and sensing, as discussed below.

3.4.1 Delivery

Routes used for the current drug administration in the body are oral passage, skin (transdermal), drug injection into blood, and inhalation. In these conventional drug administration methods, the drug reaches the site of action by systematic circulation within the body by blood and other bodily fluids. The current drug administration methods are the passive way of drug delivery, which does not have self-propulsion and controlled navigation, and tissue penetration is limited by the passive mass transport. The efficacy of oral drug administration depends upon the GI functioning and faces a slow and irregular rate of absorption. In transdermal drug administration, the stratum corneum acts as a barrier for drug diffusion. In the case of inhalation, a unique device for the purpose is needed, which is not much affordable, whereas injection of drugs directly into the bloodstream increases the drug concentration rapidly. The need to maximize the therapeutic efficacy only at some desired location of the body and to reduce the side effects at the other body parts has driven the need for TDD. Micro-robots possessing characteristics like self-propulsion, controlled navigation, cargo towing, tissue penetration, and release of cargo can be used for in situ delivery of drugs and biological components like DNA, protein, etc. For example, the microneedle [95], micropump [94], and microholder [83] are getting attention as TDD systems. A TDD system allows the rapid transport of high-concentration drugs to the diseased tissues directly and reduces the side effects on other tissues, thus increasing the efficacy of the drug. Due to the small size, a TDD system has the potential to release therapeutics in small-diameter canals like the central nervous system (especially the spine), urinary system, excretory duct of Wirsung, and ovaries. In these regions, the issue of space constraint can be addressed by providing an external power source [6] or using the surrounding [26] for the propulsion of TDD systems.

The first in vivo TDD test to treat gastric bacterial (*Helicobacter pylori*) infection in a mouse was performed by de Avila et al. [4] without any toxicological consequences. A schematic view of drug delivery by the micro-robot is shown in Fig. 3.14a. In vitro tests on T24 human urinary bladder carcinoma cells were successfully performed by Iacovacci et al. [6]. Liposomes have greatly attracted researchers' attention for potential use in TDD systems [10, 92]. Liposomes are biocompatible and have potential to be used as a drug and biocomponent carriers [103]. Mhanna et al. [10] used DOPE/DOTAP (1:3) liposomes for drug encapsulation.

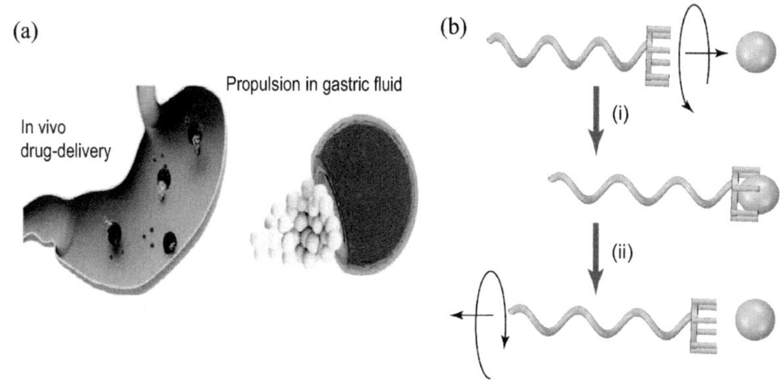

Figure 3.14 (a) Schematic view of drug delivery in the stomach by bubble-propelled micro-robots (left) and an individual micro-robot (right). Reprinted from Ref. [4] under the terms of the Creative Commons CC BY license. (b) Schematic view of cargo pick-up, transport, and drop-off by a magnetically propelled helical micro-robot attached with a microholder head.

They demonstrated in vitro, single-cell (C2C12 mouse myoblasts) delivery of the calcein drug. The liposomes were adsorbed on the surface of ABF and transported to the target location by magnetic controlling.

Iacovacci et al. [6] demonstrated in vitro controlled release of a drug at the target location by applying a compressive force on the drug-loaded hydrogel. A helical micro-robot having a microholder head can be used for drug and cargo delivery [83]. In an external rotating magnetic field, the microholder can be propelled to the cargo location (Fig. 3.14b-i). The microholder head can stably hold the cargo while transporting it via corkscrew motion. At the target location, the cargo can be dropped off by merely reversing the rotational direction of the external magnetic field, which enables the micro-robot to move in the backward direction (Fig. 3.14b-ii).

Research has shown that the TDD system is more efficient than passive drug delivery [4]. The TDD system increases the drug efficacy by direct drug supply to the diseased site, thus reducing the side effects and drug intake frequency. The TDD system has the potential to achieve therapeutic applications, which are not possible by the existing passive delivery system.

3.4.2 Surgery

Micro-robots have the potential capability to navigate through the narrowest capillaries of the human body and can perform medical procedures down to the cellular scale. Advancements in minimally invasive surgery can be done by scaling down the size of surgical tools to the microlength scale. The wireless controlling of a surgical micro-robot is essential for maneuverable, precise, non-invasive procedures in a confined space inside the body. Micro-robots can provide a variety of minimally invasive surgeries to biological entities of microlength scale or even down in the line, with flexibility, precision, and control. Micro-robots like microgrippers [2], microcannons [12], and microdrillers [15] have been tested for their capability of minimally invasive surgery. Xi et al. [15] fabricated rolled-up microdrillers and demonstrated an ex vivo drilling procedure on porcine liver tissue, as shown in Fig. 3.15a. Breger et al. [2] exploited the swelling characteristic of a hydrogel to design a soft microgripper. They fabricated a soft microgripper by sequential photolithography of swellable, soft hydrogel poly(N-isopropylacrylamide-co-acrylic acid) (pNIPAM-AAc) and a nonswellable, stiff polymer polypropylene fumarate (PPF). The microgripper was able to actuate by swelling/deswelling by absorption/desorption of water in response to the physiological temperature, as shown in Fig. 3.15b-i. They showed that combining a swellable, soft hydrogel with a nonswellable, stiff polymer endows sufficient strength to the microgripper to excise cells from a tissue clump, as shown in Fig. 3.15b-ii,iii.

The concept of a "magic bullet" is envisioned for the supply of drugs deep into the diseased tissue, in the microlength scale, at a location not accessible by conventional medical methods. Soto et al. [12] demonstrated the first use of acoustically triggered microcannons to fire nanobullets for microscale tissue penetration of the therapeutic payloads, as shown in Fig. 3.15c. They fabricated biocompatible microcannons encapsulating perfluorocarbon (PFC) emulsion with silica nanobullets. The use of a focused ultrasonic pulse triggered the vaporization of PFC, leading to rapid expulsion of gas microbubbles from one opening of the microcannon. The expulsion of microbubbles resulted in the rapid firing of the

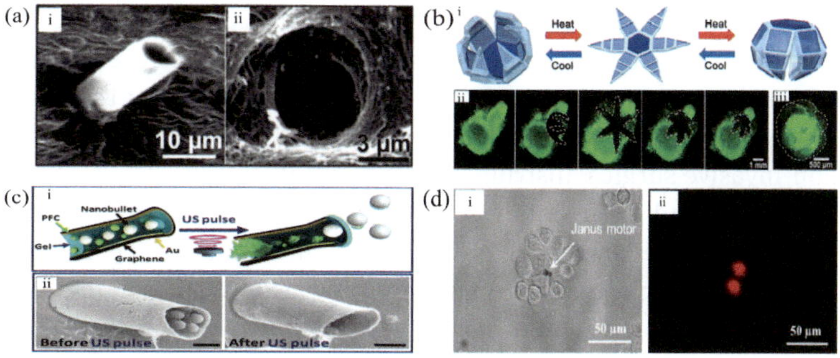

Figure 3.15 (a) SEM image of a microdriller (i) and a drilled hole (ii) after removal of the microdriller on porcine liver tissue. Reprinted with permission from Ref. [15]. Published by The Royal Society of Chemistry, under a Creative Commons Attribution 3.0 Unported Licence. (b) Microgripper fabricated with pNIPAM-AAc and PPF layers. (i) Schematic view shows micro-robot responses to the physiological temperature. (ii) Capture and excision of cells from a live fibroblast clump. (iii) Microgripper with the cell in its grip. The dashed lines show the microgripper and excised cell. Reprinted with permission from Ref. [2]. Copyright (2015) American Chemical Society (https://doi.org/10.1021/am508621s). Further permissions related to the material excerpted should be directed to the ACS. (c) Microcannon with nanobullets. (i) Schematic view showing microcannon composition before firing (left) and during firing (right) of the nanobullets. (ii) SEM image of microcannon loaded with nanobullets before firing (left) and unloaded after firing (right). Scale bar: 20 μm. Reprinted with permission from Ref. [12]. Copyright (2016) American Chemical Society. (d) Photothermal effect on Hela cells by exposing Janus micro-robot to NIR irradiation. (i) Before and (ii) after irradiation. The red color indicates dead cells. Reprinted by permission from Springer Nature Customer Service Centre GmbH: Springer Nature, *Nano Research*, Ref. [14], copyright (2016).

nanobullets and the displacement of the microcannons. The high-speed nanobullets were found to penetrate 17.5 ± 3.7 μm in a tissue phantom gel target [12]. This kind of high-speed nanobullet firing can be used for deep delivery of drugs into the diseased tissue and for delivering genetic material into cell nuclei for gene therapy.

The posterior section of an eye is crowded with physiological barriers like the vitreous humor, optic nerve, retina, etc. These barriers cause difficulty to perform ophthalmic medical procedures. To treat ophthalmic diseases, present medical procedures like topical administration of drugs, intravitreal injection, laser surgery, photodynamic therapy, etc., are used. These methods are limited with invasiveness, time consumption, and efficacy only at the early stage of the disease. Chatzipirpiridis et al. [29] electroformed non-invasive surgical instruments for ophthalmic applications: *CoNi* microtube and *CoNi*-stainless steel microneedle. They tested the magnetic maneuverability of the *CoNi* microtube in the central vitreous humor of a living lapine eye and an ex vivo porcine eye. They also performed vein puncture operation with a microneedle in a developing chicken embryo.

Stroke is a common disease causing death. Narrowing or blocking of arteries causes a reduction in blood flow to the brain, called ischemic stroke. An ischemic stroke reduces or stops oxygen supply to the brain tissues, which has fatal consequences. An ischemic stroke is primarily caused by the formation of blood clots (thrombi) in the brain arteries. There are microstirrers to generate scissor-type vibrations to dissolve thrombi [99] and thus remove clots with minimal invasion.

Wu et al. [14] demonstrated in vitro targeted photothermal therapy of HeLa cells in a cell media using a self-thermophoretic Au-PSS/PAH Janus micro-robot, as shown in Fig. 3.15d. The positively charged poly(allylamine hydrochloride) (PAH) outer layer of the Janus micro-robot attached to the negative charged HeLa cells due to electrostatic attraction. After the attachment, the sample was exposed to high-intensity near-infrared (NIR) irradiation, which reduced the viability of the HeLa cells from 100% to 57%. Their demonstration implies the combination of a Janus micro-robot and NIR irradiation can be used for precise thermal therapy and effectively induce the apoptosis of tumor cells. The use of untethered micro-robots for surgical procedures is beneficial in decreasing procedural discomfort, the sedation effect, and postprocedure complications.

3.4.3 Sensing and Diagnosis

The modern class of biosensors includes bioaffinity sensors. These bioaffinity sensors trigger specific signals after selective recognition of a specific target. The common bioaffinity sensors rely on placing of the sample drops over the receptor. The interaction between analyte and receptor is allowed over an incubation period under a quiescent condition. After the end of this incubation period, the extent of analyte–receptor interaction is measured for the assay of the analyte. This droplet-based bioassay relies solely upon the diffusion transport [104]. Therefore it is time consuming and laborious. The bioaffinity interaction can also be used to isolate a specific target from the unprocessed biological fluid. The isolation of a biological entity is a challenging task due to the complexity of biological fluids.

The development of wireless, mobile micro-robots has potential applications in sensing, diagnosis, and bioanalytics. The capability of the micro-robots does not only include the efficient alternative of the current sensing procedures but also involves the capability to sense in the previously inaccessible microscale parts of the biological environment. The biosensing micro-robots can have a surface functionalized with a bioreceptor [1, 11] or can have a built-in bioreceptor [28]. The sensing capability of the biosensors depends upon the specific interaction of the receptor with an analyte. Upon interaction with the analyte, the receptor triggers a signal, which can be traced and processed to get the qualitative and quantitative measures of the analyte. Micro-robots have potential use in "on-the-fly" sensing, capturing, and isolation of biological targets. The on-the-fly approach relies upon the continuous movement of the receptor-modified micro-robot throughout the sample domain. The continuous movement of the receptor-modified micro-robot enhances the receptor–analyte interaction as compared to that in the quiescent condition. Specifically, the continuous movement and long bubble tail generated during the propulsion of the tubular micro-robot are found to enhance the fluid convection and mixing better than that by spherical Janus and other micro-robots [105]. The enhancement in the fluid convection and mixing accelerates the bioaffinity interaction and thus the sensitivity and efficiency of the bioassay.

Micro-robotic biosensors can capture and transport a specific biological target to a clean zone for subsequent tests and analysis. The transportation of the specific biotarget requires an optimal propulsion mechanism, direction control, and large towing force of the micro-robot in the complex biological fluid. For such purpose bubble-propelled, catalytic, tubular micro-robots are widely tested [1, 9, 11, 13, 16, 28, 106]. The reasons behind the popularity of these micro-robots are the capability to tow with a large towing force and high propulsion speed, ranging between 100 and 1000 μm/s [26] in complex fluid and high-ionic-strength solution. The biosensing micro-robots with towing capability can be used to detect and isolate biological components like DNA [13], protein [11], nucleic acids [9], cancer cells [1, 16], bacteria [107], etc., in a complex biological fluid.

Nguyen and Minteer [13] demonstrated a new design strategy for "signal-on" detection of DNA targets. Their catalytic, tubular micro-robot showed movement in the presence of a DNA target only.

Circulating tumor cells (CTCs) are primarily responsible for spawning cancer metastasis. Detection of these CTCs provides diagnosis and prognosis of different kind of cancers [1]. Carcinoembryonic antigen (CEA) is the most common antigen present in cancer cells like colorectal, pancreatic, and gastric [1] cancer cells. Balasubramanian et al. [1] demonstrated in vitro isolation and transport of pancreatic cancer cells (bearing CEA + antigen) using a bubble-propelled tubular micro-robot. They functionalized the micro-robot with an anti-CEA monoclonal antibody (mAb) to selectively capture and transport the cancer cells. Selective capturing and transportation of the tumor cells without preprocessing the biological sample has great promise in the diagnosis of cancer and other diseases.

Kuralay et al. [28] demonstrated "built-in" recognition, capturing, transportation, and release of glucose and yeast cells, as shown in Fig. 3.16. They fabricated a tubular micro-robot with a poly(3-aminophenylboronic acid) (PAPBA) outer layer and a Pt catalytic inner layer. The monosaccharide recognition capability of the outer polymer layer of the micro-robot was exploited to capture the yeast cell (with sugar residues on the wall) and glucose.

Currently available flexible endoscopes and catheters provide invasive, visual diagnosis of disease for a short duration of time with discomfort to the patients. An alternative non-invasive, visual diagnosis of disease in the GI tract is facilitated by a swallowable wireless endoscopy capsule (WEC) [7]. The WECs are functionalized with illumination, a camera for image/video

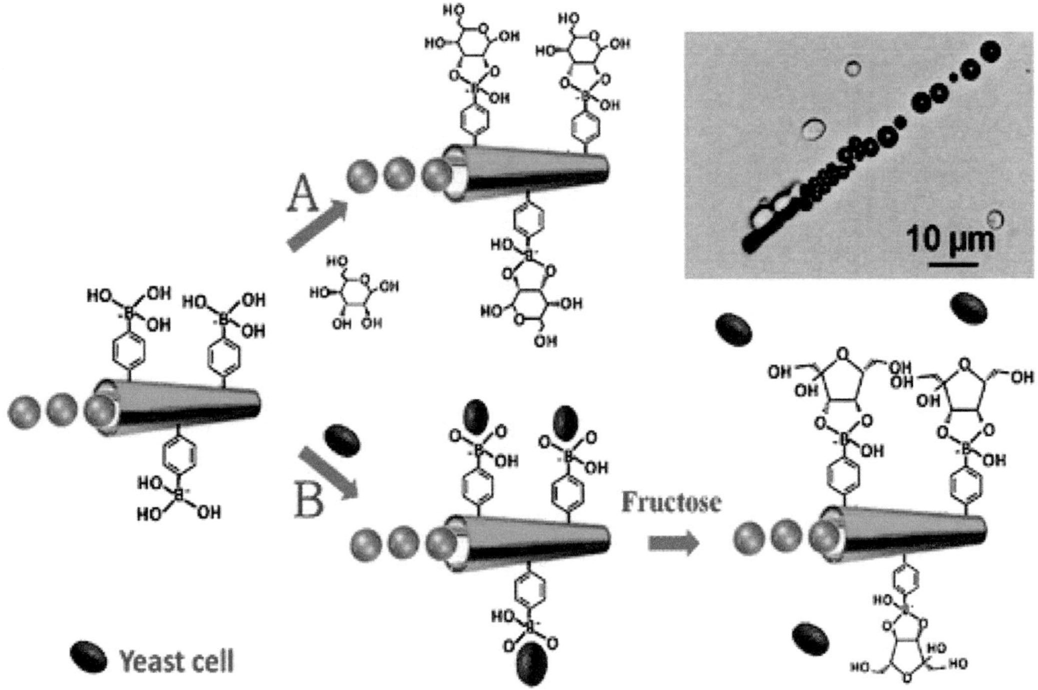

Figure 3.16 Schematic view of a bubble-propelled, tubular micro-robot with built-in boronic acid. Path (A) capturing of glucose; path (B) capturing of yeast cells and fructose-induced release of yeast cells. Inset image shows a micro-robot transporting multiple yeast cells. Reprinted with permission from Ref. [28]. Copyright (2012) American Chemical Society.

capturing, and wireless data transmission. This pill-sized WEC provides accessibility to previously difficult-to-reach parts of the GI tract (e.g., the small bowel). The commercially available WEC provides a passive diagnosis as the operator cannot control its position and orientation, and it moves by gravity and the peristalsis of the GI tract. These WECs can be used for active diagnosis purpose by controlling their position and orientation by integrating a magnetic control mechanism like a magnetic rubber shell [27]. Similar active control can also be used in the WECs for real-time sensing of pH, temperature, and pressure. The commercially available capsules are appropriate in size for the GI tract, which, with further scaling down and controlled navigation, can be employed for the active diagnosis of the circulatory, urinary, and central nervous systems. Use of micro-robots for on-the-fly sensing avoids the preprocessing of biological samples. Functionalization of the micro-robot with a receptor requires additional fabrication steps and surface manipulation.

Wu et al. [30] demonstrated the camouflage of natural red blood cells (RBCs) on a Mg biocompatible Janus micro-robot to attract, capture, and neutralize the toxins in biological fluids. They reported an enhanced detoxification efficiency of the micro-robot, attributed to the water-powered propulsion of the micro-robot and antifouling property of the RBC membrane coating. Wu et al. [30] also reported the diverse detoxification potential of the RBC-Mg micro-robot with a nerve agent simulant, methyl-paraoxon, in albumin solution. The acetylcholinesterase present in the RBC membrane can absorb and remove, and the water-powered propulsion resulted in improved speed of the decontamination process.

3.5 Discussion and Future Scope

Although there exist several sophisticated micro-robots, and many of them are currently used in practical applications, there are still many challenges to overcome in the existing micro-robots. The challenges include precise and safe locomotion and targeting and minimizing the power consumption for sensing, locomotion, data transfer, and computation. The micro-robot should not harm tissues and organs, should be biocompatible, and should have robust operation in flowing against bloodstream, respiration, peristalsis, etc. [31].

The use of a magnetic field for control and navigation of a micro-robot is prevalent as wireless transmission of high force, and torque is possible with all 6 degrees of freedom. Still, there are some limitations of the magnetic field that need to be overcome. For example, the magnetic force decays exponentially with the distance from the field source. Therefore, the use of a high-strength magnetic field is required to control the micro-robot deep inside the human body. However, the use of a maximum static magnetic field is limited to 2.5 T and for a pregnant woman to 2 T by standard regulation, which limits the use of magnetic steering of a micro-robot inside the human body. The equipment required to generate a magnetic field for medical use of a micro-robot is generally bulky and costly. This bulky equipment is required to be kept closer to the patient, which may reduce the maneuverability of the operator and patient as well. The helical tail and flexible tail micro-robots generate a flow field during their propulsion. The flow field generated by the propulsion of the flexible tail is found to influence the accuracy of positioning of the payload [33]. Therefore, further studies are required for characterizing the frequency-dependent flow field for flexible tail as well as helical tail micro-robots. Research shows that gradient pulling is efficient than the propulsion of a helical propeller, but when limitations of the magnetic field and its generation are considered, then the helical propeller is far better than gradient pulling [51].

Bubble propulsion is not suitable for in vivo applications. As H_2O_2 is the main fuel for the bubble-propelled micro-robots, there are no in vivo applications as H_2O_2 is not a biocompatible fuel. The self-propulsion of a bubble-propelled micro-robot is limited by the availability of fuel. It can only be propelled until the fuel lasts. The propulsion of a catalytic micro-robot is limited to low-ionic-strength fluids only [60]. There is a requirement of finding alternative fuels and corresponding catalytic materials and reactions. It would be beneficial if in situ fluids can be used as fuel. Water and acids are readily present in a biological media; therefore, these can be used as in situ fuels. Materials like Mg-, Al-, and Zn-based catalytic micro-robots propel by depleting protons in an acidic medium. So they can be readily used in an acidic environment without any external fuel injection. Also, these micro-robots can be used for measuring the pH of the fluid domain. However, on the other hand, the reactive material like Mg and Zn used in micro-robots dissolve in the working biological fluid; therefore, their working life is very short and cannot be used for long-time operations. The use of enzymes to decompose the fuel is also an alternative way to replace the metal catalyst from the micro-robot, to make it more efficient and biocompatible for medical applications [19, 63].

Electrophoretic propulsion is a good alternative for chemical-free, bubble-free directional propulsion of a micro-robot [79]. However, the propulsion mechanism of the phoretic effect is not clearly understood [82]. A biohybrid system faces a challenge to maintain the viability of the living cells attached to the micro-robots. Biocompatibility and proper adherence of the synthetic part to the biomicroswimmers are an essential requirement. In the case of taxis of a micro-robot, maintaining a driving gradient of the external stimuli near the micro-robot in a fluidic environment is a tough task to achieve.

Biodegradability and biocompatibility of micro-robot materials are challenges for their potential use in the medical field. If the material is biodegradable, then the micro-robot can dissolve in the biological media and disappear after completion of the operation. However, if the material is nonbiodegradable, then retrieving the micro-robot after the completion of the operation is necessary and is a major challenge. The biological environment inside the human body is very crowded; therefore, precise control of a micro-robot is required. The micro-robot components should be soft and deformable so that they do not cause any injury to the components inside the body. Rigid and hard micro-/nanorobots are generally nonbiodegradable and can also cause problems like rupture of blood vessels and tissue injury inside the body. Soft materials like synthetic polymers and plant tissues have potential use in the fabrication of a soft and biocompatible micro-robot. Also, the coupling of natural biological material with synthetic micro-robots reduces the immune evasion and biofouling effects of the biological environment, thus maintaining the efficiency and longer working life of a micro-robot.

Swimming against the dynamic blood flow is challenging with limited power. A consistent power source for continued operation is still a milestone to be achieved. On the other hand, the TDD system is challenged by the low controllability on the release rate of the therapeutics.

Now it is possible to control a micro-robot with a wireless remote or just by the mouse of a workstation [27]. Challenges for future-generation miniaturized robots are to perform collective swarming and task completion by interacting with other micro-robots. Future micro-robots will mimic the natural intelligence of their biological counterparts with adaptable and sustainable operation and group behavior with swarm intelligence. In the future, micro-robots will be able to perform any task in coordination with thousands of other micro-robots in a swarm. This collective dynamics of micro-robots is required to treat any large-size body parts, to deliver a large quantity of therapeutics to a target location, and to reduce the overall operation time. This kind of group communication and coordination is challenging, which needs to be overcome by future micro-robots. Although the collective dynamics of the micro-robots and microswimmers [108–111] is an active research area, further studies are required to understand the behavior of microswimmers, both theoretically and experimentally.

New studies on micro-robots would lead to the evolution of therapies and applications that are not yet conceived. For example, Janus micro-robots can be coated with stem cell membranes for cancer cell recognition for targeted photothermal therapy [14]. The development of micro-robots for medical applications will find its way from the lab to the real working environment if the safety of the patient is ensured. It needs to be shown that micro-robots can work in a dynamic environment inside the human body reliably and robustly.

References

1. S. Balasubramanian, D. Kagan, C.-M. Jack Hu, S. Campuzano, M. J. Lobo-Castañon, N. Lim, D. Y. Kang, M. Zimmerman, L. Zhang, J. Wang (2011). Micromachine-enabled capture and isolation of cancer cells in complex media. *Angew Chem Int Ed*, **123**(18):4247–4250.

2. J. C. Breger, C. Yoon, R. Xiao, H. R. Kwag, M. O. Wang, J. P. Fisher, T. D. Nguyen, D. H. Gracias (2015). Self-folding thermomagnetically responsive soft microgrippers. *ACS Appl Mater Interfaces*, **7**(5):3398–3405.
3. C. Chen, L. Chen, P. Wang, L.-F. Wu, T. Song (2019). Steering of magnetotactic bacterial microrobots by focusing magnetic field for targeted pathogen killing. *J Magn Magn Mater*, **479**:74–83.
4. B. E.-F. de Ávila, P. Angsantikul, J. Li, M. A. Lopez-Ramirez, D. E. Ramírez-Herrera, S. Thamphiwatana, C. Chen, J. Delezuk, R. Samakapiruk, V. Ramez (2017). Micromotor-enabled active drug delivery for in vivo treatment of stomach infection. *Nat Commun*, **8**(1):272.
5. G. Go, Z. Jin, J.-O. Park, S. Park, et al. (2018). A thermoelectromagnetically actuated microrobot for the targeted transport of therapeutic agents. *Int J Control Autom Syst*, **16**(3):1341–1354.
6. V. Iacovacci, G. Lucarini, L. Ricotti, P. Dario, P. E. Dupont, A. Menciassi (2015). Untethered magnetic millirobot for targeted drug delivery. *Biomed Microdevices*, **17**(3):63.
7. G. Iddan, G. Meron, A. Glukhovsky, P. Swain (2000). Wireless capsule endoscopy. *Nature*, **405**(6785):417.
8. A. Kafash Hoshiar, S. Jeon, K. Kim, S. Lee, J.-Y. Kim, H. Choi (2018). Steering algorithm for a flexible microrobot to enhance guidewire control in a coronary angioplasty application. *Micromachines*, **9**(12):617.
9. D. Kagan, S. Campuzano, S. Balasubramanian, F. Kuralay, G.-U. Flechsig, J. Wang (2011). Functionalized micromachines for selective and rapid isolation of nucleic acid targets fromcomplex samples. *Nano Lett*, **11**(5):2083–2087.
10. R. Mhanna, F. Qiu, L. Zhang, Y. Ding, K. Sugihara, M. Zenobi-Wong, B. J. Nelson (2014). Artificial bacterial flagella for remote-controlled targeted single-cell drug delivery. *Small*, **10**(10):1953–1957.
11. J. Orozco, S. Campuzano, D. Kagan, M. Zhou, W. Gao, J. Wang (2011). Dynamic isolation and unloading of target proteins by aptamer-modified microtransporters. *Anal Chem*, **83**(20):7962–7969.
12. F. Soto, A. Martin, S. Ibsen, M. Vaidyanathan, V. Garcia-Gradilla, Y. Levin, A. Escarpa, S. C. Esener, J. Wang (2015). Acoustic microcannons: toward advanced microballistics. *ACS Nano*, **10**(1):1522–1528.
13. K. Van Nguyen, S. D. Minteer (2015). Dna-functionalized Pt nanoparticles as catalysts for chemically powered micromotors: toward signal-onmotion-based DNA biosensor. *Chem Commun*, **51**(23):4782–4784.

14. Y. Wu, T. Si, J. Shao, Z. Wu, Q. He (2016). Near-infrared light-driven Janus capsule motors: fabrication, propulsion, and simulation. *Nano Res*, **9**(12):3747–3756.

15. W. Xi, A. A. Solovev, A. N. Ananth, D. H. Gracias, S. Sanchez, O. G. Schmidt (2013). Rolled-up magnetic microdrillers: towards remotely controlled minimally invasive surgery. *Nanoscale*, **5**(4):1294–1297.

16. X. Yu, Y. Li, J. Wu, H. Ju (2014). Motor-based autonomous microsensor for motion and counting immunoassay of cancer biomarker. *Anal Chem*, **86**(9):4501–4507.

17. W. Gao, X. Feng, A. Pei, Y. Gu, J. Li, J. Wang (2013). Seawater-driven magnesium-based Janus micromotors for environmental remediation. *Nanoscale*, **5**(11):4696–4700.

18. L. Kong, C. C. Mayorga-Martinez, J. Guan, M. Pumera (2018). Fuel-free light-powered TiO_2/Pt Janus micromotors for enhanced nitroaromatic explosives degradation. *ACS Appl Mater Interfaces*, **10**(26):22427–22434.

19. J. Orozco, V. Garcia-Gradilla, M. D'Agostino, W. Gao, A. Cortés, J. Wang (2012). Artificial enzyme-powered microfish forwater-quality testing. *ACS Nano*, **7**(1):818–824.

20. J. Orozco, D. Vilela, G. Valdés-Ramírez, Y. Fedorak, A. Escarpa, R. Vazquez-Duhalt, J. Wang (2014). Efficient biocatalytic degradation of pollutants by enzyme-releasing self-propelled motors. *Chemistry*, **20**(10):2866–2871.

21. K. Villa, J. Parmar, D. Vilela, S. Sánchez (2018). Metal-oxide-based microjets for the simultaneous removal of organic pollutants and heavy metals. *ACS Appl Mater Interfaces*, **10**(24):20478–20486.

22. C. Anthierens, C. Libersa, M. Touaibia, M. Bétemps, M. Arsicault, N. Chaillet (2000). Micro robots dedicated to small diameter canalization exploration. In *2000 IEEE/RSJ International Conference on Intelligent Robots and Systems (IROS 2000) (Cat. No. 00CH37113)*, Vol. 1, pp. 480–485.

23. K. Suzumori, T. Miyagawa, M. Kimura, Y. Hasegawa (1999). Micro inspection robot for 1-in pipes. *IEEE/ASME Trans Mechatron*, **4**(3):286–292.

24. M. Takeda, K. Namura, K. Nakamura, N. Shibaike, T. Haga, H. Takada (2000). Development of chain-type micromachine for inspection of outer tube surfaces (basic performance of the 1st prototype). In *Proceedings IEEE Thirteenth Annual International Conference on Micro Electro Mechanical Systems (Cat. No. 00CH36308)*, pp. 805–810.

25. D. Rojas, B. Jurado-Sánchez, A. Escarpa (2016). Shoot and sense Janus micromotors-based strategy for the simultaneous degradation and detection of persistent organic pollutants in food and biological samples. *Anal Chem*, **88**(7):4153–4160.
26. A. X. Lu, Y. Liu, H. Oh, A. Gargava, E. Kendall, Z. Nie, D. L. DeVoe, S. R. Raghavan (2016). Catalytic propulsion and magnetic steering of soft, patchy microcapsules: ability to pick-up and drop-off microscale cargo. *ACS Appl Mater Interfaces*, **8**(24):15676–15683.
27. F. Carpi, N. Kastelein, M. Talcott, C. Pappone (2011). Magnetically controllable gastrointestinal steering of video capsules. *IEEE Trans Biomed Eng*, **58**(2):231–234.
28. F. Kuralay, S. Sattayasamitsathit, W. Gao, A. Uygun, A. Katzenberg, J. Wang (2012). Self-propelled carbohydrate-sensitive microtransporters with built-in boronic acid recognition for isolating sugars and cells. *J Am Chem Soc*, **134**(37):15217–15220.
29. G. Chatzipirpiridis, O. Ergeneman, J. Pokki, F. Ullrich, S. Fusco, J. A. Ortega, K. M. Sivaraman, B. J. Nelson, S. Pané (2015). Electroforming of implantable tubular magnetic microrobots for wireless ophthalmologic applications. *Adv Healthcare Mater*, **4**(2):209–214.
30. Z. Wu, J. Li, B. E.-F. de Ávila, T. Li, W. Gao, Q. He, L. Zhang, J. Wang (2015). Water-powered cell-mimicking Janus micromotor. *Adv Funct Mater*, **25**(48):7497–7501.
31. M. Sitti, H. Ceylan, W. Hu, J. Giltinan, M. Turan, S. Yim, E. Diller (2015). Biomedical applications of untethered mobile milli/microrobots. *Proc IEEE*, **103**(2):205–224.
32. E. M. Purcell (1977). Life at low reynolds number. *Am J Phys*, **45**(1):3–11.
33. A. E. El-Etriby, A. Klingner, A. F. Tabak, I. S. Khalil (2018). Manipulation of non-magnetic microbeads using soft microrobotic sperm. In *IEEE 2018 International Conference on Manipulation, Automation and Robotics at Small Scales (MARSS)*, pp. 1–4.
34. I. S. Khalil, H. C. Dijkslag, L. Abelmann, S. Misra (2014). Magnetosperm: a microrobot that navigates using weak magnetic fields. *Appl Phys Lett*, **104**(22):223701.
35. I. S. Khalil, A. Fatih Tabak, A. Klingner, M. Sitti (2016). Magnetic propulsion of robotic sperms at low-Reynolds number. *Appl Phys Lett*, **109**(3):033701.

36. W. Gao, R. Dong, S. Thamphiwatana, J. Li, W. Gao, L. Zhang, J. Wang (2015). Artificial micromotors in the mouses stomach: a step toward in vivo use of synthetic motors. *ACS Nano*, **9**(1):117–123.
37. J. Li, P. Angsantikul, W. Liu, B. Esteban-Fernández de Ávila, S. Thamphiwatana, M. Xu, E. Sandraz, X. Wang, J. Delezuk, W. Gao (2017). Micromotors spontaneously neutralize gastric acid for pH-responsive payload release. *Angew Chem Int Ed*, **56**(8):2156–2161.
38. I. S. Khalil, V. Magdanz, S. Sanchez, O. G. Schmidt, L. Abelmann, S. Misra (2013). Magnetic control of potential microrobotic drug delivery systems: nanoparticles, magnetotactic bacteria and self-propelled microjets. In *2013 35th Annual International Conference of the IEEE Engineering in Medicine and Biology Society (EMBC)*, pp. 5299–5302.
39. E. Shechter, S. Martel (2010). Principles of motion control of bacterial micro-robots using oxygen gradients. In *2010 IEEE/ASME International Conference on Advanced Intelligent Mechatronics (AIM)*, pp. 848–853.
40. A. V. Singh, M. Sitti (2016). Patterned and specific attachment of bacteria on biohybrid bacteria-driven microswimmers. *Adv Healthcare Mater*, **5**(18):2325–2331.
41. J. Zhuang, M. Sitti (2016). Chemotaxis of bio-hybrid multiple bacteria-driven microswimmers. *Sci Rep*, **6**:32135.
42. L. Baraban, R. Streubel, D. Makarov, L. Han, D. Karnaushenko, O. G. Schmidt, G. Cuniberti (2013). Fuel-free locomotion of Janus motors: magnetically induced thermophoresis. *ACS Nano*, **7**(2):1360–1367.
43. H.-R. Jiang, N. Yoshinaga, M. Sano (2010). Activemotion of a Janus particle by self-thermophoresis in a defocused laser beam. *Phys Rev Lett*, **105**(26):268302.
44. W. Qin, T. Peng, Y. Gao, F. Wang, X. Hu, K. Wang, J. Shi, D. Li, J. Ren, C. Fan (2017). Catalysis-driven self-thermophoresis of Janus plasmonic nanomotors. *Angew Chem Int Ed*, **129**(2):530–533.
45. T. Qiu, T. C. Lee, A. G. Mark, K. I. Morozov, R. Münster, O. Mierka, S. Turek, A. M. Leshansky, P. Fischer (2014). Swimming by reciprocal motion at low reynolds number. *Nat Commun*, **5**:5119.
46. F. Carpi, S. Galbiati, A. Carpi (2007). Controlled navigation of endoscopic capsules: concept and preliminary experimental investigations. *IEEE Trans Biomed Eng*, **54**(11):2028–2036.
47. J.-B. Mathieu, G. Beaudoin, S. Martel (2006). Method of propulsion of a ferromagnetic core in the cardiovascular system through magnetic

gradients generated by an MRI system. *IEEE Trans Biomed Eng*, **53**(2):292–299.

48. A. Ghosh, P. Fischer (2009). Controlled propulsion of artificial magnetic nanostructured propellers. *Nano Lett*, **9**(6):2243–2245.
49. J. Li, S. Sattayasamitsathit, R. Dong, W. Gao, R. Tam, X. Feng, S. Ai, J. Wang (2014). Template electrosynthesis of tailored-made helical nanoswimmers. *Nanoscale*, **6**(16):9415–9420.
50. L. Zhang, J. J. Abbott, L. Dong, B. E. Kratochvil, D. Bell, B. J. Nelson (2009). Artificial bacterial flagella: fabrication and magnetic control. *Appl Phys Lett*, **94**(6):064107.
51. J. J. Abbott, K. E. Peyer, M. C. Lagomarsino, L. Zhang, L. Dong, I. K. Kaliakatsos, B. J. Nelson (2009). How should microrobots swim? *Int J Rob Res*, **28**(11–12):1434–1447.
52. E. Diller, J. Giltinan, P. Jena, M. Sitti (2013). Three dimensional independent control of multiplemagnetic microrobots. In *2013 IEEE International Conference on Robotics and Automation (ICRA)*, pp. 2576–2581.
53. T. Honda, K. Arai, K. Ishiyama (1996). Micro swimming mechanisms propelled by external magnetic fields. *IEEE Trans Magn*, **32**(5):5085–5087.
54. H.-W. Tung, K. E. Peyer, D. F. Sargent, B. J. Nelson (2013). Noncontact manipulation using a transversely magnetized rolling robot. *Appl Phys Lett*, **103**(11):114101.
55. R. Dreyfus, J. Baudry, M. L. Roper, M. Fermigier, H. A. Stone, J. Bibette (2005). Microscopic artificial swimmers. *Nature*, **437**(7060):862.
56. Z. Ye, S. Régnier, M. Sitti (2014). Rotating magnetic miniature swimming robots with multiple flexible flagella. *IEEE Trans Rob*, **30**(1):3–13.
57. W. Gao, S. Sattayasamitsathit, J. Orozco, J. Wang (2011). Highly efficient catalytic microengines: template electrosynthesis of polyaniline/platinum microtubes. *J Am Chem Soc*, **133**(31):11862–11864.
58. J. Li, I. Rozen, J. Wang (2016). Rocket science at the nanoscale. *ACS Nano*, **10**(6):5619–5634.
59. A. A. Solovev, Y. Mei, E. Bermúdez Ureña, G. Huang, O. G. Schmidt (2009). Catalytic microtubular jet engines self-propelled by accumulated gas bubbles. *Small*, **5**(14):1688–1692.
60. K. M. Manesh, M. Cardona, R. Yuan, M. Clark, D. Kagan, S. Balasubramanian, J. Wang (2010). Template-assisted fabrication of salt independent catalytic tubular microengines. *ACS Nano*, **4**(4):1799–1804.

61. Z. Wu, Y. Wu, W. He, X. Lin, J. Sun, Q. He (2013). Self-propelled polymer-based multilayer nanorockets for transportation and drug release. *Angew Chem Int Ed*, **52**(27):7000–7003.
62. M. Wehner, R. L. Truby, D. J. Fitzgerald, B. Mosadegh, G. M. Whitesides, J. A. Lewis, R. J. Wood (2016). An integrated design and fabrication strategy for entirely soft, autonomous robots. *Nature*, **536**(7617):451.
63. S. Sanchez, A. A. Solovev, Y. Mei, O. G. Schmidt (2010). Dynamics of biocatalytic microengines mediated by variable friction control. *J Am Chem Soc*, **132**(38):13144–13145.
64. Y. Wu, X. Lin, Z. Wu, H. Möhwald, Q. He (2014). Self-propelled polymer multilayer Janus capsules for effective drug delivery and light-triggered release. *ACS Appl Mater Interfaces*, **6**(13):10476–10481.
65. A. A. Solovev, E. J. Smith, C. C. Bof'Bufon, S. Sanchez, O. G. Schmidt (2011). Light-controlled propulsion of catalytic microengines. *Angew Chem Int Ed*, **50**(46):10875–10878.
66. T. Xu, F. Soto, W. Gao, V. Garcia-Gradilla, J. Li, X. Zhang, J. Wang (2014). Ultrasound-modulated bubble propulsion of chemically powered microengines. *J Am Chem Soc*, **136**(24):8552–8555.
67. S. Sanchez, A. N. Ananth, V. M. Fomin, M. Viehrig, O. G. Schmidt (2011). Superfast motion of catalytic microjet engines at physiological temperature. *J Am Chem Soc*, **133**(38):14860–14863.
68. J. Zhuang, R. W. Carlsen, M. Sitti (2015). pH-taxis of biohybrid microsystems. *Sci Rep*, **5**:11403.
69. D. H. Kim, U. K. Cheang, L. Kohidai, D. Byun, M. J. Kim (2010). Artificial magnetotactic motion control of tetrahymena pyriformis using ferromagnetic nanoparticles: a tool for fabrication of microbiorobots. *Appl Phys Lett*, **97**(17):173702.
70. D. Kim, A. Liu, M. Stitti (2011). Chemotactic behavior and dynamics of bacteria propelled microbeads. In *2011 IEEE/RSJ International Conference on Intelligent Robots and Systems (IROS)*, pp. 1674–1679.
71. E. Steager, C.-B. Kim, J. Patel, S. Bith, C. Naik, L. Reber, M. J. Kim (2007). Control of microfabricated structures powered by flagellated bacteria using phototaxis. *Appl Phys Lett*, **90**(26):263901.
72. E. B. Steager, M. S. Sakar, D. H. Kim, V. Kumar, G. J. Pappas, M. J. Kim (2011). Electrokinetic and optical control of bacterial microrobots. *J Micromech Microeng*, **21**(3):035001.
73. J. Yu, H. Yu, T. Xu, L. Zhang (2015). Magnetic control of amb-1 magnetotactic bacteria for micromanipulation. In *2015 IEEE International*

Conference on Automation Science and Engineering (CASE), pp. 1614–1619.

74. D. B. Weibel, P. Garstecki, D. Ryan, W. R. DiLuzio, M. Mayer, J. E. Seto, G. M. Whitesides (2005). Microoxen: microorganisms to move microscale loads. *Proc Natl Acad Sci U S A*, **102**(34):11963–11967.
75. D. Kim, A. Liu, E. Diller, M. Sitti (2012). Chemotactic steering of bacteria propelled microbeads. *Biomed Microdevices*, **14**(6):1009–1017.
76. R. W. Carlsen, M. R. Edwards, J. Zhuang, C. Pacoret, M. Sitti (2014). Magnetic steering control of multi-cellular bio-hybrid microswimmers. *Lab Chip*, **14**(19):3850–3859.
77. R. W. Carlsen, M. Sitti (2014). Bio-hybrid cell-based actuators for Microsystems. *Small*, **10**(19):3831–3851.
78. X. Lin, T. Si, Z. Wu, Q. He (2017). Self-thermophoretic motion of controlled assembled micro-/nanomotors. *Phys Chem Chem Phys*, **19**(35):23606–23613.
79. R. Dong, Q. Zhang, W. Gao, A. Pei, B. Ren (2015). Highly efficient light-driven TiO_2–Au Janus micromotors. *ACS Nano*, **10**(1):839–844.
80. F. Wong, A. Sen (2016). Progress toward light-harvesting self-electrophoretic motors: highly efficient bimetallic nanomotors and micropumps in halogen media. *ACS Nano*, **10**(7):7172–7179.
81. R. Golestanian, T. B. Liverpool, A. Ajdari (2005). Propulsion of a molecular machine by asymmetric distribution of reaction products. *Phys Rev Lett*, **94**(22):220801.
82. C. Zhou, H. Zhang, J. Tang, W. Wang (2018). Photochemically powered AgCl Janus micromotors as a model system to understand ionic self-diffusiophoresis. *Langmuir*, **34**(10):3289–3295.
83. S. Tottori, L. Zhang, F. Qiu, K. K. Krawczyk, A. Franco-Obregón, B. J. Nelson (2012). Magnetic helical micromachines: fabrication, controlled swimming, and cargo transport. *Adv Mater*, **24**(6):811–816.
84. J. Li, G. Huang, M. Ye, M. Li, R. Liu, Y. Mei (2011). Dynamics of catalytic tubular microjet engines: dependence on geometry and chemical environment. *Nanoscale*, **3**(12):5083–5089.
85. V. Luchnikov, O. Sydorenko, M. Stamm (2005). Self-rolled polymer and composite polymer/metal micro- and nanotubes with patterned inner walls. *Adv Mater*, **17**(9):1177–1182.
86. Y. Mei, G. Huang, A. A. Solovev, E. B. Ureña, I. Mönch, F. Ding, T. Reindl, R. K. Fu, P. K. Chu, O. G. Schmidt (2008). Versatile approach for integrative and functionalized tubes by strain engineering of nanomembranes on polymers. *Adv Mater*, **20**(21):4085–4090.

87. M. J. Brett, M. M. Hawkeye (2008). New materials at a glance. *Science*, **319**(5867):1192–1193.

88. L. Liu, S.-H. Yoo, S. A. Lee, S. Park (2011). Wet-chemical synthesis of palladium nanosprings. *Nano Lett*, **11**(9):3979–3982.

89. I. S. Khalil, A. F. Tabak, A. Hosney, A. Mohamed, A. Klingner, M. Ghoneima, M. Sitti (2016). Sperm-shaped magnetic microrobots: fabrication using electrospinning, modeling, and characterization. In *2016 IEEE International Conference on Robotics and Automation (ICRA)*, pp. 1939–1944.

90. D. Li, Y. Xia (2004). Electrospinning of nanofibers: reinventing the wheel? *Adv Mater*, **16**(14):1151–1170.

91. H. Ceylan, I. C. Yasa, O. Yasa, A. F. Tabak, J. Giltinan, M. Sitti (2019). 3D-printed biodegradable microswimmer for theranostic cargo delivery and release. *ACS Nano*, **13**(3):3353–3362.

92. V. H. Le, S. Zheng, J. Han, J.-O. Park (2018). Preparation of tumor targeting cell-based microrobots carrying NIR light sensitive therapeutics manipulated by electromagnetic actuating system and chemotaxis. *J Micro-Bio Rob*, **14**(3–4):69–77.

93. S. Lee, S. Kim, S. Kim, J.-Y. Kim, C. Moon, B. J. Nelson, H. Choi (2018). A capsule-type microrobot with pick-and-drop motion for targeted drug and cell delivery. *Adv Healthcare Mater*, **7**(9):1700985.

94. V. Tandon, W. S. Kang, T. A. Robbins, A. J. Spencer, E. S. Kim, M. J. McKenna, S. G. Kujawa, J. Fiering, E. E. Pararas, M. J. Mescher (2016). Microfabricated reciprocating micropump for intracochlear drug delivery with integrated drug/fluid storage and electronically controlled dosing. *Lab Chip*, **16**(5):829–846.

95. G. Traverso, C. M. Schoellhammer, A. Schroeder, R. Maa, G. Y. Lauwers, B. E. Polat, D. G. Anderson, D. Blankschtein, R. Langer (2015). Microneedles for drug delivery via the gastrointestinal tract. *J Pharm Sci*, **104**(2):362–367.

96. N. Chisnàriandini, A. P. Aji, P. D. Mariyam, J. Fuadi, Y. La Elo (2018). Measurement on real-time diagnostic of gastric tumor model using wireless endoscopy system. In *2018 5th International Conference on Information Technology, Computer, and Electrical Engineering (ICITACEE)*, pp. 105–108.

97. G. Ghosh, S. Shah, C. Maltz (2018). Ascariasis diagnosed by wireless capsule endoscopy. *Clin Gastroenterol Hepatol*, **16**(6):A23.

98. J.-Y. He, X. Wu, Y.-G. Jiang, Q. Peng, R. Jain (2018). Hookworm detection in wireless capsule endoscopy images with deep learning. *IEEE Trans Image Process*, **27**(5):2379–2392.

99. J. Yang, M. Morita, Z. Jiang (2016). Design of a novel scissoring micro-stirrer for blood clot dissolution. *Sens Actuators, A*, **248**:130–137.
100. J. S. Liu, Z. J. Gartner (2012). Directing the assembly of spatially organized multicomponent tissues from the bottom up. *Trends Cell Biol*, **22**(12):683–691.
101. S. Tasoglu, E. Diller, S. Guven, M. Sitti, U. Demirci (2014). Untethered micro-robotic coding of three-dimensional material composition. *Nat Commun*, **5**:3124.
102. Z. Lu, C. Moraes, G. Ye, C. A. Simmons, Y. Sun (2010). Single cell deposition and patterning with a robotic system. *PLoS One*, **5**(10):e13542.
103. V. P. Torchilin (2005). Recent advances with liposomes as pharmaceutical carriers. *Nat Rev Drug Discovery*, **4**(2):145.
104. J. Wang (2016). Self-propelled affinity biosensors: moving the receptor around the sample. *Biosens Bioelectron*, **76**:234–242.
105. J. Orozco, B. Jurado-Sánchez, G. Wagner, W. Gao, R. Vazquez-Duhalt, S. Sattayasamitsathit, M. Galarnyk, A., Cortés, D. Saintillan, J. Wang (2014). Bubble-propelled micromotors for enhanced transport of passive tracers. *Langmuir*, **30**(18):5082–5087.
106. L. Restrepo-Pérez, L. Soler, C. Martínez-Cisneros, S. Sánchez, O. G. Schmidt (2014). Biofunctionalized self-propelled micromotors as an alternative on-chip concentrating system. *Lab Chip*, **14**(16):2914–2917.
107. S. Campuzano, J. Orozco, D. Kagan, M. Guix, W. Gao, S. Sattayasamitsathit, J. C. Claussen, A. Merkoçi, J. Wang (2011). Bacterial isolation by lectin-modified microengines. *Nano Lett*, **12**(1):396–401.
108. D. L. Koch, G. Subramanian (2011). Collective hydrodynamics of swimming microorganisms: living fluids. *Annu Rev Fluid Mech*, **43**:637–659.
109. P. S. Mahapatra, S. Mathew (2019). Activity-induced mixing and phase transitions of self-propelled swimmers. *Phys Rev E*, **99**(1):012609.
110. J. J. Molina, Y. Nakayama, R. Yamamoto (2013). Hydrodynamic interactions of self-propelled swimmers. *Soft Matter*, **9**(19):4923–4936.
111. A. Sokolov, R. E. Goldstein, F. I. Feldchtein, I. S. Aranson (2009). Enhanced mixing and spatial instability in concentrated bacterial suspensions. *Phys Rev E*, **80**(3):031903.

Chapter 4

Microfluidics in Neuroscience

Pallavi Gupta,[a,*] Nandhini Balasubramaniam,[a,*] Kiran Kaladharan,[a] Fan-Gang Tseng,[b] Moeto Nagai,[c] Hwan-You Chang,[d] and Tuhin S. Santra[a]

[a]*Department of Engineering Design, Alumni Ave, Indian Institute of Technology, Chennai, Tamil Nadu 600036, India*
[b]*Department of Engineering and System Science, National Tsing Hua University, Hsinchu City 30013, Taiwan*
[c]*Department of Mechanical Engineering, Toyohashi University of Technology, 1-1 Hibarigaoka, Tempaku-cho, Toyohashi, Aichi 441-8580, Japan*
[d]*Department of Life Science, National Tsing Hua University, Hsinchu City 30013, Taiwan*
tuhin@iitm.ac.in, santra.tuhin@gmail.com

Microfluidic devices have shown great potential in life science research due to their designing range, real-time operations, fluid-phase diffusion gradients, programmatic switching, and flow manipulation. In most of the cases, the pressure and flow rates operating through dynamically triggered microfluidic devices induce flow-field forces (e.g., shear stresses) and their effect can be observed on the structure and function of the cell. Focused research in the field of neuroscience needs a microenvironment with spatiotemporal fluidic control, which can be provided by microfluidic devices for guiding neuron development and dynamics in vitro. Various

*These authors contributed equally to this chapter.

Microfluidics and Bio-MEMS: Devices and Applications
Edited by Tuhin S. Santra

ISBN 978-981-4800-85-3 (Hardcover), 978-1-003-01493-5 (eBook)
www.jennystanford.com

studies show that microfluidic chambers are able to meet the biological requirements for controlling mammalian neuron cell culture and development, neuron manipulation, neural stem cell (NSC) differentiation, neuropharmacology, neuroelectrophysiology, and neural implants. This chapter elaborates on microfluidic design biomimetics related to the nervous system. The re-creation of both neurodegenerative disorders and NSC behavior blueprinting microfluidic platforms is discussed, and this process can be used as a basis to construct a brain on a chip. The advantages and limitations of microfluidic devices in neuroscience are also emphasized in this chapter. Most of the devices are robust and cost effective in terms of disease modeling and drug screening.

4.1 Introduction

The human brain, consisting of hundreds of billions of neurons and trillions of connections between diversified cell types, is still a mystery to neuroscientists. The collective interactions between brain function and the physical/social environment facilitate the formation of complex physiological and social behaviors. Understanding the complex underlying mechanisms for nervous system functionality holds the key to deciphering the complex human behavior, including senses, cognition, emotions, and thinking [1]. The ability to advance this knowledge and address related clinical pathologies depends on technological innovation. Recent collaborative research initiatives, such as the Brain Research through Advancing Innovative Neurotechnologies (BRAIN) program in the United States, the Human Brain Project in Europe, Brain Mapping by Integrated Neurotechnologies for Disease Studies (Brain/MINDS) program of Japan, the China Brain Project, and Israel Brain Technologies (IBT), foster the advances in neurotechnologies and can lead to an understanding of the brain structure and function in a dynamic and systematic fashion [2]. With these joint efforts, neuroscientists can now much more efficiently address the following: (i) the structural and functional association of the brain to the complete nervous system, (ii) the basics of nervous system activities and how they are regulated, (ii) how to interpret learning mechanisms and memory on various levels, and (iv) prevention, diagnosis, and treatment of different mental and neurobiological diseases.

Traditional cell culture–based studies have offered great support in solving these tasks, yet they are able to reveal only the physiological state without any control over the cell microenvironments. Additionally, the miniaturized and automated microfluidics interface, with advancement in surface chemistry, electrochemistry, and micro-/nanotechnology, is more applicable for handling complicated chemical and biological processes. The newly emerged field of microfluidics deals with the manipulation of fluids in channels with dimensions of tens of micrometers [3]. This system can process or manipulate small (10^{-9} to 10^{-18} liters) amounts of fluids, employing channels with dimensions of tens to hundreds of micrometers. Till now, microfluidics has been applied in different disciplines, including chemistry [4–7], biology and medicine [8–12], mechanical science [13, 14], and many others [15–19]. It is noteworthy that microfluidic systems gain advantages owing to smaller amounts of reagent and sample, high sequential or parallel experimentation, similarity to the natural tissue environment, highly precise experimental control over the cellular microenvironment, and low-cost disposable devices. Indeed, it is a big task to tune these features in conventional analytical technologies. Recently, numerous functional modules, such as valves, mixing module, and pumps, have made microfluidic devices (μFDs) more precise for and easier to use in neurobiology, a critical field of life sciences [20, 21]. Studies related to neuron interactions, NSC differentiation, mechanisms of neuron-substrate interfacing, neurite growth, myelination, and the mechanisms for ion channels regulation are still in their infancy. Superior performance and easy availability of μFDs have made them a new tool to study these phenomena, which are difficult to achieve conventionally. Indeed, after the introduction of μFDs into neuroscience in 1998 [22], several studies employed μFDs to address questions related to neural cell activity and growth on various substrates and in various microenvironments, neuropharmacology, neuroelectrophysiology, NSC differentiation, neuron biosensors, and the behavior of *Caenorhabditis elegans* [21, 23–27].

The motivation of this chapter is to critically evaluate the achievements of microfluidics in terms of neurobiology. Along with efforts to integrate neuron handling and biochemical analysis on chips, the focus has been kept on microsystems providing biological insight. In the following sections, we distinguish different aspects of microfluidic-based neurobiology and discuss them in detail: (i) approaches for neuronal studies, including traditional μFDs, compartmentalized μFDs, single-cell monitoring, co-culturing, integrated microfluidic/microelectrode array (MEA), hydrogel gradient μFDs, and μFD neural probes (fiber electrodes/brain implants), (ii) neuron manipulation, (iii) axon guidance, (iv) dendritic signaling and synapse formation, (v) neural differentiation and polarity, (vi) neuropharmacology, and (vii) neuroelectrophysiology. We conclude the chapter with summary remarks and the future prospects in this promising and rapidly expanding area.

4.2 Traditional Microfluidic Devices

On the basis of the notion that the role of the nerve growth factor (NGF) is crucial in neurite development, Robert Campenot in 1970 devised the tissue culture system in which the neurites of the dissociated sympathetic neurons cultured in one fluid-filled chamber traversed the virtually impermeable partition into the second chamber (Fig. 4.1a) [28]. These pioneer compartmentalized culture devices could be used to probe, control, and culture neuronal cells under precise constraints and were later employed to study neurotrophin-mediated regulation of adult sensory neurite growth (Fig. 4.1b) (such as NGF), neurotrophin-3, and brain-derived neurotrophic factor (BDNF) [29].

Amalgamation of Campenot's inventive concept with physical sciences and engineering—involving micro-/nanotechnologies—has revolutionized the design and production of the complex, integrated, and highly controllable devices, as it facilitates local fluidic control, spatial access to distinct regions of the compartmentalized neuron, and microelectrode-based assessments of local signal transduction

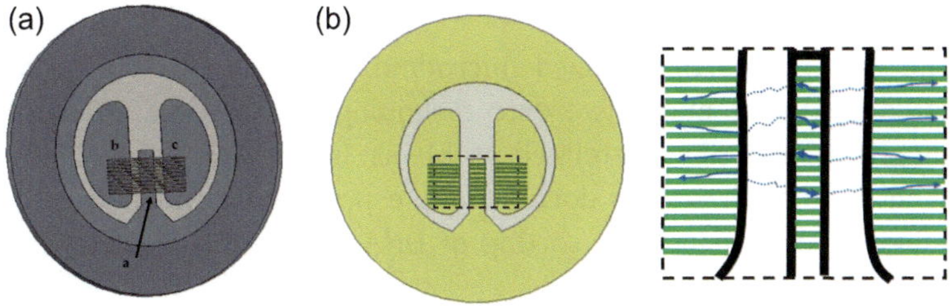

Figure 4.1 (a) The Campenot chamber was presented as a Teflon divider, creating compartments for the growth of neural processes from the soma. It is attached to a glass-bottomed, collagen-coated dish with silicon grease. Parallel scratches in the substrate provide guidance to the growing neurites into the adjacent chambers for investigating local neuron development under the influence of selective exposure of neurotrophic compounds. (b) The Campenot chamber (left) and magnification (right) of the inset showing neuron (blue) growth, which becomes compartmentalized within the device [28, 29].

events [30]. Since the 1990s, these devices have been used in life sciences—initially for molecular biology analyses (e.g., sizing and sorting of DNA fragments) [31] and proteomics and afterward via a combination of cell culture along with microfluidics—to develop miniaturized devices [32]. The pioneer research directing microfluidic technology, micropatterning, and microfabrication toward a neurobiological problem was done in 2003 by Anne M. Taylor from Noo Li Jeon's laboratory [33]. Therefore, it was the first time when a two-compartment microfabricated neuronal culture device was defined with a spatial and fluidic segregation of neuronal soma from axons [34]. This simple pioneer design acts as a key reference for numerous researchers and research labs worldwide and the backbone for the available commercialized devices.

4.3 Current Approaches

In this section, we will discuss the advancements in μFDs related to neuroscience in terms of design at the cellular/tissue organization level.

4.3.1 Compartmentalized Microfluidics

The first compartmentalized microfluidic devices (CMDs), fabricated by using microelectronic technology, were used to solve chemistry- and physics-related problems. In the late 1990s, CMDs used in biostudies were documented, especially for bioanalysis in genetics (e.g., sizing and sorting of DNA fragments) and later in proteomics [35].

Park et al. developed a CMD in which neuron cells can be cultured to improve the axonal guidance and differentiation. In this device, a radial array of microgrooves is designed to assess the axonal growth with the help of extracellular components and neurotrophic factors [36]. Similarly, this device can also be used to co-culture cortical neurons and oligodendrocytes to understand the myelination process, as shown in Fig. 4.2a [37]. Figure 4.2b-I shows two CMDs in which cortical neurons and hippocampal neurons were cultured to visualize the presynaptic localization of RNA from developing neurons, whereas Fig. 4.2b-II depicts the dimensions to maintain the volume difference between the isolated axons in a localized microenvironment. This volume difference maintains a small hydrostatic pressure, which results in a slow continuous flow through high-resistance microgrooves and the fluidic isolation of the axonal compartment [38]. Figure 4.2c [39] depicts a three-compartment μFD developed to study the synaptic connection formed in the neural networks. In this device, axons are cultured from two different lateral compartments and the central compartment is for the target neurons. In one of the lateral compartments, the neurons are cultured with inhibitors, while the other compartment is fed with the regular culture medium. The study revealed that the inhibited neurons accelerated the growth of untreated neurons such that the untreated neurons outgrew through microchannels to reach their inhibited counterparts [39, 40]. In another study, neurons and glial and skeletal muscle cells were co-cultured to create an in vitro model of the lower motor neuron (MN) and neuromuscular junction circuit [41].

Taylor et al. combined microfluidic and micropatterning techniques to fabricate a device to maintain microenvironments for cultured neuronal cells [33]. Lu et al. developed a μFD in which

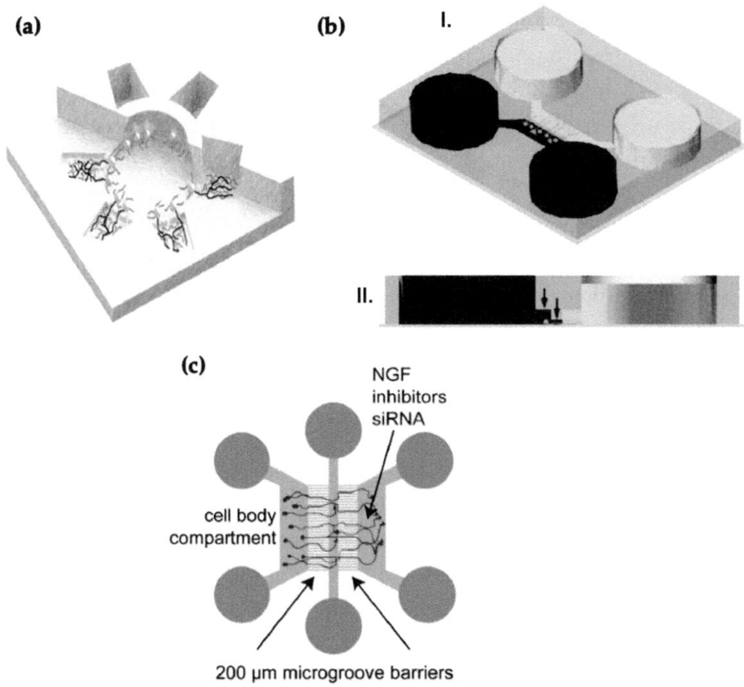

Figure 4.2 (a) A microfluidic device with a soma compartment at the center surrounded by axon compartments connected by axon-guiding microgrooves. Republished with permission of The Royal Society of Chemistry, from Ref. [37a]; permission conveyed through Copyright Clearance Center, Inc. (b, I) Microfluidic device with axonal and somal compartments and (II) dimensions depicting that the volume difference is maintained to isolate axons in a local figure-sized microenvironment. Reprinted by permission from Springer Nature Customer Service Centre GmbH: Springer Nature, *Nature Methods*, Ref. [38], copyright (2005). (c) A three-compartment microfluidic device to study synaptic formation. Reprinted by permission from Springer Nature Customer Service Centre GmbH: Springer Nature, *Nature Communications*, Ref. [39], copyright (2014).

a large somal compartment is connected to axonal compartments through a series of microchannels. This device was used to study the transport of mitochondria along single dopaminergic axons to track the transport defect that contributes to neurodegenerative diseases, like Parkinson's disease (PD) [42]. A similar µFD designed by Kerman et al. has multicompartments that can hold more cells and guides axons to extend to their targets in a sensitive environment and helps to obtain real-time imaging data of the myelination process [43, 44].

Takeuchi et al. designed a co-culture system that consists of a minichamber with a number of microcompartments fabricated on an MEA substrate. The minichamber contains two compartments with 16 microchambers and 8 microconduits used for electrophysiological recordings. The compartments were seeded with neurons and ventricular myocytes, and after 3 days, it was observed that the neurons were connected to myocytes through the microconduits [45]. A similar microfluidic co-culture device was fabricated to study the synaptic connections between neurons and myocytes [46]. Neto et al. established a co-culture system comprising neurons and osteoblasts to observe the nerve fiber formation in a bone microenvironment. Generally, the neurons extend their axons along the microgrooves if they are stuck near the microgrooves (the position of the soma defines the number of axons). The adhesion of neurons to the culture substrate depends on the neural density on the axonal side. To confine the neurons to extend their axons toward the target (to the axonal side), a microreservoir is punched on the somal side, enabling the study of the interaction between the nervous system and bone cells, as shown in Fig. 4.3 [47].

Co-culturing systems with punched holes are used to confine cells in a particular area to improve the interactions of nerve fibers with their target organs, such as developing teeth [48]. Zahavi et al. studied the mechanism of the neuromuscular junction with the help of a compartmentalized microfluidic co-culture system. This system helped to differentiate between the effect of glial-derived neurotrophic factors on motoneuron axons and that on soma and also tracked the transport of neurotrophic factors from muscle to neuron [49].

4.3.2 Single-Cell Monitoring

Neuroscience involves multiple levels of investigations, including molecular, action potential signal and propagation, cellular subsystem, and whole organism. Molecular/subcellular level investigation facilitates the study of neurotransmitters, ion channels, modulators,

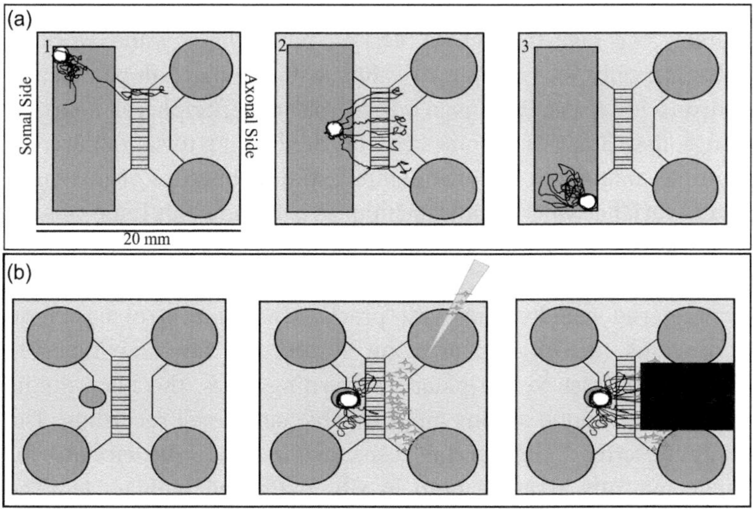

Figure 4.3 (a) An illustration of the co-culturing system depicting that the quantity of axons crossing the microgrooves is dependent on the system. (b) The co-culture system with punched holes can confine somal adhesion and guide axons to the axonal side [47].

ligands and receptors, genes and products, pathogens and microbial toxins, chemical cues, lipid membranes, synapses, axons, and growth cones. Studies at the cellular level include the analyses of neurons and glia, their interaction, and their relationships. Research at the subsystem level highlights the events of stimulation and action potential recording from neurons. Studies related to the organs and tissues of the nervous system, such as the brain, ganglia, and the spinal cord, are considered to be under this category. The next level carries out studies on animal models such as snails and mice.

As discussed earlier, compartmentalized neuron culture platforms can enable the study of a large number of neurons cultured among millimeter-sized compartments establishing substantial intercompartment connections. In such a design, for the neuron count, positions and neurite interconnections are highly random and the network is locally entangled. As a result, only a few neurons (<5%) can extend outgrowths between compartments, which acts as a limiting factor in the investigation of neurite-based biomolecule transport and reconstruction of neuronal networks. Moreover, it ends up in excessive cell usage (similar to in vitro standards), unlikely for high-throughput assays, particularly while working with rare cells, for example, cells of substantia nigra, dopaminergic neurons ideal for Parkinson's research, and peripheral neurons. Hence, data analysis becomes complicated owing to sampling from multiple animals in high-throughput studies because of interindividual variance. Microfluidic techniques overcome this issue and are well suited and capable of single-cell screening.

Therefore, Dinh et al. [50] proposed a dynamic array method for metered cell loading with precise spatial control based on a previous microfluidic arraying system by Tan and Takeuchi [51]. In contrast to the planar patterning ways, the microfluidic arraying technique attains more reliable single-cell recording. The study described differential resistance microfluidic circuits for single neurons arraying onto in situ patterned biomaterial. The biopatterning directs neurite outgrowths and maintains the spatial precision in neuronal networks. Several streamlines were produced in the fluidic circuit; the narrow streamlines exit through the neurite outgrowth channel, while others continue through the serpentine channel. Convergence takes place for most of the streamlines at the arraying site, making compact and subcellular-sized transport paths. The first cell is loaded the farthest from the inlet in the trap. After trapping of cells the streamline is disconnected for subsequent trapping of the cells serially and placing the neurons in arrays with single-cell precision (Fig. 4.4). The experiments were conducted on an array of SH-SY5Y and LUHMES cells and mouse primary neurons. The arraying process becomes fast with $\sim$100 neurons arrayed in $\sim$10 s in each culture chamber (input of $<$1000 cells) [50]. The use of multiple compartmentalized neuron array (CNA) devices can be helpful in retrieving the cells left over in the inlet chamber.

Microfluidic arraying enforces fluidic stress for less duration. On the contrary, cells under small mechanical strain for long undergo deformation, leading to altered gene expression and sometimes death. In the above-mentioned devices, this effect due to a slow array process using gravity feed for 30 min was documented. Similarly, direct trapping at microscale orifices (without the use of differential microflows) results in extreme cell body deformation, also leading to death, showing the inappropriateness for a neuron

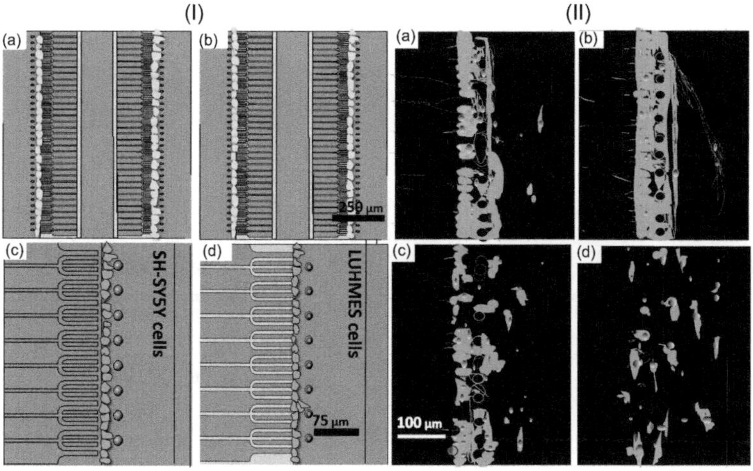

Figure 4.4 (I-a, I-b) of completely arrayed differentiated SH-SY5Y neuron-like cells inside the CNA device. The meniscus-pinning micropillars are outlined to discriminate these from the single neurons. A magnified image of the optimally arrayed SH-SY5Y cells (c) and LUHMES cells (d) imaged 4 h after arraying. The LUHMES cells have been outlined for proper visualization. (II) In situ neuron patterning. Differentiated SH-SY5Y neurons were cultured for 5 days within the CNA devices on a PL/PLL-g-PEG pattern (a, b) and on a continuous PL coating (c, d). Patterning of biomaterials enhances neurite guidance into the outgrowth microchannels [50].

culture yet establishing these systems for short-term single-cell electrophysiology [52]. Such a notion was proved by calcium imaging experiments for single neuron activity. Calcium imaging is a convenient tool to determine neuron interconnectivity and network dynamics over time, which are useful for assessing single-neuron dynamics within an arrayed neuronal network acting as a model for specific brain functions.

Replication and laser micromachining protocols were employed together with capillary pumping to optimize the single-cell-trapping efficiency of the microsieve electrode array (μSEA) platform without requiring external equipment, that is, syringes and pumps (Fig. 4.5). The capillary pumping across a microsieve generated gentle cell trapping velocities (<13.3 μm/s) and produced 80% cell trapping efficiency with 90% cell survival rates over 1 week of culture,

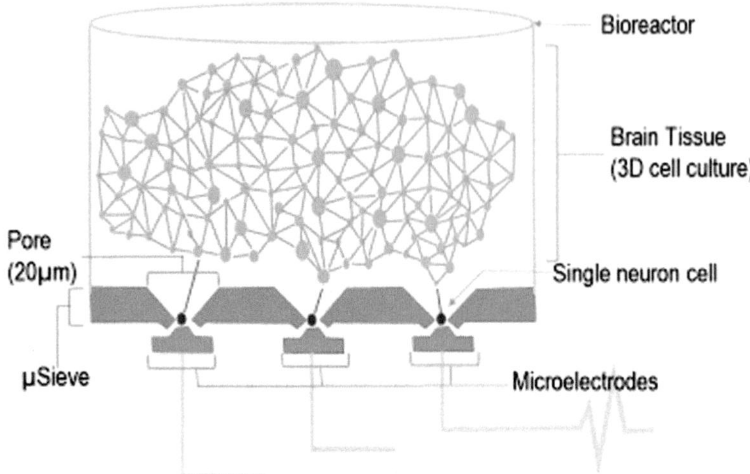

Figure 4.5 Schematic depicting the brain-on-a-chip model. A modified microsieve electrode array (μSEA) with single neurons paired to single electrodes and the 3D brain tissue construct cultured atop the μSEA inside a bioreactor [53].

allowing spatially standardized neuronal network formation. Therefore, the arraying of the growing neurons in a spatially standardized fashion would offer ease in the analysis of neurite connectivity, a hallmark neurodevelopmental end point indicator, and lead to easier correlation in the changes in connectivity to electrophysiology and biological function [53].

Another study depicted the stimulation of individual neurons in a mechanoreceptor neuron model, and the device was capable of efficiently delivering a stimulus protocol including excitation in the form of a step, as well as complex patterns, leveraging the well-characterized touch receptor neurons with an optogenetic calcium indicator. Thus the μFD is a great platform for carrying out research in the field of mechanobiology of living animals and mechanosensitive neurons [54].

A recent study by Kim et al. presented an innovative single-neuron culturing and monitoring platform with a transparent dielectrophoresis (DEP) device. The MEAs are fabricated with indium tin oxide glass and polydimethylsiloxane (PDMS) for the

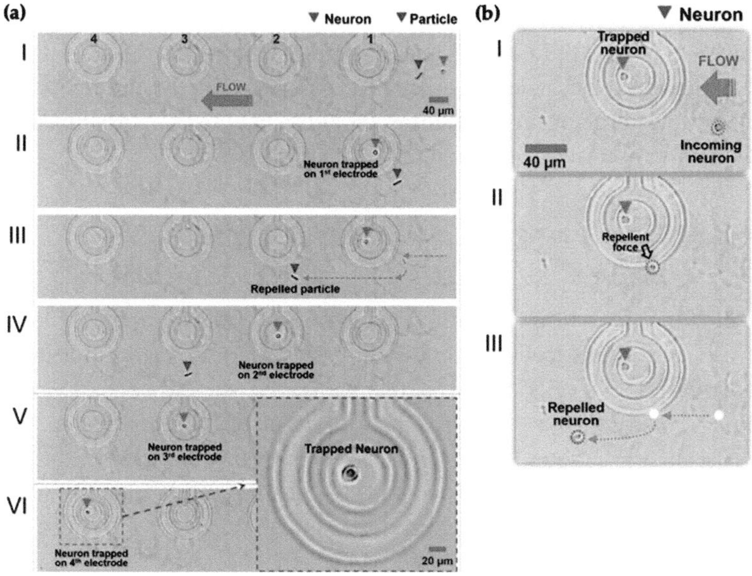

Figure 4.6 Images recorded for single-neuron manipulation on ring traps and the mechanism for precise cell trapping. (a) Incoming neuron (I) enters the first trap. (II) The neuron gets immobilized in the first trap electrode against a fluid flow. (III) Following the neuron trapping, a repelled particle keeps on moving in the flow of media. (IV) The repelled neuron is captured again in the second trap. (V, VI) The neuron is trapped in the third and fourth ring traps in turn. (b) The bouncing motion of the neuron on sensing a repulsive force. With the trapping of the target neuron in the desired electrode, the DEP force repels an incoming neuron. Later, when the incoming neuron moves outside of the electrode, the repulsive force throws the neuron out of the ring. Reprinted from Ref. [55], under the terms of the Creative Commons CC BY license. Copyright © 2018, Springer Nature.

persistent observation of neuron development and functioning. The device was fixed in a closed incubation chamber to ensure a homeostatic surrounding and fewer chances of contamination (Fig. 4.6) [55].

Single-cell analysis has gained great attention over the past decades and offers a new direction for deciphering fundamental biological phenomena, that is, cellular processes and heterogeneities [56]. Particularly in neuroscience, thorough studies of single neurons and spatially isolated neurons help in better understanding the functional neuronal network dynamics along with their fundamental molecular and cellular mechanisms [57]. Single-neuron screening has proven to be crucial to advance the personalized treatments for neurological disorders, including epilepsy, PD, Alzheimer's disease (AD), and other cognitive and motor disorders [58].

4.3.3 Co-culturing

Co-culturing refers to culturing different types of cells to understand cell-cell interactions. Interactions between cells happen in two ways, direct intercellular contact and signaling through molecules secreted by the cells. Major cellular functions, such as cell behavior, functionality, proliferation, and differentiation, are controlled by cell-cell interactions. Cell-cell interactions are a vital aspect of tissue reconstruction, and hence co-culture systems with multicellular organization are developed and studied.

In many cases, co-cultured cells are supported by cellular scaffolds developed by various methods and materials, such as electrospun fibers, hydrogels, patterned substrates, and microfluidic channels. Biocompatible synthetic polymers, that is, polycaprolactone and polylactic-*co*-glycolic acid, and natural polymers, that is, hyaluronic acid, collagen, and fibrin, are the commonly used co-culturing scaffolds. Scaffolds provide a more cell-friendly environment because of the porous structure and their micro- or nanoscale architectures.

In a conventional mixed culture system, the soluble factors cannot be differentiated from the direct-physical-contact signals between neurons and glia cells. Thus, the interaction between a single axon and a glial cell cannot be determined precisely in conventional co-culture systems. Co-culturing of cells in microfluidic channels improves the ability to control the local cellular microenvironment. This provides ease of access to supply and transfer media, eliminate the waste products from the culture system, etc.

4.3.3.1 Types of co-culture systems

Co-culture systems are classified into two types depending on the spatial arrangement of cell culture systems—direct and indirect.

4.3.3.1.1 *Direct co-culture*

One of the goals to carry out direct co-culturing is to investigate the roles of gap junctions, tight junctions, and desmosomes. Intercellular exchange of soluble molecules occurs between cells, and it is referred to as the gap junction mechanism. A tight junction occurs when cells are in close contact with one another. Tight junctions regulate the paracellular permeability, which is essential to establish compartments with different compositions in the body. Desmosomes are thread-like substances and are the main component in maintaining adhesive cell interactions. The main disadvantage of direct co-culture systems is that the effect of diffusible biomolecules secreted by one cell type on other cell types cannot be studied.

4.3.3.1.2 *Indirect co-culture*

In indirect co-culture systems, the different cell populations are physically isolated from each other (no cell-cell interactions) but cultured under the same conditions. Thus, paracrine signaling (signals produced by one cell induces a change of behavior in nearby cells) is the only way for the cells to interact with other cells in the cell culture system. The common signals generated by cells are secreted proteins (growth factors and cytokines), which influence cell behavior, proliferation, maturation, and differentiation.

4.3.3.2 Blood–brain barrier

Co-culturing of neural cells is an important approach for investigating the properties of the blood–brain barrier (BBB). The BBB comprises mainly brain endothelial cells connected by tight junctions. The paracellular diffusion of soluble factors between blood and brain is restricted by the endothelium of the cerebral blood vessels, which acts as a biochemical barrier. A μFD consisting of two separate media microchannels is employed to independently emulate highly localized microenvironments. The two microchannels are designed to supply a medium to vascular tissues on one side and neural tissues on the other side independently and maintain their microenvironments, which mimics the outside and the inside of the BBB, respectively (Fig. 4.7). Co-culturing of these tissues in

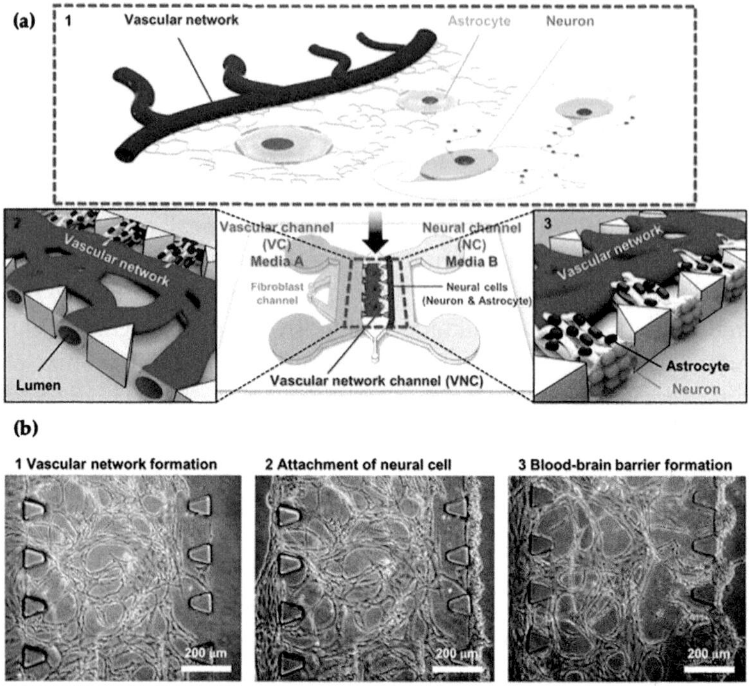

Figure 4.7 (a) Microfluidic device for investigating the BBB. (1) The neurovascular unit co-cultured with astrocytes and neurons. (2, 3) The formation of a lumen in the vascular network and neurons in the neural channel, respectively. (b) The vascular network formed in the microchannel (1), the attachment of the suspension of isolated neurons and astrocytes to the vascular networks (2), and the formation of an in vitro model of the BBB (3). Reprinted from Ref. [59], under the terms of the Creative Commons CC BY license. Copyright © 2017, Springer Nature.

a μFD helps to form a direct contact between neural and vascular tissues. Thus the neural network flows and grows into the inner lumen of the vascular network, forming an in vitro model for the BBB [59].

4.3.3.3 Co-culturing of neural cells

The vertebrate nervous system functions by rapidly conducting nerve impulses through axons insulated with multiple layers of myelin sheaths. The conventional cell culture systems, where neurons and glial cells are cultured, do not mimic the physical and biochemical environments for cell growth. Thus the interaction between axons and glia is absent when they are co-cultured in a randomly mixed form.

Microfluidic channels mimic the in vivo biological systems, which helps understand the actual system by modifying various factors. Microfluidic co-culture systems help us to understand the behavior of stem cells, cell-cell interactions of various cell lines, cell-cell signaling, and stem cell proliferation and differentiation. μFDs for co-culture systems are designed in a way such that the axons and glial cells are cultured in different compartments connected by various microchannels. The devices also help in generating and controlling shear stress, growth factor gradients, and migration assays. They also assist in eliminating the waste from the co-cultured systems.

Park et al. developed a CMD with one circular compartment in the center (for culturing soma) and one cocentric ring-shaped compartment (for co-culturing of axon/glia) connected through arrays of axon-guiding microchannels (Fig. 4.8). Because of the circular geometry, a significant number of neural somas were positioned near the axon-guiding microchannels. As the axons are smaller in dimension when compared to axon-guiding microchannels, the isolated axons extend and pass through the microchannel array to reach the glial cells. The flow of a neuron-enriched medium into the glial side is controlled by the narrow diameter of the microchannels, thus enabling the study of direct interaction between axons and glial surfaces (Fig. 4.8). The axons entering the glial compartment results in the maturation of glial cells, confirming the efficiency of the co-culturing system. Similarly, μFDs of various architectures have been designed to co-culture neuron and glial cells to improve axon-glia interactions [37].

Figure 4.9-I depicts a four-chamber co-culture device separated by three valve-enabled PDMS barriers designed to visualize and study synapse formation among neuronal populations. Out of these four cell culture chambers, the two inner chambers are filled with neurons while the outer chambers are filled with glial cells. In this device, a single pressure chamber is placed on top of the barriers

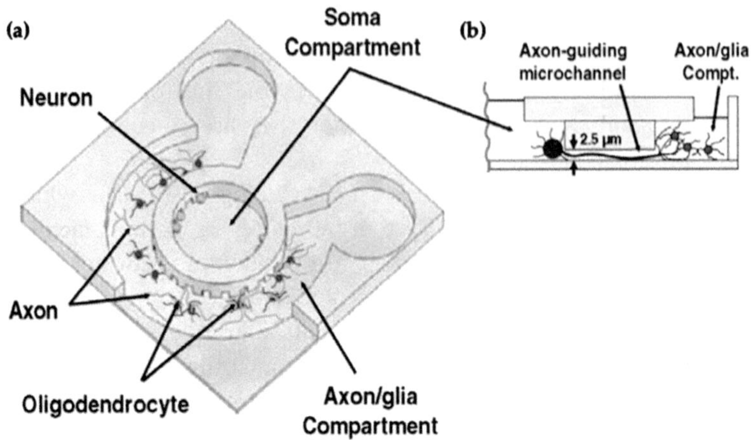

Figure 4.8 Microfluidic co-culturing system with a soma compartment and a concentric glial compartment connected by microgrooves. Reprinted by permission from Springer Nature Customer Service Centre GmbH: Springer Nature, Biomedical Microdevices, Ref. [37b], copyright (2009).

to control the fluid exchange and also to separate cell chambers. First, the glial cells are cultured on top of a PDMS surface coated with poly-L-lysine (PLL) and then the neuronal cells are loaded onto the glass surface. The glial cell chamber is vertically bonded onto the glass surface with the help of a pressure chamber. It is observed that the dendritic protrusions increase in the μFD because of the vertical layered configuration. Similarly, a parallel configuration of the neuron glia co-culture system was also designed as shown in Fig. 4.9-II. In this device, the outer two chambers coated with collagen were filled with glial cells and the inner channels coated with PLL were filled with neuronal cells. At first, glial cells were cultured and incubated to reach their confluence. Later neuronal cells were loaded into the chamber. It is noted that the dendrites grew toward the glial compartment, increasing the probability of synapse formation.

4.3.4 Integrated Microfluidic/Microelectrode Array

MEAs are neural interfacing devices consisting of a 2D arrangement of thousands of microelectrodes for the extracellular stimulation

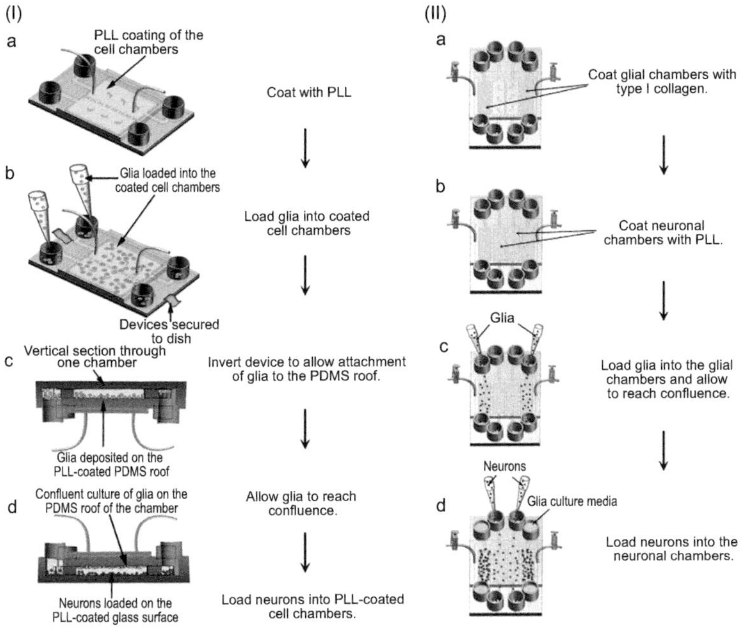

Figure 4.9 Microfluidic co-culture system with four-chamber (I) and six-chamber (II) systems, both depicting the same steps: (a) coating the substrate, (b) load neuronal cells to the coated substrate, (c) load glial cells, and (d) loading neurons to the neural chambers. Reproduced from Ref. [60], under a Creative Commons License Attribution 3.0 Unported (CC BY 3.0).

and monitoring of electrical activity in electrogenic cells. MEAs are being increasingly used for recording in parallel electrical activity from many excitable cells. These devices are useful in understanding the complex relationships between cell morphology, active channel distribution, and associated spatiotemporal patterns of neuronal activity. Spatiotemporal patterns on MEAs provide advantages in monitoring activity patterns in acute slices and dissociated neuronal networks. Metallic MEAs are used for slice culture of cells (in vivo connectivity is retained) and also for various types of dispersed primary cell cultures.

Claverol et al. developed a μFD in which the neurons are patterned on an elastomer film and a standard MEA is coupled to study the electrical recordings of the cells. Microwells in the

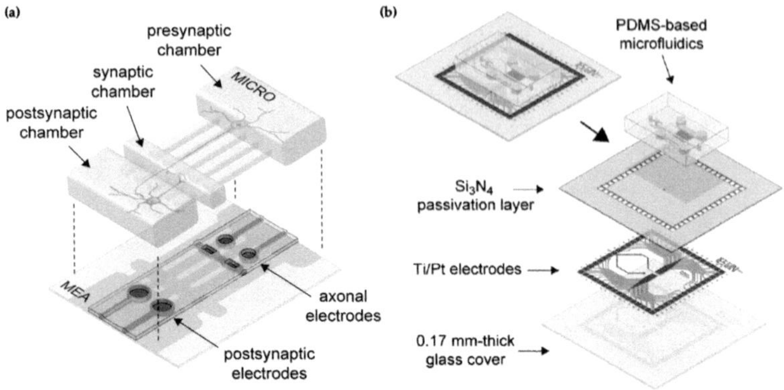

Figure 4.10 (a) Design of a patch-clamp microchannel array in which cells can be trapped by suction and cultured on the microhole array. (b) Schematic of a patch-clamp microchannel array. Republished with permission of The Royal Society of Chemistry, from Ref. [63]; permission conveyed through Copyright Clearance Center, Inc.

elastomeric film help in cell body confinement, and microchannels help in guiding the neurites for network topology specifications [61]. To obtain single-electrode recordings of extracellular potentials from confined neurons, spatially localized multisite recordings from individual microchannel are studied [62].

Morin et al. fabricated integrated microfluidic and MEA chips with predetermined connectivity to study the structure-function relationship and bioelectrical activity of neuronal cells. In this device, the microfluidic system is patterned with a number of microchambers to culture neural cells and MEAs with a planar-specified electrode layout are coupled with the device (Fig. 4.10) [63, 64]. Planar MEAs are used in conjunction with microwells to trap single neurons using DEP and study electrical patterns of neurons. DEP helps to trap each neuron into a microwell, which is connected to an electrode site of the MEA. Also, due to the electrophoresis force, glial cells are prevented from reaching the electrode site with the help of the conductivity of the buffer used in the device. Neuronal cells experience a positive dielectrophoretic force, while glial cells have a negative electrophoretic force at higher frequencies [65]. Alberti et al. developed a patch-clamp MEA system that consists of silicon/silicon dioxide (Si/SiO_2) on MEA integrated in a μFD

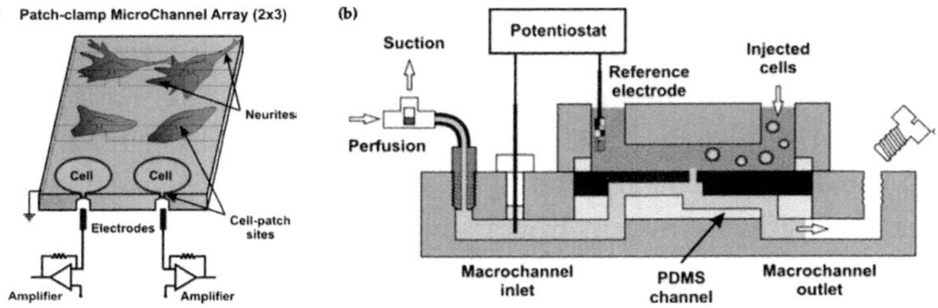

Figure 4.11 (a) Design of an MEA platform showing a microfluidic device and MEA components. (b) An exploded view of the various lithography layers for the fabrication of MEA. Reprinted by permission from Springer Nature Customer Service Centre GmbH: Springer Nature, Microfluids and Nanofluids, Ref. [66], copyright (2010).

electrical recording (Fig. 4.11). This device incorporates a number of microholes fabricated on top of microchannels. The cells are loaded from top of the system, and each cell is trapped onto the top of the microhole by suction. The electrolytic solution is loaded from the bottom, which fills the microchannels and finally reaches the microholes. A silver electrode is integrated on top of each microhole to customize the interconnections to study the electrophysiology of the neuronal cells [66].

4.3.5 Hydrogel Gradients

An important aspect of a neuronal cell culture is that the neurons form 3D structures, which are essential for exchanging information across synapses. Nevertheless, most of the in vitro substrates for microfluidics are 2D, which does not resemble the real tissue environments. The greatest challenge in tissue engineering is to create microenvironments that mimic the actual in vivo complexity of tissues and cells by incorporating gradients such as physical, chemical, temporal, and spatial factors within engineered 3D scaffolds.

Hydrogels are one of the most ideal materials for 3D tissue scaffolds that mimic the extracellular matrix (ECM). Hydrogels are also considered a replacement for the native ECM as they provide 2D or 3D solid support, a fibrous structure, nutrition, and gas permeability. Hydrogel matrices in combination with biologically active molecules can induce molecular interactions with target cells, thus enabling them to form complex geometries. Hydrogels may also aid the ability of neurons to extend new axons. The matrices of hydrogel have a low elastic modulus, matching that of a neural tissue, thus making it an ideal material for neural regeneration. To improve the directionality into hydrogel matrices, gradients are commonly introduced.

3D hydrogel matrices are placed between a source and a sink, forming gel-solution interfaces, which enables the effective diffusion of soluble molecules that generates a concentration gradient with time. Hydrogels are precisely filled in the localized area of microchannels either by the microinjection technique or by introduction through a gel-filling channel. Saadi et al. developed a μFD that generates stable gradients across 2D cell culturing microchannel and a 3D biological hydrogel. A μFD specified as a ladder device is shown in Fig. 4.12a,b. The device consists of source and sink channels connected with an array of microchannels embedded with hydrogel, which acts as a gradient generator. Since the fluidic resistance of a microgroove is high and the hydrogel matrices prevent convective inflows, diffusive transport takes place [67]. The generation of gradients with different concentration profiles can be achieved by controlling the influx and outflux areas of molecule diffusion across the gradient generation region. A linear concentration profile can be generated using a straight ladder (balanced influx and outflux area), whereas a nonlinear concentration profile can be created by a wedge ladder (unbalanced influx and outflux area). The gradient generation in these models is completely dependent on the hydrogel matrices filled in the microchannels connecting the source and the sink [67–69].

A μFD consisting of four inlet chambers and one outlet channel with high fluidic resistance is designed. The four channels are filled with cultured cells with two alternating layers of hydrogels to create a gradient. The four layers of the self-contained system can only be supplied with culture media and supplements because of the diffusion process that forces the neurites to extend. Similarly, a six-layer device with four hydrogel layers can be supplied, which stimulates a denser neurite network [70].

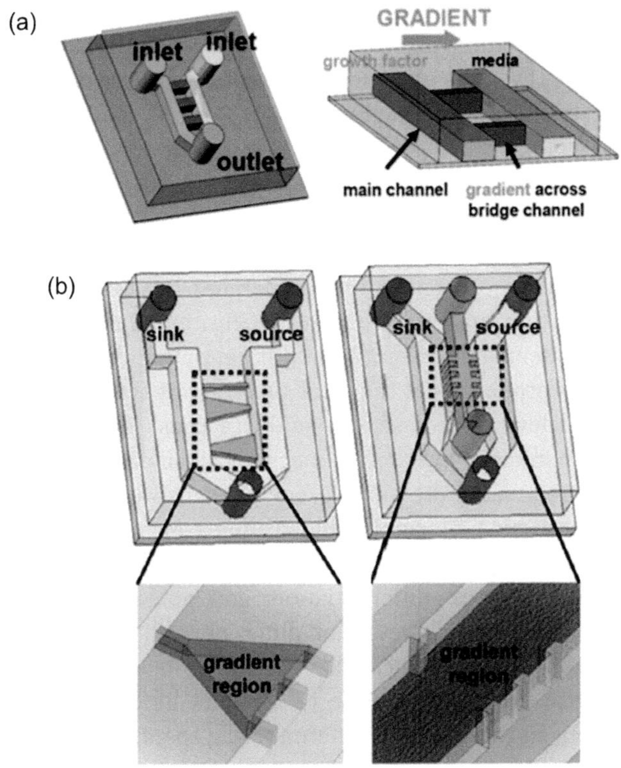

Figure 4.12 (a) Concentration gradient generated by a ladder chamber with hydrogel as a gradient. Reprinted by permission from Springer Nature Customer Service Centre GmbH: Springer Nature, Biomedical Microdevices, Ref. [67], copyright (2007). (b) Microfluidic platforms generating various concentration gradients by changing geometries of the gradient forming region. Reprinted with permission from Ref. [68]. Copyright (2007) American Chemical Society.

4.4 Applications

As discussed in earlier sections, the orientation and guided growth of neurons inside prestructured channels govern the microenvironment, providing a significant prospective for further microfluidic-based neurobiology studies. Indeed, on the basis of the understanding of above-mentioned facts, microfluidics has intruded into various applications in neuroscience already, such as NSC differentiation, polarity, axon guidance, neuropharmacology, neuroelectrophysiology, and neuron biosensors, and produced some excellent results, as explained below.

4.4.1 Neuron Differentiation and Polarity

In vitro culturing of neural cells undergoes a sequential process of differentiation, polarization, and specification of neurites (either as axons or dendrites). Neural cell differentiation is dependent on the physical properties of the local substrate as the differentiating neurons retrieve their essential guidance cues. Differentiation of NSCs can be induced by external mechanical factors, such as stress, adhesion area, substrate surface energy, substrate elasticity, and topographical cues. Biocompatible scaffolds for cell culturing are used not only to promote stem cell adhesion but also to organize cell growth, promote differentiation, and guide axonal growth. Generally, the external neuronal differentiation–inducing agents include small molecular chemicals, cytokines, and growth-inducing factors. Since the differentiation rate cannot be controlled by the optimal concentration of the differentiating agent, an alternate has to be considered [71].

Microfluidic channels exhibit a great influence in promoting neural cell differentiation as the microgrooved structure acts as a spatially defined cue for cultured cells. Micropatterned systems influence both the differentiation rate and the number of neurites extending per neuron. The dimensions and alignment of the microgroove structure greatly affect the differentiation rate, not by the adhesion behavior of the cells to the surface, but by the perturbation of the early differentiation process. For instance, the differentiation rate of NSCs grown on the microgrooves with widths smaller than the soma diameter is typically lower than that of the cells grown on a chip with microgrooves of larger widths [72]. The surface functionalization of microgrooves, such as surface coating with biomaterials such as PLL, double-walled carbon nanotubes, and gelatin, also has strong effects on the neuron differentiation rate. The result of this study clearly demonstrated that the microgrooves with coating and optimized width have significantly improved neuronal differentiation rates [73].

In the development of the nervous system, the polarity of neural cells is established by the interaction between differentiating neural cells and the extracellular environment. Topographical cues have strong effects on the polarity of the neuron culturing in a microenvironment properly maintained in microchannels. The polarity of neurons can also be induced by external factors, such as mechanical force and magnetic force. A μFD consisting of two chambers that are connected by a number of microgrooves was designed. The microgrooves were designed asymmetrically with a funnel shape to enhance the directionality of the axons (axons grow from one chamber to the other). A cortical treatment was applied in one of the chambers, which improved the differentiation of the matured cells [74].

4.4.1.1 Axon guidance

The most challenging function of neural tissue engineering is to restore the functional capabilities of damaged neural tissue (the synaptic networks have to be reformed with proper axonal guidance to reach their targets). The recent developments in the microfluidic systems with patterned and functionalized substrates have led to the control of neurite direction (axonal guidance) and enhanced neurite outgrowth. Generally, μFDs are designed in such a way that a physical barrier is incorporated with embedded microgrooves. These microgrooves connecting the compartments act as a filter by allowing the neurite cells to extend on the axonal side and trapping the cell soma on the other side of the device. The design of the μFDs can define the axon outgrowth direction and organize functional connectivity in the neural culture.

4.4.1.2 Patterned substrates

μFDs support biopatterning of substrates (micro- or nanopatterning of proteins), biomolecules, and cells. The interaction of cells with ECM, biomaterials, and other molecules can be studied by fabricating substrates with topographical and chemical cues of

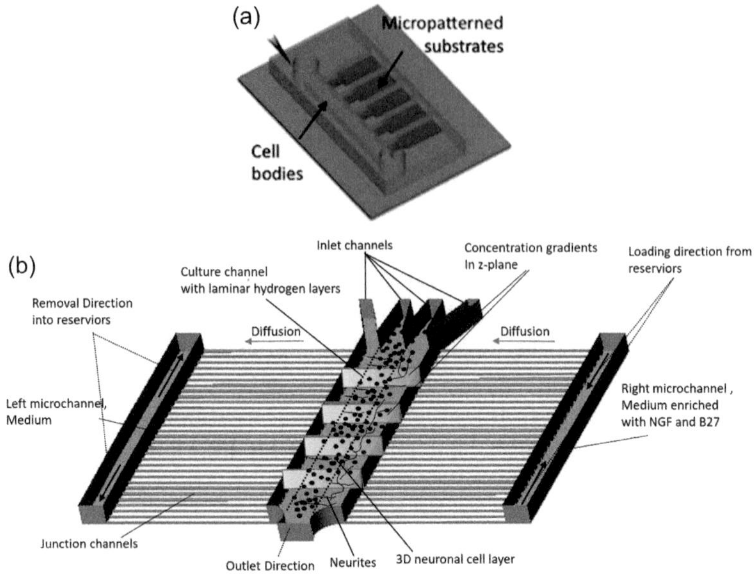

Figure 4.13 (a) A microfluidic device depicting micropatterned substrates for axon guidance and regeneration. Republished with permission of The Royal Society of Chemistry, from Ref. [75]; permission conveyed through Copyright Clearance Center, Inc. (b) Microfluidic device with a neuronal cell culture platform with a patterned hydrogel substrate and generation of a growth factor gradient through cell culture with reservoir channels [76].

cellular size. These patterned substrates can also be used to determine the ability of these materials to influence cell adhesion, proliferation, apoptosis, migration, orientation, and differentiation.

Hellman et al. designed a microfluidic-based strip assay device with a patterned substrate to guide axons, as shown in Fig. 4.13a. In this device, neural cells were selectively placed in a patterned substrate, allowing the axons to interact with biochemical cues present in the μFD. Alternating layers of permissive and inhibitory substrates were patterned on the μFD, which allows axons to extend toward their positive biochemical cues [75].

Hydrogel layers are used as a gradient generator in a μFD. In this device, hydrogel is patterned on the substrate of the μFD, where the neural cells are cultured. The media is fed to the μFD by two microchannels on either side of the culture chamber connected by an array of microchannel junctions. The concentration difference in the two microchannels creates a gradient, and the hydrogel matrix acts as a guidance cue for axon guidance, as depicted in Fig. 4.13b [76]. A four-chamber μFD in which somal and axonal compartments are connected by microchannels is designed. The soma of the neurons gets trapped in the somal compartment, and the neurons extend their axons toward the axonal compartment. The extended axons can be guided by patterning biochemical cues near the edges of the microchannels in the axonal compartment to generate the microenvironment with surface-bound inhibitory proteins [77].

4.4.2 Biochemical Gradients

In addition to patterned substrates, developing neurons respond in biochemical gradients. Gradient materials are used to screen the interactions between cells and biomaterials. External gradients that may be both long-range diffusive cues and short-range contact cues guide axonal growth and trajectory. In vivo, the biochemical gradients will never be static and thus the need to mimic and control the gradients is important. To study gradients, the μFDs are a great choice as they can generate and maintain desired gradient profiles for a long period. These devices can be categorized into two types, flow-based (diffusion occurs at the interface of laminar flows) and diffusion-based (linear gradient generated in between a source and a sink channel) gradient generators.

In flow-based gradient generators, the cell-cell communication by diffusible factors is affected because of the continuous flow, which induces higher shear stress. The diffusion-based gradient generators overcome the drawbacks of flow-based devices and hence are used to analyze paracrine signaling, which is an essential feature of the BBB. To maintain a stable gradient between the source and the sink, various methods, such as the introduction of membranes, nanofibers, hydrogels, micropillars between channels, and a diffusion chamber, are proposed.

μFDs with fluid-phase gradient cues can be used with micro-channel mixers, jets, or valves to expose cells on transient or

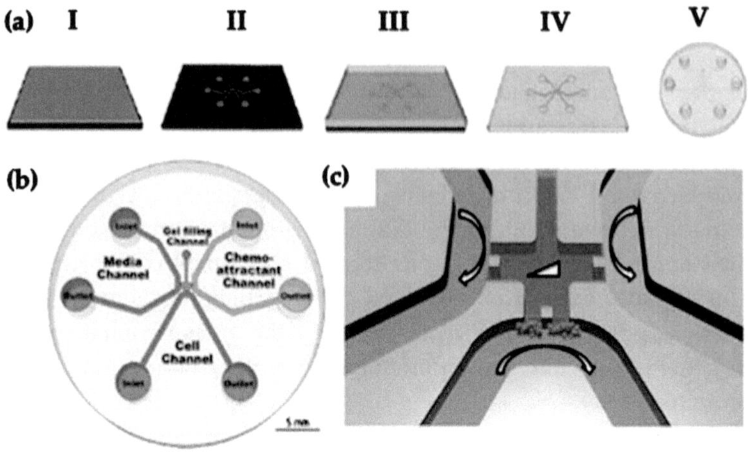

Figure 4.14 (a) Microfluidic device fabrication steps, (b) design of a three-chamber microfluidic device, and (c) media flow and growth factor gradient in the device. Republished with permission of The Royal Society of Chemistry, from Ref. [79]; permission conveyed through Copyright Clearance Center, Inc.

stable diffusive gradients. A μFD consisting of three channels interconnected by microchannels was designed along with biochemical gradients. A diffusion-mediated gradient of laminin was patterned on cover slips to guide neurons toward increasing laminin concentrations. Neuronal cells were cultured in the middle channel, and laminin-media supplements were filled into either of the side channel. It was noticed that the axons reach the diffusion zone, where the laminin is patterned and tends to grow. Then, the axons are further guided toward the laminin-filled microchannel, thus controlling the growth of axons to the target regions [78].

A μFD consisting of six channels connected at the center by a small microgroove was designed to study the surface gradients and guidance cue gradients (Fig. 4.14). A biomaterial (collagen) is filled in the microgroove, which connects all the chambers of the μFD. The media, neuronal cells, and guidance cue (chemoattractant material) are filled through the inlet chambers of the μFD. The media and the chemoattractant diffuse into the collagen by forming gradients in the microgroove. The neuronal cells are guided to grow toward the media and then are attracted by the chemoattractant, resulting in their axons getting extended [79].

Microfluidic gradient arrays are also a promising device to generate gradients and improve the guidance of axons. In this device, a large cell culture chamber is designed and an array of microjets is used to create gradients on either side of the cell culture chamber. This array of microjets is capable of generating identical concentration gradients in parallel with flow. When a biomaterial or a growth factor is supplied through a microjet inlet, a controlled flow rate is observed. Because of the surface gradient near the edges of the cell culture chamber, the neurons extend their axons toward the gradient of the chambers [80].

4.4.3 Electrophysiological Recordings

Electrophysiology is the study of the electrical properties (electric current and membrane potential) of cells and tissues, ranging from single-ion channel properties to those of whole organs. Electrophysiology using MEAs depends on the distance, strength, and stability of the interfacial contact between the electrogenic cells and an electrode. The patch-clamp technique is a classical tool for studying single- or multiple-ion channels of cells. In a traditional patch-clamp technique, a fire-polished glass pipette with a tip diameter in the range of 1–2 μm is pressed into a cell membrane using a micromanipulator. To electrically isolate cells, the membrane is sealed to the pipette with suction, enabling the recording of even single-channel ion fluxes. Although this process has a high resolution, it has very low efficiency and is time consuming.

MEAs are pretreated by using processes such as microcontact printing and substrate patterning to deposit patterned cell-adhesive proteins. This patterning results in a higher cell density at the electrodes and also increased electrical activity of the culture cells [82]. Biffi et al. fabricated a μFD in which two subcompartments were provided to culture cells and a flat MEA was coupled to the device. The two fluidically connected compartments initiated two parallel neuronal networks showing independent functionality but with similar electrophysiological activity (Fig. 4.15) [81]. The

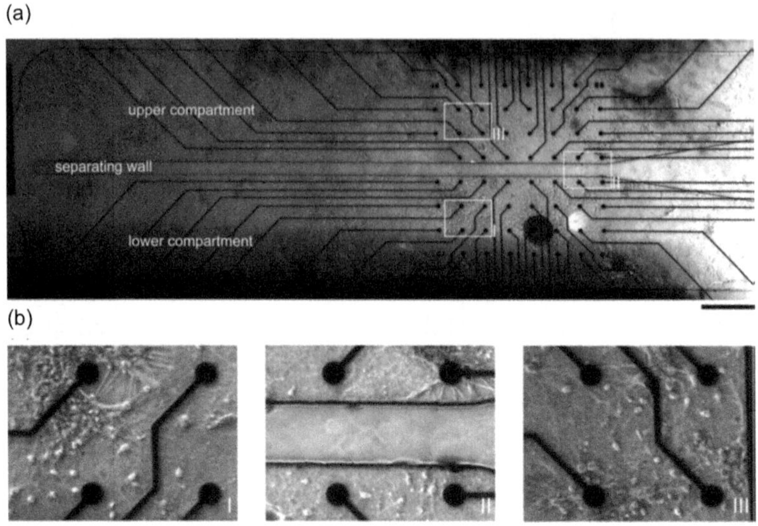

Figure 4.15 (a) Differential interference figure of a neuronal network developed in the microfluidic device. (b) Magnified images of neurons grown in devices inside the (I) lower and (III) upper compartments and (II) the region across the separation wall. Reprinted from Ref. [81], with the permission of AIP Publishing.

synchronous phenomena are uncorrelated for these two parallel subnetworks. Although the two networks basically correspond to two individual MEAs, the experimental similarity, such as cell seeding and growth in a biochemical microenvironment, decreases the experimental differences.

Goyal and Nam fabricated a μFD with planar MEAs to study the electrophysiological activity and the chemical and electrical stimulation synergism on neuronal networks (Fig. 4.16). The chemical microenvironment around the cells changes with time. Hence temporal control is important for the maintenance of healthy neural cultures. The neuronal cells were cultured in open wells, and only the neurites were under the controlled microenvironment. The electrodes underlying the neural circuits kept recording the action potentials and were used to probe the simultaneous functional alterations in the neural activity inside a controlled microenvironment [83].

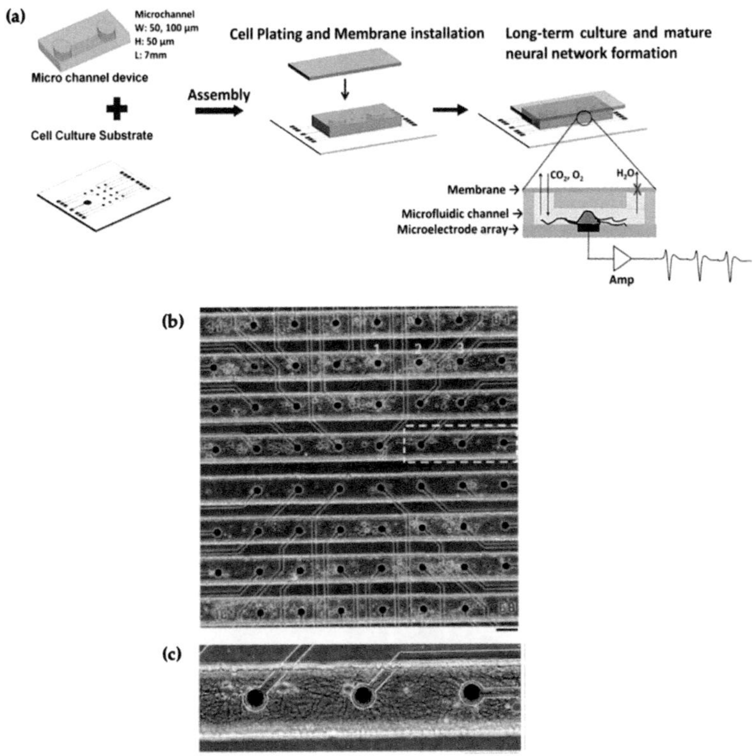

Figure 4.16 (a) Schematic of a μFD integrated with microelectrode arrays. (b) Microelectrode arrays with 100 μm wide channels and cultured neural cells. (c) Magnified view of electrodes on microelectrode arrays. Reprinted by permission from Springer Nature Customer Service Centre GmbH: Springer Nature, Biomedical Engineering Letters, Ref. [83], copyright (2011).

As discussed in the section on compartmentalized microfluidics, the μFD designed by Takeuchi allows neuronal stimulation, which helps in recording electrical signals. As shown in Fig. 4.17, the device has 64 electrodes, whose electrical activities can be measured simultaneously. The electrodes present directly under the neuronal microcompartments evoked responses. It is concluded that changes in the beat rate of the ventricular myocytes after electrically stimulating the neurons were affected by frequency and the interaction between frequency and the number of pulses [45]. A

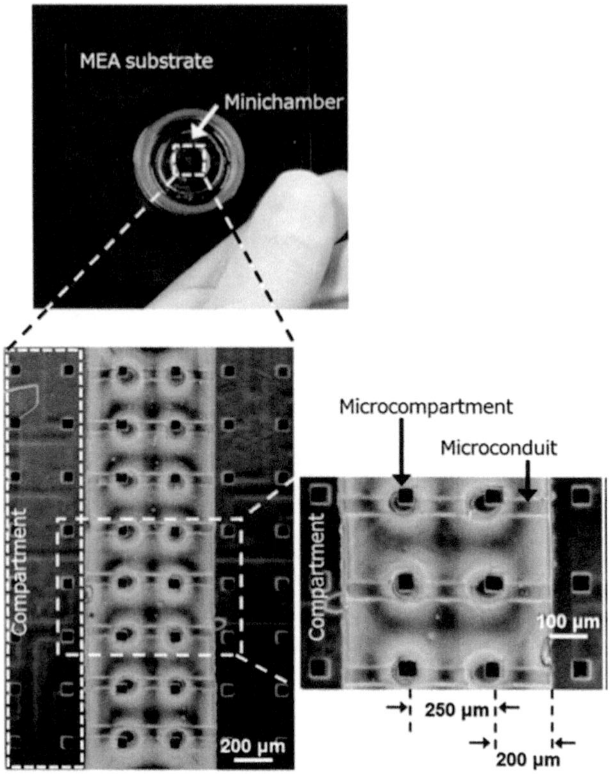

Figure 4.17 A minichamber comprising a microcompartment and microconduits on an MEA substrate. Republished with permission of The Royal Society of Chemistry, from Ref. [45], copyright (2011); permission conveyed through Copyright Clearance Center, Inc.

μFD with similar features is used to study the electrical activities of synaptically connected neurons and cardiac myocytes. It is noted that the change in the beat rate of differentiated cardiomyocytes is affected by the pulse frequency of the electrical stimulation of neurons [46].

4.4.4 Dendritic Signaling and Synapse Formation

As discussed earlier in the chapter, the polarization of neurons along with the size and density of synapses, causes difficulty in controlling the synaptic function by introducing manipulations and visualization of cell biological processes [84]. But alteration in the native neuronal network circuitry results in disordered neuron connections. Many efforts have been made to improve the cultured neuron organization and connections. Several groups have attempted to actively manipulate the dendritic outgrowth direction and polarization via chemically patterned surfaces [85–88], soluble gradients [25, 88–91], or physical structures that guide neuronal growth [92–95]. Figure 4.18 shows the images of a microfluidic chamber used to organize cultured neurons closely, which mimic the network dynamics in vivo, favoring neurophysiological studies [96–98].

Millet et al. fabricated PDMS channels to sustain low cell densities (nanoliter volume) of primary mammalian neurons, enabling the analyses of the neuron microenvironments. Such primary cultures are otherwise difficult to stabilize for longer duration. The hippocampal axons were found to be localized in the channel corners and preferred PDMS surfaces over glass in hybrid devices, as shown in Fig. 4.18a [96].

Local modifications at synapses play an important role in neuronal plasticity, but the complete information is still lacking due to the availability of limited techniques for exploring local processing within synaptic regions. Therefore, Taylor et al. developed a μFD or chamber to optimize synaptic cell biology [38]. This visualized and manipulated synapses and pre- and postsynaptic cell bodies in an independent manner [35, 99]. The device consisted of a compartmentalized microfluidic chamber that employed direct cortical growth and hippocampal axons across parallel microgrooves. This was the first attempt to direct the growth of dendrites through microgrooves and check whether there was any formation of functional synapses inside the microfluidic chamber. The two populations of neurons were cultured on either side of the microgrooves. Additionally, a small perfusion channel imposing local perturbations and probing synaptic regions of high spatial and temporal resolution was also implemented. As a proof of concept, this chamber can be used to study synapse-to-nucleus signaling in several ways using calcium imaging, study transcriptional response comparison to massed and spaced stimulation, and explore the variations in Arc transcription and mRNA localization;

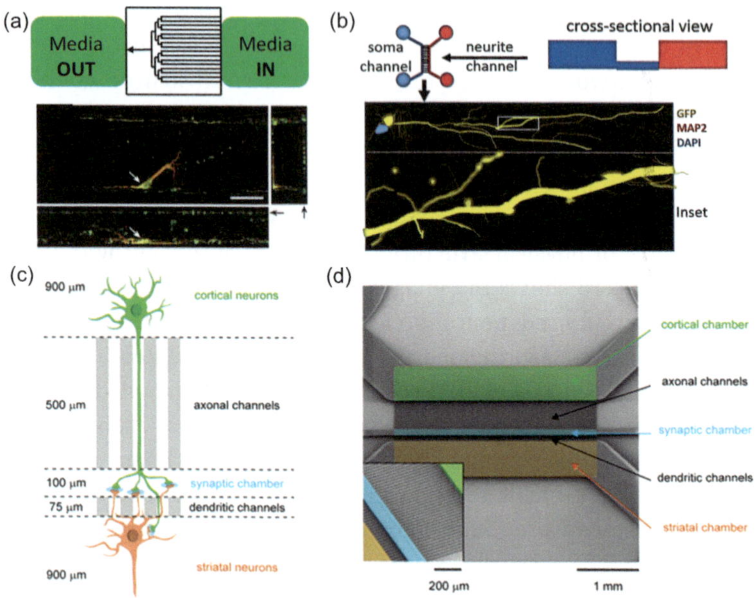

Figure 4.18 μFDs offer design flexibility to probe developing cultured neurons. (a) Ultralow densities of primary postnatal rat hippocampal neurons showing differentiation and extension of dendrites in μFDs fabricated by using PDMS. An individual neuron; the cell body is indicated by a white arrow; dendrites and axons are stained with MAP-2 (red) and tau (green); scale bar: 50 μm. Republished with permission of The Royal Society of Chemistry, from Ref. [96]; permission conveyed through Copyright Clearance Center, Inc. (b) Dendrite-to-soma signaling analysis using a multidomain μFD [97]. (c) Schematic representation of a three-compartment microfluidic device employed for the reconstruction of corticostriatal connections. Cortical neurons (upper chamber, green) connect with striatal neurons (lower chamber, orange) via microchannels and an intermediate synaptic chamber (middle chamber, blue). The microchannel length facilitates cortical axons (>450 μm) and striatal dendrites (<450 μm) to extend and connect in the synaptic chamber. (d) Microphotograph of the silicon template for PDMS-based microfluidic device fabrication [98]. (c and d) Reprinted from Ref. [98], under the terms of the Creative Commons CC BY license (CC by 4.0).

subsequent to the local introduction of the mGluR1 group I agonist, dihydroxyphenylglycine was also demonstrated. Later, in 2017, Nagendran et al. studied and assessed the progress of intrinsic events occurring in the long projection pyramidal neurons after distal axotomy, hence remodeling synapses [100]. For this study, microfluidic chambers containing microgroove-embedded barriers of $\sim$900 μm in length were used. The results show that axotomized pyramidal neurons undergo dendritic spine loss followed by a trans-synaptic enhancement in presynaptic excitability. They also found that directly injured neurons show excitability and loss of inhibitory inputs. Axotomy shows considerable detrimental effects on netrin-1. But the exogenous netrin-1 protein several hours after axotomy re-establishes spine density and brings presynaptic excitability and inhibitory inputs onto the injured neurons to optimal levels [100].

Dendrites act as conduits of information influx and integration, from synapse to nucleus and vice versa. Interconnected compartments with μFDs also control dendrite probing in subregions. To mimic the synaptic release, neurotransmitters were administered inside ultrathin fluid streams of chemoattractants, placed or relocated across neurons in μFDs [99]. The advancements in controlling the microenvironments of dendrites have helped resolve the fundamental synapse formation and regulation mechanism [101, 102]. This approach has proven useful in resolving mechanisms of BDNF induction of transcription-mediated synapse strengthening. Compartmentalization and fluid isolation of dendrites from the somatic body of the rat embryonic cortical neurons in μFDs has helped discover a novel, dendrite-localized neurotrophin signaling pathway (Fig. 4.18b). Treatment by BDNF and other pharmacological agents can be restricted to the dendritic somatic part inside the above-described device. The study revealed that the dendrite-to-nucleus induction of *c-Fos* expression induced by BDNF is dependent on Ca^{2+}, Trk-activity independent in dendrites, and also mediated via mitogen-activated protein kinase (MAPKK or MEK1/2). However, dendritic signal processing is different from MEK5-mediated axonal retrograde signaling, and it is dependent on intradendritic mRNA translation for the expression of the immediate early gene. All these factors separate dendritic pathways from the retrograde neurotrophin signaling existing in axons [97].

It is a difficult task to discriminate mediators of axodendritic synaptic differentiation and degeneration in the corticostriatal pathway of neurons in vivo. Plume dynamics, source micropipette diffusion, and static-fluidic conditions make it challenging to exactly control the exogenous cues in conventional cell culture plates. But μFDs offer easy control over neurite directionality and also enable *en passant* analysis of axodendritic connections and corticostriatal synapses. Tapered interconnecting channels in a μFD direct cortical axons toward a neighboring chamber culturing striatal neurons [103]. The approach overcomes the shortcomings of conventional methods by identifying the specific types of neurons in a co-culture population and separating two different populations [104], along with allowing abundant *en passant* axodendritic interactions. Moreover, the μFD microenvironment favors striatal neuron differentiation, consequently generating longer dendrites, higher spine density, enhanced phosphorylated extracellular signal–regulated kinase activation, and spontaneous Ca^{2+} oscillation synchronous with cortical neuron activity [105–108]. Recently, Moutaux et al. developed a microfluidic-based approach for the reconstruction of a time and space-controlled neuronal network suitable for fast spinning confocal video microscopy. This μFD is composed of two fluidically isolated neuronal chambers that are connected via a set of thin microchannels through which neurites can grow and come in contact each other in an intermediate synaptic compartment (Fig. 4.18c,d). This configuration allowed network development synchronization between various conditions, thereby facilitating systematic analyses and reproducibility. The spatiotemporal features allow cross-compared axonal trafficking of two motile organelles, dense core vesicles, and mitochondria throughout network maturation [98].

4.4.5 Developmental Study at Cell Population/Tissue/ Organ-on-a-Chip (Brain-on-a-Chip) Level

Human brain development is a complex process—a controlled environment hosting interactions among intracellular signaling, secreted signaling molecules, and cell surface receptors [109]. The modern neurobiological methods have provided overwhelming evidence for brain development and factors affecting it—the interaction of genetic, epigenetic, and environmental factors. To achieve complete insight into the signals shaping the developing brain, cultures would require to include several cell types in three dimensions. The central nervous system (CNS) consists of mixed populations of neurons, glia, microglia, and endothelial cells interacting mutually. Topographical neural structures can simply be achieved by re-creating the brain microenvironment. But this needs a detailed study of the critical environmental cues. To construct organized 3D cultures, the culturing medium ought to have the material properties, design and fabrication, and imaging processes to be optimized for the cell type. Rat hippocampal neurons prefer PDMS microchannels over glass substrates, and neurites have a strong affinity for topographic cues (walls meeting at right angles) [96]. Given the condition of competitive substrate cues, the hippocampal neurons prefer physical contact guidance cues over chemical ligands. Therefore, cellular tension and chemical substrate cues must access different competitive signaling pathways [110].

To preserve the developmental organization in vitro, Smith et al. fabricated a μFD for re-creating the controlled microenvironment of chemical morphogens similar to developing tissue. The microdevice works on simple Fickian diffusion principles for creating the concentration landscapes of crucial morphogens, mimicking the in vivo environment to differentiate naïve stem cells into their specialized cell fates. A few microfluidic platforms have studied spinal cord patterning [111–114]; however, only Demers et al. could show the recapitulation of the spatial organization of neural tube development in vitro [113]. During spinal cord development (Fig. 4.19), chemical cues released from organizing centers around the neural tube, that is, notochord, paraxial mesoderm, and roof/floor plates, direct the differentiation of neural precursor cells into mature neurons [113].

MN differentiation experiments were also conducted inside the two-channel and four-channel microdevices. The results showed that the differentiation happened on crossing the threshold retinoic acid (RA) concentration of 0.1 μM and as an intangible function of RA gradients in a two-channel device. However, in a four-channel device

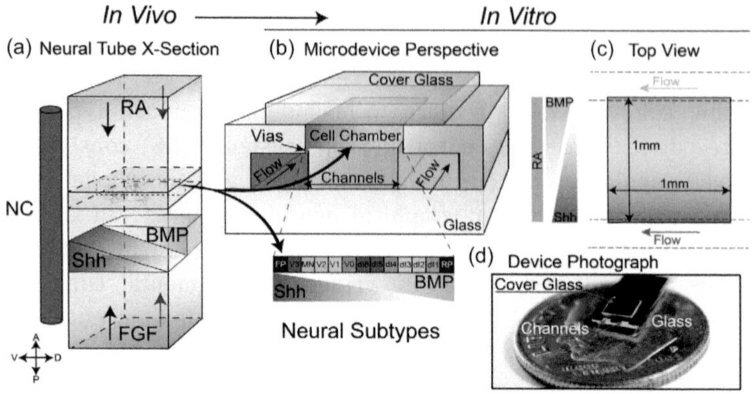

Figure 4.19 Graphics showing microfluidic reconstruction of the neural tube. (a, b) Neural tube schematic re-created inside the device with four 1° signals (RA, SHH, BMP, and FGF) used for bulk neural tube (NC, notochord) patterning (a). The SHH gradient directs the ventral neural progenitor differentiation into discrete domains of neurons (b). Flow channels supplying nutrients along with desired guidance molecules into the cell culture chamber. (c) A top (rotated) view of (b) as visualized by the cover glass. (d) Macroscale indication of the device on top of a dime. FP, floor plate; RP, roof plate. Reproduced from Ref. [113] with permission of the Company of Biologists.

(more extensive and realistic environments), the MNs preferred a particular purmorphamine/RA signaling combination. Likewise, an anterior-to-posterior RA gradient generally existing in vivo could be explored to be employed in the four-channel microdevice. Ultimately, the authors showed the effect of an opposing bone morphogenetic protein (BMP) gradient on an MN band. It induces a more defined MN band sharpening by inhibiting neuroectodermal lineages [115].

Microfluidic platforms have shown remarkable advancement over existing tools as they create better biomimicking surroundings, which helps biologists to resolve complex developmental questions, such as subtype specification mechanisms, interactions of signaling pathways, and transcription factor. The current modified version with dual source/sink channels along with simple variants [116] can help in deciphering cellular differentiation in a user-defined concentration landscape, where no more attention needs to be paid, unlike in traditional cell culture experiments.

In vitro MN differentiation and neural development inside microfluidic devices can be helpful in therapeutics. The generation of specific spatially defined populations of neural subtypes in vitro can lead to the treatment of neural disorders, such as amyotrophic lateral sclerosis (ALS) and spinal muscular atrophy, where preferential neural subtype deaths occur [117]. However, in order to re-create the early developmental events forming distinct neural subtypes, answers to the unanswered fundamental questions about neural tube development are required. Coincidentally, this simple yet powerful device showed the potential to develop neural tube surroundings from the ground space. Scientists successfully generated a single "slice" of the neural tube using only three morphogens. Therefore, in the future, there is a lot of scope for combinatorial investigation using different chemical morphogens, like fibroblast growth factor (FGF) and transforming growth factor β families, temporal effects, and live-cell imaging of patterning events [113].

A closer approximation of the physical 3D structure of the developing brain can be achieved by combining the layered hydrogel scaffolds into the microfluidic channels. Cerebral cortex layers were mimicked at the same scale by fabricating four agarose hydrogel layers. Layers with embedded neurons are intercalated in the adjacent layers. The encapsulated growth factors (serum, B27 supplement, and NGF) are released diffusively and direct developing neurons [118]. Consequently, organization of the laminar brain tissue occurs with embedded neurons, axon growth, and synaptogenesis within the 3D matrixes, which is useful for studying corticogenesis in vitro. Multichannel μFDs can facilitate the formation of the large, interrelating populations of dissociated cells (e.g., neurons, glia, and microglia) and organotypic tissue slices. Organotypic slices via microchannels can interconnect and show synchronized neural activity [119]. These complex tissue-level constructions can help in the precise manipulation of neural circuits and enhance the understanding of the neural plasticity development, circadian-clock coupling, and neuropathologies.

The application of μFDs in the developmental interactions of neurons, glia, and other cells of interrelating brain regions is still in its infancy. For example, AD-like pathological conditions are present only in a subset of neuronal networks grown within μFDs; similar methods can be used to investigate developing neuronal networks. Therefore, the flexibility of μFD design makes it compliant with isolated neurons in compartmentalized microdomains and 3D tissue constructs for probing several developmental questions [120].

4.4.6 Neurodegenerative Studies

The progressive loss of neural function, because of protein aggregation, oxidative stress, and misfolding in the CNS and the peripheral nervous system, leads to neurodegenerative diseases like PD, AD, and ALS. Till now, animal models are the gold standard for understanding the neurodegenerative disease pathology and physiology of cell behavior in vivo. Nonetheless, commonly, it is not possible to predict drug efficacy in humans using animal models owing to the species-specific variations between humans and animal models. Again, all animal models do not reflect human disease accurately [121–124]. Additionally, the global trend related to "the 3Rs" (reduction, refinement, and replacement of animal testing) is pushing the scientific community to find an alternative for animal testing. Among the alternatives, the 3D in vitro models eliminate the limitations of 2D cell culture but also offer limited similarity between animal models and human physiology as co-culturing of various cell types is not feasible. For instance, in order to decipher the highly organized microenvironment present in the human brain and spinal cord, we need to consider the physical compartmentalization of neuronal networks and vascular network. Differences in the ECM composition, special biochemicals for particular cell culture systems, and dynamic mechanical stresses (i.e., fluidic stresses developing from pressure gradients and physical strains because of muscular contraction) need to be included in the design. Figure 4.20 provides the potential of the in vitro models to help understand neurodegenerative disease. Moreover, the traditional 3D models are costly and have low throughput and the protocols are often irreproducible [123].

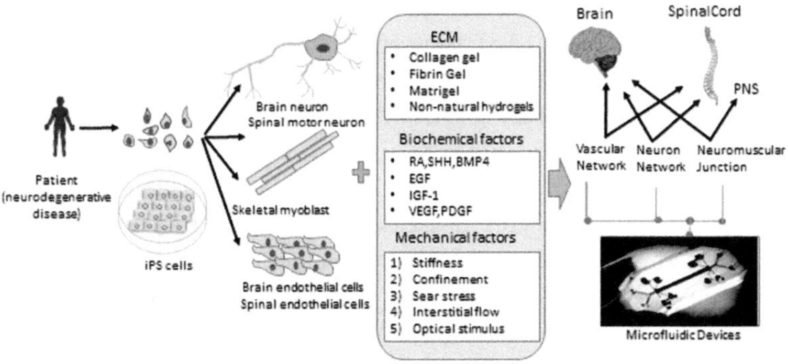

Figure 4.20 Potential of in vitro models to help understand neurodegenerative disease [124].

Therefore, as discussed earlier, μFDs, which are highly customizable, culture fewer cells in less cell culture medium in comparison to traditional cell culture, have low contamination risk, and provide competent high-throughput experimentation, lessen the cell population up to a few hundreds or even to the a level of single cell. This makes screen perturbations possible even at the individual cell level [124]. A microfluidic cell culture has the capacity to integrate with systematic biosensors with a culture platform for detecting physiological parameters, analyzing external stimuli in situ, and non-invasively probing cellular behavior [125]. For example, the induced pluripotent stem (iPS)-derived and embryonic stem cell–derived cells differentiate into mature neurons in approximately 1 month in a traditional cell culture, while fluidic devices use fewer cells, consume less reagents, have a lower contamination risk, and provide high-throughput experimentation along with allowing the capture perturbations at the level of individual cells. The experimental setup needs costly growth factors, like basic FGF, and activation of necessary requirements to differentiate and maintain iPS-derived mature cells, along with increasing the spatial and temporal resolution, in lower quantities [124]. In recent times, 3D microenvironment reconstruction of stem cell differentiation (iPS cells and NSCs) led to the development of neurodegenerative disease models. For example, a culture of neurons differentiated from iPS cells derived from fibroblasts of a familial AD patient generated an AD model. Pathological events of AD were reviewed by employing amyloid-β (Aβ) expressing neuronal cells, attained after differentiation of AD mutation NSCs [126]. Pathogenesis of ALS was studied the by co-culturing astrocyte neurons; genetically modified astrocytes overexpress either a human wild-type (WT) or a super oxide dismutase enzyme 1 (SOD1) mutation [127]. Overexpression of SOD-WT in astrocytes decreased oxidative stress on cortical neurons, those in close metabolic contact. On the other hand, after severe glutamate treatment, cortical neurons metabolically related to SOD-mutant astrocytes diminished protein expression of synapsin, which indicates that the SOD-mutant enzyme has potential toxic effects, mimicking ALS.

To decipher the pathogenesis mechanisms in neurodegenerative diseases, for example, AD, PD, and MN disorders (ALS), in vivo–like in vitro systems are required. Therefore, vascular network and neuronal network assembly into a neurovascular unit in the brain or neuromuscular junctions is necessary to mimic the brain and the spinal cord. Emerging studies state that the BBB is involved in the neurodegenerative disease pathogenesis, that is, AD, PD, and Huntington's disease (HD), and also works as the primary determinant for drug delivery to the brain [128–130]. Therefore, various attempts have been made to model the BBB: (a) 2D monolayer–based BBB model, (b) a lumen-based BBB model, and (c) lumen-based BBB model with neural cells. The detailed structures are presented in Fig. 4.21.

After the failure of the amyloid hypothesis for AD cure at clinical levels, various attempts have been made in order to completely understand the complex AD pathogenesis mechanisms and therapeutic interventions. Choi et al. studied the neurotoxicity of Aβ oligomeric assemblies by creating a gradient of Aβ oligomeric assemblies within microchannels on a microfluidic platform [133]. Park et al. from the same group developed a 3D neurosphere AD model in a microfluidic system, which closely mimics the brain tissue, as shown in Fig. 4.22a. Aβ neurotoxicity was studied on 3D neurospheres via an osmotic pump described earlier [134]. Deleglise et al. investigated the neuronal degeneration pattern with a Aβ peptide by employing the somato-dendritic compartments [135]. Soma dendrites and axons were isolated in each separate

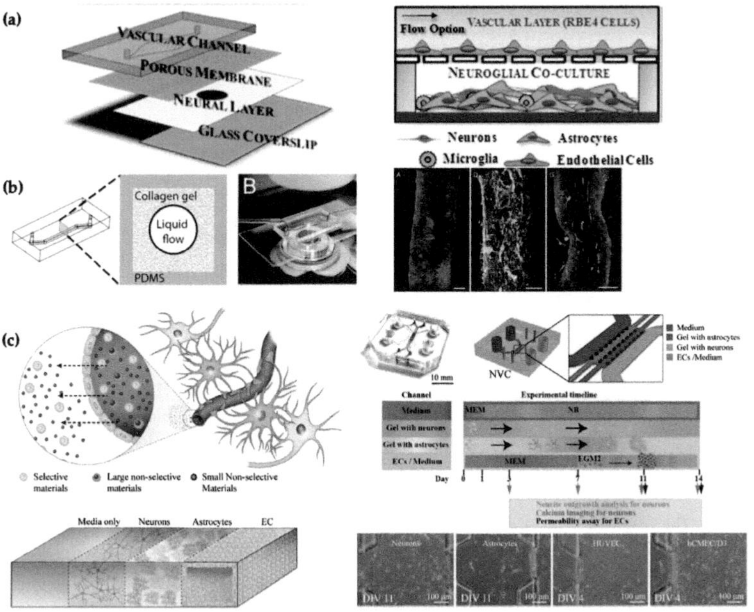

Figure 4.21 In vitro models of the BBB. (a) 2D monolayer–based BBB model. Republished with permission of The Royal Society of Chemistry, from Ref. [130], copyright (2017); permission conveyed through Copyright Clearance Center, Inc. (b) Lumen-shaped BBB model; Reproduced from Ref. [131] under a Creative Commons Attribution 4.0 International (CC BY 4.0) license. (c) Lumen-shaped BBB model with neural cells. Republished with permission of The Royal Society of Chemistry, from Ref. [132], copyright (2017); permission conveyed through Copyright Clearance Center, Inc.

fluidic chamber, which allows separate Aβ peptide treatment on each cell type. The results showed the connection of soma dendrites with Aβ peptide–induced synaptic loss, but not with axons. Song et al. also investigated the role of the neuronal network in Aβ transmission, contributing to the plaque spreading in neurodegeneration [136], as shown in Fig. 4.22b.

Cho et al. introduced a microfluidic system to demonstrate the migration and accumulation of microglia under an Aβ gradient, as shown in Fig. 4.22c [137]. Initially, the gradients of soluble Aβ alters the microglial morphology, followed by favoring migration toward the soluble Aβ–filled chamber. Later, this group conducted the

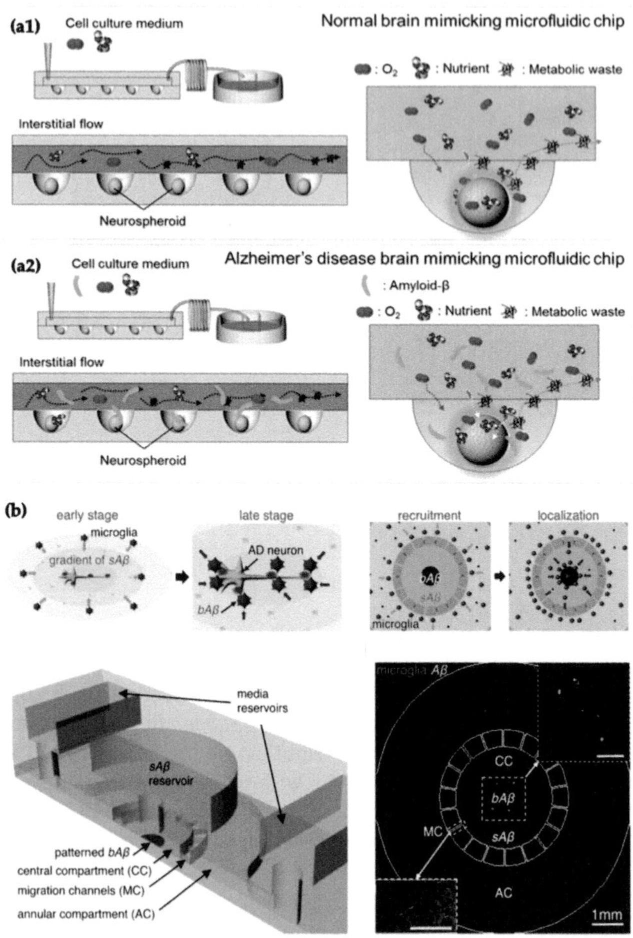

Figure 4.22 In vitro model of Alzheimer's disease. (a1) Schematic diagrams of a normal brain-mimicking microfluidic chip and (a2) an Alzheimer's disease brain-mimicking microfluidic chip; both are under dynamic conditions with a flow of normal medium containing oxygen and nutrients without and with 5 μM synthetic amyloid-*β* (1–42) for 3 days, respectively, for normal brain and an AD model. Republished with permission of The Royal Society of Chemistry, from Ref. [134], copyright (2015); permission conveyed through Copyright Clearance Center, Inc. (b) Replication of the AD-A*β* environment in a microfluidic chemotaxis platform. Reprinted by permission from Springer Nature Customer Service Centre GmbH: Springer Nature, Scientific Reports, Ref. [137], copyright (2013).

experiments on Aβ fibril–coated surfaces. They found that the surface-patterned Aβ selectively decreased microglial motility. Taking the cue from the earlier experiments, Cho et al. showed the synergistic effect of both surface-patterned Aβ and the soluble Aβ on microglial accumulation. Bianco et al. shifted the paradigm toward the study of cellular interactions of astrocytes and neurons occurring during neuroinflammatory events developed inside a microfluidic system and investigated the mechanisms of AD pathogenesis [138]. They observed that astrocytes from the hippocampus (but not cortical) of the brain impose an unfavorable effect on neurons under exposure to Aβ fibrils. The study shows the role of astrocytes in modeling of AD pathogenesis.

Moving on to the other aspect of AD pathogenesis, BBB dysfunction plays an important role. Various well-controlled microfluidic studies are already documented in this aspect. The 3D microfluidic-system-based BBB model considers many BBB-related pathological events occurring in AD. For instance, changes in BBB permeability can be visually identified under the influence of treatment with Aβ- or AD-related chemokines. Along with that, complex microfluidics-based BBB models designed with endothelial cells with astrocytes and pericytes help in understanding the well-controlled cell-cell interactions via co-culturing of Aβ neurons. These systems monitor real-time physical and chemical events among cells.

Similarly, to date, few in vivo–like closely mimicking models are present for analyzing PD. Freundt et al. [139] demonstrated a primary neuron culture model in the presence of fibrils of α-Syn in μFDs equipped with microgrooves. Lewy bodies get spread to the various parts by α-Syn fibril internalization by neurons, followed by transportation along the axons in a characteristic pattern. Therefore, to increase the resemblance of these models to the PD pathophysiology, α-Syn transport tracking can be integrated with compartmentalized microfluidic systems and cells isolated from a PD patient. The chambers for neuron patterning facilitate neural activity quantitatively for PD pathogenesis. The integration of microvalves to regulate fluid routing and reconstitution of paracrine signaling among neural cells could make these models more realistic for PD pathogenesis.

Till now, limited knowledge of the HD pathogenesis mechanism has constrained the development of a realistic HD model because cells in a Petri dish lack multiple critical features of neural cells developing in vivo for HD: (a) role of Htt in altering cellular function, (b) role of supporting brain cells (other than neurons) in neurodegenerative diseases. In spite of being a promising technology, no 3D microfluidic model exists for HD.

4.5 Future Prospects

The neuroscience community is determined to get a breakthrough related to the fundamentals of brain functionality and pathology. Many of these tasks require advanced microfluidic systems. These systems should be based on biologically and biomechanically compatible materials, multimodal or multifunctional modules, and switchable neuronal function in freely moving animals and should have high spatiotemporal resolution via wireless or optical controls. An innovative microfluidic interface offers tremendous possibilities for in vivo pharmacology, chemogenetics, optogenetics, and the rather new field of optopharmacology (optical control of molecules for stimulation or inhibition in a particular cell with high temporal and spatial precision). The highly localized pharmacological infusion for targeted therapies without affecting surrounding tissues may prove to be invaluable for clinical theranostics for brain tumors, neural injury, and many neurodegenerative diseases (PD, AD, Creutzfeldt–Jakob disease, etc.). Therefore, beyond fundamental research, the advancement in the reconstruction of the human microenvironmental systems using human-derived cells and novel artificial intelligence analysis methods will reduce the gap between preclinical and clinical research via in vitro tools. The high-throughput screening of brain pathologies with nervous system on a chip would be helpful in addressing the underlying mechanisms and facilitating the advancement of therapies. Till now, many studies have been carried out to investigate MN degeneration, but the entire degeneration pathogenesis has still not been deciphered. The future scope of this research will lie in understanding the interactions of vascular networks and MNs (along with skeletal myotubes). The combination of a high-throughput μFD with brain organoids could offer novel ways for basic development studies and drug screening and tissue engineering applications [140].

The capabilities of advanced lab-on-a-chip (LOC) devices can be predicted in the fields of optogenetics [141] to monitor and manipulate neuron-level events in neuroscience. In the future, LOCs involving on-chip light microscopes for advanced optogenetics are expected to be designed. Then the system can operate using tiny animals, such as worms, for transfection, transduction, and optical manipulation. Connectomics is one more advanced field that could benefit from LOCs. It includes management of data produced via synaptic neuron circuits and network connectivities to comprehend forms and functions of the brain [142].

Single-cell genetic manipulation is estimated substantively in the field of system neuroscience. A study by Rancz et al. has paved the way for providing a biophysical profile of a cell prior to some explicit genetic perturbation. This combinatorial scheme apprehends the synaptic and anatomical receptive field of a neuron, and the LOC approach offers an improved technique to perform the same activity; integrating and automating would be comparatively easier [143].

Future neuroscience LOCs would also be useful for following applications but not restricted to them:

- LOCs can be modularized with standard intermodule interfacing, leading to the creation of novel instruments offering multitasking in an automated manner. It may also be possible to create biodegradable chips capable of emulating real animal brains by patterning neurons and glial cells on them. Besides, in vitro brains along with in silico brains (artificial neural networks implemented on high-performance computing devices), or in vitro–in silico brains, can act as alternatives to animal models.
- Commercialization of the disposable LOCs will help neuroscientists to order LOCs from retail websites as per need to carry out experiments. Neuroscientists will get an opportunity to switch and upgrade instrumentation. For example, in the future, it may be feasible to extend and integrate highly sophisticated digital holographic microscopy (DHM) with

LOCs. The upgraded DHM would not need labeling dyes and microelectrodes for anatomical and physiological study of neurons in culture. It would use the distortion of laser directed toward the object to determine the shape of the neurons [23]. With the integration of the haptics [144] and virtual reality [145] into neuroscience with LOCs, we could observe configurations and dynamics of neuron clusters in holographs. This advancement will help researchers to get the sense of touching neurons while observing the response evoked by mechanical and optical stimuli, both electrophysiological and morphological.

Owing to the formation of random connections in promiscuous neurospheres, specific spatial signal detection is still a challenge in 3D neural network signal detection, which brings forward the need of 3D MEA development.

Coming to the gradient of neurotrophic factors in microfluidic chambers, future experiments could be undertaken to maintain constant gradients via continuous flow of media and chemoattractants for longer periods by employing external pumps. Apart from known guidance cues, such as netrin-1, brain pulp, and slit-2, new guidance molecules, factors affecting growth cone morphology, or axon guiding genes could also be studied using microfluidic gradient channels.

Approximately 90% of single-cell trapping has been achieved in a "deterministic" manner on μFDs; therefore, the device offers great potential to conduct advanced cell studies at the single-cell level, such as metabolism, drug toxicity, pharmacokinetics, shear stress activation, chemical signaling pathway activation, and inhibition.

4.6 Summary

In the current chapter, we demonstrated the studies of neuroscience or neuroengineering with emphasis on μFDs, which are effective tools in therapeutics and diagnostics. With the emergence of μFDs, neuroscientists have gained unprecedented success in manipulating the microenvironments surrounding neurons and therefore monitoring the structural and functional changes in cells. The unmatched controlling capacity of the local physical, chemical, and fluidic environments, while accessing and probing microdomains of the developing neurons, offers extraordinary prospects in various fields. It indicates the implications of microfluidics in decoding neuronal polarity, exploring axonal guidance and dendritic function, and understanding the developmental processes shaping the nervous system. Single-cell functions in a neural network are also outlined in this chapter. Microfluidics has been demonstrated to be a significant, perhaps even revolutionary, platform for the field of neurobiology. It has the capacity to produce data in ways that are not possible for conventional technologies. LOC-based studies for neuroscience were discussed, along with their features and future. Current LOC prototypes, when they can be delivered as commercial off-the-shelf products, have the potential to become essential systems for integrated molecular and cellular neuroscience research in the near future. The brain on a chip using human cells can potentially predict efficacy and off-target toxicity of drugs during preclinical drug development studies.

However, in order to widen the practical applicability of microfluidics in neurobiology, there is still a great deal of work to be done on μFDs, which are still in their infancy. Though current applications of microfluidics in neurobiology are still facing various problems, μFDs offer huge prospects to provide superior performance by integrating multifunctional modules in a single device, speed automation, and high throughput and efficiently contend with laboratory-scale technologies in neurobiology research.

References

1. National Science Foundation (n.d.). https://www.nsf.gov/ (accessed May 30, 2019).
2. Investigative Human Neuroscience, Brain Initiative (n.d.). https://braininitiative.nih.gov/brain-programs/investigative-human-neuroscience (accessed May 30, 2019).
3. G. M. Whitesides (2006). The origins and the future of microfluidics. *Nature*, **442**:368–373; doi:10.1038/nature05058.

4. E. E. Coyle, M. Oelgemöller (2008). Micro-photochemistry: photochemistry in microstructured reactors. The new photochemistry of the future? *Photochem Photobiol Sci*, **7**:1313; doi:10.1039/b808778d.
5. K. Jahnisch, V. Hessel, H. Löwe, M. Baerns (2004). Chemistry in microstructured reactors. *Angew Chem Int Ed*, **43**:406–446; doi:10.1002/anie.200300577.
6. C. C. Lee, G. Sui, A. Elizarov, C. J. Shu, Y.-S. Shin, A. N. Dooley, J. Huang, A. Daridon, P. Wyatt, D. Stout, H. C. Kolb, O. N. Witte, N. Satyamurthy, J. R. Heath, M. E. Phelps, S. R. Quake, H. R. Tseng (2005). Multistep synthesis of a radiolabeled imaging probe using integrated microfluidics. *Science (80)*, **310**:1793–1796; doi:10.1126/science.1118919.
7. Y. Kikutani, T. Kitamori (2004). Micro-flow reaction systems for combinatorial syntheses. *Macromol Rapid Commun*, **25**:158–168; doi:10.1002/marc.200300192.
8. E. K. Sackmann, A. L. Fulton, D. J. Beebe (2014). The present and future role of microfluidics in biomedical research. *Nature*, **507**:181–189; doi:10.1038/nature13118.
9. C. M. B. Ho, S. H. Ng, K. H. H. Li, Y.-J. Yoon (2015). 3D printed microfluidics for biological applications. *Lab Chip*, **15**:3627–3637; doi:10.1039/c5lc00685f.
10. D. Holmes, S. Gawad (2010). The application of microfluidics in biology. *Methods Mol Biol*, **583**:55–80; doi:10.1007/978-1-60327-106-6_2.
11. W. C. Tian, E. Finehout (2009). Microfluidic systems for cellular applications. In *Microfluidics for Biological Applications*, Springer US, Boston, MA, pp. 185–221; doi:10.1007/978-0-387-09480-9_6.
12. W.-C. Tian, E. Finehout (2009). Microfluidic systems for protein separations. In *Microfluidics for Biological Applications*, Springer US, Boston, MA, pp. 165–184; doi:10.1007/978-0-387-09480-9_5.
13. T. S. Santra, F. G. Tseng (2021). *Handbook of Single Cell Technology*, Springer Nature, Singapore, ISBN 978-981-10-8952-7.
14. P. Shinde, L. Mohan, A. Kumar, K. Dey, A. Maddi, A. N. Patananan, F.-G. Tseng, H.-Y. Chang, M. Nagai, T. S. Santra (2018). Current trends of microfluidic single-cell technologies. *Int J Mol Sci*, **19**:3143; doi:10.3390/ijms19103143.
15. W. C. Tian, E. Finehout (2009). Microfluidic systems for engineering vascularized tissue constructs. In *Microfluidics for Biological Applications*, Springer US, Boston, MA, pp. 223–240; doi:10.1007/978-0-387-09480-9_7.

16. W. C. Tian, E. Finehout (2009). Introduction to microfluidics. In *Microfluidics for Biological Applications*, Springer US, Boston, MA, pp. 1–34; doi:10.1007/978-0-387-09480-9_1.
17. F. G. Tseng, T. S. Tuhin (2015). *Micro/Nanofluidic Devices for Single Cell Analysis*, European Networks, MDPI AG.
18. F. G. Tseng, T. S. Santra (2016). *Single Cell Analysis in Biotechnology and Systems Biology*, European Networks, MDPI AG.
19. F. G. Tseng, T. S. Santra (2016). *Essentials of Single-Cell Analysis: Concepts, Applications and Future Prospects*, Springer-Verlag, Berlin Heidelberg, Germany, ISBN 978-3-662-49116-4.
20. B. J. Kirby, T. J. Shepodd, E. F. Hasselbrink (2002). Voltage-addressable on/off microvalves for high-pressure microchip separations. *J Chromatogr A*, **979**(1–2):147–154; doi:10.1016/S0021-9673(02)01453-X.
21. J. Wang, L. Ren, L. Li, W. Liu, J. Zhou, W. Yu, D. Tong, S. Chen (2009). Microfluidics: a new cosset for neurobiology. *Lab Chip*, **9**:644–652; doi:10.1039/b813495b.
22. M. O. Heuschkel, L. J. Guérin, B. Buisson, D. Bertrand, P. Renaud (1998). Buried microchannels in photopolymer for delivering of solutions to neurons in a network. *Sens Actuators, B*, **48**(1–3):356–361; doi:10.1016/S0925-4005(98)00071-9.
23. A. K. Soe, S. Nahavandi, K. Khoshmanesh (2012). Neuroscience goes on a chip. *Biosens Bioelectron*, **35**:1–13; doi:10.1016/j.bios.2012.02.012.
24. T. Osaki, Y. Shin, V. Sivathanu, M. Campisi, R. D. Kamm (2018). In vitro microfluidic models for neurodegenerative disorders. *Adv Healthc Mater*, **7**:1–29; doi:10.1002/adhm.201700489.
25. L. J. Millet, M. U. Gillette (2012). New perspectives on neuronal development via microfluidic environments. *Trends Neurosci*, **35**:752–61; doi:10.1016/j.tins.2012.09.001.
26. M. Badiola-Mateos, A. Hervera, J. A. del Río, J. Samitier (2018). Challenges and future prospects on 3D in-vitro modeling of the neuromuscular circuit. *Front Bioeng Biotechnol*, **6**:194; doi:10.3389/fbioe.2018.00194.
27. M. Badiola-Mateos, A. Hervera, J. A. del Río, J. Samitier (2018). Challenges and future prospects on 3D in-vitro modeling of the neuromuscular circuit, *Front Bioeng Biotechnol*, **6**:1–9; doi:10.3389/fbioe.2018.00194.
28. R. B. Campenot (1977). Local control of neurite development by nerve growth factor. *Proc Natl Acad Sci*, **74**(10):4516–4519; doi:10.1073/pnas.74.10.4516.

29. L. J. Millet, M. U. Gillette (2012). Over a century of neuron culture: from the hanging drop to microfluidic devices. *Yale J Biol Med*, **85**:501–521.
30. M. U. G. Larry, J. Millet (2013). New perspectives on neuronal development via microfluidic environments. *Trends Neurosci*, **35**:752–761; doi:10.1016/j.tins.2012.09.001.New.
31. H. Chou, C. Spence, A. Fu, A. Scherer, S. Quake (1998). Disposable microdevices for DNA analysis and cell sorting technology. *Proc. Solid state Sensor and Actual Workshop*, Hilton Head, SC, June 8–11, pp. 11–14.
32. G. W. Bishop (2016). 3D printed microfluidic devices. *Microfluid Biol Fundam Appl*, 103–113; doi:10.1007/978-3-319-40036-5_4.
33. A. M. Taylor, S. W. Rhee, C. H. Tu, D. H. Cribbs, C. W. Cotman, N. L. Jeon (2003). Microfluidic multicompartment device for neuroscience research. *Langmuir*, **19**:1551–1556; doi:10.1021/la026417v.
34. J. W. Park, B. Vahidi, A. M. Taylor, S. W. Rhee, N. L. Jeon (2006). Microfluidic culture platform for neuroscience research. *Nat Protoc*, **1**:2128–2136; doi:10.1038/nprot.2006.316.
35. E. Neto, L. Leitão, D. M. Sousa, C. J. Alves, I. S. Alencastre, P. Aguiar, M. Lamghari (2016). Compartmentalized microfluidic platforms: the unrivaled breakthrough of in vitro tools for neurobiological research. *J Neurosci*, **36**:11573–11584; doi:10.1523/jneurosci.1748-16.2016.
36. J. Park, S. Kim, S. I. Park, Y. Choe, J. Li, A. Han (2014). A microchip for quantitative analysis of CNS axon growth under localized biomolecular treatments. *J Neurosci Methods*, **221**:166–174; doi:10.1016/j.jneumeth.2013.09.018.
37. (a) J. Park, H. Koito, J. Li, A. Han (2012). Multi-compartment neuron-glia co-culture platform for localized CNS axon-glia interaction study. *Lab Chip*, **12**:3296–3304; doi:10.1039/c2lc40303j. (b) J. Park, H. Koito, J. Li, et al. (2009). Microfluidic compartmentalized co-culture platform for CNS axon myelination research. *Biomed. Microdevices*, **11**(6).
38. A. M. Taylor, M. Blurton-jones, S. W. Rhee, D. H. Cribbs, W. Carl (2006). A microfluidic culture platform for CNS axonal injury, regeneration and transport. *Nat Methods*, **2**:599–605; doi:10.1038/nmeth777.
39. G. Yang, C. Sau, W. Lai, J. Cichon, W. Li (2015). Sleep promotes branch-specific formation of dendritic spines after learning. *HHS Public Access*, **344**:1173–1178; doi:10.1126/science.1249098.
40. A. Coquinco, L. Kojic, W. Wen, Y. T. Wang, N. L. Jeon, A. J. Milnerwood, M. Cynader (2014). A microfluidic based in vitro model of synaptic competition. *Mol Cell Neurosci*, **60**:43–52; doi:10.1016/j.mcn.2014.03.001.

41. K. A. Southam, A. E. King, C. A. Blizzard, G. H. McCormack, T. C. Dickson (2013). Microfluidic primary culture model of the lower motor neuron-neuromuscular junction circuit. *J Neurosci Methods*, **218**:164–169; doi:10.1016/j.jneumeth.2013.06.002.

42. X. Lu, J. S. Kim-Han, K. L. O'Malley, S. E. Sakiyama-Elbert (2012). A microdevice platform for visualizing mitochondrial transport in aligned dopaminergic axons. *J Neurosci Methods*, **209**:35–39; doi:10.1016/j.jneumeth.2012.05.021.

43. B. E. Kerman, H. J. Kim, K. Padmanabhan, A. Mei, S. Georges, M. S. Joens, J. A. J. Fitzpatrick, R. Jappelli, K. J. Chandross, P. August, F. H. Gage (2015). In vitro myelin formation using embryonic stem cells. *Development*, **142**:2213–2225; doi:10.1242/dev.116517.

44. J. Park, H. Koito, J. Li, A. Han (2009). A multi-compartment CNS neuron-glia co-culture microfluidic platform. *J Vis Exp*, 6–8; doi:10.3791/1399.

45. A. Takeuchi, S. Nakafutami, H. Tani, M. Mori, Y. Takayama, H. Moriguchi, K. Kotani, K. Miwa, J. K. Lee, M. Noshiro, Y. Jimbo (2011). Device for co-culture of sympathetic neurons and cardiomyocytes using microfabrication. *Lab Chip*, **11**:2268–2275; doi:10.1039/c0lc00327a.

46. A. Takeuchi, K. Shimba, M. Mori, Y. Takayama, H. Moriguchi, K. Kotani, J. K. Lee, M. Noshiro, Y. Jimbo (2012). Sympathetic neurons modulate the beat rate of pluripotent cell-derived cardiomyocytes in vitro. *Integr Biol (United Kingdom)*, **4**:1532–1539; doi:10.1039/c2ib20060k.

47. E. Neto, C. J. Alves, D. M. Sousa, I. S. Alencastre, A. H. Lourenço, L. Leitão, H. R. Ryu, N. L. Jeon, R. Fernandes, P. Aguiar, R. D. Almeida, M. Lamghari (2014). Sensory neurons and osteoblasts: close partners in a microfluidic platform. *Integr Biol (United Kingdom)*, **6**:586–595; doi:10.1039/c4ib00035h.

48. P. Pagella, E. Neto, L. Jiménez-Rojo, M. Lamghari, T. A. Mitsiadis (2014). Microfluidics co-culture systems for studying tooth innervation. *Front Physiol*, **5**:1–8; doi:10.3389/fphys.2014.00326.

49. E. E. Zahavi, A. Ionescu, S. Gluska, T. Gradus, K. Ben-Yaakov, E. Perlson (2015). A compartmentalized microfluidic neuromuscular co-culture system reveals spatial aspects of GDNF functions. *J Cell Sci*, **128**:1241–1252; doi:10.1242/jcs.167544.

50. N. D. Dinh, Y. Y. Chiang, H. Hardelauf, J. Baumann, E. Jackson, S. Waide, J. Sisnaiske, J. P. Frimat, C. Van Thriel, D. Janasek, J. M. Peyrin, J. West (2013). Microfluidic construction of minimalistic neuronal co-cultures. *Lab Chip*, **13**:1402–1412; doi:10.1039/c3lc41224e.

51. W.-H. Tan, S. Takeuchi (2007). A trap-and-release integrated microfluidic system for dynamic microarray applications. *Proc Natl Acad Sci*, **104**(4):1146–1151; doi:10.1073/pnas.0606625104.
52. C. Ionescu-Zanetti, R. M. Shaw, J. Seo, Y.-N. Jan, L. Y. Jan, L. P. Lee (2005). Mammalian electrophysiology on a microfluidic platform. *Proc Natl Acad Sci*, **102**(26):9112–9117; doi:10.1073/pnas.0503418102.
53. E. Moonen, R. Luttge, J. P. Frimat (2018). Single cell trapping by capillary pumping using NOA81 replica moulded stencils. *Microelectron Eng*, **197**:1–7; doi:10.1016/j.mee.2018.04.010.
54. A. L. Nekimken, H. Fehlauer, A. A. Kim, S. N. Manosalvas-Kjono, P. Ladpli, F. Memon, D. Gopisetty, V. Sanchez, M. B. Goodman, B. L. Pruitt, M. Krieg (2017). Pneumatic stimulation of C. elegans mechanoreceptor neurons in a microfluidic trap. *Lab Chip*, **17**(6):1116–1127; doi:10.1039/c6lc01165a.
55. H. Kim, I. K. Lee, K. Taylor, K. Richters, D. H. Baek, J. H. Ryu, S. J. Cho, Y. H. Jung, D. W. Park, J. Novello, J. Bong, A. J. Suminski, A. M. Dingle, R. H. Blick, J. C. Williams, E. W. Dent, Z. Ma (2018). Single-neuronal cell culture and monitoring platform using a fully transparent microfluidic DEP device. *Sci Rep*, **8**:13194; doi:10.1038/s41598-018-31576-2.
56. T. Konry, S. Sarkar, P. Sabhachandani, N. Cohen (2016). Innovative tools and technology for analysis of single cells and cell–cell interaction. *Annu Rev Biomed Eng*, **18**:259–284; doi:10.1146/annurev-bioeng-090215-112735.
57. J. H. Marshel, T. Mori, K. J. Nielsen, E. M. Callaway (2010). Targeting single neuronal networks for gene expression and cell labeling in vivo. *Neuron*, **67**(4):562–574; doi:10.1016/j.neuron.2010.08.001.
58. A. Gross, J. Schoendube, S. Zimmermann, M. Steeb, R. Zengerle, P. Koltay (2015). Technologies for single-cell isolation. *Int J Mol Sci*, **16**(8):16897–16919; doi:10.3390/ijms160816897.
59. S. Bang, S. R. Lee, J. Ko, K. Son, D. Tahk, J. Ahn, C. Im, N. L. Jeon (2017). A low permeability microfluidic blood-brain barrier platform with direct contact between perfusable vascular network and astrocytes. *Sci Rep*, **7**:1–10; doi:10.1038/s41598-017-07416-0.
60. M. Shi, D. Majumdar, Y. Gao, B. Brewer, C. R. Goodwin, J. A. Mclean, D. Li, D. J. Webb (2014). Glia co-culture with neurons in microfluidic platforms promotes the formation and stabilization of synaptic contacts. *Lab Chip*, **13**:3008–3021; doi:10.1039/c3lc50249j.Glia.
61. E. Claverol-Tinturé, M. Ghirardi, F. Fiumara, X. Rosell, J. Cabestany (2005). Multielectrode arrays with elastomeric microstructured

overlays for extracellular recordings from patterned neurons. *J Neural Eng*, **2**(2):L1–7; doi:10.1088/1741-2560/2/2/L01.

62. E. Claverol-Tinturé, J. Cabestany, X. Rosell (2007). Multisite recording of extracellular potentials produced by microchannel-confined neurons in-vitro. *IEEE Trans Biomed Eng*, **54**:331–335; doi:10.1109/TBME.2006.880903.
63. E. Moutaux, B. Charlot, A. Genoux, F. Saudou, M. Cazorla (2018). An integrated microfluidic/microelectrode array for the study of activity-dependent intracellular dynamics in neuronal networks. *Lab Chip*, **18**:3425–3435; doi:10.1039/C8LC00694F.
64. F. Morin, N. Nishimura, L. Griscom, B. LePioufle, H. Fujita, Y. Takamura, E. Tamiya (2006). Constraining the connectivity of neuronal networks cultured on microelectrode arrays with microfluidic techniques: a step towards neuron-based functional chips. *Biosens Bioelectron*, **21**:1093–1100; doi:10.1016/j.bios.2005.04.020.
65. F. T. Jaber, F. H. Labeed, M. P. Hughes (2009). Action potential recording from dielectrophoretically positioned neurons inside micro-wells of a planar microelectrode array. *J Neurosci Methods*, **182**:225–235; doi:10.1016/j.jneumeth.2009.06.013.
66. M. Alberti, D. Snakenborg, J. M. Lopacinska, M. Dufva, J. P. Kutter (2010). Characterization of a patch-clamp microchannel array towards neuronal networks analysis. *Microfluid Nanofluidics*, **9**:963–972; doi:10.1007/s10404-010-0619-y.
67. W. Saadi, S. W. Rhee, F. Lin, B. Vahidi, B. G. Chung, N. L. Jeon (2007). Generation of stable concentration gradients in 2D and 3D environments using a microfluidic ladder chamber. *Biomed Microdevices*, **9**:627–635; doi:10.1007/s10544-007-9051-9.
68. B. Mosadegh, C. Huango, J. W. Park, H. S. Shin, B. G. Chung, S. K. Hwang, K. H. Lee, H. J. Kim, J. Brody, N. L. Jeon (2007). Generation of stable complex gradients across two-dimensional surfaces and three-dimensional gels. *Langmuir*, **23**:10910–10912; doi:10.1021/la7026835.
69. S. Kim, H. J. Kim, N. L. Jeon (2010). Biological applications of microfluidic gradient devices. *Integr Biol*, **2**(11–12):584–603; doi:10.1039/c0ib00055h.
70. A. Kunze, A. Bertsch, M. Giugliano, P. Renaud (2009). Microfluidic hydrogel layers with multiple gradients to stimulate and perfuse three-dimensional neuronal cell cultures. *Procedia Chem*, **1**:369–372; doi:10.1016/j.proche.2009.07.092.

71. M. R. Lee, K. W. Kwon, H. Jung, H. N. Kim, K. Y. Suh, K. Kim, K. S. Kim (2010). Direct differentiation of human embryonic stem cells into selective neurons on nanoscale ridge/groove pattern arrays. *Biomaterials*, **31**:4360–4366; doi:10.1016/j.biomaterials.2010.02.012.
72. A. Béduer, C. Vieu, F. Arnauduc, J. C. Sol, I. Loubinoux, L. Vaysse (2012). Engineering of adult human neural stem cells differentiation through surface micropatterning. *Biomaterials*, **33**:504–514; doi:10.1016/j.biomaterials.2011.09.073.
73. A. Béduer, L. Vaysse, E. Flahaut, F. Seichepine, I. Loubinoux, C. Vieu (2011). Multi-scale engineering for neuronal cell growth and differentiation. *Microelectron Eng*, **88**:1668–1671; doi:10.1016/j.mee.2010.12.049.
74. J.-M. Peyrin, B. Deleglise, L. Saias, M. Vignes, P. Gougis, S. Magnifico, S. Betuing, M. Pietri, J. Caboche, P. Vanhoutte, J.-L. Viovy, B. Brugg (2011). Axon diodes for the reconstruction of oriented neuronal networks in microfluidic chambers. *Lab Chip*, **11**:3663–3673; doi:10.1039/c1lc20014c.
75. A. N. Hellman, B. Vahidi, H. J. Kim, W. Mismar, O. Steward, N. L. Jeon, V. Venugopalan (2010). Examination of axonal injury and regeneration in micropatterned neuronal culture using pulsed laser microbeam dissection. *Lab Chip*, **10**:2083–2092; doi:10.1039/b927153h.
76. A. Kunze, R. Meissner, P. Renaud (2010). Neurite guidance through 3D hydrogel layers in a microfluidic environment. In *Proceedings of the 14th International Conference on Miniaturized Systems for Chemistry and Life Sciences (MicroTAS 2010)*. **1**:187–189, ISBN:978-0-9798064-3-8.
77. H. J. Kim, J. W. Park, J. W. Park, J. H. Byun, B. Vahidi, S. W. Rhee, N. L. Jeon (2012). Integrated microfluidics platforms for investigating injury and regeneration of CNS axons. *Ann Biomed Eng*, **40**:1268–1276; doi:10.1007/s10439-012-0515-6.
78. L. J. Millet, M. E. Stewart, G. Nuzzo, M. U. Gillette (2010). Guiding neuron development with planar surface gradients of substrate cues deposited using microfluidic devices, *Lab Chip*, **10**(12):1525–1535; doi:10.1039/c001552k.
79. C. R. Kothapalli, E. Van Veen, S. De Valence, S. Chung, I. K. Zervantonakis, F. B. Gertler, R. D. Kamm (2011). A high-throughput microfluidic assay to study neurite response to growth factor gradients. *Lab Chip*, **11**:497–507; doi:10.1039/c0lc00240b.

80. N. Bhattacharjee, A. Folch (2017). Large-scale micro fluidic gradient arrays reveal axon guidance behaviors in hippocampal neurons. *Nat Publ Gr Microsyst Nanoeng*, **3**:17003; doi:10.1038/micronano.2017.3.

81. E. Biffi, F. Piraino, A. Pedrocchi, G. B. Fiore, G. Ferrigno, A. Redaelli, A. Menegon, M. Rasponi (2012). A microfluidic platform for controlled biochemical stimulation of twin neuronal networks. *Biomicrofluidics*, **6**:1–10; doi:10.1063/1.3699975.

82. P. G. Gross, E. P. Kartalov, A. Scherer, L. P. Weiner (2007). Applications of microfluidics for neuronal studies. *J Neurol Sci*, **252**:135–143; doi:10.1016/j.jns.2006.11.009.

83. G. Goyal, Y. Nam (2011). Neuronal micro-culture engineering by microchannel devices of cellular scale dimensions. *Biomed Eng Lett*, **1**:89–98; doi:10.1007/s13534-011-0014-y.

84. T. Takano, C. Xu, Y. Funahashi, T. Namba, K. Kaibuchi (2015). Neuronal polarization. *Development*, **142**:2088–2093; doi:10.1242/dev.114454.

85. M. Scholl, C. Sprössler, M. Denyer, M. Krause, K. Nakajima, A. Maelicke, W. Knoll, A. Offenhäusser (2000). Ordered networks of rat hippocampal neurons attached to silicon oxide surfaces. *J Neurosci Methods*, **104**:65–75; doi:1016/S0165-0270(00)00325-3.

86. D. A. Stenger, J. J. Hickman, K. E. Bateman, M. S. Ravenscroft, W. Ma, J. J. Pancrazio, K. Shaffer, A. E. Schaffner, D. H. Cribbs, C. W. Cotman (1998). Microlithographic determination of axonal/dendritic polarity in cultured hippocampal neurons. *J Neurosci Methods*, **82**:167–173; doi:10.1016/S0165-0270(98)00047-8.

87. A. K. Vogt, F. D. Stefani, A. Best, G. Nelles, A. Yasuda, W. Knoll, A. Offenhäusser (2004). Impact of micropatterned surfaces on neuronal polarity. *J Neurosci Methods*, **134**:191–198; doi:10.1016/j.jneumeth.2003.11.004.

88. A. K. Vogt, G. Wrobel, W. Meyer, W. Knoll, A. Offenhäusser (2005). Synaptic plasticity in micropatterned neuronal networks. *Biomaterials*, **26**:2549–2557; doi:10.1016/j.biomaterials.2004.07.031.

89. S. K. W. Dertinger, X. Jiang, Z. Li, V. N. Murthy, G. M. Whitesides (2002). Gradients of substrate-bound laminin orient axonal specification of neurons. *Proc Natl Acad Sci USA*, **99**:12542–12547; doi:10.1073/pnas.192457199.

90. L. J. Millet, M. E. Stewart, R. G. Nuzzo, M. U. Gillette (2010). Guiding neuron development with planar surface gradients of substrate cues deposited using microfluidic devices. *Lab Chip*, **10**:1525; doi:10.1039/c001552k.

91. L. Kam, W. Shain, J. N. Turner, R. Bizios (2001). Axonal outgrowth of hippocampal neurons on micro-scale networks of polylysine-conjugated laminin. *Biomaterials*, **22**:1049–1054; doi: 10.1016/S0142-9612(00)00352-5.
92. A. Tooker, E. Meng, J. Erickson, Y.-C. Tai, J. Pine (2004). Development of biocompatible parylene neurocages. In *International Conference of the IEEE Engineering in Medicine and Biology Society*, pp. 2542–2545; doi:10.1109/IEMBS.2004.1403731.
93. A. M. Taylor, M. Blurton-Jones, S. W. Rhee, D. H. Cribbs, C. W. Cotman, N. L. Jeon (2005). A microfluidic culture platform for CNS axonal injury, regeneration and transport. *Nat Methods*, **2**:599–605; doi:10.1038/nmeth777.
94. A. M. Taylor, S. W. Rhee, N. L. Jeon (2006). Microfluidic chambers for cell migration and neuroscience research. In *Microfluidic Techniques*, Humana Press, New Jersey, pp. 167–178; doi:10.1385/1-59259-997-4:167.
95. A. Tooker, J. Erickson, G. Chow, Y.-C. Tai, J. Pine (2006). Parylene neurocages for electrical stimulation on silicon and glass substrates. In *2006 International Conference of the IEEE Engineering in Medicine and Biology Society*, pp. 4322–4325; doi:10.1109/IEMBS.2006.260472.
96. L. J. Millet, M. E. Stewart, J. V. Sweedler, R. G. Nuzzo, M. U. Gillette (2007). Microfluidic devices for culturing primary mammalian neurons at low densities. *Lab Chip*, **7**:987–994; doi:10.1039/b705266a.
97. M. S. Cohen, C. B. Orth, H. J. Kim, N. L. Jeon, S. R. Jaffrey (2011). Neurotrophin-mediated dendrite-to-nucleus signaling revealed by microfluidic compartmentalization of dendrites. *Proc Natl Acad Sci*, **108**:11246–11251; doi:10.1073/pnas.1012401108.
98. E. Moutaux, W. Christaller, C. Scaramuzzino, A. Genoux, B. Charlot, M. Cazorla, F. Saudou (2018). Neuronal network maturation differently affects secretory vesicles and mitochondria transport in axons. *Sci Rep*, **8**:1–14; doi:10.1038/s41598-018-31759-x.
99. A. M. Taylor, D. C. Dieterich, H. T. Ito, S. A. Kim, E. M. Schuman (2010). Microfluidic local perfusion chambers for the visualization and manipulation of synapses. *Neuron*, **66**:57–68; doi:10.1016/j.neuron.2010.03.022.
100. A. M. Taylor, T. Nagendran, R. L. Bigler, R. S. Larsen, S. B. Frost, B. D. Philpot, R. J. Nudo, R. L. Bigler, S. B. Frost, B. D. Philpot, R. J. Nudo, A. M. Taylor (2017). Distal axotomy enhances retrograde presynaptic excitability onto injured pyramidal neurons via trans-synaptic signaling. *Nat Commun*, **8**:625; doi:10.1038/s41467-017-00652-y.

101. E. J. Botzolakis, A. Maheshwari, H. J. Feng, A. H. Lagrange, J. H. Shaver, N. J. Kassebaum, R. Venkataraman, F. Baudenbacher, R. L. Macdonald (2009). Achieving synaptically relevant pulses of neurotransmitter using PDMS microfluidics. *J Neurosci Methods*, **177**:294–302; doi:10.1016/j.jneumeth.2008.10.014.

102. A. Jain, M. U. Gillette (2015). Development of microfluidic devices for the manipulation of neuronal synapses. *Microfluid Compart Platforms Neurobiol Res*, **103**:127–137; doi:10.1007/978-1-4939-2510-0_7.

103. J. M. Peyrin, B. Deleglise, L. Saias, M. Vignes, P. Gougis, S. Magnifico, S. Betuing, M. Pietri, J. Caboche, P. Vanhoutte, J. L. Viovy, B. Brugg (2011). Axon diodes for the reconstruction of oriented neuronal networks in microfluidic chambers. *Lab Chip*, **11**:3663–3673; doi: 10.1039/c1lc20014c.

104. A. M. Kaufman, A. J. Milnerwood, M. D. Sepers, A. Coquinco, K. She, L. Wang, H. Lee, A. M. Craig, M. Cynader, L. A. Raymond (2012). Opposing roles of synaptic and extrasynaptic NMDA receptor signaling in cocultured striatal and cortical neurons. *J Neurosci*, **32**:3992–4003; doi:10.1523/jneurosci.4129-11.2012.

105. A. M. Taylor, D. C. Dieterich, H. T. Ito, S. A. Kim, E. M. Schuman (2010). Microfluidic local perfusion chambers for the visualization and manipulation of synapses. *Neuron*, **66**:57–68; doi:10.1016/j.neuron.2010.03.022.

106. M. J. Geden, S. E. Romero, M. Deshmukh (2019). Apoptosis versus axon pruning: molecular intersection of two distinct pathways for axon degeneration. *Neurosci Res*, **139**:3; doi:10.1016/J.NEURES.2018.11.007.

107. V. S. Van Laar, B. Arnold, S. B. Berman (2019). Primary embryonic rat cortical neuronal culture and chronic rotenone treatment in microfluidic culture devices. *Bio-Protoc*, **9**(6):pii.e3192; doi:10.21769/BioProtoc.3192.

108. S. Yokoyama, A. Otomo, S. Hadano, H. Kimura (2019). An open-type microdevice to improve the quality of fluorescence labeling for axonal transport analysis in neurons. *Biomicrofluidics*, **13**:034104; doi:10.1063/1.5090968.

109. D. S. Bassett, E. Bullmore (2006). Small-world brain networks. *Neuroscientist*, **12**:512–523; doi:10.1177/1073858406293182.

110. N. Gomez, S. Chen, C. E. Schmidt (2007). Polarization of hippocampal neurons with competitive surface stimuli: contact guidance cues are preferred over chemical ligands. *J R Soc Interface*, **4**:223–233; doi:10.1098/rsif.2006.0171.

111. O. C. Amadi, M. L. Steinhauser, Y. Nishi, S. Chung, R. D. Kamm, A. P. McMahon, R. T. Lee (2010). A low resistance microfluidic system for the creation of stable concentration gradients in a defined 3D microenvironment. *Biomed Microdevices*, **12**:1027–1041; doi:10.1007/s10544-010-9457-7.

112. J. Y. Park, S. K. Kim, D. H. Woo, E. J. Lee, J. H. Kim, S. H. Lee (2009). Differentiation of neural progenitor cells in a microfluidic chip-generated cytokine gradient. *Stem Cells*, **27**:2646–2654; doi:10.1002/stem.202.

113. C. J. Demers, P. Soundararajan, P. Chennampally, G. A. Cox, J. Briscoe, S. D. Collins, R. L. Smith (2016). Development-on-chip: *in vitro* neural tube patterning with a microfluidic device. *Development*, **143**:1884–1892; doi:10.1242/dev.126847.

114. C. Yang, X. Li, Q. Q. Li, Q. Q. Li, H. Li, B. Zhang, J. Lin, C. J. Demers, P. Soundararajan, P. Chennampally, G. A. Cox, J. Briscoe, S. D. Collins, R. L. Smith, M. Plass, J. Solana, F. Alexander Wolf, S. Ayoub, A. Misios, P. Glažar, B. Obermayer, F. J. Theis, C. Kocks, N. Rajewsky, I. K. Suzuki, D. Gacquer, R. Van Heurck, D. Kumar, M. Wojno, A. Bilheu, A. Herpoel, N. Lambert, J. Cheron, F. Polleux, V. Detours, P. Vanderhaeghen, M. Placzek, J. Briscoe (2018). Development-on-chip: *in vitro* neural tube patterning with a microfluidic device. *Development*, **143**:1884–1892; doi:10.1242/dev.126847.

115. C. D. Stern (2006). Neural induction: 10 years on since the "default model." *Curr Opin Cell Biol*, **18**:692–697; doi:10.1016/j.ceb.2006.09.002.

116. R. L. Smith, C. J. Demers, S. D. Collins (2010). Microfluidic device for the combinatorial application and maintenance of dynamically imposed diffusional gradients. *Microfluid Nanofluidics*, **9**:613–622; doi:10.1007/s10404-010-0574-7.

117. K. C. Kanning, A. Kaplan, C. E. Henderson (2010). Motor neuron diversity in development and disease. *Annu Rev Neurosci*, **33**:409–440; doi:10.1146/annurev.neuro.051508.135722.

118. A. Kunze, M. Giugliano, A. Valero, P. Renaud (2011). Biomaterials micropatterning neural cell cultures in 3D with a multi-layered scaffold. *Biomaterials*, **32**:2088–2098; doi:10.1016/j.biomaterials.2010.11.047.

119. Y. Berdichevsky, K. J. Staley, M. L. Yarmush (2010). Building and manipulating neural pathways with microfluidics. *Lab Chip*, **10**:999–1004; doi:10.1039/b922365g.

120. L. J. Millet, M. U. Gillette (2012). New perspectives on neuronal development via microfluidic environments. *Trends Neurosci*, **35**:752–761; doi:10.1016/j.tins.2012.09.001.
121. W. M. S. Russell, R. L. Burch (1959). *The Principles of Humane Experimental Technique*. Methuen, London; doi:10.1017/CBO97811074 15324.004.
122. L. Brichta, W. Shin, V. Jackson-Lewis, J. Blesa, E. L. Yap, Z. Walker, J. Zhang, J. P. Roussarie, M. J. Alvarez, A. Califano, S. Przedborski, P. Greengard (2015). Identification of neurodegenerative factors using translatome-regulatory network analysis. *Nat Neurosci*, **18**:1325–1333; doi:10.1038/nn.4070.
123. K. H. Nam, A. S. T. Smith, S. Lone, S. Kwon, D. H. Kim (2015). Biomimetic 3D tissue models for advanced high-throughput drug screening. *J Lab Autom*, **20**:201–215; doi:10.1177/2211068214557813.
124. S. I. Nishimura, M. Ueda, M. Sasai (2009). Cortical factor feedback model for cellular locomotion and cytofission. *PLoS Comput Biol*, **5**:e1000310; doi:10.1371/journal.pcbi.1000310.
125. Q. Liu, C. Wu, H. Cai, N. Hu, J. Zhou, P. Wang (2014). Cell-based biosensors and their application in biomedicine. *Chem Rev*, **114**:6423–6461; doi:10.1021/cr2003129.
126. C. Sh, K. Yh, H. M, S. C, L. S, D. C, C. H, H. B, A. C, M. J, K. Jb, Z. C, W. Bj, P. M, K. Dm, W. Cj, W. Sl, T. Re, K. Dy, S. H. Choi, Y. H. Kim, M. Hebisch, C. Sliwinski, S. Lee, C. D'Avanzo, H. Chen, B. Hooli, C. Asselin, J. Muffat, J. B. Klee, C. Zhang, B. J. Wainger, M. Peitz, D. M. Kovacs, C. J. Woolf, S. L. Wagner, R. E. Tanzi, D. Y. Kim (2014). A three-dimensional human neural cell culture model of Alzheimer's disease. *Nature*, **515**:274–278; doi:10.1038/nature13800.
127. A. Kunze, S. Lengacher, E. Dirren, P. Aebischer, P. J. Magistretti, P. Renaud (2013). Astrocyte-neuron co-culture on microchips based on the model of SOD mutation to mimic ALS. *Integr Biol (United Kingdom)*, **5**:964–975; doi:10.1039/c3ib40022k.
128. B. V. Zlokovic (2011). Neurovascular pathways to neurodegeneration in Alzheimer's disease and other disorders. *Nat Rev Neurosci*, **12**:723–738; doi:10.1038/nrn3114.
129. R. Kortekaas, K. L. Leenders, J. C. H. Van Oostrom, W. Vaalburg, J. Bart, A. T. M. Willemsen, N. H. Hendrikse (2005). Blood-brain barrier dysfunction in Parkinsonian midbrain in vivo. *Ann Neurol*, **57**:176–179; doi:10.1002/ana.20369.
130. A. K. H. Achyuta, A. J. Conway, R. B. Crouse, E. C. Bannister, R. N. Lee, C. P. Katnik, A. A. Behensky, J. Cuevas, S. S. Sundaram (2013). A modular

approach to create a neurovascular unit-on-a-chip. *Lab Chip*, **13**:542–553; doi:10.1039/c2lc41033h.

131. A. Herland, A. D. Van Der Meer, E. A. FitzGerald, T. E. Park, J. J. F. Sleeboom, D. E. Ingber (2016). Distinct contributions of astrocytes and pericytes to neuroinflammation identified in a 3D human blood-brain barrier on a chip. *PLoS One*, **11**:1–21; doi:10.1371/journal.pone.0150360.
132. G. Adriani, D. Ma, A. Pavesi, R. D. Kamm, E. L. K. Goh (2017). A 3D neurovascular microfluidic model consisting of neurons, astrocytes and cerebral endothelial cells as a blood-brain barrier. *Lab Chip*, **17**:448–459; doi:10.1039/c6lc00638h.
133. Y. J. Choi, S. Chae, J. H. Kim, K. F. Barald, J. Y. Park, S. H. Lee (2013). Neurotoxic amyloid beta oligomeric assemblies recreated in microfluidic platform with interstitial level of slow flow. *Sci Rep*, **3**:1921; doi:10.1038/srep01921.
134. J. Park, B. K. Lee, G. S. Jeong, J. K. Hyun, C. J. Lee, S. H. Lee (2015). Three-dimensional brain-on-a-chip with an interstitial level of flow and its application as an in vitro model of Alzheimer's disease. *Lab Chip*, **15**:141–150; doi:10.1039/c4lc00962b.
135. B. Deleglise, S. Magnifico, E. Duplus, P. Vaur, V. Soubeyre, M. Belle, M. Vignes, J. L. Viovy, E. Jacotot, J. M. Peyrin, B. Brugg (2014). β-amyloid induces a dying-back process and remote trans-synaptic alterations in a microfluidic-based reconstructed neuronal network. *Acta Neuropathol Commun*, **2**:145; doi:10.1186/s40478-014-0145-3.
136. H. L. Song, S. Shim, D. H. Kim, S. H. Won, S. Joo, S. Kim, N. L. Jeon, S. Y. Yoon (2014). β-Amyloid is transmitted via neuronal connections along axonal membranes. *Ann Neurol*, **75**:88–97; doi:10.1002/ana.24029.
137. H. Cho, T. Hashimoto, E. Wong, Y. Hori, L. B. Wood, L. Zhao, K. M. Haigis, B. T. Hyman, D. Irimia (2013). Microfluidic chemotaxis platform for differentiating the roles of soluble and bound amyloid-β on microglial accumulation. *Sci Rep*, **3**:1823; doi:10.1038/srep01823.
138. F. Bianco, N. Tonna, R. D. Lovchik, R. Mastrangelo, R. Morini, A. Ruiz, E. Delamarche, M. Matteoli (2012). Overflow microfluidic networks: application to the biochemical analysis of brain cell interactions in complex neuroinflammatory scenarios. *Anal Chem*, **84**:9833–9840; doi:10.1021/ac302094z.
139. E. C. Freundt, N. Maynard, E. K. Clancy, S. Roy, L. Bousset, Y. Sourigues, M. Covert, R. Melki, K. Kirkegaard, M. Brahic (2012). Neuron-to-neuron transmission of α-synuclein fibrils through axonal transport. *Ann Neurol*, **72**:517–524; doi:10.1002/ana.23747.

140. F. Yu, W. Hunziker, D. Choudhury (2019). Engineering microfluidic organoid-on-a-chip platforms. *Micromachines*, **10**:165; doi:10.3390/mi10030165.

141. A. R. Adamantidis, F. Zhang, A. M. Aravanis, K. Deisseroth, L. De Lecea (2007). Neural substrates of awakening probed with optogenetic control of hypocretin neurons. *Nature*, **450**:420–424; doi:10.1038/nature06310.

142. P. Hagmann, L. Cammoun, X. Gigandet, R. Meuli, C. J. Honey, J. Van Wedeen, O. Sporns (2008). Mapping the structural core of human cerebral cortex. *PLoS Biol*, **6**:e159; doi:10.1371/journal.pbio.0060159.

143. E. A. Rancz, K. M. Franks, M. K. Schwarz, B. Pichler, A. T. Schaefer, T. W. Margrie (2011). Transfection via whole-cell recording in vivo: bridging single-cell physiology, genetics and connectomics. *Nat Neurosci*, **14**:527–533; doi:10.1038/nn.2765.

144. M. O. Ernst, M. S. Banks (2002). Humans integrate visual and haptic information in a statistically optimal fashion. *Nature*, **415**:429–433; doi:10.1038/415429a.

145. S. Faumont, G. Rondeau, T. R. Thiele, K. J. Lawton, K. E. McCormick, M. Sottile, O. Griesbeck, E. S. Heckscher, W. M. Roberts, C. Q. Doe, S. R. Lockery (2011). An Image-Free Opto-Mechanical system for creating virtual environments and imaging neuronal activity in freely moving caenorhabditis elegans. *PLoS One*, **6**:e24666; doi:10.1371/journal.pone.0024666.

Chapter 5

Vascularized Microfluidic Organ on a Chip and Its Applications

Qiyue Sun,[a] **Jianghua Pei,**[a] **Qinyu Li,**[a] **and Xiaolin Wang**[a,b]
[a]*Department of Micro/Nano Electronics, Shanghai Jiao Tong University, Shanghai 200240, China*
[b]*Institute of Medical Robotics, Shanghai Jiao Tong University, Shanghai 200240, China*
xlwang83@sjtu.edu.cn

The microvasculature plays a critical role in the human physiology and is closely associated with various human diseases. By combining advanced microfluidic-based techniques, engineered 3D vascularized microtissue models provide a precise and reproducible platform to study the microvasculature in vitro, which is an essential and primary component for engineering versatile organs on chips and achieving greater biological relevance. In this chapter, we discuss current strategies for engineering microvessels in vitro. By closely simulating relevant factors found in vivo, such as the biomechanical, biochemical, and biological microenvironment, it is possible to create more accurate organ-specific models, including both healthy and pathological vascularized microtissues with their respective vascular barrier properties. We further discuss the integration of tumor cells/spheroids into the engineered microvasculature to model the vascularized tumor on a chip and its potential

Microfluidics and Bio-MEMS: Devices and Applications
Edited by Tuhin S. Santra

ISBN 978-981-4800-85-3 (Hardcover), 978-1-003-01493-5 (eBook)
www.jennystanford.com

applications in the study of cancer metastasis and anticancer drug screening. We conclude with our commentaries on current progress in and future perspective of on-chip vascularization techniques for fundamental and clinical/translational research.

5.1 Introduction

The circulatory system plays a vital role in maintaining homeostasis in the human body. It comprises a closed network of arteries, veins, and capillaries that allow the blood to circulate throughout the body, not only for waste product removal, but also for gas exchange and nutrient transportation, all of which are essential for organ viability. Besides participating in metabolic function, the microvasculature in different organ microenvironments has unique biological functions and physical properties, such as maintaining the solute and water balance between the blood and tissue compartments or responding to different deformations and stress fluctuations [1]. Recently, the concept of "organ on a chip" has been proposed to establish in vitro models that can mimic the microphysiological function and 3D microstructure of human organs more accurately and specifically compared to the traditional 2D cultures and animal models [2]. In addition to supplying nutrients and oxygen to the cultured tissue by perfusing the culture medium, vascularization of the organ on a chip can also contribute to the establishment of organ-specific microenvironments and microphysiological function by having selective barrier functions similar to the in vivo organ. In other words, to better mimic the characteristics and functions of specific human organs in vitro, it is necessary to integrate a perfusable and functional 3D microvasculature with different organ-on-a-chip systems. Microfluidic technologies have emerged as useful tools for the development of organs on chips, which can offer precise control over various aspects of the cellular microenvironment, such as a different profile of fluid flow, gradient of various growth factors, and mechanical properties of versatile biomaterials. All these advantages can help biomimic in vitro vascularized microtissue models.

Furthermore, besides the substantial physiological research that can be carried out after constructing and characterizing the microvasculature in homeostatic conditions in these systems, there is a great potential to model pathological conditions to study vascular-related diseases [3]. Especially for cancer biology, the tumor vasculature plays a critical role in several key events in the metastatic cascade, such as intravasation and extravasation. Engineered microvessels can be well suited to the study of mechanisms of tumor growth and metastasis, drug screening, and cancer therapies by establishing the vascularized microtumor models in vitro.

In this chapter, we focus on the generation of microvascular networks (μVNs) in 3D engineered tissue constructs and their integration into vascularized microtumor models by combining microfluidics, microfabrication, biomaterials, and tissue engineering technologies. We first discuss the current strategies for tissue vascularization. Next, we highlight relevant factors that induce vascularization inside microfluidic systems. We then provide a brief introduction to selective vascular barrier properties in different human organs and methods to construct and characterize these properties. Then, we review the current state-of-the-art in vitro vascularized tumor-on-a-chip models in various disease stages and their potential applications in anticancer drug screening. We conclude with our vision to improve the current approaches for creating vascularized microtissues. This review will provide a better understanding of the vascularization process for organ-on-a-chip systems and its applications in cancer research.

5.2 In vitro Vascularization Strategies

In many early studies to understand the basis of vascular biology, 2D models were constructed by plating endothelial cells (ECs) on a flat surface, such as a Petri dish [4], a porous membrane [5], or patterned hydrogel [6], to form a confluent monolayer to mimic the blood vessel wall. However, these 2D models cannot properly replicate the exact physical structure of blood vessels in vivo, and of note is their circular shape and polarized luminal/abluminal surfaces. It is known that the blood vessels in the human body have circular cross sections. To mimic the microvascular in vitro, it is necessary to replicate not only their microphysiological function but also their 3D microstructure (i.e., circular cross section). In addition, blood vessel polarization in the apical (luminal) to basal (abluminal) axis is important for directed secretion of proteins, which can induce cord hollowing by changing the cell shape to form a lumen [7]. Thus, it is challenging to use these 2D models to replicate luminal flow in vivo and study its effect on vasculature. Due to the advantages of highly regulated spatial and temporal control over cell patterning, chemical gradients, and mechanical stimuli, microfluidics-based techniques have been widely utilized to create 3D models in vitro [8, 9]. Besides that, microfluidic technologies also enable the continuous perfusion of the cell culture medium to supply the oxygen and nutrients as well as to remove waste, which is the necessary condition to realize the long-term survival of these vascularized microtissues. In addition, with the combination of versatile biomaterials and tissue engineering techniques, many 3D in vitro vascularization strategies have been developed, which can be broadly categorized into three main types: (i) EC lining–based methods, (ii) vasculogenesis- and angiogenesis-based methods, and (iii) hybrid methods.

5.2.1 EC Lining–Based Methods

Briefly, EC lining methods are achieved by allowing the ECs to form a confluent monolayer on the inner walls of the microfluidic channel. The major advantage of these methods is that the microvasculature geometry and dimensions can be easily controlled using different microfabrication techniques with high flexibility. Furthermore, the shear stress imposed on the microvessels can be precisely controlled and calculated on the basis of the channel dimensions and the applied flow rate. However, these methods cannot mimic vascular formation in vivo, which mainly relies on the natural process of vasculogenesis and angiogenesis to form the complex μVNs. Since it is difficult to evenly distribute a high density of ECs inside the microfluidic channels with small diameters, which may induce a microchannel blockage, EC lining methods are only suitable for constructing large blood vessels with diameters greater than 50 μm

[10]. The hollow microstructure (i.e., a single channel or a channel network) can be made of either hydrogel or polydimethylsiloxane (PDMS). For hydrogel-based microchannels, various micromolding methods reported in the literature so far can be grouped into three main types: (i) the microneedle-based removable method for a single microchannel, (ii) the micropatterned, planar hydrogel slab bonding method for a single-layer microchannel network, and (iii) the dissolvable material–based sacrificial micromolding method for a multilayer microchannel network. For an EC lining inside a PDMS-based microfluidic channel, since cells might not adhere tightly to the PDMS surface, a thin coating of basement membrane proteins (laminin, fibronectin, collagen IV, etc.) on the microchannel inner walls is needed to enhance cell adherence.

5.2.1.1 Microneedle-based removable method

The basic procedure of this method is to insert a cylindrical object (a microneedle, a wire, a capillary tube, etc.) into a hydrogel solution and remove it after the hydrogel is fully polymerized to create a single microchannel. Then ECs are seeded onto the inner surface of the microchannel to form a confluent EC monolayer [11–13]. Briefly, this method typically requires two steps: creating a microchannel and lining it with ECs. Furthermore, Sadr et al. proposed a new method to realize the rapid one-step engineering of microtubular constructs that combined self-assembled monolayer-based cell transfer and hydrogel photo-cross-linking techniques [14]. After ultraviolet cross-linking, the gold-sputtered rod modified with oligopeptide was removed from the polymerized hydrogel by applying an electrical stimulation, which would transfer the layer of human umbilical vein endothelial cells (HUVECs) to 3D geometrically defined vascular-like structures in hydrogels. Moreover, a bilayer vascular structure of smooth muscle cells (SMCs) and HUVECs could be formed by a sequential deposition of SMCs, fibronectin as the cell adhesive layer, and HUVECs, which exhibited a stronger barrier function than the EC monolayer [15]. However, since the cylindrical objects need to be removed from hydrogel, this method is limited to a simple, single, straight microvascular channel [12, 16].

5.2.1.2 Micropatterned planar hydrogel slab bonding method

Rather than a single microchannel, a single-layer microchannel network can also be fabricated by using the standard lithographic patterning method. Mu et al. developed an interconnected microvascular array by using the additive bonding between a micropatterned collagen gel slab and the liquid molding under the effect of fibrillogenesis (Fig. 5.1A) [17]. Briefly, they placed a gelled and structured hydrogel layer (HL) on hydrogel solution (HS) held by PDMS barriers (PB) at 4°C, and realized the bonding by increasing the temperature to 37°C as well as adding calcium chloride solution. Then, ECs were seeded in hydrogel microchannels to achieve the confluent monolayer with the EC lining method. Using this model, several vascular biology studies were performed, including vessel sprouting, interaction between ECs and mural cells, endothelial barrier function under different flow conditions, and EC response to various biochemical signals [17]. Compared with the microneedle-based removable method, although 2D planar networks can be easily formed, they usually have a rectangular cross section due to the inherent characteristics of templates produced by lithographic methods.

5.2.1.3 Dissolvable material–based sacrificial micromolding method

In this method, a multilayer microchannel network can be fabricated by dissolving or melting the dissolvable gel or solid material embedded in a gel matrix. With the development of 3D printing technology, more complex multilayer microchannel networks can be interconnected using a wide range of template materials, such as carbohydrate glass [18], Pluronic F127 [19], agarose [20], gelatin [21], sodium alginate [22], and synthetic poly(ethylene glycol) [23]. Miller et al. created a multilayer microvessel network from a carbohydrate glass template that can provide sufficient mechanical support in an open lattice (Fig. 5.1B) [18]. After dissolving the carbohydrate glass in the cell culture medium, the perfusable, hollow, cylindrical network structure was left inside fibrin gel. HUVECs were then seeded into these hollow microstructures to form a 3D vascular network. While this strategy has the potential to create a complex 3D μVN, it is time intensive and still limited to larger vessels determined by the resolution of the 3D printer.

5.2.1.4 EC lining inside a PDMS-based microfluidic channel

PDMS-based microfluidic chips enable a single-layer microchannel network design with diameters ranging from 60 μm to 200 μm, and different microenvironment factors (e.g., flow profile and shear stress) can be precisely controlled with the sophisticated microfluidic technologies [24]. In vivo, ECs are attached to a basement membrane 40–120 nm in thickness [25]. Therefore, for ECs to adhere to a PDMS-based microchannel, extracellular matrix (ECM) proteins like laminin, fibronectin, or collagen IV need to be deposited onto the inner wall of the microchannel prior to lining it with ECs (Fig. 5.1C-i) [26, 27]. Moreover, microchannels fabricated through soft lithography technology have an inherent rectangular or square cross section, which cannot closely mimic in vivo microvessels, with a circular cross section. However, Esch et al. showed that ECs could be grown within both square and semicircular microchannels [28]. It was found that shear force has a greater impact on these endothelial structures than geometry and the channel geometry only influences the final cross-sectional profile of the vascular lining.

To solve the problem of noncircular cross sections, Bischel et al. developed a "viscous finger patterning" method that took advantage of basic fluidic principles to create 3D lumens with circular cross sections inside a PDMS-based microchannel [29, 30]. Briefly, the culture medium was pumped into a partially polymerized hydrogel at the center of the microchannel, creating a circular channel at the center while filling the rectangular corners with hydrogel. Similarly, a circular PDMS microchannel can also be fabricated by introducing a pressurized air stream inside a partially solidified rectangular microchannel filled with liquid PDMS or liquid silicone oligomer and then baking the device to fully cure the coated layer [31, 32]. Furthermore, Zhang et al. developed a hollow PDMS tube with adjustable diameter and wall thickness to mimic the elastomeric free-form blood vessels with the EC lining method,

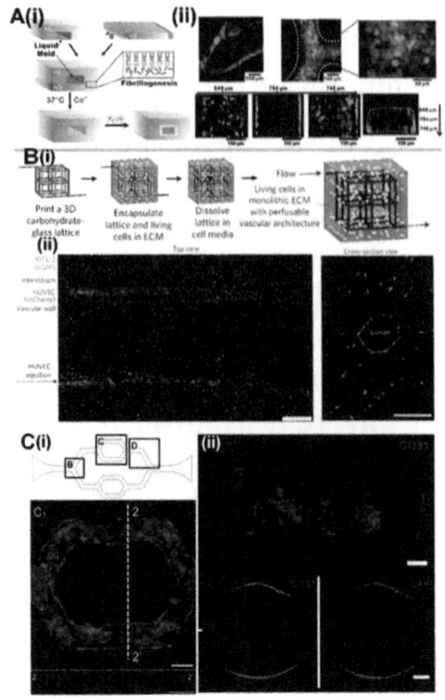

Figure 5.1 EC lining–based methods for in vitro vascularization. (A) Micropatterned planar hydrogel slab bonding method. (i) Schematic representations of engineering the vascular network in hydrogel by using the additive bonding between a micropatterned collagen gel slab and a flat hydrogel layer. (ii) The monolayer of HUVEC seeded on the inner wall of branched hydrogel microchannels and it fluorescent images at different depth. Republished with permission of The Royal Society of Chemistry, from Ref. [17], copyright (2013); permission conveyed through Copyright Clearance Center, Inc. (B) Dissolvable material-based sacrificial micromolding method. (i) Schematic of a 3D interconnected microvessel network formed by the rapid casting of a carbohydrate glass lattice as the sacrificial element with a 3D printer. The lattice is encapsulated in the ECM along with living cells and dissolved in minutes in cell media without damaging nearby cells, which can generate a monolithic tissue construct with a vascular architecture. (ii) The micrograph (left) shows HUVECs expressing mCherry attached to the hydrogel wall to generate the microvessel network. Scale bar: 1 mm. The micrograph (right) shows an endothelial monolayer-lined vascular lumen surrounded by 10T1/2 cells after 9 days in culture. Scale bar: 200 μm. Reprinted by permission from Springer Nature Customer Service Centre GmbH: Springer Nature, *Nature Materials*, Ref. [18], copyright (2012). (C) EC lining inside a PDMS-based microfluidic channel. (i) The schematic of a PDMS-based microfluidic channel and the confocal image of an endothelial cell at location C of the channel with a rectangle cross section. Scale bar: 100 μm. Adapted from Ref. [27], under a Creative Commons Attribution License. Copyright © 2015 Li et al. (ii) A confocal reconstruction image showing the complete lumen formed by the HUVECs inside the PDMS tube and confocal fluorescence micrographs of the cross-sectional views of an endothelialized PDMS tube stained with CD31/nuclei. Scale bar: 100 μm (top) and 200 μm (bottom). Adapted with permission of The Royal Society of Chemistry, from Ref. [33], copyright (2016); permission conveyed through Copyright Clearance Center, Inc.

which could potentially replace inert plastic tubes to integrate multiple organoids into a single microfluidic circuitry (Fig. 5.1C-ii) [33].

5.2.2 Vasculogenesis- and Angiogenesis-Based Methods

Vasculogenesis and angiogenesis are the fundamental processes of vascular development in vivo [34, 35], as shown in Fig. 5.2A. Vasculogenesis gives rise to the formation of the first primitive vascular plexus during early embryogenesis, while angiogenesis is responsible for the remodeling and expansion of the vascular network [36]. In adults, new blood vessels may form via vasculogenesis through the recruitment of ECs from bone marrow [37], as well as through angiogenesis stimulated by distress signals from the parenchymal tissue under certain physiological and pathological conditions, such as wound healing, exercise, and tumor growth [38]. Different from EC lining–based methods, vasculogenesis- and angiogenesis-based methods do not require additional microstructures to guide ECs into forming a vascular structure. They allow the seeded cells inside the ECM to self-assemble into a 3D μVN de novo. Therefore, these methods can create a more natural vascular structure than the EC lining methods.

5.2.2.1 Vasculogenesis

Vasculogenesis is a process by which blood vessels are formed from endothelial progenitor cells (EPCs) de novo. Within a microfluidic device, Hsu et al. demonstrated the formation of μVNs using endothelial colony-forming cell-derived ECs and normal human lung fibroblasts (NHLFs) mixed in fibrin gel and embedded inside a series of diamond-shaped tissue chambers (Fig. 5.2B) [39–41]. Under a physiological level of interstitial flow, a continuous and perfusable microvessel network formed after 3 weeks by perfusing the cell culture medium in the absence of vascular endothelial growth factor (VEGF) and basic fibroblast growth factor (bFGF) [39–41]. It was reported that the vascularized microtissue could be cultured, remained viable for up to 40 days, and showed disruption and shedding over time [41]. Furthermore, it was found that a combination of fibroblast-derived proteins would promote EC sprouting and was necessary for EC lumen formation [42]. VEGF and bFGF can also be utilized as chemical factors to stimulate vasculogenesis. Raghavan et al. presented a method to control the microvessel geometry by spatially patterning ECs within micromolded collagen gels and demonstrated that lumenized microvessels formed within 24–48 h on being stimulated by VEGF and bFGF [43]. With this method, microvessels of various sizes and shapes can be achieved by modifying the collagen gel concentration and the seeded cell density. It was found that both the average branch length and the effective diameter decreased with an increase in fibrinogen concentrations, while the branch length, diameter, and area fraction of the vascularized region increased with an increase in the EC seeding density [44]. Beyond these, paracrine signaling by co-culturing ECs with other types of cells can also influence vascular morphology, such as the number of branches, the average branch length, the percent vascularized area, and the average vessel diameter. It was found that a larger coverage area and network stability can be achieved by co-culturing ECs with fibroblasts, while the diameter and the average vessel length are reduced under the effect of angiogenic growth factors [45]. Furthermore, besides fibroblasts, stromal cells like pericytes, SMCs, and mesenchymal stem cells (MSCs) can contribute to vascular stabilization [46]. These findings will provide mechanisms to fine-tune the μVN in vitro with specified morphological properties.

5.2.2.2 Angiogenesis

Angiogenesis is the formation of new blood vessels from pre-existing vessels, which is an important mechanism for vascular network remodeling. Angiogenesis occurs through a series of defined steps, including ECM degradation by matrix metalloproteinases (MMPs), angiogenic stimulus, sprouting, elongation and branching, lumen formation, anastomosis, and stabilization or regression [47]. Within a microfluidic device, angiogenic sprouting from the EC monolayer can also be induced by a VEGF gradient [49, 50]. In addition, Kim et al. developed a 3D microfluidic in vitro model to investigate angiogenic sprouting from vasculogenesis-derived μVNs in the presence of interstitial flow (Fig. 5.2C) [51]. It was found that angiogenic sprouting was promoted only in the direction opposite to that of the interstitial flow and suppressed in the direction of flow.

To form a perfusable μVN, these sprouts need to connect with each other through a process called anastomosis. Yeon et al. established an array of perfusable microvessels from the EC monolayer lined on both sidewalls of fibrin gel confined inside a microfluidic channel (Fig. 5.2D) [52]. By co-culturing with lung fibroblasts, the angiogenic sprouts were stimulated and fused together after 3–4 days. The spacing of the angiogenic sprouts can be well controlled by adjusting the spacing of the microstructures that confine the ECM during the microfabrication process. Similarly, Song et al. demonstrated a microfluidic device that could accurately reproduce the dynamics of vascular anastomosis on treatment by VEGF [53]. All these devices will enable a new generation of studies to investigate the mechanisms of angiogenesis and provide a novel and practical platform for drug screening.

5.2.2.3 Hybrid methods

By controlling different factors within the microenvironment, different mechanisms to stimulate vascularization can be achieved on a single microfluidic device. Kim et al. presented a microfluidic device with five parallel microchannels that could generate a perfusable microvessel network through either vasculogenesis or angiogenesis by seeding the cells and fibrin gel into different channels (Fig. 5.3A) [54]. For the vasculogenesis process, HUVECs and NHLFs were mixed in a fibrinogen solution and introduced separately into the central channel and the stromal cell culture channels. For the angiogenesis process, the central channel and the stromal cell culture channels were solely filled with fibrin gel and HUVECs were attached onto the gel wall inside one medium channel contralateral to the NHLF seeding. Recently, Wang et al. developed a versatile microfluidic device design with a robust construction methodology to establish an interconnected, perfused vascular network from artery to capillary beds to vein (Fig. 5.3B) [26]. This method combined multiple strategies and vascular development processes such as vasculogenesis, EC lining, angiogenesis, and anastomosis

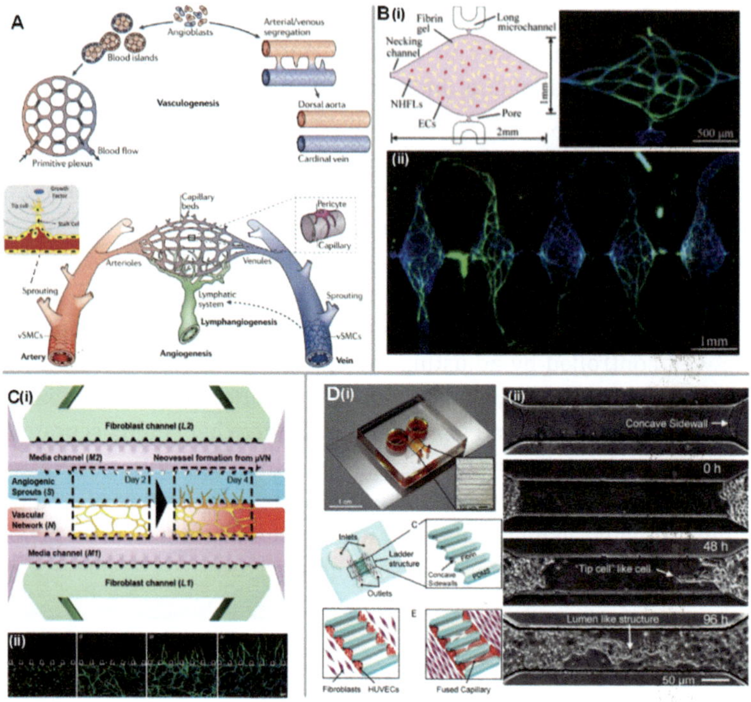

Figure 5.2 Vasculogenesis- and angiogenesis-based methods for in vitro vascularization. (A) Schematic of vasculogenesis and angiogenesis in vivo. Adapted by permission from Springer Nature Customer Service Centre GmbH: Springer Nature, *Nature Reviews Molecular Cell Biology*, Ref. [36], copyright (2011). (B) 3D microvascular network formation by vasculogenesis. (i) Schematic of a 3D microtissue chamber seeded with ECs and NHLFs in fibrin gel and the microvascular network formed, labeled with CD31 (green) and 4′,6-diamidino-2-phenylindole (DAPI) (blue) for nuclei. (ii) Fluorescent image of 5 interconnected microvascular networks. Adapted with permission of The Royal Society of Chemistry, from Ref. [39], copyright (2013); permission conveyed through Copyright Clearance Center, Inc. (C) Angiogenesis from a 3D microvascular network. (i) Configuration of the microfluidic device and schematic of the angiogenesis assay from a microvascular network. (ii) Temporal sequence of interconnected vascular network formation and angiogenic sprouting from the network formed inside the adjacent microchannel. Scale bar: 100 μm. Adapted with permission of The Royal Society of Chemistry, from Ref. [51], copyright (2016); permission conveyed through Copyright Clearance Center, Inc. (D) Anastomosis of sprouts into interconnected and perfusable blood vessels. (i) Schematic of microfluidic device design and self-organized capillary networks formed by HUVECs attached on the concave sidewalls from both sides. (ii) Images of sprouting HUVECs from both sides and fused capillary formation over time. Adapted with permission of The Royal Society of Chemistry, from Ref. [52], copyright (2012); permission conveyed through Copyright Clearance Center, Inc.

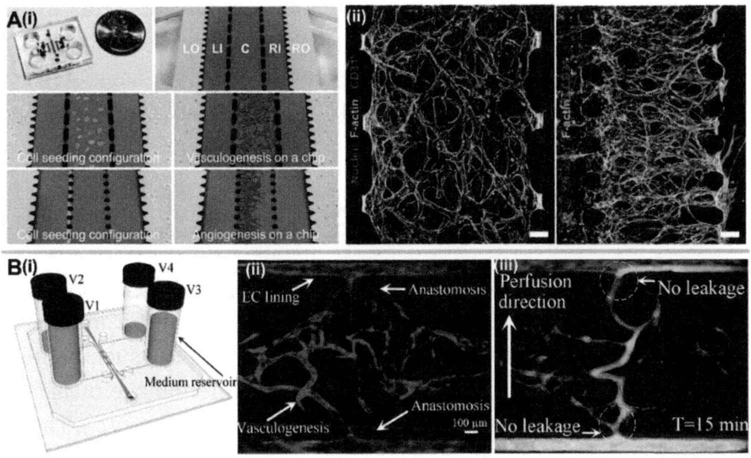

Figure 5.3 Hybrid methods for in vitro vascularization. (A) Vasculogenesis or angiogenesis realized on a single microfluidic device. (i) Schematic of microfluidic chip design and cell seeding configurations for vasculogenesis and angiogenesis at different microfluidic channels. (ii) Confocal micrographs showing the overall architectures of vascular networks established by vasculogenic and angiogenic processes on day 4. Scale bar: 100 μm. Republished with permission of The Royal Society of Chemistry, from Ref. [54], copyright (2013); permission conveyed through Copyright Clearance Center, Inc. (B) Advanced vascularization model for generating an intact and perfusable 3D microvascular network. (i) Schematic of microfluidic device design by using four medium reservoirs with different hydrostatic pressures. (ii) Intact microvascular network formation that incorporates different stages of vascular development, including vasculogenesis, EC lining, sprouting angiogenesis, and anastomosis, in sequential order. (iii) 70 kDa FITC-dextran perfusion confirms physiologic tightness of the EC junctions and completeness of the interconnections between artery/vein and the capillary network. From Ref. [26]—Reproduced by permission of The Royal Society of Chemistry.

to take place on a single platform in the proper sequence. After the formation of a capillary network inside the tissue chamber via vasculogenesis, the EC lining along the microfluidic channels adjacent to the tissue chamber served as an artery and a vein. To promote a tight interconnection between the artery/vein and the capillary network, sprouting angiogenesis was induced, leading to anastomosis of the μVN inside the tissue chamber and the EC lining along the microfluidic channels. More importantly, this work demonstrated that nonphysiological leakage from the vasculature into the interstitial space could be prevented, which was validated by 70 kDa fluorescein isothiocyanate (FITC)-labeled dextran perfusion.

5.3 Vascular-Inducing Factors

Microfluidic systems have become an emerging tool for developing microvasculature in vitro due to their advantages in terms of the precise control of factors within the microenvironment, such as fluid flow at the physiological levels, distribution of different chemical factors, and different ECM properties (stiffness, orientation, etc.). These inducing factors can be divided into three main types: biomechanical factors, extracellular (or diffusible) signaling molecules, and cell source and cell–cell interaction. Table 5.1 shows a summary of the main factors that influence different vessel parameters.

5.3.1 Biomechanical Factors

In a microfluidic device, biomechanical factors are various forces generated by fluid flow through the microfluidic channel or the ECM. Shear stress is the dominant biomechanical factor that highly depends on the fluid flow. It was found that shear stress would enhance barrier functions by decreasing vascular permeability and narrowing vascular wall to improve stability [12, 55–57]. For EC lining, ECs were elongated and aligned with the direction of the flow under the effect of shear stress (Fig. 5.4A-ii). In addition, shear stress has been shown to attenuate HUVEC invasion through nitric oxide signaling, irrespective of the interstitial flow direction and the VEGF gradient [58, 59]. Galie et al. further concluded that intraluminal shear stress and transmural flow through the endothelium above 10 dyn/cm^2 can trigger ECs to sprout and invade into the underlying matrix, independent of cell–cell junction maturation and the pressure gradient across the monolayer [60].

Interstitial flow is another important regulator of various cell behaviors both in vitro and in vivo [61], and the force it imposes on ECs can promote capillary formation. In vivo interstitial flow

Table 5.1 Main factors that influence different vessel parameters

Factors	Vessel parameters	Effect
Shear stress	Barrier function	Decreases permeability [12, 55–57]
	Cell orientation	Elongates and aligns cells along the flow direction in the EC lining [57]
	EC invasion	Attenuates invasion [58, 59]
Interstitial flow	Vasculogenesis	Promotes capillary formation regardless of flow direction [40]
	Angiogenesis	Ensures that angiogenesis is more active in the direction opposite to the interstitial flow [51]
Stiff ECM	Vessel lumen	Generates vessel lumens of small diameters [63]
	Sprouting invasion	Significantly reduces sprouting invasion [70]
Fiber arrangements of the ECM	Vessel alignment	Ensures alignment along the fiber orientation [64]
VEGF concentration	Endothelial sprouting	Induces sprouting at low concentrations (2.5–5 ng/mL) [67]
VEGF gradient	Vessel morphology	Induces sprouting if positive and vessel dilation if negative [59]
Hypoxia	Angiogenesis	Promotes angiogenesis, such as tumor angiogenesis [78]
Co-culture with fibroblasts	Perfusable vascular formation	Promotes EC sprouting and lumen formation [40, 52]
Co-culture with mural cells	Vessel stabilization	Ensures more stability [88–91]

in the 0.1–1 μm/s range can modulate vessel formation [62]. Inside the microfluidic device, the convective velocity in the range of 1.7–11 μm/s with high a Péclet number (Pe $>$ 10), defined as the ratio of convective to diffusive transport, can stimulate vasculogenesis regardless of the interstitial flow direction (Fig. 5.4A-i) [40]. However, for angiogenesis, new sprouts from the existing vascular network are more active in the direction opposite to that of the interstitial flow [51]. Moreover, it was found that the basal-to-apical transendothelial flow could trigger the transition of ECs from a quiescent to an invasive phenotype to undergo angiogenesis [48].

In addition to fluid forces, hydrogel/ECM mechanical properties such as gel permeability, pore size, fiber diameter and stiffness, and degradation speed also play an important role in regulating vascular formation [45]. Collagen and fibrin are the most commonly used natural hydrogels, and their mechanical properties can be flexibly fine-tuned. Collagen's mechanical properties can be adjusted by changing temperature or pH, while fibrin's mechanical properties can be adjusted by changing the fibrinogen and thrombin concentrations. It was found that stiffer gels would generate vessel lumens with smaller diameters [63] and the invasion of sprouting tubular structures into the stiff collagen gel was significantly reduced compared with that into the soft gel (Fig. 5.4A-iii) [64]. Furthermore, fiber arrangements inside the ECM can also affect vessel alignment in response to contact guidance [65]. In addition to these passive mechanical forces generated by the ECM with different mechanical properties, the active mechanical force imposed on the ECM has a profound effect on the microvasculature response. For example, the free-floating scaffolds result in randomly orientated vessels. However, the cyclic stretching applied on uniaxially fixated seeded constructs results in diagonal vessels, whereas the static stretching results in vertical vessels [66].

5.3.2 Extracellular (or Diffusible) Signaling Molecules

Growth factors are cell-secreted substances capable of stimulating cellular growth, proliferation, and differentiation. VEGF is the most important growth factor that regulates EC migration and proliferation through VEGF receptor-2 signaling [67]. It was found that VEGF at lower concentrations (2.5–5 ng/mL) induces endothelial sprouting rather than at higher concentrations (15–35 ng/mL) [68]. In addition, negative VEGF gradients result in a process resembling vessel dilation whereas positive VEGF gradients induce sprouting [59]. Besides VEGF, other angiogenic growth factors, such as angiopoietin 1, transforming growth factor β1, and bFGF, can direct angiogenic sprouting from a low to high concentration gradient via endothelial tip cell filopodia and align the sprouts into well-organized structures [64, 69, 70]. Therefore, establishing a concentration gradient of growth factors is crucial for tissue vascularization through either vasculogenesis or angiogenesis in vitro (Fig. 5.4B). This can be achieved by embedding either alginate spheroids that contain fibroblasts or the microparticles that encapsulate growth factors into the hydrogel to create a local concentration gradient or by perfusing cell culture media containing growth factors through the microfluidic device [71, 72]. Compared to single growth factors acting on the μVN, a combination of multiple growth factors can highly enhance vascular density and stability [73, 74]. Growth factors with different gradient profiles can also be adjusted by tuning flow rate and designing different microfluidic configurations [75]. For example, a simple device with one central gel channel positioned between two medium channels can generate a linear gradient profile across the gel channel [76]. Besides a linear gradient profile in one direction, two gradient profiles can also be generated orthogonally, which could be used to examine cellular morphogenesis under the influence of multiple well-defined gradients [49]. In addition to the spatial effect of growth factors, temporal change of VEGF concentration could also affect angiogenesis. It was found that an initial gradient of a high VEGF concentration, with a programmed concentration decrease over time, yielded optimal angiogenic sprouting, as compared to a constant dose of VEGF [77]. Therefore, controlled release of growth factors, optimal combination of growth factors, dosage, gradient profile, and appropriate exposure time will facilitate proper vascular formation and stability.

In addition to the direct influence of different growth factors, oxygen tension may be considered as an indirect method that affects the secretion of VEGF. It is known that hypoxia inducible factor 1 (HIF-1) secreted in a hypoxic environment can initiate the transcription of VEGF, which plays a critical role in inducing the expression of antiapoptotic protein, especially in pathological angiogenesis models such as tumor angiogenesis [78]. In microfluidics, the desired oxygen microenvironment can be generated through different methods. For example, due to the high gas permeability of PDMS used in microfluidics, both hypoxia microenvironment and oxygen tension gradient can be flexibly created by allowing different gases with specific concentrations to flow, perfusing gas-equilibrated medium, or loading oxygen scavenging or generation chemicals [79, 80]. Furthermore, with the combination of poor gas-permeable thermoplastic, interference from the external microenvironment can be inhibited, which can improve the oxygen control inside the PDMS-thermoplastic hybrid microfluidic devices [81, 82].

5.3.3 Cell Source and Cell–Cell Interaction

ECs are the main cellular component for tissue vascularization. HUVECs have been widely used for microvascular construction in vitro due to their relatively simple isolation process and well-established literature references. However, other EC sources may be used instead of HUVECs for better outcomes or for certain applications that require tissue-specific ECs. For example, it was found that EPCs exhibited a substantial proliferative capacity, which is different from HUVECs, which have limited proliferation in the absence of proangiogenic factors [83, 84]. Recently, induced pluripotent stem cell (iPSC)-derived ECs have become a new EC source for personalized medicine research [85–87]. To create a mature and stable μVN in vitro, it is necessary to co-culture ECs with mural cells, such as fibroblasts, pericytes, or MSCs. Fibroblasts can facilitate EC sprouting and lumen formation by synthesizing and maintaining essential matrix proteins (e.g., collagen I), as well as secreting various angiogenic growth factors [40, 52]. In addition, pericytes and SMCs wrapping around capillary and larger vessels (e.g., arterioles, arteries, venioles, and veins) play an important role in stabilizing newly formed vessels (Fig. 5.4C) [88–91]. MSCs also exhibit a functional role in vascular stabilization. For example, a co-culture of HUVECs and MSCs derived from either bone marrow or human embryonic stem cells can increase sprouting branches and maintain structural integrity of the μVN in vitro [76, 92].

5.4 Selective Vascular Barrier

ECs exhibit different cellular structures and barrier properties depending on specific organs in the human body. For example,

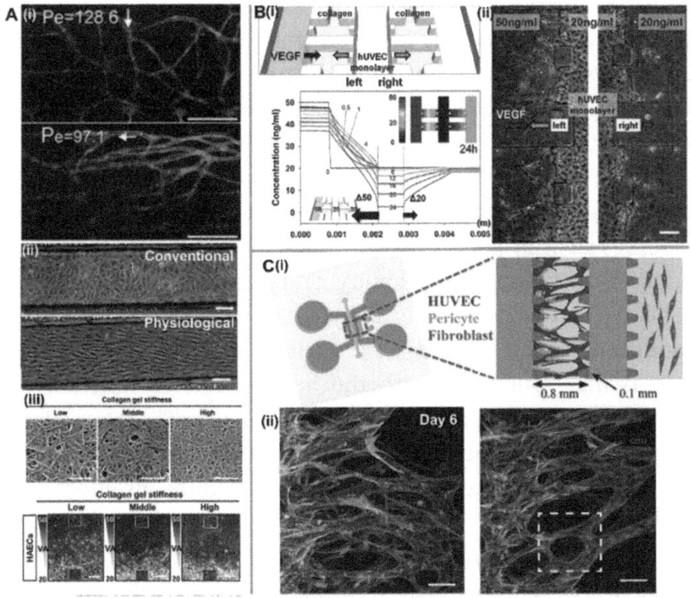

Figure 5.4 Three major vascular-inducing factors in vascularization on a chip. (A) Biomechanical factors. (i) The influence of interstitial flow with a high Péclet number on vasculogenesis and the orientation of the microvascular network under the effect of the interstitial flow in transverse and longitudinal directions. Scale bar: 500 μm. Adapted with permission of The Royal Society of Chemistry, from Ref. [40], copyright (2013); permission conveyed through Copyright Clearance Center, Inc. (ii) The effect of shear stress on EC lining; the ECs were elongated and aligned with the direction of the flow under the physiological conditions compared with the conventional conditions. Scale bar: 100 μm. Adapted by permission from Springer Nature Customer Service Centre GmbH: Springer Nature, *Scientific Reports*, Ref. [57], copyright (2014). (iii) The effect of gel stiffness on angiogenesis; the invasion of tubular structures into the stiff collagen gel was significantly reduced compared with their invasion in the soft gel. Scale bar: 1 μm (top) or 150 μm (bottom). Adapted from Ref. [64], under a Creative Commons Attribution 4.0 International License (https://creativecommons.org/licenses/by/4.0/). (B) Extracellular (or diffusible) signaling molecules. (i) Schematic of a chip design where HUVECs were cultured in the center channel and the simulation result of a VEGF gradient. (ii) Apparent sprouting angiogenesis on the left-hand side, with a steeper gradient, compared with the right-hand side, with a gentle gradient. Scale bar: 150 μm. Adapted with permission from Ref. [50]. Copyright (2011) American Chemical Society. (C) Cell source and cell–cell interactions. (i) Schematic of a microfluidic device composed of a central vessel channel, two adjacent media channels, and the outermost fibroblast channel. (ii) Confocal images showing matured pericytes covering the perfusable EC network on day 6. Scale bar: 100 μm. Adapted from Ref. [91], under the terms of the Creative Commons Attribution License (https://creativecommons.org/licenses/by/4.0/).

ECs at the blood-brain barrier (BBB) have very tight cell–cell junctions, resulting in very low permeability, to protect the central nervous system from toxins and plasma fluctuations (Fig. 5.5A) [93–95]. To develop an organ-on-a-chip system with tissue-specific characteristics, it is necessary to incorporate a μVN with tissue-specific barrier functions. Besides improving the barrier function by imposing shear stress, an increase in the circumferential stretch on the vessel wall perpendicular to the direction of the flow will enhance endothelial permeability and vascular tone due to the overloaded pressure that is associated with hypertension [96]. In addition to the stimuli with various mechanical factors, the microvascular will react with the loss of barrier function upon simulation with inflammatory cytokines [11]. Therefore, factors such as cell phenotype, perfusion condition, vessel geometry, and growing factors and cytokines will have a significant effect on the vessel permeability.

In versatile organs on chips, usually, the design consists of a vascular barrier made from a porous membrane sandwiched between the blood vessel compartment containing ECs and the other tissue/organ compartment containing other cell types [97, 98]. This porous membrane could facilitate the creation of the cellular-tissue interface as well as cross talk between these two compartments. The porous membrane can be fabricated from a variety of materials using different methods, like the PDMS through template-based soft lithography [99] and polymeric track etched membranes [100]. The pore size of the porous membrane is likely the most important parameter to affect cell transmigration, physical contact with other cells, and paracrine signaling. In addition, the mechanical properties of the porous membrane, like stiffness and strain, would affect the organization and function of organs, as successfully demonstrated in the lung on a chip [101] and the gut on a chip (Fig. 5.5B) [102]. Furthermore, the surface properties of the porous membrane, functioning as an equivalent to the ECM and more specifically the cell basement membrane, would directly stimulate cell response. It was found that a porous membrane with higher roughness would increase cell adhesion due to more adsorbed protein, while a porous membrane with moderate hydrophilicity would lead to the highest cell adhesion,

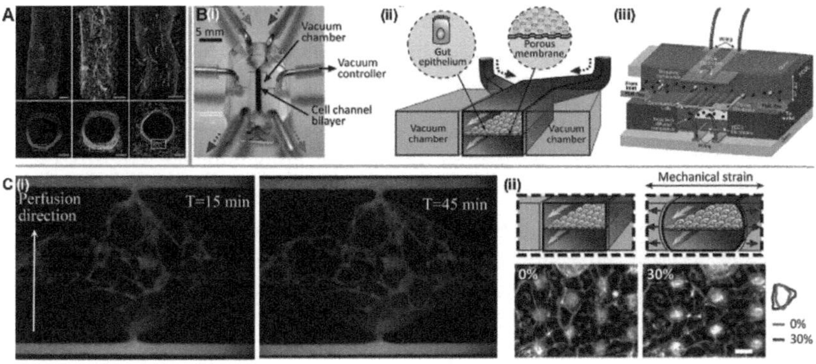

Figure 5.5 Selective vascular barrier properties of different organ chips, constructed with a porous membrane and permeability characterization methods. (A) Microvasculature with different structures and selective barrier functions in different organs on chips. The BBB, with tight cell–cell junctions. Fluorescence confocal micrographs of the brain microvessel engineered from human brain microvascular endothelial cells (left) and co-culture with pericytes (middle) and astrocytes (right). Scale bar: 200 μm. Adapted from Ref. [95], under the Creative Commons Attribution License. Copyright © 2016 Herland et al. (B) Human gut on a chip. (i) Photographic image of the chip prototype. (ii) A monolayer of gut epithelial cells is cultured on the flexible porous ECM-coated membrane in the middle of the central microchannel. Vacuum chambers are designed parallel to the channels that can apply mechanical strain to the cell layer. (iii) Schematics (top) and phase contrast images (bottom) of intestinal monolayers cultured within the gut on a chip in the absence (left) or presence (right) of mechanical strain (30%; arrow indicates the direction) exerted by applying suction to the vacuum chambers. Scale bar: 20 μm. Adapted with permission of The Royal Society of Chemistry, from Ref. [102], copyright (2012); permission conveyed through Copyright Clearance Center, Inc. (C) Vascular permeability characterization. (i) Permeability coefficient calculation by measuring perfusion fluorescent solutes across the vessel lumen wall over time. Scale bar: 100 μm. From Ref. [26]—Reproduced by permission of The Royal Society of Chemistry. (ii) Schematic of a multilayered microfluidic device design with incorporated electrode for internal TEER measurement on a BBB model. Adapted with permission of The Royal Society of Chemistry, from Ref. [93], copyright (2012); permission conveyed through Copyright Clearance Center, Inc.

spreading, and growth. Therefore, to better mimic organ-specific vascular barrier properties, it is necessary to integrate a porous membrane with suitable microstructures (pore size, porosity, etc.), mechanical properties (substrate stiffness, strain, etc.), and surface properties (surface roughness and topography, surface chemistry, etc.).

As is known, the tighter endothelial junctions of an established tissue barrier correspond to the lower solute permeability. Immunostaining of junctional markers, such as platelet EC adhesion molecule-1 and vascular endothelial cadherin, and the deposition of basement membrane proteins, such as laminin and collagen IV, are commonly used to validate vascular integrity qualitatively [65]. For quantitative analysis, fluorescent-tagged dextran or other biologically relevant molecules are typically utilized to characterize the barrier function by measuring their permeability coefficient across the vessel wall (Fig. 5.5C-i) [26, 103, 104]. Since albumin (MW = 66.5 kDa) is the most common protein in blood plasma, fluorescent-tagged bovine serum albumin or 70 kDa dextran is commonly used [104]. The vascular permeability coefficient of microvessels in vitro is typically in around 10^{-6} cm s^{-1}, which is comparable to the value of the tumor microvasculature in vivo [11, 17]. In addition to permeability measurement, transendothelial electrical resistance (TEER) is a widely accepted method to quantify vascular integrity by characterizing the electric impedance across the endothelium (Fig. 5.5C-ii) [93, 105]. Compared to the molecular permeability test, TEER can be performed easily and in real time. However, TEER measurement in self-assembled microvascular models remains a challenge compared to TEER measurement in EC monolayer or single-vessel models.

5.5 Application of Engineered Microvascular Networks to Cancer Biology

Besides a better understanding of the basic mechanisms of vascularization, the other promising application of engineered microvessels is in modeling human diseases in vitro. Endothelial dysfunction is a major physiological mechanism that can cause various vascular diseases, such as thrombosis, atherosclerosis, and inflammation [106]. Especially within the field of cancer biology, in vitro tumor models have provided important tools for cancer research and served as low-cost anticancer drug screening platforms. On the basis of the tumor formation type and vascular integration, these models can be broadly classified into four categories: transwell based [107], spheroid based [108], hydrogel droplet embedded culture [109], and vascularized tumor [110]. More than 90% of cancer-related mortality is attributed to cancer metastasis [111], which often involves multiple steps closely associated with vascular pathology, such as tumor angiogenesis [112], intravasation [113], and extravasation [114]. Therefore, engineered vascularized tumor models play an important role in studying cancer metastasis.

5.5.1 Tumor Angiogenesis

Since tumor growth can be suppressed by cutting out the nutrient and oxygen supply through tumor vessels, it is essential to have a complete understanding of tumor angiogenesis to develop new cancer therapies. It is known that the rapidly growing angiogenic vasculature around a tumor consequently leads to irregular sprouting, tortuous microstructure, dysfunctional or absent perivascular cells, and leaky barrier properties, all of which are thought to promote tumor metastasis [115]. Several in vitro studies have been performed to study tumor angiogenesis by co-culturing cancer cells and ECs in microfluidic devices. A study is normally performed by establishing a vertical 2D endothelial monolayer on the sidewalls, which can facilitate better imaging of angiogenic sprouting into the 3D ECM on the axial plane. By attaching ECs to the sidewall of fibrin gel, 3D sprouting was promoted by the factors secreted by the highly malignant human glioblastoma, which exhibited aberrant morphology compared to the NHLF-induced sprouts [54]. Similarly, Chung et al. engineered a platform to evaluate and quantify capillary growth and EC migration from an intact EC monolayer attached onto a collagen gel wall by co-culturing with MTLn3 cancer cells [63]. They observed that cancer cells could either attract ECs and induce capillary formation or have minimal effect, while SMCs

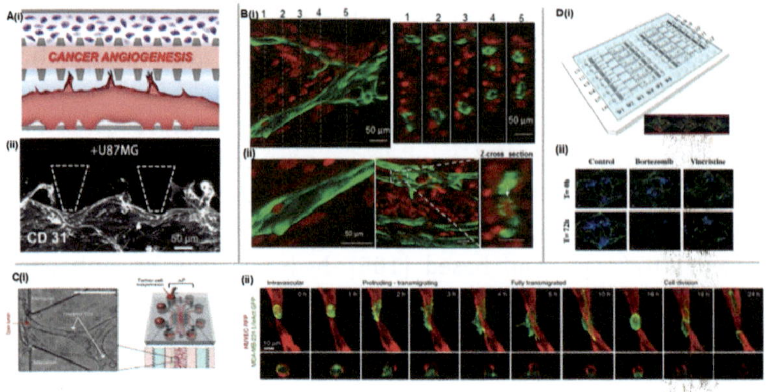

Figure 5.6 Vascularized tumor on a chip for studying tumor metastasis and anticancer drug screening. (A) Tumor angiogenesis. (i) Schematic of a chip design for tumor angiogenesis. (ii) The 3D confocal image (CD31) shows the angiogenic sprouts stemming from the microvessel wall growing toward the upper channel stimulated by the proangiogenic factors from U87MG cancer cell lines. Adapted from Ref. [116], with the permission of AIP Publishing. (B) Tumor intravasation. (i) MDA-MB-231 cells (red) were implanted in a prevascularized (green) device and the tumors were grown for four days without intravasated tumor cells. (ii) A tumor cell in the process of intravasating into the vessel. Scale bar: 50 μm. Republished with permission of The Royal Society of Chemistry, from Ref. [120], copyright (2018); permission conveyed through Copyright Clearance Center, Inc. (C) Tumor extravasation. (i) Microvessel bed formation and tumor cell perfusion procedure. (ii) Time-lapse sequence of an arrested tumor cell extravasating from a lumen and subsequently undergoing cell division over 24 h. Reprinted by permission from Springer Nature Customer Service Centre GmbH: Springer Nature, *Nature Protocols*, Ref. [122], copyright (2017). (D) Large-scale anticancer drug screening. (i) Schematic of a microfluidic platform design custom-fitted into a standard 96-well plate format and the formation of vascularized microtumor tissue inside three tissue chambers. (ii) Drug screening validation with bortezomib and vincristine characterized by the effect on the morphology microvascular network as well as the tumor size. Scale bar: 50 μm. From Ref. [132]—Reproduced by permission of The Royal Society of Chemistry.

suppress endothelial activity. Furthermore, Lee et al. presented a tumor angiogenesis model to quantify angiogenic sprouting from vasculogenesis-derived microvessels (Fig. 5.6A) [116]. It was found that low flow velocities commonly exist in tumor vasculature, suggesting that high shear stress regulation of angiogenic activity was absent in these vessels, thereby driving tumor angiogenesis due to the increased tumor-expressed angiogenic factors and endothelial permeability [117]. To further study the complex multicellular interactions during the tumor angiogenesis, a prevascularized tumor spheroid model was developed by co-culturing tumor cells, ECs, and fibroblasts embedded in fibrin gel, which exhibited robust sprouting angiogenesis after 7 days of culture [118]. In addition, different from physiological angiogenesis, tumor angiogenesis is mainly stimulated by the release of HIF-1 due to excessive cell proliferation and dysfunctional vasculature, which initiates transcription of VEGF to stimulate vascular outgrowth toward the tumor tissue [78]. DelNero et al. developed an oxygen-controlled 3D alginate-based tumor model to study tumor angiogenesis in hypoxic conditions, which demonstrated that proinflammatory pathways are a critical regulator for tumor angiogenesis under a hypoxia microenvironment [119].

5.5.2 Tumor Intravasation

Tumor intravasation is the invasion of cancer cells through the basal membrane into blood or lymphatic vessels near the tumor stroma, which can be triggered by a chemotactic gradient, oxygen tension, and impaired endothelial barrier function [113]. Shirure et al. developed a platform to simultaneously and dynamically observe hallmark features of tumor progression, including cell proliferation, angiogenesis, cell migration, and tumor cell intravasation (Fig. 5.6B). They found that enhancing the permeability of the vessels by treating with VEGF or thrombin does not significantly increase the rate of tumor intravasation [120]. Similarly, Wong and Searson presented a platform that positioned metastatic cancer cells next to artificial, EC-lined microvessels embedded in the ECM. Using real-time live cell fluorescence microscopy, they observed a variety of tumor-endothelium interactions, including invasion, intravasation, angiogenesis, and tumor cell dormancy [121].

5.5.3 Tumor Extravasation

Tumor extravasation refers to circulating tumor cells transmigrating through the endothelium and lodging at the secondary organs.

Chen et al. developed a microfluidic platform to study tumor cell extravasation from vasculogenesis-derived μVNs (Fig. 5.6C) [122]. They showed how to engineer an in vitro microfluidic-based assay featuring perfusable μVNs for real-time observation and quantification of tumor cell extravasation. Furthermore, Jeon et al. developed a microfluidic 3D in vitro model to analyze organ-specific human breast cancer cell extravasation into bone and muscle-mimicking microenvironments through a μVN concentrically wrapped in mural cells [123]. They found that the extravasation rates of breast cancer cells were significantly higher in the bone-mimicking microenvironment compared to those in the control matrices without stromal cells or the muscle-mimicking microenvironments. This result validated a widely accepted phenomenon that specific tumor types preferentially metastasize to specific organ systems.

5.5.4 Tumor Microenvironment

The tumor microenvironment (TME) plays an important role during the tumor metastasis process, which can be broadly classified into biophysical cues (e.g., intramural flow, interstitial flow, and ECM) and biochemical cues (e.g., molecular gradient of growth factors, cytokines, oxygen, and metabolites). For intramural flow in the bioengineered microvessel models, a critical method is to develop a perfusable and functional μVN with either EC lining or vasculogenesis-/angiogenesis-based methods. The tumor intravasation/extravasation processes are mainly regulated by the alteration of vessel wall permeability and integrity under the effect of a shear stress induced by the intramural flow inside the vessel lumen. Therefore, tumor invasion and metastasis events are unlikely to occur in large vessels like arteries and veins due to the high shear stress along the tight and thick vessel walls. In contrast, the smaller capillaries would easily trap the tumor cells, which are more likely to penetrate through the capillaries due to the thinner walls and lower shear stresses [124]. The interstitial flow in the TME can lead to a spatial gradient of secreted cytokines, which would contribute to the directional tumor cell migration along/against the interstitial flow direction [125]. Besides the biological flow, the properties of the ECM in the TME are the other major biophysical factors to affect the tumor cell behavior, including stiffness, pore size, and fiber alignment [126]. It was found that the ECM that is both stiff and with a large pore size would facilitate the tumor cell migration to promote its invasion events. Furthermore, the hollow microchannel inside the ECM remodeled by the MMP can serve as a highway to speed up tumor cell migration [127]. Similarly, the well-aligned hydrogel fibers can also be regarded as hollow microsized tunnels, which can be realized by using the shear stress on the unpolymerized hydrogel inside a narrow microfluidic microstructure or directly applying the mechanical stretching force on the polymerized hydrogel.

Biochemical factors play a paramount role in tumor progression. Both tumor cells and stromal cells elaborate chemokines and cytokines. These biochemical factors act either by autocrine or by paracrine mechanisms to sustain tumor cell growth and induce angiogenesis and immune response in the tumor [128]. By using microfluidic systems, well-controlled biochemical gradient profiles can be established. Furthermore, due to the diminished supply of nutrients and oxygen caused by the defective vascular with heterogeneous perfusion in fast-growing solid tumor tissue, the TME is often hypoxic. Therefore, oxygen tension is another critical chemical factor in the TME.

5.5.5 Application of Vascularized Tumor on a Chip

5.5.5.1 Anticancer drug screening

The main application of in vitro vascularized tumor models is anticancer drug screening. Current drug screening methods still rely heavily on 2D cell culture assays or animal models. While 2D cell culture assays fail to replicate the 3D structure and complexity of tissues in vivo, animal models fail to capture the human-specific response to drugs due to intrinsic differences in anatomy and physiology [129]. While 3D tumor spheroids can address some of these shortcomings, many tumor cell types, especially those with a highly invasive phenotype, cannot readily form spheroids. Since most drugs are delivered to the targeted tissue through the circulatory system rather than being directly imposed on the cultured tissue in vivo, incorporating the vasculature into tumor models can better mimic drug delivery to provide precise information like drug dosage as a reference in further clinical trials. Besides the vascularized tumor tissue, a co-culture with other stromal cells is another important aspect to mimic the drug delivery microenvironment. It was found that a co-culture of stromal cells and tumor cells showed a significantly higher drug resistance as compared to the culture only with tumor cells [130]. Sobrino et al. developed a 3D microtumor model supported by perfused human microvasculature for anticancer drug screening [131]. They found that the microvessels in this platform are sensitive to the effects of both antiangiogenic (pazopanib, linifanib, cabozantinib, axitinib, etc.) and vascular disrupting agents (vincristine, taxol, etc.). Therefore, this vascularized tumor model can be used to identify drugs that target cancer cells directly or indirectly by inducing the vasculature to starve off tumor growth. On the basis of this work, Phan et al. further developed an arrayed vascularized microtumor platform for large-scale drug screening applications and successfully demonstrated its capacity to identify both antiangiogenic and anticancer drugs from a small library of compounds (Fig. 5.6D) [132].

5.5.5.2 Different tumor therapies

The widely used strategies for antiangiogenic therapies are to destroy the tumor vasculature, thereby depriving the tumor of nutrients. However, these methods will also severely compromise the delivery of oxygen to the solid tumor, which induces a hypoxic microenvironment to often make chemotherapeutics and radiation less effective. In addition, the interstitial hypertension and non-uniform blood flow induced by the abnormalities and leakiness of tumor vasculature also push the drugs and oxygen to accumulate in the concentrated regions far from the solid tumor, no matter how much therapeutics and oxygen are pumped into the tumor. Therefore, another concept for tumor antiangiogenic therapy is to study the normalization of tumor vasculature, which can temporarily "normalize" the abnormal structure and function of tumor vasculature to function like normal blood vessels by using certain antiangiogenic agents to make it more efficient for oxygen and drug delivery [133]. The microfluidics-based vascularized tumor model can provide a better understanding of tumor vessel normalization at the molecular, cellular, and functional organ levels. Besides antiangiogenic therapies, immunotherapies against cancer have shown enormous potential, leading to the recent Food and Drug Administration approval of several drugs that reduce cancer progression [134]. Therefore, more studies on cancer-immune cell interactions would be better characterized by using vascularized tumor models perfused with immune cells [135].

5.6 Conclusions and Future Perspectives

In this chapter, we have provided an overview of different vascularization strategies based on microfluidic technology. These strategies can be classified into three categories: EC lining–based methods, vasculogenesis- and angiogenesis-based methods, and hybrid methods. It is apparent that each strategy carries both advantages and limitations. While the EC lining–based methods enable immediate media perfusion with well-controlled biomechanical factors, they are inefficient and suitable only for creating larger blood vessels based on predefined microchannel dimensions. In contrast, while vasculogenesis- and angiogenesis-based methods provide very limited microenvironment control, they are suitable for creating capillary networks down to a few micrometers in diameter by utilizing the natural morphogenic properties of ECs. Therefore, an ideal strategy for proper tissue vascularization may lie in the combination of more than one technique to realize hybrid methods. Moreover, multiple vascular-inducing factors, such as biomechanical factors, extracellular (or diffusible) signaling molecules, cell source, and cell–cell interaction, should be integrated into more complex microfluidic designs simultaneously to closely mimic the native microenvironments in different organs with specific vascular barrier functions. Furthermore, engineered microvessels also play an important role in modeling various human diseases, especially for cancer research. A vascularized tumor on a chip can serve as a powerful tool to study the mechanisms of cancer metastasis at different stages, like tumor angiogenesis, intravasation, and extravasation, as well as their potential applications in anticancer drug screening and different tumor therapies.

While many proof-of-principle studies have demonstrated versatile vascularization strategies and their applications in tumor models, challenges still remain. For example, there is an urgent need to establish more precise and geometrically controlled methods with more dynamic and flexible on-chip microenvironment control to recapitulate all aspects of vasculature in vivo. To construct a functional vascularized microtissue, interactions between ECs, supporting mural cells, and organ-specific cells with in vivo–relevant cell densities are required. Additionally, current culture media are optimized for specific cell types when cultured in monolayers. Once these cells are in a complex microenvironment, it is necessary to have a universal blood substitute to equally supplement multiple cell types. Another major consideration is to develop thick, implantable vascularized tissue constructs from microns to millimeters with synthetic biocompatible and degradable scaffold materials for tissue repair or regeneration, as described in the recent AngioChip design [136]. An exciting future application of engineered microvessels is their potential use in the field of precision medicine by creating patient-specific microvessel models using ECs derived from iPSCs. Future work on vascularized tumor models may rely on a better understanding of the TME and how to manipulate different microenvironment factors to influence cancer progression. Besides vascular-inducing factors, the solid tumor stress induced by the abnormally high amounts of collagen and other proteins is also a key factor in designing physiologically relevant in vitro vascularized tumor models. For drug screening applications, it is imperative to develop automated, high-throughput, reproducible, cost-effective, and easy-to-use microfluidic 3D cell culture systems. In terms of device design, it is necessary to integrate multiple on-chip detection sensors and analysis components for real-time monitoring and analysis of growing tissues. Finally, while the majority of organ-on-a-chip platforms are fabricated using PDMS and its derivatives due to their low cost, ease of use, and biocompatibility, they can absorb small molecules with transient mechanical properties, making them ineffective for long-term cell/tissue culture and drug studies. Therefore, there has been a push toward thermoplastic materials (like polystyrene, cyclic olefin copolymer, and polycarbonate) to replace PDMS for organ-on-a-chip platforms [137].

To summarize, although significant progress has been made toward tissue vascularization and related vascularized tumor models in the past decades, numerous challenges remain and will be addressed in future studies with a combination of microfluidics, microfabrication, biomaterials, tissue engineering, and other related technologies. Thus, a multidisciplinary, collaborative approach between engineers and biologists is required to develop more novel microphysiological systems that can closely mimic organ-specific physiology in vivo.

Acknowledgments

This work was supported by grants from the National Natural Science Foundation of China (No. 31600781), the Science and Technology Commission of Shanghai Municipality (17JC1400202 and 19ZR1475000), the Interdisciplinary Program of Shanghai Jiao Tong University (YG2016MS06), and the Fundamental Research Funds for the Central Universities.

References

1. R. Jain (2003). Molecular regulation of vessel maturation. *Nat Med*, **9**:685–693.
2. S. Bhatia, D. Ingber (2014). Microfluidic organs-on-chips. *Nat Biotechnol*, **32**:760–772.
3. S. Rayner, Y. Zheng (2016). Engineered microvessels for the study of human disease. *J Biomech Eng*, **138**:110801.
4. R. Kramer, G. Nicolson (1979). Interactions of tumor cells with vascular endothelial cell monolayers: a model for metastatic invasion. *Proc Nat Acad Sci USA*, **76**:5704–5708.
5. E. Young, M. Watson, S. Srigunapalan, A. Wheeler, C. Simmons (2010). Technique for real-time measurements of endothelial permeability in a microfluidic membrane chip using laser-induced fluorescence detection. *Anal Chem*, **82**:808–816.

6. L. Wang, Z. Zhang, J. Wdzieczak-Bakala, D. Pang, J. Liu, Y. Chen (2011). Patterning cells and shear flow conditions: convenient observation of endothelial cell remoulding, enhanced production of angiogenesis factors and drug response. *Lab Chip*, **11**:4235–4240.
7. J. C. Pelton, C. E. Wright, M. Leitges, V. L. Bautch (2014). Multiple endothelial cells constitute the tip of developing blood vessels and polarize to promote lumen formation. *Development*, **141**:4121–4126.
8. S. Halldorsson, E. Lucumi, R. Gómez-Sjöberg, R. Fleming (2015). Advantages and challenges of microfluidic cell culture in polydimethylsiloxane devices. *Biosens Bioelectron*, **63**:218–231.
9. A. Hasan, A. Paul, N. E. Vrana, X. Zhao, A. Memic, Y. S. Hwang, M. R. Dokmeci, A. Khademhosseini (2014). Microfluidic techniques for development of 3D vascularized tissue. *Biomaterials*, **35**:7308–7325.
10. J. Tien, K. H. K. Wong, J. G. Truslow (2013). Vascularization of microfluidic hydrogels. In *Microfluidic Cell Culture Systems*, C. Bettinger, J. T. Borenstein, S. L. Tao, eds. Elsevier, Oxford, UK, pp. 205–221.
11. K. Chrobak, D. Potter, J. Tien (2006). Formation of perfused, functional microvascular tubes in vitro. *Microvasc Res*, **71**:185–196.
12. G. Price, K. Wong, J. Truslow, A. Leung, C. Acharya, J. Tien (2010). Effect of mechanical factors on the function of engineered human blood microvessels in microfluidic collagen gels. *Biomaterials*, **31**:6182–6189.
13. K. Wong, J. Truslow, A. Khankhel, K. Chan, J. Tien (2012). Artificial lymphatic drainage systems for vascularized microfluidic scaffolds. *J Biomed Mater Res Part A*, **101A**:2181–2190.
14. N. Sadr, M. Zhu, T. Osaki, T. Kakegawa, Y. Yang, M. Moretti, J. Fukuda, A. Khademhosseini (2011). SAM-based cell transfer to photopatterned hydrogels for microengineering vascular-like structures. *Biomaterials*, **32**:7479–7490.
15. H. Yoshida, M. Matsusaki, M. Akashi (2012). Multilayered blood capillary analogs in biodegradable hydrogels for in vitro drug permeability assays. *Adv Funct Mater*, **23**:1736–1742.
16. C. Buchanan, E. Voigt, C. Szot, J. Freeman, P. Vlachos, M. Rylander (2014). Three-dimensional microfluidic collagen hydrogels for investigating flow-mediated tumor-endothelial signaling and vascular organization. *Tissue Eng Part C*, **20**:64–75.
17. X. Mu, W. Zheng, L. Xiao, W. Zhang, X. Jiang (2013). Engineering a 3D vascular network in hydrogel for mimicking a nephron. *Lab Chip*, **13**:1612–1618.

18. J. Miller, K. Stevens, M. Yang, B. Baker, D. Nguyen, D. Cohen, E. Toro, A. Chen, P. Galie, X. Yu, et al. (2012). Rapid casting of patterned vascular networks for perfusable engineered three-dimensional tissues. *Nat Mater*, **11**:768–774.
19. D. Kolesky, R. Truby, A. Gladman, T. Busbee, K. Homan, J. Lewis (2014). 3D bioprinting of vascularized, heterogeneous cell-laden tissue constructs. *Adv Mater*, **26**:3124–3130.
20. L. Bertassoni, M. Cecconi, V. Manoharan, M. Nikkhah, J. Hjortnaes, A. Cristino, G. Barabaschi, D. Demarchi, M. Dokmeci, Y. Yang, et al. (2014). Hydrogel bioprinted microchannel networks for vascularization of tissue engineering constructs. *Lab Chip*, **14**:2202–2211.
21. A. Golden, J. Tien (2007). Fabrication of microfluidic hydrogels using molded gelatin as a sacrificial element. *Lab Chip*, **7**:720–725.
22. X. Wang, Z. Jin, B. Gan, S. Lv, M. Xie, W. Huang (2014). Engineering interconnected 3D vascular networks in hydrogels using molded sodium alginate lattice as the sacrificial template. *Lab Chip*, **14**:2709–2716.
23. V. Tsang, A. Chen, L. Cho, K. Jadin, R. Sah, S. DeLong, J. West, S. Bhatia (2007). Fabrication of 3D hepatic tissues by additive photopatterning of cellular hydrogels. *FASEB J*, **21**:790–801.
24. Y. Xia, G. Whitesides (1998). Soft lithography. *Annu Rev Mater Sci*, **28**:153–184.
25. A. McGuigan, M. Sefton (2007). The influence of biomaterials on endothelial cell thrombogenicity. *Biomaterials*, **28**:2547–2571.
26. X. Wang, D. Phan, A. Sobrino, S. George, C. Hughes, A. Lee (2016). Engineering anastomosis between living capillary networks and endothelial cell-lined microfluidic channels. *Lab Chip*, **16**:282–290.
27. X. Li, S. Xu, P. He, Y. Liu (2015). In vitro recapitulation of functional microvessels for the study of endothelial shear response, nitric oxide and (Ca^{2+})i. *PLoS One*, **10**:e0126797.
28. M. Esch, D. Post, M. Shuler, T. Stokol (2011). Characterization of in vitro endothelial linings grown within microfluidic channels. *Tissue Eng Part A*, **17**:2965–2971.
29. L. Bischel, E. Young, B. Mader, D. Beebe (2013). Tubeless microfluidic angiogenesis assay with three-dimensional endothelial-lined microvessels. *Biomaterials*, **34**:1471–1477.

30. L. Bischel, S. Lee, D. Beebe (2012). A practical method for patterning lumens through ECM hydrogels via viscous fingering patterning. *J Lab Autom*, **17**:96–103.
31. M. Abdelgawad, C. Wu, W. Chien, W. Geddie, M. Jewett, Y. Sun (2011). A fast and simple method to fabricate circular microchannels in polydimethylsiloxane (PDMS). *Lab Chip*, **11**:545–551.
32. L. Fiddes, N. Raz, S. Srigunapalan, E. Tumarkan, C. Simmons, A. Wheeler, E. Kumacheva (2010). A circular cross-section PDMS microfluidics system for replication of cardiovascular flow conditions. *Biomaterials*, **31**:3459–3464.
33. W. Zhang, Y. Zhang, S. Bakht, J. Aleman, S. Shin, K. Yue, M. Sica, J. Ribas, M. Duchamp, J. Ju, et al. (2016). Elastomeric free-form blood vessels for interconnecting organs on chip systems. *Lab Chip*, **16**:1579–1586.
34. S. Patan (2004). Vasculogenesis and angiogenesis. *Cancer Treat Res*, **117**:3–32.
35. M. Potente, T. Makinen (2017). Vascular heterogeneity and specialization in development and disease. *Nat Rev Mol Cell Biol*, **16**:477–494.
36. S. P. Herbert, D. Y. Stainier (2011). Molecular control of endothelial cell behaviour during blood vessel morphogenesis. *Nat Rev Mol Cell Biol*, **12**:551–564.
37. O. M. Tepper, J. M. Capla, R. D. Galiano, D. J. Ceradini, M. J. Callaghan, M. E. Kleinman, G. C. Gurtner (2005). Adult vasculogenesis occurs through in situ recruitment, proliferation, and tubulization of circulating bone marrow-derived cells. *Blood*, **105**:1068–1077.
38. P. McLoughlin, M. P. Keane (2011). Physiological and pathological angiogenesis in the adult pulmonary circulation. *Compr Physiol*, **1**:1473–1508.
39. Y. Hsu, M. Moya, C. Hughes, S. George, A. Lee (2013). A microfluidic platform for generating large-scale nearly identical human microphysiological vascularized tissue arrays. *Lab Chip*, **13**:2990–2998.
40. Y. Hsu, M. Moya, P. Abiri, C. Hughes, S. George, A. Lee (2013). Full range physiological mass transport control in 3D tissue cultures. *Lab Chip*, **13**:81–89.
41. M. Moya, Y. Hsu, A. Lee, C. Hughes, S. George (2013). In vitro perfused human capillary networks. *Tissue Eng Part C*, **19**, 730–737.
42. A. C. Newman, M. N. Nakatsu, W. Chou, P. D. Gershon, C. C. Hughes (2011). The requirement for fibroblasts in angiogenesis: fibroblast-derived matrix proteins are essential for endothelial cell lumen formation. *Mol Biol Cell*, **22**:3791–3800.

43. S. Raghavan, C. Nelson, J. Baranski, E. Lim, C. Chen (2010). Geometrically controlled endothelial tubulogenesis in micropatterned gels. *Tissue Eng Part A*, **16**:2255–2263.
44. C. M. Ghajar, X. Chen, J. W. Harris, V. Suresh, C. C. Hughes, N. L. Jeon, A. J. Putnam, S. C. George (2008). The effect of matrix density on the regulation of 3-D capillary morphogenesis. *Biophys J*, **94**:1930–1941.
45. J. Whisler, M. Chen, R. Kamm (2014). Control of perfusable microvascular network morphology using a multiculture microfluidic system. *Tissue Eng Part C*, **20**:543–552.
46. L. F. Mendes, R. P. Pirraco, W. Szymczyk, A. M. Frias, T. C. Santos, R. L. Reis, A. P. Marques (2012). Perivascular-like cells contribute to the stability of the vascular network of osteogenic tissue formed from cell sheet-based constructs. *PLoS One*, **7**:e41051.
47. E. Logsdon, S. Finley, A. Popel, F. Gabhann (2013). A systems biology view of blood vessel growth and remodelling. *J Cell Mol Med*, **18**:1491–1508.
48. V. Vickerman, R. Kamm (2012). Mechanism of a flow-gated angiogenesis switch: early signaling events at cell–matrix and cell–cell junctions. *Integr Biol*, **4**:863–874.
49. Y. Shin, J. Jeon, S. Han, G. Jung, S. Shin, S. Lee, R. Sudo, R. Kamm, S. Chung (2011). In vitro 3D collective sprouting angiogenesis under orchestrated ANG-1 and VEGF gradients. *Lab Chip*, **11**:2175–2181.
50. G. Jeong, S. Han, Y. Shin, G. Kwon, R. Kamm, S. Lee, S. Chung (2011). Sprouting angiogenesis under a chemical gradient regulated by interactions with an endothelial monolayer in a microfluidic platform. *Anal Chem*, **83**:8454–8459.
51. S. Kim, M. Chung, J. Ahn, S. Lee, N. Jeon (2016). Interstitial flow regulates the angiogenic response and phenotype of endothelial cells in a 3D culture model. *Lab Chip*, **16**:4189–4199.
52. J. Yeon, H. Ryu, M. Chung, Q. Hu, N. Jeon (2012). In vitro formation and characterization of a perfusable three-dimensional tubular capillary network in microfluidic devices. *Lab Chip*, **12**:2815–2822.
53. J. Song, D. Bazou, L. Munn (2012). Anastomosis of endothelial sprouts forms new vessels in a tissue analogue of angiogenesis. *Integr Biol*, **4**:857–862.
54. S. Kim, H. Lee, M. Chung, N. Jeon (2013). Engineering of functional, perfusable 3D microvascular networks on a chip. *Lab Chip*, **13**:1489–1500.

55. E. Tzima, M. Irani-Tehrani, W. Kiosses, E. Dejana, D. Schultz, B. Engelhardt, G. Cao, H. DeLisser, M. Schwartz (2005). A mechanosensory complex that mediates the endothelial cell response to fluid shear stress. *Nature*, **437**:426–431.
56. O. Khan, M. Sefton (2010). Endothelial cell behaviour within a microfluidic mimic of the flow channels of a modular tissue engineered construct. *Biomed Microdevices*, **13**:69–87.
57. H. Abaci, Y. Shen, S. Tan, S. Gerecht (2014). Recapitulating physiological and pathological shear stress and oxygen to model vasculature in health and disease. *Sci Rep*, **4**:4951.
58. D. Nguyen, S. Stapleton, M. Yang, S. Cha, C. Choi, P. Galie, C. Chen (2013). Biomimetic model to reconstitute angiogenic sprouting morphogenesis in vitro. *Proc Nat Acad Sci USA*, **110**:6712–6717.
59. J. Song, L. Munn (2011). Fluid forces control endothelial sprouting. *Proc Nat Acad Sci USA*, **108**:15342–15347.
60. P. Galie, D. Nguyen, C. Choi, D. Cohen, P. Janmey, C. Chen (2014). Fluid shear stress threshold regulates angiogenic sprouting. *Proc Nat Acad Sci USA*, **111**:7968–7973.
61. M. Swartz, M. Fleury (2007). Interstitial flow and its effects in soft tissues. *Annu Rev Biomed Eng*, **9**:229–256.
62. C. Ng, C. Helm, M. Swartz (2004). Interstitial flow differentially stimulates blood and lymphatic endothelial cell morphogenesis in vitro. *Microvasc Res*, **68**:258–264.
63. S. Chung, R. Sudo, P. Mack, C. Wan, V. Vickerman, R. Kamm (2009). Cell migration into scaffolds under co-culture conditions in a microfluidic platform. *Lab Chip*, **9**:269–275.
64. H. Seo, H. Jeong, H. Joo, S. Choi, C. Park, J. Kim, J. Choi, L. Cui, S. Hong, S. Chung, et al. (2016). Intrinsic FGF2 and FGF5 promotes angiogenesis of human aortic endothelial cells in 3D microfluidic angiogenesis system. *Sci Rep*, **6**:28832.
65. F. Laco, M. Grant, R. Black (2012). Collagen-nanofiber hydrogel composites promote contact guidance of human lymphatic microvascular endothelial cells and directed capillary tube formation. *J Biomed Mater Res A*, **101A**:1787–1799.
66. D. Rosenfeld, S. Landau, Y. Shandalov, N. Raindel, A. Freiman, E. Shor, Y. Blinder, H. H. Vandenburgh, D. J. Mooney, S. Levenberg (2016). Morphogenesis of 3D vascular networks is regulated by tensile forces. *Proc Natl Acad Sci USA*, **113**:3215–3220.

67. N. Ferrara (2009). VEGF-A: a critical regulator of blood vessel growth. *Eur Cytokine Netw*, **20**:158–163.

68. M. Nakatsu, R. Sainson, J. Aoto, K. Taylor, M. Aitkenhead, S. Pérez-del-Pulgar, P. Carpenter, C. Hughes (2003). Angiogenic sprouting and capillary lumen formation modeled by human umbilical vein endothelial cells (HUVEC) in fibrin gels: the role of fibroblasts and Angiopoietin-1. *Microvasc Res*, **66**:102–112.

69. H. Gerhardt, M. Golding, M. Fruttiger, C. Ruhrberg, A. Lundkvist, A. Abramsson, M. Jeltsch, C. Mitchell, K. Alitalo, D. Shima, et al. (2003). VEGF guides angiogenic sprouting utilizing endothelial tip cell filopodia. *J Cell Biol*, **161**:1163–1177.

70. C. Del Amo, C. Borau, R. Gutiérrez, J. Asín, J. García-Aznar (2016). Quantification of angiogenic sprouting under different growth factors in a microfluidic platform. *J Biomech*, **49**:1340–1346.

71. A. Perets, Y. Baruch, F. Weisbuch, G. Shoshany, G. Neufeld, S. Cohen (2003). Enhancing the vascularization of three-dimensional porous alginate scaffolds by incorporating controlled release basic fibroblast growth factor microspheres. *J Biomed Mater Res A*, **65A**:489–497.

72. S. Lim, C. Kim, A. Aref, R. Kamm, M. Raghunath (2013). Complementary effects of ciclopirox olamine, a prolyl hydroxylase inhibitor and sphingosine 1-phosphate on fibroblasts and endothelial cells in driving capillary sprouting. *Integr Biol*, **5**:1474-1484.

73. X. Hao, E. Silva, A. Manssonbroberg, K. Grinnemo, A. Siddiqui, G. Dellgren, E. Wardell, L. Brodin, D. Mooney, C. Sylven (2007). Angiogenic effects of sequential release of VEGF-A165 and PDGF-BB with alginate hydrogels after myocardial infarction. *Cardiovasc Res*, **75**:178–185.

74. S. Nillesen, P. Geutjes, R. Wismans, J. Schalkwijk, W. Daamen, T. van Kuppevelt (2007). Increased angiogenesis and blood vessel maturation in acellular collagen–heparin scaffolds containing both FGF2 and VEGF. *Biomaterials*, **28**:1123–1131.

75. B. Baker, B. Trappmann, S. Stapleton, E. Toro, C. Chen (2013). Microfluidics embedded within extracellular matrix to define vascular architectures and pattern diffusive gradients. *Lab Chip*, **13**:3246–3252.

76. J. Jeon, S. Bersini, J. Whisler, M. Chen, G. Dubini, J. Charest, M. Moretti, R. Kamm (2014). Generation of 3D functional microvascular networks with human mesenchymal stem cells in microfluidic systems. *Integr Biol*, **6**:555–563.

77. E. Silva, D. Mooney (2010). Effects of VEGF temporal and spatial presentation on angiogenesis. *Biomaterials*, **31**:1235–1241.

78. B. Krock, N. Skuli, M. Simon (2011). Hypoxia-induced angiogenesis good and evil. *Genes Cancer*, **2**:1117–1133.
79. M. Brennan, M. Rexius-Hall, L. Elgass, D. Eddington (2014). Oxygen control with microfluidics. *Lab Chip*, **14**:4305–4318.
80. P. Oomen, M. Skolimowski, E. Verpoorte (2016). Implementing oxygen control in chip-based cell and tissue culture systems. *Lab Chip*, **16**:3394–3414.
81. G. Mehta, J. Lee, W. Cha, Y. Tung, J. Linderman, S. Takayama (2009). Hard top soft bottom microfluidic devices for cell culture and chemical analysis. *Anal Chem*, **81**:3714–3722.
82. C. Chang, Y. Cheng, M. Tu, Y. Chen, C. Peng, W. Liao, Y. Tung (2014). A polydimethylsiloxane–polycarbonate hybrid microfluidic device capable of generating perpendicular chemical and oxygen gradients for cell culture studies. *Lab Chip*, **14**:3762–3772.
83. J. Kim, K. Yang, H. Park, S. Cho, S. Han, Y. Shin, S. Chung, J. Lee (2014). Implantable microfluidic device for the formation of three-dimensional vasculature by human endothelial progenitor cells. *Biotechnol Bioprocess Eng*, **19**:379–385.
84. X. Dai, S. Cai, Q. Ye, J. Jiang, X. Yan, X. Xiong, Q. Jiang, A. Wang, Y. Tan (2011). A novel in vitro angiogenesis model based on a microfluidic device. *Chin Sci Bull*, **56**:3301–3309.
85. R. Samuel, L. Daheron, S. Liao, T. Vardam, W. Kamoun, A. Batista, C. Buecker, R. Schafer, X. Han, P. Au, et al. (2013). Generation of functionally competent and durable engineered blood vessels from human induced pluripotent stem cells. *Proc Nat Acad Sci USA*, **110**:12774–12779.
86. D. Reed, G. Foldes, S. Harding, J. Mitchell (2013). Stem cell-derived endothelial cells for cardiovascular disease: a therapeutic perspective. *Br J Clin Pharmacol*, **75**:897–906.
87. Z. Wang, P. Au, T. Chen, Y. Shao, L. Daheron, H. Bai, M. Arzigian, D. Fukumura, R. Jain, D. Scadden (2007). Endothelial cells derived from human embryonic stem cells form durable blood vessels in vivo. *Nat Biotechnol*, **25**:317–318.
88. G. Bergers (2005). The role of pericytes in blood-vessel formation and maintenance. *Neuro Oncol*, **7**:452–464.
89. A. Van der Meer, V. Orlova, P. ten Dijke, A. van den Berg, C. Mummery (2013). Three-dimensional co-cultures of human endothelial cells and embryonic stem cell-derived pericytes inside a microfluidic device. *Lab Chip*, **13**:3562–3568.

90. Y. Liu, S. Rayatpisheh, S. Chew, M. Chan-Park (2012). Impact of endothelial cells on 3D cultured smooth muscle cells in a biomimetic hydrogel. *ACS Appl Mater Interfaces*, **4**:1378–1387.
91. J. Kim, M. Chung, S. Kim, D. Jo, J. Kim, N. Jeon (2015). Engineering of a biomimetic pericyte-covered 3D microvascular network. *PLoS One*, **10**:e0133880.
92. N. Boyd, S. Nunes, J. Jokinen, L. Krishnan, Y. Chen, K. Smith, S. Stice, J. Hoying (2011). Microvascular mural cell functionality of human embryonic stem cell derived mesenchymal cells. *Tissue Eng Part A*, **17**:1537–1548.
93. R. Booth, H. Kim (2012). Characterization of a microfluidic in vitro model of the blood-brain barrier (μBBB). *Lab Chip*, **12**:1784–1792.
94. J. Brown, V. Pensabene, D. Markov, V. Allwardt, M. Neely, M. Shi, C. Britt, O. Hoilett, Q. Yang, B. Brewer, et al. (2015). Recreating blood-brain barrier physiology and structure on chip: a novel neurovascular microfluidic bioreactor. *Biomicrofluidics*, **9**:054124.
95. A. Herland, A. van der Meer, E. FitzGerald, T. Park, J. Sleeboom, D. Ingber (2016). Distinct contributions of astrocytes and pericytes to neuroinflammation identified in a 3D human blood-brain barrier on a chip. *PLoS One*, **11**:e0150360.
96. A. Birukova, S. Chatchavalvanich, A. Rios, K. Kawkitinarong, J. Garcia, K. Birukov (2006). Differential regulation of pulmonary endothelial monolayer integrity by varying degrees of cyclic stretch. *Am J Pathol*, **168**:1749–1761.
97. H. Chung, M. Mireles, B. Kwarta, T. Gaborski (2018). Use of porous membranes in tissue barrier and co-culture models. *Lab Chip*, **18**:1671–1689.
98. T. Pasman, D. Grijpma, D. Stamatialis, A. Poot (2018). Flat and microstructured polymeric membranes in organs-on-chips. *J R Soc Interface*, **15**:20180351.
99. D. Huh, H. J. Kim, J. P. Fraser, D. E. Shea, M. Khan, A. Bahinski, G. A. Hamilton, D. E. Ingber (2013). Microfabrication of human organs-on-chips. *Nat Protoc*, **8**:2135–2157.
100. V. V. Abhyankar, M. Wu, C. Y. Koh, A. V. Hatch (2016). A reversibly sealed, easy access, modular (SEAM) microfluidic architecture to establish in vitro tissue interfaces. *PLoS One*, **11**:e0156341.
101. D. Huh, B. Matthews, A. Mammoto, M. Montoya-Zavala, H. Hsin, D. Ingber (2010). Reconstituting organ-level lung functions on a chip. *Science*, **328**:1662–1668.

102. H. Kim, D. Huh, G. Hamilton, D. Ingber (2012). Human gut-on-a-chip inhabited by microbial flora that experiences intestinal peristalsis-like motions and flow. *Lab Chip*, **12**:2165–2174.

103. W. Yuan, Y. Lv, M. Zeng, B. Fu (2009). Non-invasive measurement of solute permeability in cerebral microvessels of the rat. *Microvasc Res*, **77**:166–173.

104. L. Shi, M. Zeng, Y. Sun, B. Fu (2014). Quantification of blood-brain barrier solute permeability and brain transport by multiphoton microscopy. *J Biomech Eng*, **136**:031005.

105. B. Srinivasan, A. Kolli, M. Esch, H. Abaci, M. Shuler, J. Hickman (2015). TEER measurement techniques for in vitro barrier model systems. *J Lab Autom*, **20**:107–126.

106. A. Barakat (2013). Blood flow and arterial endothelial dysfunction: mechanisms and implications. *C R Phys*, **14**:479–496.

107. K. Hulkower, R. Herber (2011). Cell migration and invasion assays as tools for drug discovery. *Pharmaceutics*, **3**:107–124.

108. J. Friedrich, C. Seidel, R. Ebner, L. Kunz-Schughart (2009). Spheroid-based drug screen: considerations and practical approach. *Nat Protoc*, **4**:309–324.

109. H. Yabushita, M. Ohnishi, M. Komiyama, T. Mori, M. Noguchi, T. Kishida, Y. Noguchi, K. Sawaguchi, M. Noguchi (2004). Usefulness of collagen gel droplet embedded culture drug sensitivity testing in ovarian cancer. *Oncol Rep*, **12**:307–311.

110. J. Butler, H. Kobayashi, S. Rafii (2010). Instructive role of the vascular niche in promoting tumour growth and tissue repair by angiocrine factors. *Nat Rev Cancer*, **10**:138–146.

111. C. Chaffer, R. Weinberg (2011). A perspective on cancer cell metastasis. *Science*, **331**:1559–1564.

112. S. M. Weis, D. A. Cheresh (2011). Tumor angiogenesis: molecular pathways and therapeutic targets. *Nat Med*, **17**:1359–1370.

113. S. P. Chiang, R. M. Cabrera, J. E. Segall (2016). Tumor cell intravasation. *Am J Physiol Cell Physiol*, **311**:C1–C14.

114. K. Stoletov, H. Kato, E. Zardouzian, J. Kelber, J. Yang, S. Shattil, R. Klemke (2010). Visualizing extravasation dynamics of metastatic tumor cells. *J Cell Sci*, **123**:2332–2341.

115. D. Fukumura, D. Duda, L. Munn, R. Jain (2010). Tumor microvasculature and microenvironment: novel insights through intravital imaging in pre-clinical models. *Microcirculation*, **17**:206–225.

116. H. Lee, W. Park, H. Ryu, N. Jeon (2014). A microfluidic platform for quantitative analysis of cancer angiogenesis and intravasation. *Biomicrofluidics*, **8**:054102.
117. C. Buchanan, S. Verbridge, P. Vlachos, M. Rylander (2014). Flow shear stress regulates endothelial barrier function and expression of angiogenic factors in a 3D microfluidic tumor vascular model. *Cell Adh Migr*, **8**:517–524.
118. S. M. Ehsan, K. M. Welch-Reardon, M. L. Waterman, C. C. Hughes, S. C. George (2014). A three-dimensional in vitro model of tumor cell intravasation. *Integr Biol*, **6**:603–610.
119. P. DelNero, M. Lane, S. S. Verbridge, B. Kwee, P. Kermani, B. Hempstead, A. Stroock, C. Fischbach (2015). 3D culture broadly regulates tumor cell hypoxia response and angiogenesis via pro-inflammatory pathways. *Biomaterials*, **55**:110–118.
120. V. S. Shirure, Y. Bi, M. B. Curtis, A. Lezia, M. M. Goedegebuure, S. P. Goedegebuure, R. Aft, R. C. Fields, S. C. George (2018). Tumor-on-a-chip platform to investigate progression and drug sensitivity in cell lines and patient-derived organoids. *Lab Chip*, **18**:3687–3702.
121. A. Wong, P. Searson (2014). Live-cell imaging of invasion and intravasation in an artificial microvessel platform. *Cancer Res*, **74**:4937–4945.
122. M. Chen, J. Whisler, J. Fröse, C. Yu, Y. Shin, R. Kamm (2017). On-chip human microvasculature assay for visualization and quantification of tumor cell extravasation dynamics. *Nat protocols*, **5**:865.
123. J. Jeon, S. Bersini, M. Gilardi, G. Dubini, J. Charest, M. Moretti, R. Kamm (2015). Human 3D vascularized organotypic microfluidic assays to study breast cancer cell extravasation. *Proc Nat Acad Sci USA*, **112**:214–219.
124. P. Guo, B. Cai, M. Lei, Y. Liu, B. M. Fu (2014). Differential arrest and adhesion of tumor cells and microbeads in the microvasculature. *Biomech Model Mechanobiol*, **13**:537–550.
125. W. J. Polacheck, J. L. Charest, R. D. Kamm (2011). Interstitial flow influences direction of tumor cell migration through competing mechanisms. *Proc Natl Acad Sci USA*, **108**:11115–11120.
126. S. P. Carey, C. M. Kraning-Rush, R. M. Williams, C. A. Reinhart-King (2012). Biophysical control of invasive tumor cell behavior by extracellular matrix microarchitecture. *Biomaterials*, **33**:4157–4165.
127. H. D. Foda, S. Zucker (2001). Matrix metalloproteinases in cancer invasion, metastasis and angiogenesis. *Drug Discov Today*, **6**:478–482.

128. D. Raman, P. J. Baugher, Y. M. Thu, A. Richmond (2007). Role of chemokines in tumor growth. *Cancer Lett*, **256**:137–165.

129. E. W. Esch, A. Bahinski, D. Huh (2015). Organs-on-chips at the frontiers of drug discovery. *Nat Rev Drug Discov*, **14**:248–260.

130. Z. Xu, Y. Gao, Y. Hao, E. Li, Y. Wang, J. Zhang, W. Wang, Z. Gao, Q. Wang (2013). Application of a microfluidic chip-based 3d co-culture to test drug sensitivity for individualized treatment of lung cancer. *Biomaterials*, **34**:4109–4117.

131. A. Sobrino, D. Phan, R. Datta, X. Wang, S. Hachey, M. Romero-López, E. Gratton, A. Lee, S. George, C. Hughes (2016). 3D microtumors in vitro supported by perfused vascular networks. *Sci Rep*, **6**:31589.

132. D. Phan, X. Wang, B. Craver, A. Sobrino, D. Zhao, J. Chen, L. Lee, S. George, A. Lee, C. Hughes (2017). A vascularized and perfused organ-on-a-chip platform for large-scale drug screening applications. *Lab Chip*, **17**:511–520.

133. R. Jain (2005). Normalization of tumor vasculature: an emerging concept in antiangiogenic therapy. *Science*, **307**:58–62.

134. A. Boussommier-Calleja, R. Li, M. B. Chen, S. C. Wong, R. D. Kamm (2016). Microfluidics: a new tool for modeling cancer-immune interactions. *Trends Cancer*, **2**:6–19.

135. F. Mattei, G. Schiavoni, A. De Ninno, V. Lucarini, P. Sestili, A. Sistigu, A. Fragale, M. Sanchez, M. Spada, A. Gerardino, et al. (2014). A multidisciplinary study using in vivo tumor models and microfluidic cell-on-chip approach to explore the cross-talk between cancer and immune cells. *J Immunotoxicol*, **11**:337–346.

136. B. Zhang, M. Montgomery, M. Chamberlain, S. Ogawa, A. Korolj, A. Pahnke, L. Wells, S. Massé, J. Kim, L. Reis, et al. (2016). Biodegradable scaffold with built-in vasculature for organ-on-a-chip engineering and direct surgical anastomosis. *Nat Mater*, **15**:669–678.

137. P. M. Van Midwoud, A. Janse, M. T. Merema, G. M. Groothuis, E. Verpoorte (2012). Comparison of biocompatibility and adsorption properties of different plastics for advanced microfluidic cell and tissue culture models. *Anal Chem*, **84**:3938–3944.

Chapter 6

DNA Gene Microarray Biochip and Applications

Jian-Chiun Liou
School of Biomedical Engineering, Taipei Medical University, Taipei 11031, Taiwan
jcliou@tmu.edu.tw

With the completion of the sequencing of human genetic maps, we gradually understood the structure of the genes. Next, scientists need to understand the significance and interrelationship of tens of thousands of genes (genetic decoding) and to study the function of proteomics, combined with microelectronics, micromechanics, life sciences, and bioinformation products, such as "biochips" (a tool to speed up decoding). Biochips can be applied in genetic function research, new drug development, clinical testing, strain screening, and environmental control.

Liquid droplets are printed and used to form DNA microarray biochips. An array is a well-ordered arrangement. Gene chips are DNA microarrays. Microchip technology, developed by computer science and technology, uses probes to immobilize thousands of kinds of DNA in a high-density lattice on a nonporous glass slide or nylon membrane. On the surface, the tested sample is a cDNA target. The glass slide or nylon membrane is then hybridized with the sample. The target nucleic acid in the sample is hybridized and fixed

Microfluidics and Bio-MEMS: Devices and Applications
Edited by Tuhin S. Santra

ISBN 978-981-4800-85-3 (Hardcover), 978-1-003-01493-5 (eBook)
www.jennystanford.com

on the cDNA microarray glass and contains the probe points of the complementary nucleic acid sequence. After washing, the sample nucleic acid without hybridization is removed and the position of the hybridization reaction can be recorded. Therefore, only one experiment is required. The cDNA microarray can record thousands of gene expression patterns. The applications of gene chips are quite extensive, for example, in the analysis of DNA sequences, embryogenesis studies, prediction of the effect and toxicity of drugs on different individuals, search for tissue-specific genes, gene expression profiles of diseases, and gene expression among species.

6.1 Introduction

The life phenomenon of an organism is presented through proteins. A variety of different proteins are derived from the expression of deoxyribonucleic acid (DNA) and genetic material ribonucleic acid (RNA). The DNA and the RNA contain all the genes of the organism. So to understand the mysteries of biology, you must start with the DNA or the RNA, which stores the genes. DNA is a linear, giant molecule composed of four different nuclear oxalic acids. It combines a specific group of nuclear oxalic acids with two linear molecules by a base pair. It forms a double helix. If we have a small piece of DNA, the molecules on this DNA are irradiated or fluorescently labeled as probes. Once the base sequence above it is complementary to the DNA sequence to be tested or the message RNA, both can be attracted by base pair attraction. This process is called a hybridization reaction, in which the DNA to be tested is linked to the mark and can be observed by means of fluorescent light on the mark. DNA microarray wafers, commonly referred to as biochips, have been commercialized since the early 1990s. It has revolutionized biochemical analysis. DNA microarray wafers are nuclear oxalic acid fragments that have been sequenced and prelabeled using the microelectromechanical system technology. DNA sequence droplets are implanted in tens of thousands of squares as small as micrometer size in the wafer. It is then

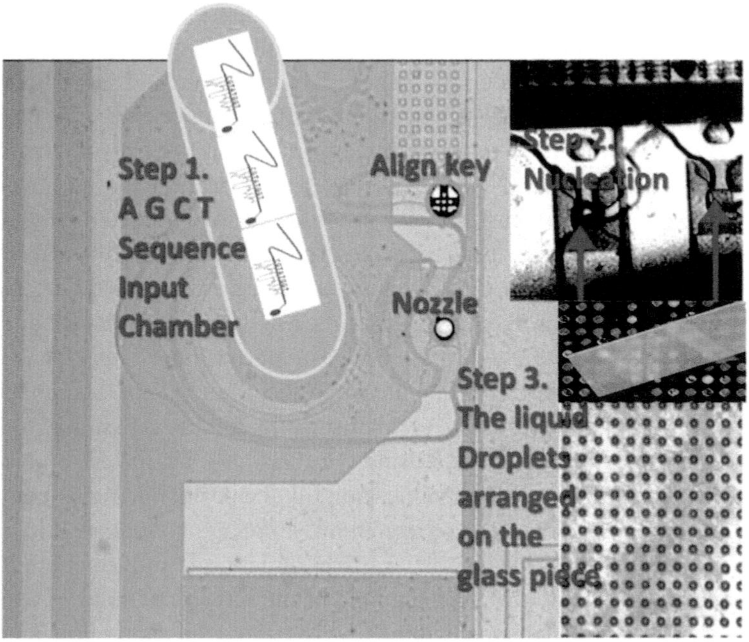

Figure 6.1 DNA sequence.

hybridized to the nuclear oxalic acid fragment to be detected. Using a specific correspondence between the base pairs, by microscopic imaging techniques, it is possible to predict the ordering of the successfully ligated nuclear oxalic acid fragments from the known sorted DNA fragments on the probe [1–12]. In addition, since the sequencing of the human genetic map has been completed, the next phase of biochemical analysis technology development will focus on to the identification of "protein bodies." Protein bodies are thousands of specific protein products that are expressed by cells that are altered by the external environment. The type of these products is inextricably linked to the original DNA sequence of the cell, as shown in Fig. 6.1. Therefore, microarray-based protein wafers have also become the focus of the next wafer technology development [13–18].

Two new biochips have been commercialized, Agilent's microfluidic wafers and Zyomyx's microarray protein wafers. Microfluidic wafers typically use electro-osmotic flow to drive fluid movement between microchannels that connect the components. Electro-osmotic flow is also commonly used in capillary electrophoresis of DNA sequencing. The construction and operation of this novel instrument is similar to that used in the literature by the Renes laboratory earlier. On a wafer the size of a credit card, a photolithography method commonly used in a general semiconductor device process is first used. It is a micropipe that cuts two crosses (the width and depth are usually 50–100 microns). A small container tank is made at each of the four ends of the pipe, and the other side of the pipe is covered with another wafer. In operation, the sample is first injected into one of the container slots and then a voltage is applied. It uses electro-osmotic flow to introduce and fill one of the microchannels. After the original voltage is turned off, a voltage is applied across the microchannels in the other vertical direction to induce another electro-osmotic flow. A small sample of the intersection of the two cross-shaped micropipes is driven by the electro-osmotic flow and moves over the pipe that is not filled with the sample.

When the sample is driven by electro-osmotic flow, molecules with different charge properties will have different rates of movement. Therefore, the time for each component in the sample to move to the end of the pipe is also different. The sample for electrochemically detecting a thin electrode sheet adheres to the tube wall near the end of the tube. Or a transparent window is formed on the wafer covered thereon, and the light source is detected by fluorescence. In addition, the sample is taken out of the wafer by a tiny catheter, and the ionization interface device is connected to convert the sample into ions, which are then detected by a mass spectrometer. Agilent's microfluidic wafer instrument is a fluorescent technique that separates DNA samples by electrophoresis. In operation, the sample and reagents are simply added to the designated sample container on the wafer and the wafer is placed in a slot of the instrument size of a CD-ROM. The instrument operates, detects, and performs data analysis. It has a wide range of uses in new drug development, clinical medical testing, and forensic science [19–22].

6.2 Combining Nanotechnology's Biochips

Following the profound impact of DNA microarray wafer technology on biochemical analysis, various new-generation wafer devices are being developed on a daily basis. Microfluidic electrophoresis wafers and protein microarray wafers are two of the most representative new technologies. Although microfluidic electrophoresis wafers have matured, the application of DNA sequencing–related technology is still the most important. The practicality of protein microarray wafers needs to be further investigated. In recent years, more and more scientists have applied nanotechnology to the development of other new biochips. It is expected that in the near future, its progress will soar and more new biochips will be available.

The ejection performance of the DNA jet monomer needs to be calculated first by heating the heater inside the wafer. The fluid is brought to a superheating state at a brief moment of heating, thereby generating hot bubbles. At the same time, thermal bubbles used to generate instantaneous starting high pressure provide a source of thrust. The fluid is ejected out of the orifice of the spray monomer to achieve atomization.

The self-developed calculation program calculates the properties of the thermal bubbles generated by the heater inside the wafer, including the initial temperature, pressure, and volume of the thermal bubbles as a function of time. At the same time, the mean exit velocity of the fluid at the outlet of the orifice is calculated. It provides the source of thrust required for the FLOW-3D computing software. Experimental observations were made on the inkjet head by Rembe et al. [11], as shown in Fig. 6.2, and the experimental observation conditions were an operation voltage of 15 volts and a heater heating time of 3 μs, equivalent to 6.5 watts of power for heating water. The time it takes to produce thermal bubbles is defined by the experimental observations as 3.1 μs. To verify the accuracy and reliability of the self-developed calculation program, the object of the simulation is calculated for the inkjet head. The calculation result of Fig. 6.3 is a function of the initial temperature, pressure, and volume of the thermal bubble as a function of time when the same operating voltage and heating time are given. At the same time, the obtained hot bubble generation time was 3 μs. It uses

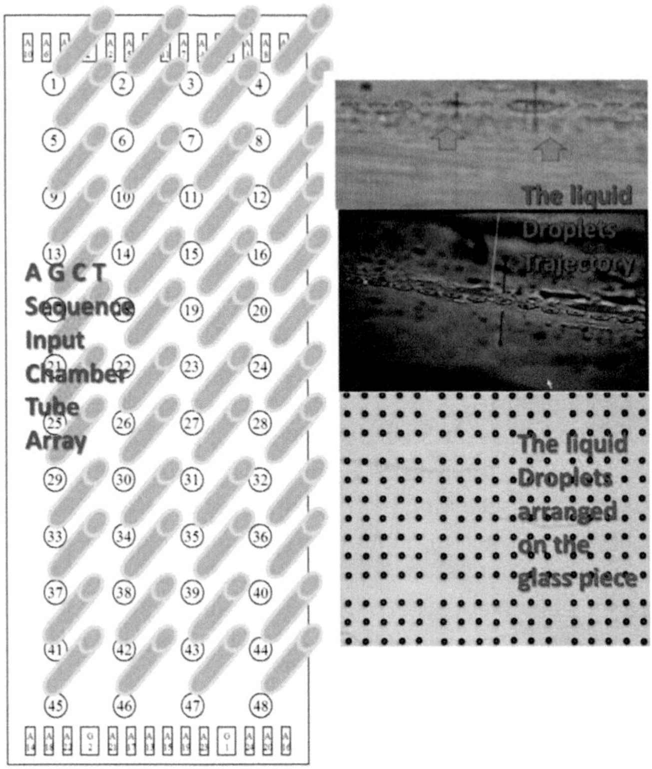

Figure 6.2 Experimental observation conditions.

the calculation of the average exit velocity function of the fluid in the orifice section to provide the source of the thrust required for FLOW-3D simulation software calculations.

6.3 Simulation of the Injection Performance of a Single-Channel Injection Chamber

As a result of the study of the maximum flow rate of the single injection, the maximum flow rate of the monomer designed with 50 orifices was 4.18 cc/min when the working fluid was water. However, the maximum flow rate when spraying DNA is only 2.9 cc/min. Spraying wafers uses different working fluids, and

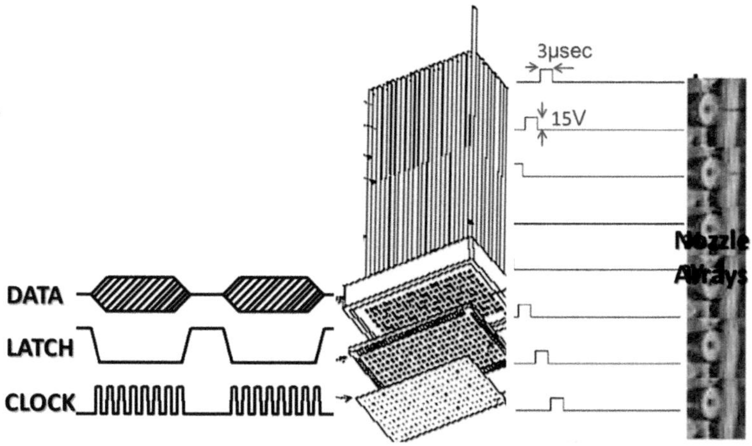

Figure 6.3 Operation voltage of 15 V and heater heating time of 3 μs as a function of the volume, temperature, and pressure of hot bubbles as a function of time.

the characteristics of the jets will vary. There are three physical properties of fluids that affect the size of the jet flow, fluid density, viscosity coefficient, and surface tension, which for water are 1000 kg/m^3, 1 cps, and 73 dyne/cm, respectively. For DNA these are 756 kg/m^3, 0.7 cps, and 20.5 dyne/cm, respectively. The above measurement results indicate that different fluids have heating times and maximum flow rates in the same spray monomer. It is a question worthy of further discussion.

The first computational simulation is done for a conventional type of firing chamber that supplies fluid in a single channel. The geometry of the ejection cavity is an operating frequency of 5 KHz for the fluid to be ejected (Figs. 6.4 and 6.5). Since the calculation range of the droplet flight direction (Z direction) is about 400 μm and the droplet passes through the top section of the calculation area at about 40 μs, the average velocity of the front end of the droplet can be estimated to be about 10 m/s. The trailing length of the tail is about 300 μm, which is roughly consistent with the observations in the relevant literature. From the calculation results of the X-Z cross section at 200 μs, the liquid level of the fluid has reached the outlet of the nozzle and tends to be stable. It means that the fluid

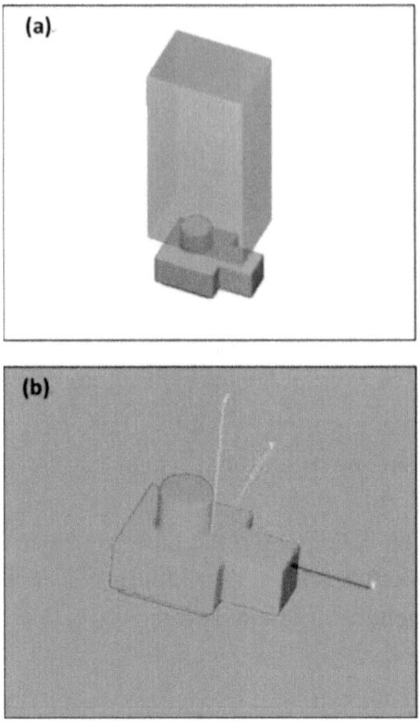

Figure 6.4 Computational simulation established; (a) calculation area range and (b) 3D view of a microfluidic single channel.

has been almost backfilled, and the result shows that the injection chamber is designed to operate at a frequency of 5 kHz and there is no ink supply. In addition, since the single-channel injection cavity is not completely symmetrical to the *Z* axis direction, the tail of the droplet is easily deflected by the asymmetrical flow field force due to the extremely small mass, which deflects the flight direction. It can also be seen from the calculation result of 40 μs instantaneously that the tail of the droplet is biased toward the direction of the supply liquid passage. As the flight time increases, the amount of lateral displacement of the deflection becomes more pronounced. Satellite droplets are formed as the droplets reach the target of the jet. In jet printing processes for inkjet printers or other industrial applications, it is necessary to redesign the geometry so it reduces the asymmetry of the firing chamber relative to the direction of the spray (*Z* axis direction). Its purpose is to reduce the generation of

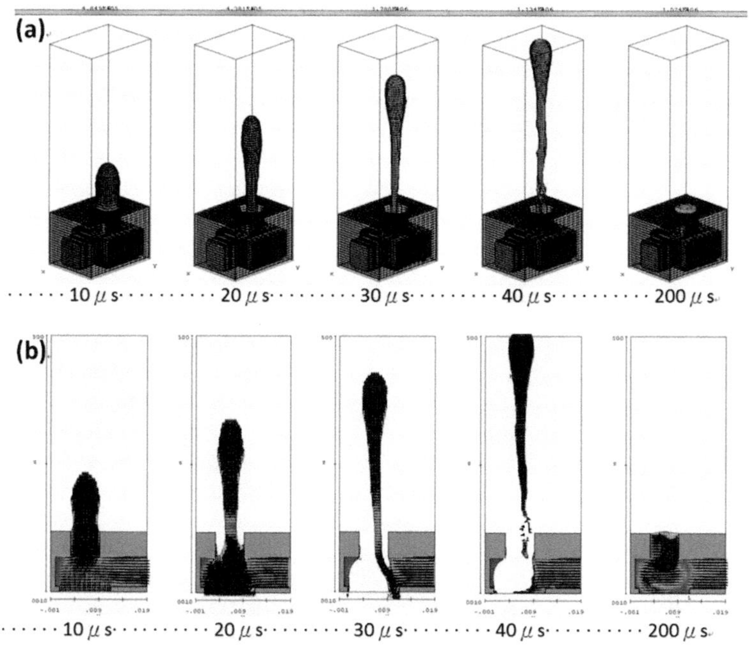

Figure 6.5 Calculation a single-channel injection cavity with a simulated water diameter of 60 μm and thickness of 50 μm at an operating frequency of 5 kHz in (a) 3D side view and (b) X-Z 2D cross section.

satellite droplets. For the application of this study, the research goal is the degree of atomization and the increase of injection frequency and flow rate. Therefore, the directional problem does not affect the achievement of the plan objectives. Figure 6.6 shows a diameter of 60 μm and thickness of 30 μm. This 2D section clearly shows the flow field of the fluid in the ejection chamber immediately before and after the injection. The results show that at 10 μs, the hot bubble is still at the growth stage. In addition to ejecting droplets from the orifice, the fluid is pushed from the ejection chamber into a single channel and discharged from it at the same time. At 20 μs, it can be seen that the thermal bubble has entered the dissipating phase, the liquid has begun to backfill, and the velocity field vector is directed by the channel to the interior of the ejection chamber. At the 30 μs and 40 μs stages, a certain amount of fluid volume is carried

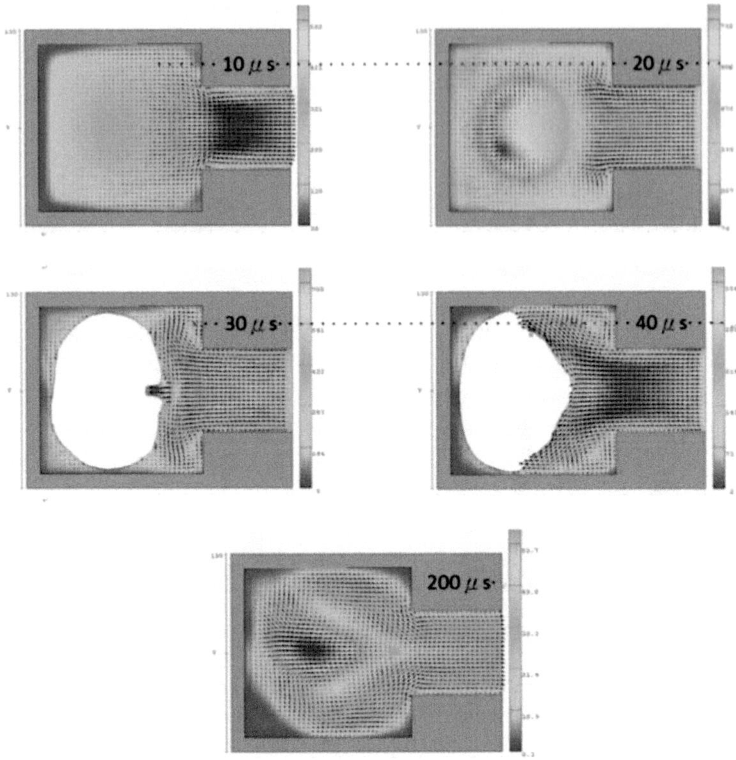

Figure 6.6 Calculation of a single-channel injection cavity with a simulated water diameter of 60 μm and thickness of 50 μm, with an operating frequency of 5 KHz, in the X-Y 2D section.

away as the liquid has been ejected from the orifice. As a result, no fluid can completely fill the interior of the injection chamber. At this time, the outside air enters the injection cavity through the injection hole. It simultaneously waits for the backflow of the fluid through the flow channel and then discharges the air from the orifice. If the air is not completely exhausted, it may cause air bubbles to block the passage. Because the bubbles formed by the air occupy the ejection chamber, the next ejection action occurs. It does not spray a proper volume of liquid smoothly. Since the fluid takes away a large amount of the heat generated by the heater, it has the effect of dissipating heat so the heat is accumulated inside the wafer where the ejection chamber is located. The vicious cycle will cause the efficiency of the jet to begin to drop, and eventually the droplets will not be ejected. Due to an excessive internal temperature of the wafer, therefore, the design of the microfluidic channel of the poor ejection chamber may cause bubbles to remain inside, and the ejection efficiency is lowered. If the heater heating operation is not stopped immediately, the wafer is eventually destroyed because the temperature is too high.

For the purpose of this project, the shape of the droplet of the second shot and the length of the tail drag are different from those of the first shot. The second shot represents the flow of the fluid in the injection chamber that has not been backfilled and stabilized, that is, the second injection action has started, which will influence the second injection effect (Fig. 6.7). By analogy, the subsequent jetting effect will be worse. At the same time, the calculation results of the end of the injection cycle of 100 μs and 200 μs show that there is no satellite liquid formed by the tail dragging at 100 μs. There is a drop in the calculation area. At 200 μs, a satellite droplet is at the top of the calculation area and is about to leave the calculation area. It can be seen that the second injection causes the tail of the droplet to be long. The droplet easily breaks into several small droplets and finally forms satellite droplets. Figure 6.8 shows the results of the X-Z 2D section. From this result, it is clear that at 100 μs and 200 μs, the fluid has not completely filled the nozzle hole of the injection chamber. The single-cavity injection cavity size design is not suitable for operating frequencies up to 10 KHz and the subsequent calculation simulation should be based on the operating frequency of 5 KHz so that the result of the calculation simulation is an unoperated frequency. Figure 6.9 is still the calculation result of the X-Y 2D section with a jet cavity height of 60 μm and a height of 30 μm. Since the fluid is not completely filled with the injection cavity at 100 μs and 200 μs, it can be seen that the velocity vector of the internal flow field is still directed to the direction of the injection cavity by the single-channel direction of the supply fluid, indicating that the fluid backfilling phase is still in progress and the flow field has not stabilized.

Through simulation of the liquid in the 3D model, it is possible to feed back the calculated total amount of liquid discharged and

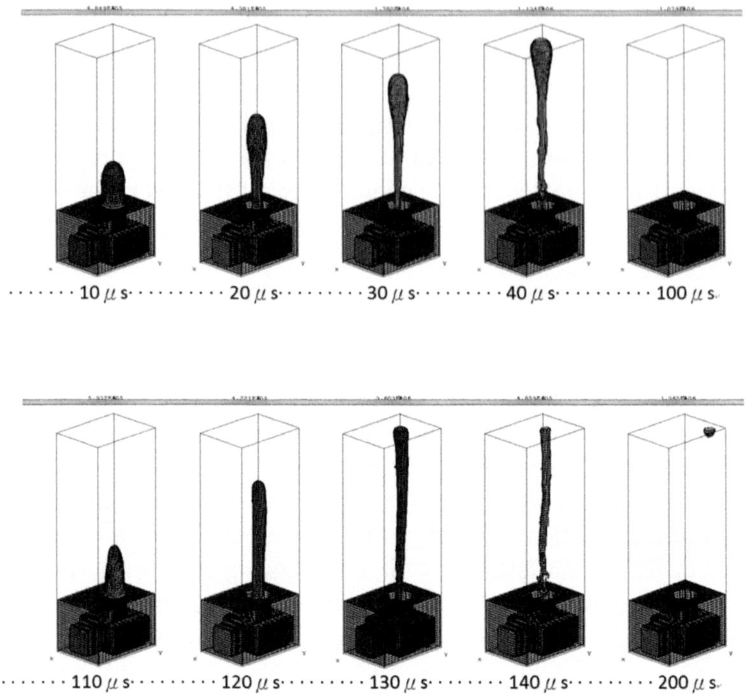

Figure 6.7 Calculation of a single-channel injection chamber with a simulated water diameter of 60 μm and thickness of 50 μm at a 10 KHz operating frequency in a 3D side view.

the speed at which the orifice is ejected. Therefore, the time of the control port must be controlled between each nozzle hole. It is used to prevent the spoiled foreign bodies from interfering with each other and not accurately falling to the coordinate position of the glass pick.

6.4 Experiment and Results

DNA detection usually takes several steps to complete. In traditional array instruments, a micropipette is operated by means of a module with a robotic arm and moved back and forth between different reagents or sample containers to complete the detection step.

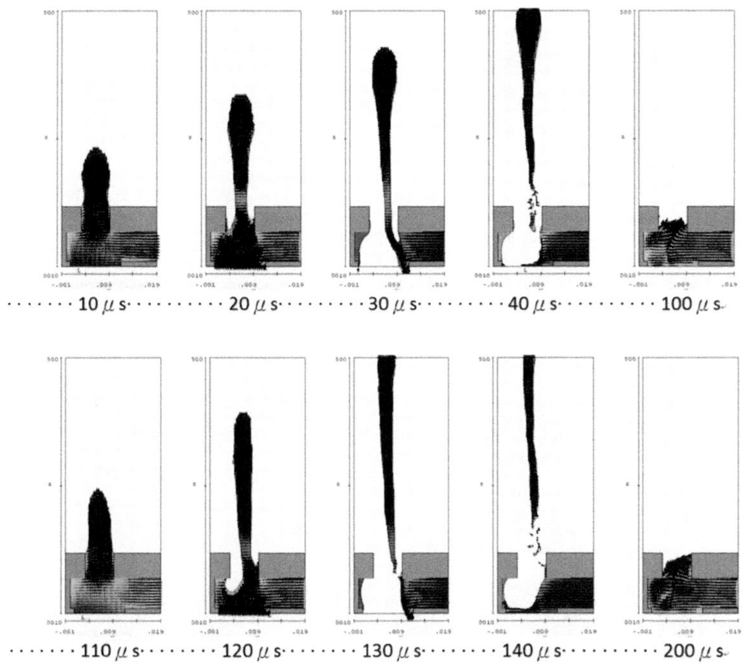

Figure 6.8 *X-Z* 2D section (calculation of the single-channel injection cavity with a simulated water diameter of 60 μm and thickness of 50 μm, with an *X-Z* 2D cross section at 10 a KHz operating frequency).

Thanks to the help of the first generation of biochips, the genetic ordering of many things and organisms, such as rice, mice, and humans, has been successively decoded. Biochips are also evolving to focus on rapid, singular operations, large-molecular-weight analysis, and high technology. To simplify the operating procedure, a microfluidic wafer was developed. The microfluidic wafer is characterized by various components that are required to be used in the detection process, such as a mixing reaction tank, a heating reaction tank, a separation pipe, and a detection tank. They are all concentrated on the same wafer. It then uses the electro-osmotic flow generated by the applied voltage or uses a miniaturized pump or centrifugal force to drive the sample or reagent to move in the micropipes connected between the components to complete the test. This one-piece multifunction wafer is also called a lab-on-a-chip.

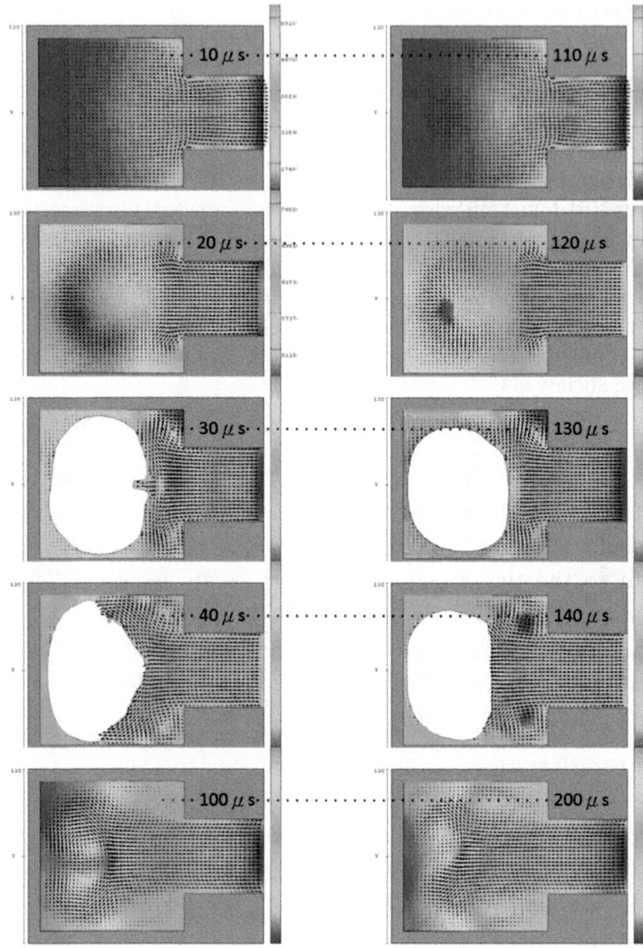

Figure 6.9 Calculation of the single-channel injection cavity with a simulated water diameter of 60 μm and thickness of 50 μm, with an X-Y 2D cross section at a 10 KHz operating frequency.

The concept of detecting DNA fragments using microarray wafers can also be applied to the detection of proteins. Therefore, the design and fabrication of protein microarray wafers is quite similar to that of DNA wafers. Thousands of proteins are first implanted in a grid of several micrometers square, and various proteins in the sample are detected and reacted with specific proteins immobilized on the microarray. As with DNA microarray detection methods, proteins in the sample have been previously labeled with fluorescent functional groups for color development. The microscope is then used to magnify the image to complete the detection. The reaction between proteins is usually done by using a special identification mechanism between the antibody and the antigen. This is different from the sequence in which the DNA fragment utilizes a specific correspondence between various base pairs. So far, for all proteins involved in the development and progression of disease, the antigen-to-antibody recognition response is not fully understood. In addition, the synthesis and purification techniques of these antibodies are not sophisticated. The microarray protein wafer technology will be widely used in the research regarding a variety of new drugs once there is a breakthrough in antigen protein research and antibody preparation technology. In addition to the above-described microarray protein wafers produced by Georgex, there are other wafer devices that use other working principles to determine proteins. For example, Ciphergen Biosystems developed a matrix-assisted laser desorption method using a mass spectrometer technology to identify protein-structured wafers and Biacore International developed a product using surface plasma.

In the bubble-type inkjet technique, a heating electrode is disposed on the wall of the tube at the position of the inkjet head and the inkjet head is instigated to generate DNA liquid droplets by heating the selected heating element by electron pulses, as shown in Fig. 6.10. When the electrode is heated to a certain temperature, the ink droplets become tiny bubbles, burst, and then are ejected through the heating chamber through the inkjet head to adhere to the surface of the substrate. The amount of ink droplets on the substrate depends on the temperature control of the heating device.

The research and development of droplet-type microfluidic wafers is based on the characteristics of the original single droplets and the different droplets generated by the structural design as shown in Fig. 6.11, which are applied to medical disease detection, biological culture, and chemical blending. To match different usage methods, different types of droplets are needed, and there are many

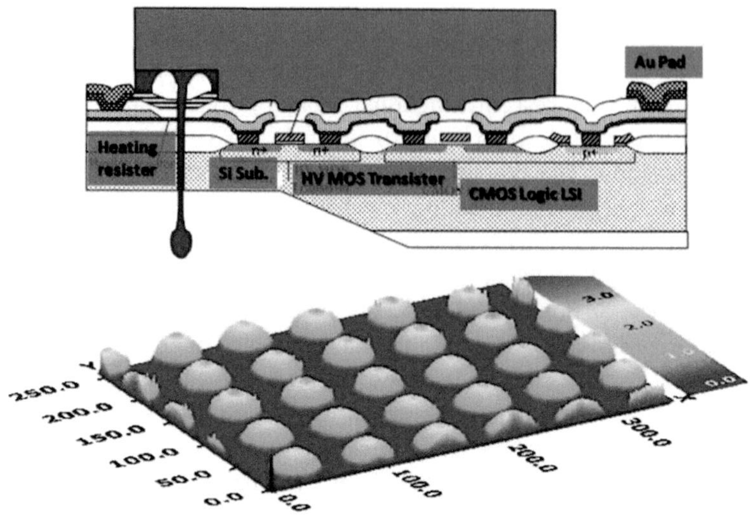

Figure 6.10 Generation of DNA liquid droplets.

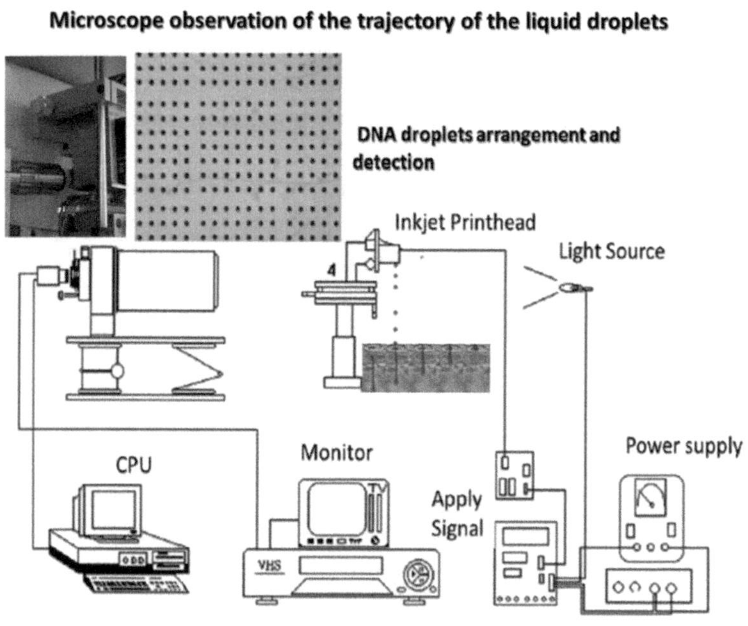

Figure 6.11 Original single droplets.

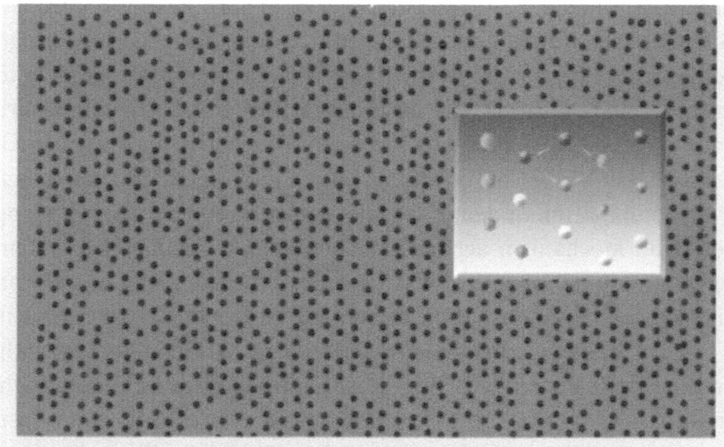

Figure 6.12 Small volume of droplets on a glass substrate.

ways to produce droplets. The mature method is to change the structural design using the structural angle and the carrier fluid flow rate. Different factors help to change the size of the droplets produced.

The importance of droplet research is that the main droplets have the advantages of a small volume (10^{-15}–10^{-9} liters), no diffusion of the sample (content stability), stable reaction conditions (no evaporation or reaction with other substances), no cross contamination (droplets are arranged at fixed distances in the tube), and quick mixing (by internal fluid circulation to achieve mixing), as shown in Fig. 6.12. Droplet technology mainly uses structure and flow rate to control droplet size. It is a simple and effective way. Most of the droplets are developed on the basis of this.

In this DNA gene microarray biochip and applications, we can design cavities of different shapes. Its design concept is how much liquid can be ejected from the liquid in the pool.

6.5 Conclusions

DNA microarrays, also known as DNA arrays or DNA wafers, place thousands or tens of thousands of nucleic acid probes over a few square centimeters. The time of the control port must be controlled between each nozzle hole. It is used to prevent the spoiled foreign bodies from interfering with each other and not accurately falling to the coordinate position. The DNA, cDNA, RNA, etc., in the sample are bound to the probe and detected by fluorescence. It is a very quick replacement of the composition of the liquid ingredients. In such a huge combination of genes, it is necessary to control the matching detection mechanism accurately. This technology includes system signal manipulation and a base frequency mechanism. The test can provide a lot of information related to the genetic sequence. It is a tool for genomics technology and genetics technology research. Researchers can use gene chips to simultaneously and quantitatively analyze large (tens of thousands) gene expressions with fast, accurate, and low-cost bioanalytical testing capabilities. Each of the sprayed DNA beads is capable of meeting precise coordinate positions. Such a small volume of DNA liquid can interfere with the movement during the ejection. How to overcome this interference is an important research topic in the future.

Acknowledgments

This work was financially supported by the Higher Education Sprout Project, Ministry of Education (MOE), Taiwan (DP2-108-21121-01-0-05-04).

References

1. J.-C. Liou, W.-D. Lin (2016). Monolithic CMOS/MEMS spray chip for addressing medical DNA gene sequence. In *3rd International Conference on Green Technology and Sustainable Development (GTSD)*, pp. 82–85.
2. K. Ugsornrat, T. Pogfai, T. Maturos, A. Wisitsoraat, A. Tuantranont (2012). Experimental study of digital microfluidic biochip. In *4th 2011 Biomedical Engineering International Conference*, pp. 207–211.
3. L. M. Cannon, A. Jagarapu, C. A. Vargas-Garcia, M. J. Piovoso, R. Zurakowski (2017). Implications of measurement assay type in design of HIV experiments. In *IEEE 56th Annual Conference on Decision and Control (CDC)*, pp. 4106–4111.

4. M. W. Royal, N. M. Jokerst, R. B. Fair (2013). Droplet-based sensing: optical microresonator sensors embedded in digital electrowetting microfluidics systems. *IEEE Sens J*, **13**(12):4733–4742.
5. J.-Y. Lee, H.-C. Park, J.-Y. Jung, H.-Y. Kwak (2003). Bubble nucleation on micro line heaters. *ASME J Heat Transfer*, **121**:687–692.
6. J.-H. Park, Y. S. Oh (2005). Investigation to minimize heater burnout in thermal thin film print heads. *Microsyst Technol*, **11**(1):16–22.
7. H. Akamatsu, N. Matsukubo, Y. Inokuma, T. Aso, H. K. Shindo (2007). Maximum performance of thermal printhead. In *23rd International Conference on Digital Printing Technologies, Technical Program and Proceedings, Digital Fabrication 2007*, pp. 142–147.
8. T. Lindemann, H. Ashauer, T. Goettsche, H. Sandmaier, Y. Yu, R. P. Peters, D. Sassano, A. Bellone, A. Scardovi, R. Zengerle, P. Koltay (2005). Bubble jet printhead with integrated polyimide nozzle plate. In *18th IEEE International Conference on Micro Electro Mechanical Systems (IEEE Cat. No. 05CH37610)*, pp. 560–563.
9. F.-G. Tseng, C.-J. Kim, C.-M. Ho (2002). A high-resolution high-frequency monolithic top-shooting microinjector free of satellite drops—part II: fabrication, implementation, and characterization. *J Microelectromech Syst*, **11**(5):437–447.
10. F.-G. Tseng, C.-J. Kim, C.-M. Ho (2002). A high-resolution high-frequency monolithic top-shooting microinjector free of satellite drops—part I: concept, design, and model. *J Microelectromech Syst*, **11**(5):427–436.
11. C. Rembe, S. aus der Wiesche, M. Beuten, E. P. Hofer (1999). Investigations of nonreproducible phenomena in thermal ink jets with real high speed cine photomicrography. *J Imaging Sci Technol*, **43**(4):325–331.
12. A. Ashkin, J. M. Dziedzic (1981). Observation of optical resonances of dielectric spheres by light scattering. *Appl Opt*, **20**(10):1803–1814.
13. H.-B. Lin, A. L. Huston, B. L. Justus, A. J. Campillo (1986). Some characteristics of a droplet whispering-gallery-mode laser. *Opt Lett*, **11**(10):614–616.
14. S. C. Hill, D. H. Leach, R. K. Chang (1993). Third-order sum-frequency generation in droplets: model with numerical results for third-harmonic generation. *JOSA B*, **10**(1):16–33.
15. J.-C. Liou, C.-J. Chen, M.-H. Chuang (2004). Lower power supply thermal bubble printhead chip with MEMS technology increasing thermal energy effect. In *Final Program and Proceedings of IS&T's NIP20: International Conference on Digital Printing Technologies*, pp. 863–866.

16. M. Murata, M. Kataoka, R. Nayve, A. Fukugawa, Y. Ueda, T. Mihara, M. Fujii, T. Iwamori (2001). High resolution long array thermal ink jet printhead with on-chip LSI heater plate and micromachined Si channel plate. *IEICE Trans Electron*, **E84C**(12):1792–1800.
17. J.-C. Liou, F.-G. Tseng (2007). Reduced 30% scanning time 3D multiplexer integrated circuit applied to large array format 20KHZ frequency inkjet print heads. *DTIP 2007*, Stresa, Italy, pp. 106–111.
18. J.-C. Liou, F.-G. Tseng (2008). Three-dimensional architecture of multiplexing data registration integrated circuit for large-array ink jet printhead. *J Imaging Sci Technol*, **52**(1):1-1-7.
19. S. J. Shin, K. Kuk, J. W. Shin, C. S. Lee, Y. S. Oh, S. O. Park (2004). Thermal design modifications to improve firing frequency of back shooting inkjet printhead. *Sens Actuators, A*, **A114**(2–3):387–391.
20. S. J. Shin, K. Kuk, J. W. Shin, C. S. Lee, Y. S. Oh, S. O. Park (2003). Firing frequency improvement of back shooting inkjet printhead by thermal management. In *TRANSDUCERS '03. 12th International Conference on Solid-State Sensors, Actuators and Microsystems. Digest of Technical Papers* (*Cat. No. 03TH8664*), pp. 380–383.
21. J.-C. Liou, C.-H. Liu, C.-J. Chen (2004). A bubble-jet MEMS printhead integrated with CMOS de-multiplexer. In *2004 IEEE/SEMI Advanced Semiconductor Manufacturing Conference and Workshop* (*IEEE Cat. No. 04CH37530*), pp. 413–419.
22. R. Nayve, M. Fujii, A. Fukugawa, T. Takeuchi, M. Murata, Y. Yamada, M. Koyanagi (2004). High-resolution long-array thermal ink jet printhead fabricated by anisotropic wet etching and deep Si RIE. *J Microelectromech Syst*, **13**(5):814–821.

Chapter 7

Microneedles: Current Trends and Applications

Hima Manoj,[a,b,*] Pallavi Gupta,[a,*] Loganathan Mohan,[a] Moeto Nagai,[c] Syrpailyne Wankhar,[a] and Tuhin S. Santra[a]

[a]*Department of Engineering Design, Alumni Ave, Indian Institute of Technology, Chennai, Tamil Nadu 600036, India*
[b]*Department of Bioengineering, Christian Medical College, IDA Scudder Rd, Vellore, Tamil Nadu 632004, India*
[c]*Department of Mechanical Engineering, Toyohashi University of Technology, 1-1 Hibarigaoka, Tempaku-cho, Toyohashi, Aichi 441-8580, Japan*
tuhin@iitm.ac.in, santra.tuhin@gmail.com

Transdermal drug delivery facilitates the direct transport of drugs across the skin, deep into the systemic circulation. Particularly since the last decade, the role of microneedles has remarkably improved in combinatorial approaches for drug delivery across the skin. The field of microneedles is an amalgamation of technologies and pharmaceutical sciences research enabling transdermal drug delivery and has tremendous prospects with the emergence of new formulation and fabrication techniques for particles and microneedles. The advancements in microneedles have gained tremendous attention among researchers owing to the various

*These authors contributed equally to this chapter.

Microfluidics and Bio-MEMS: Devices and Applications
Edited by Tuhin S. Santra

ISBN 978-981-4800-85-3 (Hardcover), 978-1-003-01493-5 (eBook)
www.jennystanford.com

biomedical advantages of microneedles. After employing several approaches to overcome the barrier nature of the stratum corneum, transdermal delivery via microneedles has proved to be of significance. It offers a convenient route for drug delivery in various clinical implications. The shortcomings of oral drug delivery and the pain associated with hypodermic needles have shifted the drug delivery orientation toward the transdermal route. This chapter focuses on the history, types, mechanisms, and applications of drug delivery, along with the recent innovations in microneedle drug delivery systems. This chapter also highlights the delivery of different therapeutic agents efficiently via microneedles of various designs and compositions, ultimately showing the potential role of microneedles as a transdermal drug delivery system.

7.1 Introduction

In recent years, drug delivery research has extensively pursued how to transport proteins, DNA, genes, antibodies, and vaccines into cells or tissues safely and efficiently [1–3]. The conventional drug delivery approaches, for example, oral administration and hypodermic injection, exhibit limited functionality owing to the drug inactivation through phase I metabolism during oral delivery and the requirement of a well-trained person for a hypodermic injection. Transdermal drug delivery overcomes these drawbacks, facilitating a safe, easy-to-access, and patient friendly approach. The unfavorable hydrophilicity and large size of biomolecules do not allow passive diffusion across the dermal layers of the skin. The stratum corneum (SC) layer of the skin acts as a barrier against drugs that are lipophobic and large in molecular size, creating complications in the designing of topical formulations. Therefore, various possible options for topical or transdermal delivery systems have been explored to improve drug permeation through the skin, like nanocarrier-loaded topical creams, transdermal patches, and microneedles.

Microneedles generally involve an array of 3D microstructures having lengths of less than 1000 μm. Such dimensions and arrangements make them ideal for transdermal drug delivery [4, 5]. Microneedles can easily pierce the SC, creating transient microchannels that help in the passive diffusion of the drug into the skin. Microneedles can be designed to have superficial penetration depths that do not touch nerve receptors in the lower reticular dermis (i.e., up to 200 μm). This results in a painless administration process while ensuring optimal drug delivery as required. Faster onset of action, better patient compliance, possibility of self-administration, and improved permeability and efficacy are the key characteristics of this technology [6]. Drugs applied onto the epidermis or dermis layers of the skin are readily absorbed via diffusion and a lymph-mediated uptake mechanism in the regional capillaries and lymphatic networks [10–12]. Additionally, the main function of the skin is to act as an immunological barrier for the body [13–16]. Therefore, from the immunological perspective, improved local immune responses can be attained by the microneedle administration route because of the higher number of resident antigen-presenting cells (APCs), including Langerhans cells and subsets of T lymphocytes, present in the dermal layer [17–20]. Microneedle-based vaccination has also shown a dose-sparing effect in comparison to the traditional intramuscular immunization, stimulating adequate humoral, cellular, and mucosal immune responses [21, 22].

Till date, microneedles have been applicable for the delivery of diverse cargos, ranging from native protein therapeutics to formulations based on nano- or microparticles [12, 23–28]. Among different material-based microneedles, the biocompatible polymer-based microneedle devices have got leverage of addressing several concerns arising in transdermal protein delivery [7, 29–32]. Another advantage of polymeric microneedles is the elimination of the sharp biohazard wastes, which was not possible in the initial microneedle development period. They avoid the detrimental effects on drug stability with a tailorable drug loading capacity, unlike silicon or metal microneedles [33–35]. The polymeric matrix also maintains bioactive protein in a dried state for a long term without a cold chain requirement, reducing the cost and transportation constraints [36–38]. Further tuning of the design has helped in optimizing the physiochemical properties and spatiotemporal release profiles of the drugs over a period ranging from hours to days [39–42]. This chapter summarizes the delivery of different types of cargos through representative types of microneedles for a variety of applications, along with a discussion on the current state of the art, and highlights the future potential of microneedles for drug delivery. The challenges for clinical translation of microneedles are also briefly discussed in the end.

7.2 History of Microneedles

Traditionally, topical drug therapy refers to the application of medicinal substances or drug-laden substances to the diseased or infected skin as this was supposed to be the only condition when the skin barrier is compromised. The exposed skin was considered to be the only route for direct absorption of drug into viable skin. But with the understanding of the fact that intact skin is a semipermeable barrier opened prospects for systemic delivery of medicines. During the early twentieth century, the skin was reported to be more permeable to lipophilic agents. But a study by Smith et al. in 1919 revealed that the barrier properties of the skin had been mainly attributed to the top layers [7]. Subsequently, Scheuplein explored a three-layered laminate model of the skin, composed of the SC, the epidermis, and the dermis [8]. They also investigated in vitro skin permeability for a wide spectrum of drugs and drug permeation exhibited by Fickian diffusion. The outermost 10–15 μm of tissue was supposed to offer the maximum resistance against drug absorption after separation of the SC from the lower layers of the skin. Subsequently, by the 1960s, systemic delivery of drugs applied topically to the intact SC started to attract attention.

A 3-day patch delivering scopolamine for motion sickness treatment was first used for transdermal systemic delivery, which received approval for use in the United States in 1979 [9]. Microneedles were first conceptualized in the 1970s [10], but after the late 1990s only, significant research advancements were made with improvement in microfabrication manufacturing technology [11]. In the first published paper on microneedles, Henry et al. highlighted silicon microneedle use to assist in calcein (model drug) delivery through excised human skin in vitro [11]. An array of solid microneedles embedded in cadaver skin increased calcein delivery by 3 orders of magnitude. Improved transport occurred through leakage pathways present between the needles and skin. Later, skin permeability was enhanced by 1 more order of magnitude due to the unplugged pores left on the skin after removal of the needles. A follow-up study was performed on cadaver skin by McAllister et al. [12]. Their group studied the permeability of various compounds, and the results assessed that insulin, bovine serum albumin (BSA), and latex nanoparticles (NPs, up to 100 nm in diameter) were able to cross the skin after microneedle treatment [13]. Mathematical modeling of the data showed that the transport of the above-mentioned compounds occurs by a simple diffusion process. In these studies, microneedles were fabricated from a silicon wafer using lithography and reactive ion etching techniques. Since then, extensive investigation has been done on these devices [3, 5, 11–13]. In 2004, microneedle arrays were considered to be permitted for the transdermal transport of macromolecules and possibly supramolecular complexes and microparticles [17]. But in the last decades, microneedle technology has improved tremendously owing to the extensive research that has been carried out using a wide range of materials, fabrication techniques, and microneedle designs [18].

7.3 Mechanism of Drug Delivery via Microneedles

There are two major factors determining the mechanism of delivery via microneedles: (i) mechanical disruption of the skin and (ii) release of the drug or vaccine within the epidermis layer. From the epidermis layer, the drug or vaccine can more readily reach its target for action. Most researchers have focused on creating microscopic holes in the skin via insertion of solid microneedles fabricated from silicon or metal. There exist some approaches for the delivery of substances via microneedles. The "poke with patch" approach is based on creating pores with microneedles followed by the application of a transdermal patch (or some prototype) to the skin surface [19]. Drug transportation takes place by diffusion

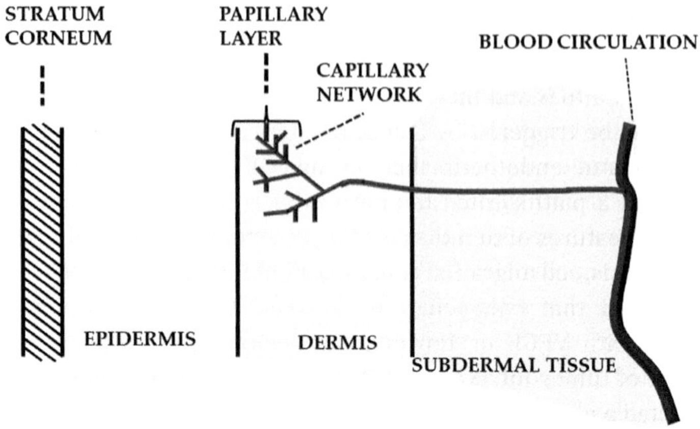

Figure 7.1 Schematic showing the human skin model for a mechanistic analysis of skin permeation [21].

or probably iontophoresis under the influence of an electric field. The "coat and poke" approach uses needles coated with the drug and then inserted into the skin [20]. The required amount of drug is loaded onto the microneedle itself, thereby avoiding any drug reservoir on the skin surface. "Dip and scrape" is a variation of the coat-and-poke approach, where microneedles are first dipped into a drug solution and then scraped across the skin surface to leave behind the drug within microabrasions created by the needles [20]. Hollow microneedles have been designed and studied, which are more reminiscent of an injection than a patch. Though it was not easy to fabricate hollow needles, active fluid flow takes place through the needle bore and into the skin. This results in much faster rates of delivery, which can be modulated over time [19].

Moving onto the uptake of the drug via a microneedle, a systemically active drug must possess some physicochemical properties to reach a target tissue. These properties facilitate the absorption of the drug through the skin. The uptake of the drug by the capillary network in the dermal papillary layer is shown in Fig. 7.1. The rate of permeation, $\mathrm{d}Q/\mathrm{d}t$, across various layers of skin tissues can be expressed as

$$\mathrm{d}Q/\mathrm{d}t = P_{\mathrm{S}}(C_{\mathrm{d}} - C_{\mathrm{r}}), \tag{7.1}$$

where C_d and C_r are the concentrations of the skin penetrant in the donor phase and the receptor phase, respectively; P_S is the overall permeability coefficient of the skin, which can be expressed as

$$P_S = K_S D_{SS}/h_S; \tag{7.2}$$

K_S is the partition coefficient of the penetrant; D_{SS} is the apparent diffusivity of the penetrant; and h_S is the thickness of the skin.

This mathematical relationship deciphers the rate of drug permeation across the skin layers and, therefore, helps in designing the microneedle, the drug dosage, etc.

The next section represents the biological and chemical events occurring at the microneedle insertion physically.

For microneedle application, as discussed earlier, after the microneedle device is inserted into the skin and removed, a micron-size pore is left in the skin. Pore resealing might take some time. This delayed pore resealing increases the chances of infection, which makes the time taken by the skin to recover its barrier properties crucial. Electrical impedance measurement is a practical way for studying pore resealing, which can take 2–40 h, depending on whether the skin is occluded and on needle geometry. Briefly, the breaking of collagen strands and controlled stimulation of the healing response are the basis for the microneedle working principle. It consists of three steps: (i) initiation/inflammation, (ii) proliferation, and (iii) remodeling. In Phase I, after the microneedle penetrates the skin, growth factors are released and the fibroblasts are activated and start proliferating, resulting in the initiation and production of proteins at the injury site for action. Apart from activating fibroblast proliferation, the release of growth factors stimulates the release of cytokines and also facilitates the propagation of intracellular matrix proteins. Phase II involves the replacement of the dominant leucocytes by monocytes; fibroblast proliferation dominates and monocyte, keratinocytes, and fibroblasts continue to release the growth factors. In the final phase, that is, Phase III, the conversion of collagen occurs through tissue remodeling and the maturation of vascular tissues results in the tightening of the skin, as explained in the schematic in Fig. 7.2.

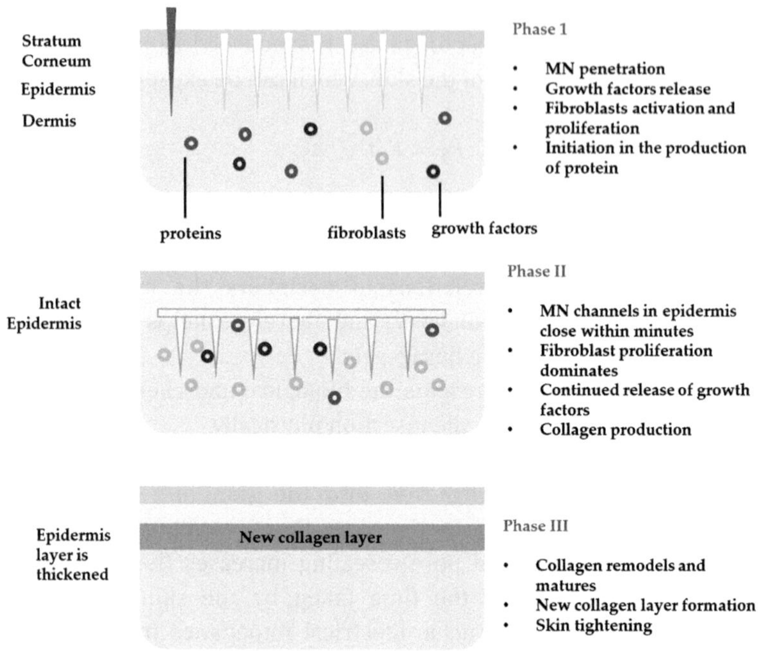

Figure 7.2 Schematic representation of the principle of action of microneedle-based devices [22].

How the drug gets released from the microneedle coating into the blood can be explained as follows: First between the needle wall and the surrounding skin, the skin fluid fills up the annular space. Second, the dissolved drug diffuses across the epidermis and crosses the epidermal or dermal junction into the dermis, where it is absorbed by the blood microcirculation. Third, until the solubility limit is reached or all the drug on the coating is diffused, the drug keeps dissolving into the surrounding skin fluid. Hence, proper knowledge of the annular gap width and the solubility of the coated drug can help in avoiding wastage from coating the needle with excess drug. Fourth, the annular gap width is filled with fluid into which the drug dissolves and the total area of gap width affects permeability. This proves that geometrical parameters such as radius and needle length can be related to permeability.

7.4 Types of Microneedles

7.4.1 Solid Microneedles

Solid microneedles are designed to make pores in the outermost skin layer for easy delivery of drugs. These are able to deliver drugs through passive diffusion across skin layers [5, 23] and also have local effects [24]. There are mainly two types of solid microneedles:

7.4.1.1 Solid, durable microneedles

These are solid microneedles coated with the drug and can be provided to the patient in two ways: either after the removal of the microneedle, the patch containing drugs will remain on the skin, or the solid microneedle itself will be coated with the drug so that the drug will be dissolved into the skin after the insertion of the microneedle into skin [25].

7.4.1.2 Solid, degradable microneedles

Solid microneedles are formed by using drug-infused biodegradable and water-soluble polymers, which increases the drug volume capacity of the microneedle [30].

As discussed in the previous section, Prausnitz et al. first reported a fourfold increase in the permeability of the model drug, calcein, into the human skin after treatment with a silicon microneedle [11]. After that, the first generation of microneedles for transdermal applications included mainly solid silicon, metals (e.g., stainless steel and titanium), and ceramics for fabrication [11]. Later, Narayanan et al. fabricated long and tapered solid silicon microneedles using a tetramethyl ammonium hydroxide etching process with an average height of 158 μm and a base width of 110.5 μm, as shown in Fig. 7.3 [26]. To show improved bioavailability and mechanical strength of solid microneedles, the same group fabricated gold-coated solid silicon microneedles 250 μm in height and 52.8 in base width, with an aspect ratio of 4.73, a tip angle 24.5°, and a diameter of 45 μm [27]. Li et al. studied biodegradable solid polymer microneedles fabricated using polylactic acid (PLA) with mechanical strength sufficient enough

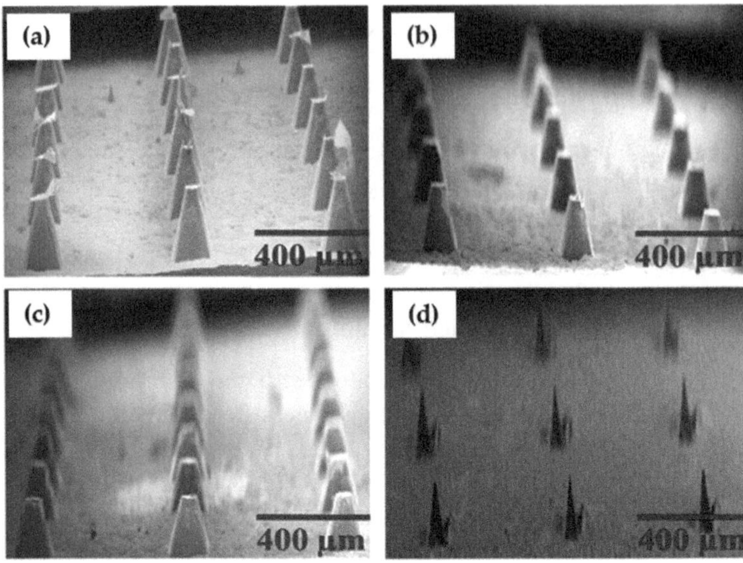

Figure 7.3 Solid Si microneedle array with a SiO_2 layer (a). Partially (b) and fully removed SiO_2 layer (c). Solid Si microneedles (following 2 h 45 min of etching) (d). Reprinted by permission from Springer Nature Customer Service Centre GmbH: Springer Nature, *International Journal of Advanced Manufacturing Technology*, Ref. [26], copyright (2016).

for rupturing the SC and piercing it. Microneedles fabricated with 800 μm depth and 256 microneedles per cm^2 density increased drug permeation [28]. Stainless-steel microneedles and their arrays are also studied by various researchers and can enhance captopril and metoprolol tartrate delivery into skin [1].

7.4.2 Hollow Microneedles

Hollow microneedles contain hollow tips connected to drug reservoirs. They work in a manner to the way a conventional hypodermic injection works. They puncture the skin and then infuse the drugs that are in a liquid formulation through the hollow needle into the tissue. There is no limitation of drug capacity, unlike solid microneedles; therefore, arbitrary quantities of drugs in liquid formulation can be loaded. Generally, there are two types of

Figure 7.4 SEM micrograph of the microneedles. (a) The height of a microneedle array was found to be 300 μm, with an intermicroneedle distance of 500 μm. (b) A single microneedle has outer diameters of 110 μm and 150 μm at the tip and the base, respectively. Reprinted from Ref. [30]. © IOP Publishing. Reproduced with permission. All rights reserved.

hollow microneedle designs as well, one mimicking the traditional hypodermic needle with a single microneedle and the other an array of multiple hollow microneedles [25, 29].

Hollow microneedles provide an easy path for drug delivery into the skin or other tissues, allowing pressure-driven flow of large-dosage solutions. As in the case of conventional hypodermic injections, the pressure applied to hollow microneedles can modulate the drug flow rate and thus enable a rapid bolus injection, a slow infusion, or a time-varying delivery rate with a single flexible microneedle patch and also can deliver a liquid formulation to a wide area [23].

Generally, hollow microneedles are prepared using a rigid substrate by standard microelectromechanical system (MEMS) techniques like deep reactive ion etching, wet chemical etching, and laser micromachining. Yet Vinayakumar et al. fabricated a hollow stainless-steel microneedle array by the following method, as shown in Fig. 7.4. First, circular pores and circular steps were made on either side of an acetone and isopropyl alcohol clean, circular stainless-steel plate (30 × 30 mm) using electron discharge machining. After that, a femto-second laser machining system fabricated the hollow microneedle array of 4 × 4 with 300 μm height, 500 μm intermicroneedle distance, and an outer diameter of 110 μm and 150 μm at the tip and base, respectively [30].

Innovative fabrication methods are developed to make polymer hollow microneedles for flexible devices. Polyphenylsulfone polymer is used to drill and mill to create a bevel-shaped tip and a hollow structure [29]. A digital micromirror stereo lithography instrument was also used recently to fabricate hollow polymer microneedles from liquid resin polymerization. The lithography, electroplating, and molding (LIGA) technique helps fabricate hollow microneedles via X-ray exposure onto polymethyl methacrylate (PMMA) through a mask [31, 32]. A laser-based direct two-photon polymerization rapid prototyping system is also used to form hollow polymer microneedles with a wide variety of geometries. The process involves two-photon polymerization and polydimethyl siloxane (PDMS) micromolding for the fabrication of hollow microneedles with exceptional mechanical properties [33]. Figure 7.5 shows a schematic of microneedles of different designs for drug delivery.

Mishra et al. developed hollow microneedles that work as reservoirs or micropumps while integrating with the drug [35]. The novelty of this fabrication is the effective fluidic interaction between the reservoir and the microneedles present on the substrate. SU-8 microneedles are fabricated using two steps: development by direct laser writing and alignment on the silicon substrate with microfluidic ports pre-etched by wet chemical etching. The microneedles are 500–600 μm in length, with an outer diameter of 100 μm and an inner diameter of 40 μm and an aspect ratio of 5. The microneedles show much higher compressive and bending forces (0.27 N and 0.022 N per needle, respectively) than resistive forces offered by the skin. Successful penetration was achieved for mice and rat skin without any breakage, and a microfluidic analysis showed that its lumen depicts a flow rate of around 0.93 $\mu L\ s^{-1}$ at a 2 KPa pressure difference at the inlet [35]. To increase the efficiency Maaden et al. prepared hollow microneedles with fused silica using hydrofluoric acid etching. These microneedles efficiently injected even small traces of vaccine into the skin via automation mode, nullifying the problems associated with hypodermic needles [36].

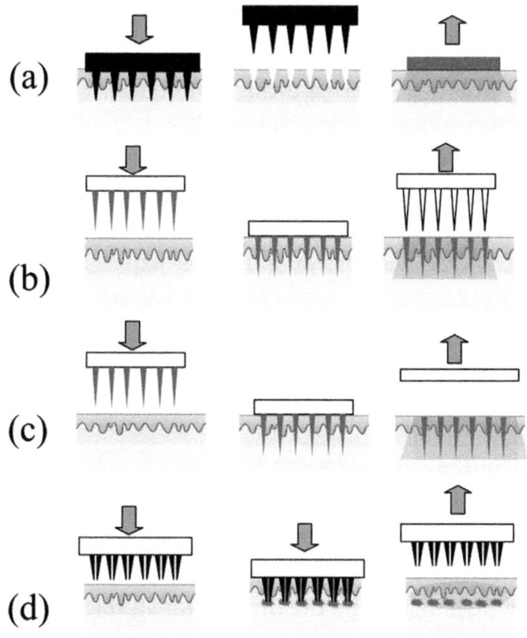

Figure 7.5 Schematic showing drug delivery using various microneedles. (a) Solid microneedles form micron-sized pores across the stratum corneum to permeabilize the skin. After application of a patch containing the drug, the needle patch is removed from the site. (b) Solid microneedles coated with dry drugs or vaccine for faster dissolution in the skin. (c) Encapsulation of the drug or vaccine in polymeric microneedles shows faster and controlled release in the skin. (d) Hollow microneedles facilitate injection of the drug solution. Reprinted from Ref. [34] under a Creative Commons Attribution License (CC BY) (https://creativecommons.org/licenses/by/4.0/).

7.4.3 Polymer Microneedles

Polymers are the key component used for fabricating polymeric microneedles. According to their in vivo performance, they are dissolvable, swellable, and biodegradable [37]. The polymer matrix protects the embedded drugs from mechanical shear and biological degradation. Once a microneedle penetrates the skin and forms a pore, it dissolves or degrades in the skin to release the drugs. To provide sufficient mechanical strength to the polymer microneedle, materials like PLA, PMMA, SU-8, and maltose are being used.

Solid polymer microneedles are normally patterned using an unconventional process due to their tapered 3D structure. The most common way to fabricate this tapered structure is inclined ultraviolet (UV) exposure technology. When a photomask and a negative, thick photoresist-coated substrate are fixed together, the holding stage is tilted and rotated near the UV source. During UV light illumination, the incident and reflected UV at the interface between the resist and the substrate is exploited and thus the polymer photoresist can be patterned into sharp tips. The desired tip can be fabricated by controlling the tilt degree and the exposure dosage. Apart from UV exposure, deep X-ray exposure is also deployed to fabricate microneedle structures, using the LIGA technique [23].

Kim et al. proposed the droplet-borne air-blowing (DAB) method for mild (4°C–25°C) and fast ($\leq$10 min) microneedle fabrication without any drug wastage. The pressure and time of droplet dispenser govern the amount of drug reinforced in the microneedle, and air blowing helps shape the droplet into a microneedle. During fabrication, it has to be ensured that the microneedle provides the force sufficient enough to penetrate skin. The base structure of the two-layered DAB-microneedle maintains the delivery of the drug without wastage. A similar outcome was observed for a DAB-based insulin-loaded microneedle and subcutaneous injection in terms of bioavailability (96.6 $\pm$ 2.4%) and downregulation of the glucose level. Air blowing shaped this droplet into a microneedle and provided the microneedle sufficient force to penetrate the skin [38]. Later, Lee et al. found that pyramid-shaped polymeric microneedles possess greater mechanical strength than conical ones owing to bigger cross section areas at the same base width/diameter. Increment in the base width/diameter (i.e., decrease in the aspect ratio) improves the mechanical strength of the microneedles of the same shape [39].

7.4.3.1 Dissolving microneedles

Dissolvable polymers are made up of water-soluble polymers for quick and complete dissolution in the skin after insertion without producing biohazardous waste, as shown in Fig. 7.6 [40]. The

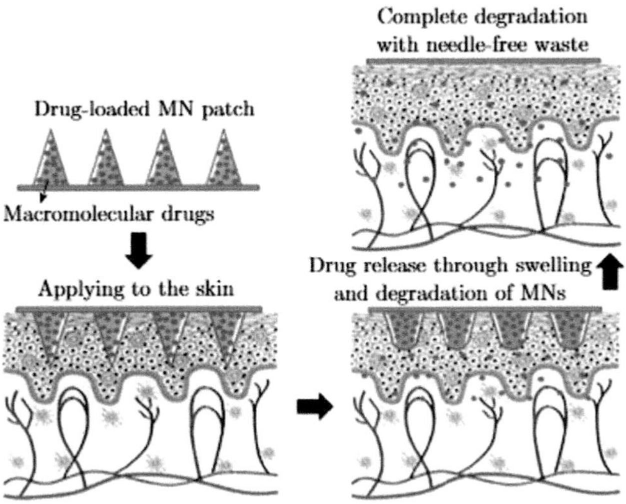

Figure 7.6 The common mechanism of degradable microneedles. Reprinted with permission from Ref. [40]. Copyright (2012) American Chemical Society.

degraded microneedle made up of chitosan of a suitable molecular weight is cleared by the kidney in vivo, whereas high-molecular-weight chitosan chains undergo protease degradation for renal clearance.

To shorten the time of dissolution, microneedles were optimized in an arrowhead design, separating from the shaft within seconds, staying embedded in the skin, and rapidly releasing the encapsulated drug [37]. The skin penetration of microneedles, efficiency of the drug release, and the bioactivity and stability of embedded drugs are also dependent on the properties of the polymers [41]. Most of the dissolvable polymers are water-soluble polysaccharides, such as dextran, sodium chondroitin sulfate, hydroxypropyl cellulose, hydroxypropyl methylcellulose, carboxymethyl cellulose (CMC), sodium alginate, hyaluronic acid (HA), and amylopectin [42].

Dissolving microneedles are usually fabricated by micromolds. Micromolds with specified tapered structures are filled with melted polymers. After these polymers are solidified or polymerized in the micromolds, the sharp tips of the microneedles are integrated with drug formulations. Various materials, including CMC [43], chondroitin sulfate, polyvinylpyrrolidone (PVP), PLA copolymers (polylactic-*co*-glycolic acid [PLGA]) [44], and fibroin, have been deployed to fill into the mold cavities to form microneedles. Since a microneedle may not be entirely inserted into the skin, the drug needs to be encapsulated in multiple layers in the microneedle or an air bubble needs to be added at the microneedle base tip to gain higher efficiency. This layered or air bubble formation can be achieved by a particle-based molding method. The method is based on filling micromolds with polymer microparticles to yield microstructures with multiple reinforcements and complex geometries under mild process conditions. Polymer (PLA, polyglycolic acid [PGA], and PLGA) microparticles 1–30 μm in size were made by spray-drying and emulsion techniques either with or without encapsulation of model drug compounds. These polymer microparticles were packed inside PDMS micromolds at room temperature and melted or bonded to create microstructures according to various protocols [45].

Dissolving polymer microneedle patches of 3-aminophenylboronic acid–modified alginate (Alg-APBA) and hyaluronate dissolve quickly inside the interstitial fluid of the skin after insertion. Naturally occurring, biocompatible, and biodegradable polysaccharide alginate (Alg) and hyaluronate are suitable candidates for sustainable material development in the pharmaceutical industry [46]. Figure 7.7 shows the fabrication of Alg-APBA/hyaluronate microneedles and insulin-loaded Alg-APBA/hyluronate.

Briefly, the master mold consists of around 100 (10×10) pyramid-shaped microneedles ($L \times W \times H \approx 300$ μm × 10 μm × 700 μm, respectively, and the space between the microneedles is ~600 μm). For fabricating Alg-APBA/hyaluronate microneedle patches, first a composite paste was prepared by mixing Alg-APBA (WAlg-APBA:WHA = 1:1) with sodium hyaluronate. Later, a 1% $CaCl_2$ aqueous solution chelated the alginate and hyaluronate, increasing the mechanical strength of the microneedles. Rhodamine B (RB), insulin, or fluorescein isothiocyanate insulin was mixed in the composite paste. This drug-loaded paste (~100 mg) was filled inside a PDMS mold, followed by centrifugation (8000 rpm for 3 min). The centrifugation sediments the composites in the micropores existing in the mold. After discarding the excess drug-

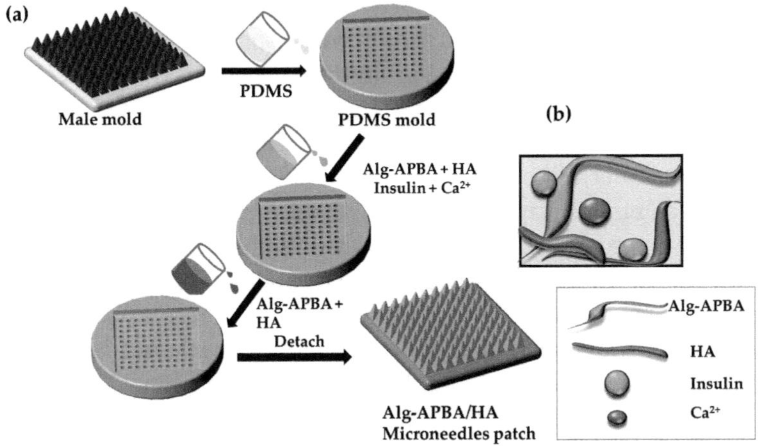

Figure 7.7 Fabrication of Alg-APBA/hyaluronate microneedles by the micromold method (a) and the internal structure of the prepared insulin-loaded Alg-APBA/hyaluronate microneedle (b) [46].

loaded composite, the same centrifugation process was done with drug-deficient composites (~300 mg) inside the mold (for 5 min). After drying overnight in an ambient atmosphere, the Alg-APBA/hyaluronate microneedle patches were stored in dark inside a desiccator until they were peeled off from the molds [46].

7.4.3.2 Biodegradable polymers

Biodegradable polymers are polymers that degrade over a period of time. These polymers are PLA, chitosan, PGA, and PLGA, which can form the matrix and then degrade and release the incorporated drugs for sustained delivery over a long period of time after application [47]. The polymers degrade into nontoxic components, excreted from the body through natural metabolic pathways later. However, these microneedles should be inserted and remain intact in the skin for many days for proper utilization of the degradation property of the biodegradable polymer. The degradation rate of the polymer regulates the drug release rate and duration [37].

These biodegradable microneedles can be divided into two classes: (i) synthetic, that is, PLA, PLGA, PLA copolymers, PGA, polycarbonate, polystyrene (PS), and polycaprolactone (PCL), and

(ii) natural, that is, chitosan, HA, etc. Hong et al. fabricated solid biodegradable microneedles with varying aspect ratios from melted PLA. The microneedle height had no significance in terms of altering their constant degradation rate (25 μm h^{-1}) and surface erosion, even under the influence of an accelerated degradation environment (i.e., alkaline condition) [48]. To improve the properties of microneedles, Kim et al. encapsulated hydrogel microparticles composed of poly-*N*-isopropyl acrylamide (PNIPAAm) in the microneedles fabricated from melted PLGA. The author indicated that the encapsulated PNIPAAm microparticles can quickly swell after insertion into the skin, triggering crack formation and partial mechanical failure of the microneedle at the time it leaves the skin (1 h after insertion). The remnant needle tip kept releasing the drug for days and up to a few months. The drug release occurred via passive diffusion and slow degradation of the PLGA microneedle matrix [47]. Synthetic polymers such as PLGA are soluble in organic solvents or melt at high temperatures. Later, Chen et al. also made an attempt to enhance the mechanical strength of the chitosan with an insertion depth of 200–250 μm, with sustained release of BSA with chitosan degradation for up to at least 8 days [40]. Jin et al. fabricated biodegradable microneedles by using mechanically robust and physiologically inert chitin for a tuberculosis skin test [49]. The water-insoluble chitin microneedles did not show any substantial swelling or degradation. Furthermore, they were able to achieve a 0.2 N of tip buckling force for a single chitin microneedle showing significant skin insertion, as a 0.1 N force punctures human skin. Figure 7.8 shows the steps for

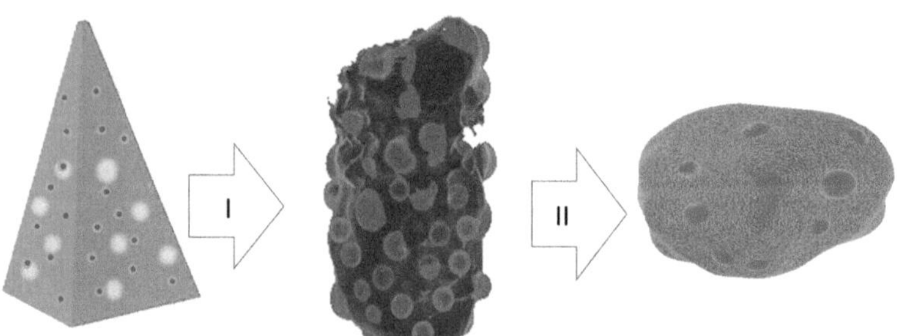

Figure 7.8 Microneedle separation from the skin mediated through hydrogel swelling [50].

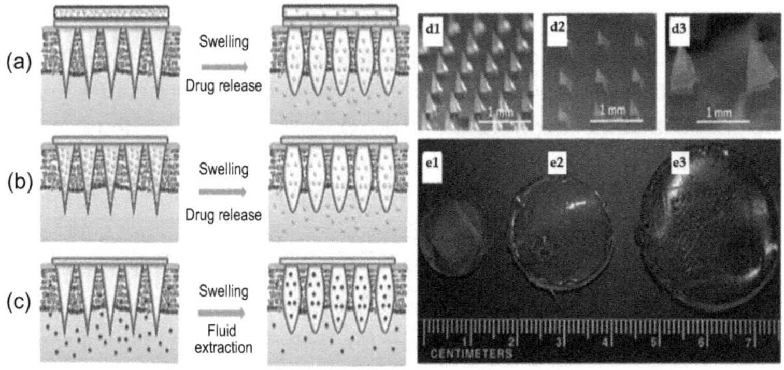

Figure 7.9 (a, b) Swellable microneedles for transdermal drug delivery; (c) interstitial fluid extraction. Following insertion into the skin, the microneedles swell up and release the drugs, preloaded with adhesive patches (a) or microneedle tips (b). On the contrary, interstitial fluid–containing biomarkers can be imbibed into microneedles (c). (d, e) The swelling ability of swellable microneedles fabricated via cross-linking of an aqueous solution of PMVE/MA and PEG. In dry conditions, each microneedle is 600 μm in height and 300 μm in width at the base (d1, e1). After 24 h incubation in water, microneedles swell to various sizes corresponding to the degree of hydrogel cross-linking during molding: cross-linking for 24 h (d2, e2) and 72 h (d3, e3). Reprinted with permission of The Royal Society of Chemistry, from Ref. [37], copyright (2017); permission conveyed through Copyright Clearance Center, Inc.

the separation of the microneedle from skin using hydrogel swelling. These microneedles are not readily dissolvable and therefore would ensure the long-term delivery of targets during application to the skin [50].

7.4.3.3 Swellable polymers

Swellable microneedles are made with swellable polymers/cross-linked hydrogels. The swelling of the polymer allows the extraction of the interstitial fluid from the skin while the preloaded drugs are released, as shown in Fig. 7.9a–c [37]. In contrast to dissolving microneedles, swellable microneedles show two characteristic features: (i) they do not deposit microneedle tip materials in the skin, and (ii) they allow termination of treatment in the case of adverse drug reactions or overdose [51]. The cross-linking density of polymers governs the swelling capacity of the polymer [40], therefore tuning the drug release rate in transdermal drug delivery. Lastly, it is worth considering that swellable microneedles must be mechanically strong enough in the dry state for penetration into the skin. The polymers suited for such microneedles are poly(methyl vinyl ether maleic anhydride) (PMVE/MA)-PEG, poly(2-hydroxyethyl methacrylate), polystyrene-*b*-polyacrylic acid, and polyvinyl alcohol (PVA).

Donnelly et al. fabricated swellable microneedles from aqueous blends containing Gantrez S-97, PEG, and Na_2CO_3 [13]. Lau et al. introduced the advancement route of insulin delivery by fabricating multilayered polymeric microneedle patches with flexible swellable pedestals. In their fabrication, three layers were packed in each patch: silk fibroin + insulin as the outer layer, silk fibroin as the inner layer, and PVA as the flexible pedestal. The composite multilayered patch exhibited higher mechanical strength and thus deeper skin penetration in comparison to the pure silk fibroin microneedle patch. Moreover, the multilayered patch facilitated insulin release at a very high rate, inducing maximal decrement in blood glucose (BG) levels after 2 h of application. The outcome has shown great clinical potential for the pain-free and sustained treatment of diabetes mellitus [52].

All the above-mentioned microneedles are swellable microneedles, which are fabricated using superswelling polymers. The hydrophilic structure allows large uptake of water in the 3D polymeric network. Owing to the interstitial fluid in the skin, channels can form between capillary circulation and the drug patch. This causes swelling when the microneedle is inserted. Before punching into the skin, the microneedles cause disruption in the skin barrier and upon swelling, act as a rate controlling membrane. These swellable microneedles are flexible in size and shape, with easy disinfection and intact removal from the skin [53]. Migdadi et al. used hydrogel forming microneedles for delivering metformin transdermally. The study targets decrease in the gastrointestinal (GI) side effects associated with oral delivery. The results showed improvement in the permeation and bioavailability of the drug with such microneedles [54].

7.5 Microneedle Material and Its Properties

The microneedle was first fabricated using silicon [11, 55]. Many other materials have been used since for microneedle fabrication, such as stainless steel, dextrin [56], glass [57], ceramic [58], maltose [59, 60], galactose [61], and various polymers [12, 62, 63]. We will now discuss microneedle fabrication using different materials (Fig. 7.10).

7.5.1 Silicon

The first microneedle was made from silicon in the 1990s [4]. Silicon has a crystalline structure with an anisotropic nature, and it has elastic moduli of 50 to 180 GPa depending on the alignment of the crystal lattice [9, 47, 64]. It has attractive properties, like flexibility, which allows the fabrication of needles with different sizes and shapes. Yet the major limitations of this material are high cost, a time-consuming complex fabrication process, biocompatibility issues, and material brittleness, so some part may break and remain in the skin, causing some serious health issues [9].

7.5.2 Metal

Stainless-steel, titanium, palladium, nickel, and palladium-cobalt alloys are the metals mainly used for the fabrication of microneedles [65, 66]. These metals exhibit excellent mechanical properties and good biocompatibility. They are strong enough to avoid breaking on penetrating the skin and thus are more suitable than silicon. The first metal used in the production of microneedles was stainless steel by Hashmi et al. in 1995 [55]. Later, titanium was also used as a good alternative to stainless steel [9, 67].

7.5.3 Ceramic

Ceramic microneedles have been fabricated using ceramic micromolding and sintering, PDMS microneedle mold, and ceramic sintering. The major types of ceramics used are calcium sulfate dihydrate (gypsum [$CaSO_4 0.2H_2O$]) and calcium phosphate

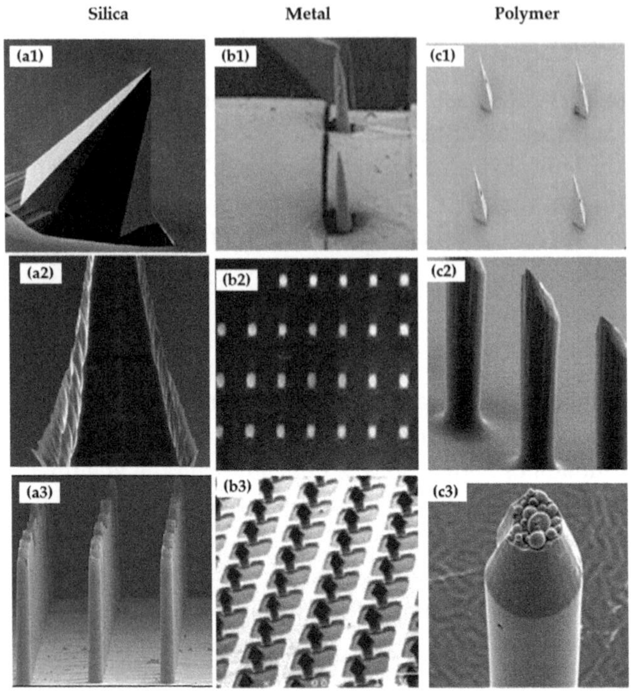

Figure 7.10 Solid microneedles made of silicon, metal and polymer. Silica microneedles: (a1) Reprinted by permission from Springer Nature Customer Service Centre GmbH: Springer Nature, *Pharmaceutical Research*, Ref. [1], copyright (2009). (a2) Reprinted from Ref. [2] under a Creative Commons Attribution 4.0 International (CC BY 4.0) License (https://creativecommons.org/licenses/by/4.0/). (a3) Reprinted by permission from Springer Nature Customer Service Centre GmbH: Springer Nature, *Microsystem Technologies*, Ref. [3], copyright (2017). Metal microneedles: (b1) Reprinted by permission from Springer Nature Customer Service Centre GmbH: Springer Nature, *AAPS Journal*, Ref. [4], copyright (2011). Reprinted by permission from Springer Nature Customer Service Centre GmbH: Springer Nature, *International Journal of Advanced Manufacturing Technology*, Ref. [5], copyright (2016). (b3) Reprinted by permission from Springer Nature Customer Service Centre GmbH: Springer Nature, *AAPS Journal*, Ref. [4], copyright (2011). Polymer microneedles: (c1) Reprinted by permission from Springer Nature Customer Service Centre GmbH: Springer Nature, *Microsystem Technologies*, Ref. [6], copyright (2005). (c2) Reprinted by permission from Springer Nature Customer Service Centre GmbH: Springer Nature, *Pharmaceutical Research*, Ref. [7], copyright (2006). (c3) Reprinted by permission from Springer Nature Customer Service Centre GmbH: Springer Nature, Ref. [7], copyright (2006).

dihydrate (brushite [$CaHPO_4.2H_2O$]) [68]. In recent years, an organically modified ceramic called Ormocer® has been widely used to fabricate microneedles. Alumina (Al_2O_3) is also employed owing to its chemical properties; for example, the highly energetic ionic and covalent bonds between Al and O atoms make strong bonds of this material. These types of ceramic contain three-dimensionally cross-linked copolymers [69], and thus the material has good mechanical properties.

7.5.4 Silica Glass

Silica glass is a physiologically inert and pure compound made from silicon and oxygen atoms. It has the capability to reduced brittle characteristics, which makes it suitable for the fabrication of microneedles [70]. Borosilicate glass made from silica and boron trioxide is more elastic in nature but because a manual fabrication process is used to fabricate microneedles, it is time consuming [12] and has limited application for experimental purposes [9].

7.5.5 Carbohydrate

The most common sugars used are maltose, mannitol, trehalose, sucrose, xylitol, galactose, and polysaccharides [39, 71]. The drug-loaded carbohydrate mixture is cast to carbohydrate slurries, which are molded by using silicon or metal templates for the fabrication of the microneedles [60]. They release the drug by time-based dissolution of the carbohydrate. It is one of the suitable materials for the fabrication of microneedles because of cost-effectiveness and safety for human health. Yet the fabrication process becomes challenging due to their degradation at high temperatures [9].

7.5.6 Polymer

A wide variety of polymers, including PMMA [72], PLA [73], PLGA [63], PGA [12], polycarbonate [74], cyclic olefin copolymer, PVP [75], PVA [75], PS [76], PMVE/MA [77], and SU-8 photoresist [78], are reported for microneedle fabrication. Mostly, dissolving or biodegradable and hydrogel-forming microneedle arrays are made from these polymers. Microneedles fabricated with these polymers have less mechanical strength than other materials but are tougher than glass and ceramics [9, 47].

Therefore, the choice of materials for a microneedle must satisfy multiple criteria that depend on the patch design and patch components. The microneedles need to be sufficiently strong to puncture the skin. They should also be biocompatible, suitable for manufacturing processes using current manufacturing practice, and ideally inexpensive. For nondissolving microneedles, this is often accomplished by using metals such as stainless steel and titanium, polymers such as polycarbonate and PMMA, and silicon. The drug is typically present in a water-soluble matrix that dissolves into the skin. This matrix must be not only biocompatible but also suitable for delivery into and clearance from the body. The material should also be compatible with the drug as the drug remains encapsulated in it during manufacturing, storage, and delivery. Materials that are often used as water-soluble matrices are polymers that can help to provide mechanical strength and viscosity during coating, such as CMC, methyl cellulose, PVP, PVP, HA, chondroitin sulfate, dextran, dextrin, sodium alginate, and sugars.

7.6 Microneedle Patch Design Parameters and Evaluation

7.6.1 Microneedle and Patch Geometry

The geometry of individual microneedles and the overall patch can affect microneedle patch performance. Microneedles are typically 100–1000 μm long. Wider microneedles and smaller aspect ratios can increase the strength and prevent microneedle fracture or, more typically, deformation [35]. But wider microneedles find it harder to penetrate deeply into the skin and can increase pain. Some microneedles taper continuously from a wide base to a sharp tip, which increases the mechanical strength and is needed to support polymer microneedles.

Microneedle tips are often in the range of 1 to 10 μm, which is sufficiently sharp to enable skin insertion. Tips may be less sharp, and in some cases tip sharpness can be well below 1 μm (notably for etched silicon microneedles). Microneedles are assembled into arrays that typically range from tens to tens of thousands of microneedles. The array area is often in the range of 1–10 cm^2. This can result in microneedles with a wide range of densities. A microneedle with a large density can suffer from a bed-of-nails effect, where the insertion force is distributed among too many microneedles, the result being that none can penetrate the skin [9].

This may be overcome by having sharper tips and stronger insertion forces. A lower microneedle density makes skin insertion easier and accommodates a wider area, but the smaller number of microneedles typically reduces the amount of drug that can be delivered. Increasing the microneedle array area accommodates more microneedles but can increase pain. More importantly, a greater array area of microneedle insertion into skin surfaces makes insertion difficult and typically nonplanar and deformable. The overall size of the microneedle is typically larger than the microneedle array, and it can provide additional surface area for skin adhesive and make the patch sufficiently large for easy handling by patients.

7.6.2 Drug and Vaccine Stability in Microneedles

Microneedles need to be stored under conditions that maintain drug activity and ensure microneedle integrity, like sharpness, strength, and sterility. This is accomplished not only through microneedle design but also via packaging that can provide physical protection from mechanical damage and environmental contamination. Storage at appropriate temperature and humidity is also essential. Several studies have shown that trehalose-stabilized influenza vaccines coated onto microneedles lose vaccine activity, which correlates with the crystallization of the sugar over time and the fact that it may have been nucleated at the coating-microneedle interface. In another study, using two-component formulations containing saccharides and/or amino acids in dissolving microneedle patches, influenza vaccine was stabilized for at least 2 years at 25°C as well as through multiple freeze cycles. Measles and hepatitis B vaccines have also been stabilized using microneedle formulations [79, 80].

7.6.3 Mechanical Strength of Microneedles

Microneedles are supposed to possess adequate adhesion property between tip and microtubes and sufficient stiffness of the microtubes for desired penetration. The mechanical strength of microneedles is assessed to estimate the above-mentioned properties. A study showed that during mechanical testing, the microneedles broke only at the interface, when the exerting load was larger than the threshold value. However, the SU-8 microtubes were strong enough to withstand the pressure. After characterization of 20 samples, the average threshold value was calculated to be 7.36 ± 0.48 N for the microneedles (300 μm base and 1000 μm height) since the minimal force required for a successful penetration was reported to be less than 1 N with similar microneedle dimensions and the device was reliable during the penetration process [23].

7.6.4 Characterization of Penetration

Permeation of the drug through the skin is recorded by a diffusion cell apparatus, as shown in Fig. 7.11. Pig ear skin is the most common model for such experiment. The skin is mounted between the receptor and donor compartments. Then the cumulative permeation profiles of microneedle-treated and microneedle-untreated skin are compared [81].

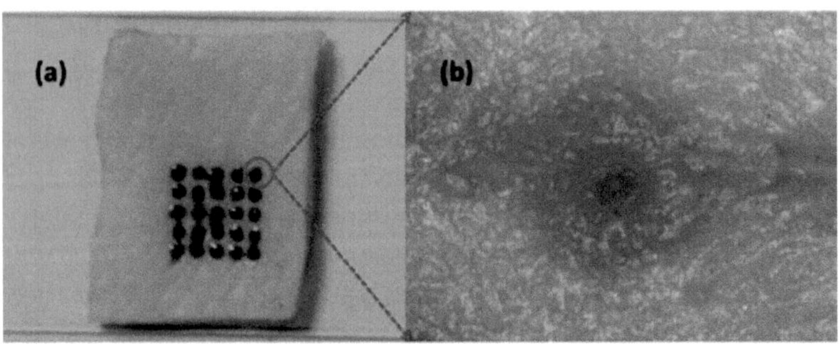

Figure 7.11 (a) Penetration testing results on porcine cadaver skin. (b) In vitro skin permeation studies. Reprinted by permission from Springer Nature Customer Service Centre GmbH: Springer Nature, *AAPS PharmSciTech*, Ref. [81], copyright (2017).

7.6.5 In vivo Animal Model Studies

In vivo studies were performed on mice using microneedles and their therapeutic potentials further evaluated. Hairless rats can be used for the study. The animals should be anesthetized with a suitable technique. To keep a check on the transepidermal water loss, water loss before and after the insertion of the microneedle also needs to be measured. A Delfin VapoMeter is employed to record this parameter [82].

7.7 Applications of Microneedles

Though solid and hollow microneedles were developed to increase the skin permeability with their advanced functionality over hypodermic needles, with advancement in research, now these are applicable in transdermal, ocular, and intracellular delivery.

7.7.1 Delivery to the Skin

The skin is the primary barrier protecting the human body from pathogens and excess water loss. Owing to the highly effective barrier properties of the skin, transdermal drug delivery for local or systemic effects becomes a major challenge. The outer layer of the skin, the SC, made up of dead keratinocytes and intracellular matrix rich in cholesterol, triglycerides, and ceramides, constitutes a lipophilic and dense structure (1.4 g/cm^3) [5]. Though the thickness of the SC is 10–15 μm, it is generally impermeable to drugs with molecular weights higher than 500 Da and Log P out of the 1–3 range. Log P, or the partition coefficient, is a component of Lipinski's rule of 5 and a measure of the lipophilicity of a drug. Moreover, sometimes, the transdermal flux is too low to be considered as successful percutaneous permeation, causing a problem for drugs with a high therapeutically effective dose [83]. As the SC thickness is in microns, the microneedles have the ability to overcome the limitations without going deep into the skin. They also overcome the SC barrier without inflicting pain and avoid damaging the blood vessels present in the reticular dermis of the skin. Therefore, there has been a tremendous increase in microneedle-based drug administration across the skin, including administration of low-molecular-weight drugs, biotherapeutics, vaccines, etc.

Low-molecular-weight drugs are administered using microneedle pretreatment of the skin to increase its permeability. As drugs for systemic indications generally require high doses, initial studies involving this approach were performed for the delivery of small marker compounds, that is, calcein (molecular weight 623 Da) [11], galanthamine (molecular weight 287 Da) [84], and docetaxel in elastic liposomes [85]. Pretreatment with microneedles followed by drug application was used for the delivery of naltrexone across guinea pig and human skin. Microneedle patches were used for drug delivery to human subjects, and the subjects showed elevated and prolonged naltrexone plasma levels for 3 days after microneedle application [86]. Microneedle patches were also employed in topical photodynamic therapy using 5-aminolevulinic acid or 5-aminolevulinic acid methyl ester delivery through skin [87]. The production of the photosensitizer protoporphyrin IX was measured to be higher after delivery of the drug combined with skin pretreatment using a microneedle both in humans and rats. This approach was also used for successfully delivering the preformed photosensitizer *meso*-tetra(*N*-methyl-4-pyridyl) porphinetetratosylate [88].

With the current progress, the most recently randomized clinical trial has been conducted on 21 men to investigate the enhanced delivery of lidocaine, a local anesthesia, after pretreatment with microneedles. With pretreatment, the anesthesia was introduced in half the duration when compared to the topical 4% lidocaine cream [89].

7.7.2 Cosmetology

Microneedles, with features like overcoming the barrier of the SC and being minimally invasive, painless and atraumatic, and bacterial contamination- and long-term-irritation proof, have gained application in cosmetology [90]. Skin restoration also takes place within 1–3 days after treatment, which allows the application of microneedles safely on delicate skin, mainly face, for cosmetology.

Conventionally, cosmetically active ingredients are delivered intradermally through a hypodermic needle (causing pain) or a cream formulation (decreasing the desirable effects). Microneedles are also able to facilitate the transdermal delivery of skin-impermeable, high-molecular-weight, and hydrophilic peptides and proteins [6, 18]. Till now, microneedle application in cosmetology has been aimed at improvement of the skin appearance; skin restoration and rejuvenation [91]; blemish and scar treatment [92–94]; delivery of cosmetically active ingredients like ascorbic acid (AA), eflornithine, and retinyl retinoate (RR) [8]; hair growth regulation [95, 96]; and wrinkle treatment (using a dissolvable microneedle composed of hydrophilic biodegradable polymers PVP and PVP copolymerized with poly(ethylene glycol) dimethacrylate, ascorbic acid, AA, and RR) [97–101].

Mohammed et al. used solid, removable microneedles for increased permeation of lysine-threonine-threonine-lysine-serine (KTTKS), melanostatin, and rigin, the peptides for skin aging and melanogenesis study. Confocal laser scanning microscopy showed that melanostatin reduces melaninogenesis, rigin has an anti-inflammatory effect, and KTTKS helps in collagen production. The melanostatin and KTTKS delivery and distribution were significantly increased over the skin in contrast to rigin, that is, with very slight difference to passive delivery [91]. Further, in vivo experiments revealed that a microneedle roller (192 needles, 500 μm length) applied to the dorsal skin surface of mouse, trimmed via an electric clipper, visually enhanced the results of topically applied VANIQA® (eflornithine hydrochloride, 13.9%) cream. Therefore, preliminary results prove the effectiveness of this technique of eflornithine delivery with a microneedle roller for facial hirsutism treatment [95]. The converse effect was achieved with skin pretreatment via a microneedle roller followed by the topical application of minoxidil, resulting in hair growth stimulation in the clinical trials of androgenic alopecia treatment [96].

With the advancement in cosmetology, solid microneedles are mostly employed as rollers (as shown in Fig. 7.12) and derma pens. Commercially available microneedle rollers have been successfully treating large areas of skin, increasing skin restoration and

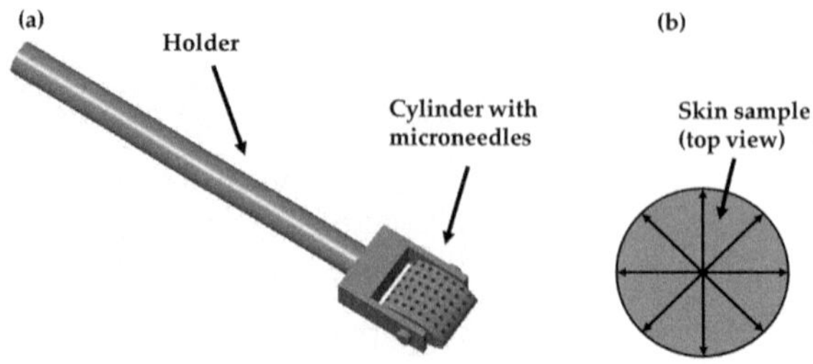

Figure 7.12 Schematic showing solid microneedle–based dermarollers for cosmetology application [102].

permeation and inducing collagen production for both cosmetic and drug applications.

The first Food and Drug Administration (FDA)-approved microneedle product was the Dermaroller®, a cylindrical roller with a surface completely occupied by solid metal microneedles (0.2–2.5 mm in length). The shorter needles are designed especially for home use by patients to improve skin texture, and the longer ones are used in clinics to treat scars and hyperpigmentation. In other studies, microneedle rollers were used to increase optical skin clearing by glycerol [103, 104] and enhanced the skin permeability of chondroitin sulfate [105]. MicroHyala®, a patch covered with dissolving microneedles containing HA to treat wrinkles, and LiteClear® for acne treatment are commercially available worldwide. Other FDA-approved microneedle-based cosmetology applications include C-8 (cosmetic type), CIT-8 (a collagen induction therapy), MF-8 type, MS-4, Soluvia®, and h-patch for cosmetology [18]. In spite of several advantages, solid removable microneedles may cause side effects, such as pain, skin irritation, and inflammation, mainly in the process of percutaneous collagen induction. To overcome this issue, solid microneedles are combined with topical drug treatment in a multiple-step process [105].

7.7.3 Microneedles for Ocular Drug Delivery

In the field of ocular drug delivery, a major challenge is the inaccessibility and extremely soft and delicate nature of the ocular tissues. Ocular drug delivery therapies are useful for diseases such as glaucoma, bacterial/fungal keratitis, dry-eye syndrome, corneal neovascularization, diabetic retinopathy, and other eye-related diseases. In particular, the continuous application of hypodermic needles increases the risk of ocular complications and poor patient compliance [106]. In such a scenario, microneedles have some advantages related to overcoming the ocular barriers (bypassing ocular barrier function, e.g., the epithelium and the sclera), thus allowing the localized delivery of drug molecules within the ocular tissue. This results in minimized pain, lower tissue damage, and reduced risk of infection. It also increases patient compliance owing to the almost invisible length of microneedles, and the ability to achieve targeted drug delivery to the eye [107].

As mentioned earlier, the corneal epithelium decreases the ocular drug bioavailability to very low levels, acting as a barrier. To avoid this disadvantage, Jiang et al. developed fluorescein-coated microneedles, increasing the drug's bioavailability by 2 orders of magnitude. Similarly, pilocarpine-coated microneedles showed greater and longer lasting pupil constriction in comparison to topical drug treatment. This study showed minimally invasive drug delivery to eyes via both intrascleral and intracorneal routes with the aid of coated microneedles. In vitro studies conducted on human cadaveric sclera proposed the designing of microneedles in such a way that it can be inserted into but does not cross the sclera, with fast diffusion of coated drug formulations, small molecules, proteins, and DNA [108]. In vivo studies performed on rabbit cornea showed rapid, sustained fluorescein delivery for several hours, with approximately 70% bioavailability [108].

Glaucoma is an ophthalmic problem caused due to the increase of intraocular pressure, which results in defects in the visual pathways, such as the degeneration of the optic nerve. The conventional glaucoma therapy, including drops and ointments, are tough to manage in the long term due to hinderance by the eye's natural defenses, like blinking and tears. Eye injections are extremely painful and carry the chances of infection and damage to the eyes [109]. Laser iridotomy for glaucoma treatment results in short-term side effects like elevated eye pressure and long-term side effects like seeing a glare or halos. To overcome these shortcomings, Yadav et al. developed microneedles comprising a glass pipette or a metallic

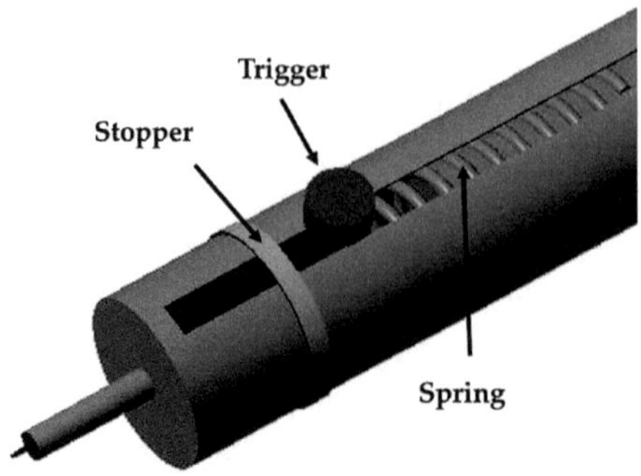

Figure 7.13 Microneedle patches attached with a spring for controlled ocular drug delivery [111].

type, where one end is attached to one end of a tubing while the other end is beveled, helping in insertion into the small blood vessels, particularly the aqueous type of vessels in the eye network [110].

The suprachoroidal space, which is present between the sclera and the choroid, is one of the best pathways for targeted drug delivery into the ocular zone. Here, biocompatible and bioresponsive microneedles play a key role in transdermal delivery [107, 111]. Park et al. developed microneedle patches equipped with a spring and an insertion profile, enabling insertion into the scleral tissues, as shown in Fig. 7.13, and transscleral delivery of RB. This system, with microneedles with variable insertion depths, would also have the possibility to reach deeper ocular segments in a minimally invasive way [111].

Later, Jiang et al. reported that for ocular treatment it is better to use hollow glass microneedles prepared from borosilicate cylindrical glass micropipette tubes with outer and inner diameters of 1.5 mm and 0.86 mm, respectively. Needle insertion up to 700–1080 μm deep into tissue along with sulforhodamine infusion at a pressure of 15 psi resulted in the retraction of the tissue in 60 μm increments. Here, titanium deep reactive ion etching was employed to avoid the shortcomings of large array sizes of microneedles, delivering therapeutically relevant dosages with complex thickness fenestrations (i.e., windows), which are reservoirs for passive delivery [108]. Thus the carrying capacity gets enhanced by up to five times in comparison to solid microneedles of the same dimensions, along with greater subsurface deposition [112].

For ocular drug delivery Parker et al. fabricated microneedles 500–1500 μm long, 50–200 μm wide, 50 μm thick, and with a needle tip angle of 60° using 50-μm-thick Grade 1 CP Ti and titanium inductively coupled plasma deep etch (TIDE) process [113]. The fabrication process entailed the following steps: (i) the deposition of a 3 μm plasma-enhanced-chemical-vapor-deposition SiO_2 etch mask, (ii) lithographic patterning, (iii) dry etching of the mask, (iv) anisotropic through-etching of the underlying Ti using the TIDE process, and (v) mask removal by dry etching. Thus the device had the ability for fenestrations, increasing its drug-loading capacity by up to five times in comparison to that of solid microneedles of identical shank dimensions in ocular drug delivery [114].

To increase the proportion of microneedle insertion, Song et al. fabricated a solid SU-8 resin–based microneedle and then combined it in a macroscale applicator to work as a microneedle pen with a base area of 200 × 200 μm^2 and a height of 140 μm, along with RB, Evans blue or sunitinib malate, and a polymer carrier, which are considered as model drugs for dip coating of a microneedle. The microneedle pen can precisely localize the drug within the stromal membrane of the cornea, otherwise difficult to do when given topically because of the corneal epithelium, present as a barrier [115].

Even though methacrylated hyaluronic acid (MeHA) microneedles are less stiff than HA microneedles [116], HA dissolves quickly into a wet surface like that of the cornea, leading to a burst release of the drug [117]. On the other hand, MeHA shows more resistivity against dissolution and in turn a slower rate of drug release [118, 119]. Than et al. assayed an eye-contact double-layered microneedle patch for controlled ocular drug delivery (Fig. 7.14a) using a simple micromolding method (Fig. 7.14b). The microneedles have an HA inner core and a MeHA as the outer layer. The highly dissolvable HA is coated by MeHA so that the microneedles can penetrate into the wet surface of the cornea. The fabricated patch has a pyramidal-shaped microneedle array with a tip diameter ≈10 μm, height ≈500 μm, base width ≈250 μm, and intermediate spacing

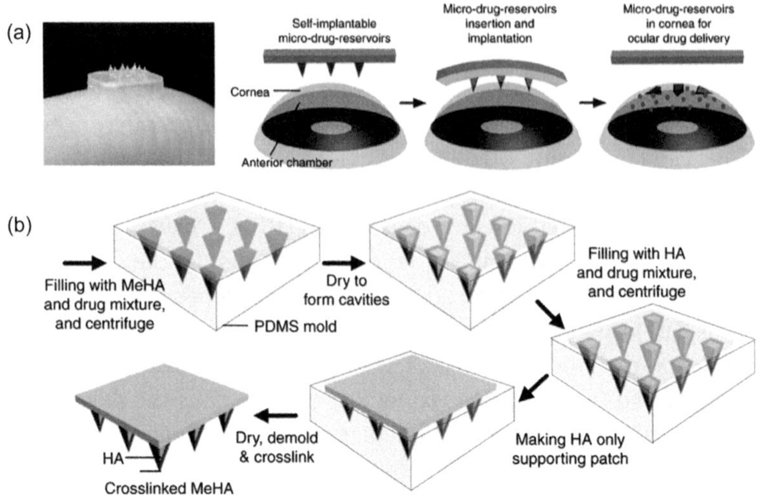

Figure 7.14 Illustration of an eye-contact patch equipped with an array of self-implantable micro-drug-reservoirs for ocular drug delivery; (a) schematic and (b) fabrication process of the polymeric patch with an array of needle-shaped and double-layered microdrug reservoirs. Reprinted from Ref. [109] under a Creative Commons Attribution 4.0 International License (https://creativecommons.org/licenses/by/4.0/).

≈400 μm. The final microneedle dimensions are smaller than those of stainless-steel templates (300 μm base, 600 μm height) owing to the shrinkage of PDMS and HA/MeHA during the fabrication process [109].

Effective ocular drug delivery is challenging because of various eye barriers. In this context, microneedles will play a major role in the field of ocular drug delivery and neither cause discomfort to patients nor require skilled persons. When compared with conventional treatment in the form of topical eye drops (need of high doses at specific intervals and with side effects) and painful, risky intraocular injections, microneedle patches and their various designs can help patients to conveniently and efficiently treat eye disorders at home. Hence, optimization of microneedle designs, injection volumes, method of injection or retraction, forces of injection, pressure of infusion, and tissue damage need to be thoroughly investigated. Also, microneedles can be integrated with sustained drug delivery formulations, such as nano-/microparticles, in situ forming injectable implants, and drug suspensions/solutions, in order to allow targeted and long-term delivery of the formulation within the desired ocular tissue. Finally, microneedles can achieve a major role in revolutionizing ocular drug delivery, like transdermal drug delivery.

7.7.4 Insulin Delivery

Type 1 diabetes and to some extent type 2 diabetes are currently dependent on painful and highly inconvenient hypodermic or subcutaneous insulin injections for medical interventions. Topical insulin delivery does not meet the dose quantity owing to the large molecular weight of insulin (5808 Da) [120]. Therefore, to help diabetic patients, in 1997, a skin perforating device facilitating transdermal delivery of insulin was patented [121]. However, Martanto et al. formulated a more consistent and fundamental method for insulin delivery [122]. Solid removable microneedles were inserted into a diabetic hairless rat model in vivo for insulin delivery, increasing the skin permeability of insulin, which in turn decreased BG by 80%, consistent with 0.05–0.5 U of insulin injected subcutaneously [122].

Till now various microneedles, such as solid removable [122], dissolvable, hollow [70, 123–125], polymer, phase transition, and bioresponsive, have been fabricated for insulin delivery. Hollow microneedles have made progress to human studies for insulin delivery to type 1 diabetes subjects to increase insulin pharmacokinetics by almost two times, which may afford better control over postprandial plasma glucose levels [70, 123]. Hollow metal microneedles induced faster insulin absorption accompanied by rapid onset and offset of metabolic effect on BG levels than subcutaneous injection in larger cohorts of type I diabetes patients [124–128]. Davis et al. used a hairless rat model for insulin delivery with arrays of hollow microneedles [129]. Using the robust fabrication method suitable for inexpensive mass production, the modified-LIGA process, microneedles with mechanical strength sufficient enough to pierce living skin without causing any breakage were developed. There was a 47% decline in the BG levels observed with insulin delivery via microneedles [129]. Later, hollow glass microneedles by Wang et al. [130] were precisely inserted with the assistance of a rotary drilling device to a controlled depth and delivered controlled amounts of insulin. In in vitro experiments conducted on hairless rat skin, the epidermis showed less fluorescence after injections of fluorescently labelled insulin. The results suggest a slower penetration rate of insulin across the dermal-epidermal junction, therefore mostly staying in the dermis with a ± 60 μm resolution. It was concluded that hollow microneedles are useful for precise microinjection into skin, particularly when single needle insertion is assisted by rotary drilling and then retracted before infusion or a microneedle array is inserted by mechanical vibration [12]. In 2006, Ito et al. investigated dissolvable dextrin microneedles loaded with various doses of insulin (0.5, 1, and 2.5 IU/kg). The percutaneous administration into mice in vivo demonstrated the pharmacological availability of 97.7% (0.5 IU/kg), 93.3% (1 IU/kg), and 91.3% (2.5 IU/kg) for all doses. Investigation of the stability of insulin in the microneedle preparations showed that the insulin remaining after 1 month of storage at −80°C was 98.2%, at 20°C was 98.9%, and at 40°C was 99%. Overall this work demonstrated the usefulness of self-dissolving microneedles for the transdermal delivery of insulin [131]. Further, thermal changes in the fabrication process were addressed by droplet-borne air blowing to the polymer droplet. Thus the droplet solidifies into a microneedle shape, facilitating insulin delivery [71]. The high bioavailability (96.6% $\pm 2.4\%$) was supported via regulated BG levels for type 1 diabetic mice models. Migalska et al. fabricated dissolvable microneedles from an aqueous blend of PMVE/MA for insulin delivery. After treatment, the diabetic mice had significantly reduced BG levels [132]. Similarly, various other approaches were used to employ dissolvable microneedles for insulin delivery, for example, encapsulation of insulin into a dissolvable starch/gelatin microneedle patch [133]. Gelatin as a matrix provides biodegradability, biocompatibility, non-immunogenicity, and film-forming ability, while starch forms a durable and strong composite blend suitable for the microneedle matrix, as shown in Fig. 7.15 [133].

Further, an insulin-loaded patch consisting of 100 chondroitin sulphate dissolvable microneedles was used and two to four

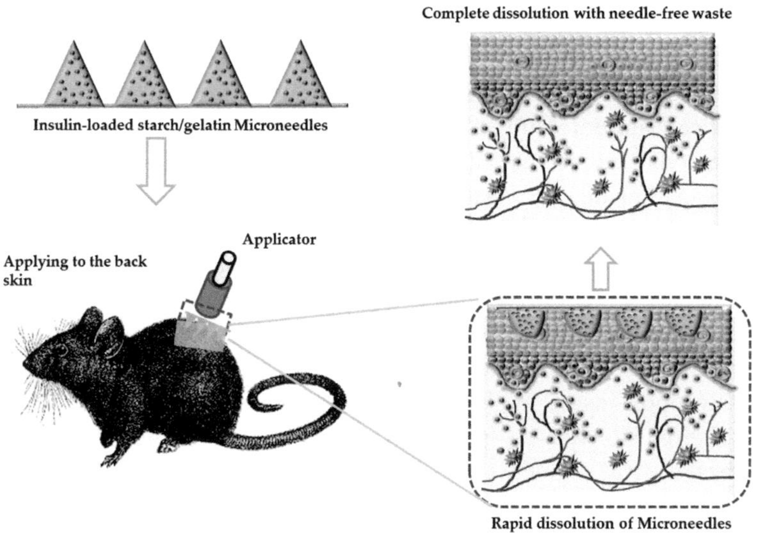

Figure 7.15 Schematic showing the transdermal delivery of insulin using starch/gelatin microneedles; the microneedles rapidly dissolve into the skin to release encapsulated insulin [133].

patches were applied on the shaved abdominal skin of dogs. The authors found 70% more bioavailability of insulin from patches. In addition, 99% recovery was observed owing to the excellent stability even after 1 month [134], proving the usefulness of dissolvable microneedles for immediate insulin delivery in an economical and minimum invasive manner. The results were also favorable for larger animal models, providing insight into the chances of scalability of the technique for human applications. Therefore, for the first time, Gupta et al. conducted insulin delivery with microneedles in 2 adults with type 1 diabetes. The study was conducted to analyze the resealing of the skin's barrier properties after microneedle insertion and to determine the duration of increased skin permeability as a function of microneedle geometry and skin occlusion. Results indicated that in the absence of occlusion, all microneedle-treated sites recovered barrier properties within 2 h, while the occluded sites resealed more slowly, with resealing windows ranging from 3 to 40 h, depending on the microneedle geometry. Upon removal of occlusion, the skin barrier resealed rapidly [135]. Similar studies were conducted to investigate a pharmacokinetic and pharmacodynamic response for insulin injection in type 1 diabetes [136–139]. More recently, a commercially available microneedle device (MicronJet600) was used to inject insulin into 17 patients with type 2 diabetes [6]. Other interesting approaches of insulin delivery include (i) positive nanovesicles driven by iontophoresis through the skin with microneedle-induced microchannels [140], (ii) patches loaded with hypoxia-sensitive vesicles, triggering fast automatic glucose-responsive "closed loop" insulin delivery [132], (iii) stretchable devices with biodegradable microneedles [141], (iv) phase-transition microneedles (hard glassy state to hydrogel state upon insertion) [142], and (v) bioresponsive microneedles, triggered to release protein therapeutics in response to the physiological signals [143–146]. This platform often integrates with the polymeric matrix or submicrometer- or nanometer-sized particles in which a therapeutic protein of interest is encapsulated [147].

7.7.5 Oral and Gastrointestinal Drug Delivery

Traverso and his team developed a pill coated with microneedles with a reservoir in which the drug delivery occurs directly from the internal core of the pill to the target tissue. The pill is capable of selective delivery at different levels of the GI tract because the microneedles are covered with selective bioadhesive pH-sensitive coatings. Such selectivity is due to the exposure to the different environments encountered along the GI tract, which vary according to pH, protease content, bacterial content, and mucosal characteristics. This selective targeting is attractive to study different disease processes with effect in different portions of the GI tract [148].

Seon-Woo et al. fabricated a microneedle patch system for the purpose of improving the adhesive strength and delivery of triamcinolone acetonide (TA) into the mucosa. This system consisted of an adhesive layer (50 μm thickness and 300 μm length) containing 18% TA and a backing layer. After patches were attached to the tape-stripped porcine skin for 10 s, the adhesive strength was observed to be 1.69 N mm/cm^2 with microneedles and 0.41 N mm/cm^2 without microneedles. It showed a delivery rate of 42% more TA into the skin compared to an oral mucosal patch without microneedles (TA AL). The addition of microneedles can increase the oral mucosal patch adhesion to the mucosa, thereby improving the efficiency of drug deliver [149]. Ma et al. developed coated microneedles for direct and minimally invasive intratumoral delivery of anticancer drugs. It was prepared by PLGA NP-encapsulated doxorubicin (DOX) on in-plane (1D) microneedles. Out-of-plane (2D) microneedle arrays (1 cm^2 area) with 57 microneedles coated with free DOX demonstrated uniform DOX delivery in porcine cadaver buccal tissue >3 mm in depth. It was confirmed that the hypodermic injection in to the porcine buccal tissue had leakage of the injected volume (about 25% of the injected 80 μL) [150].

One of the effective treatments for oral cancer was demonstrated by Chen et al. using a light-activated microneedle system that continuously provided photothermal therapy and chemotherapy to superficial tumors and had an anticancer effect. This system has PCL microneedles that contains a photosensitive material such as lanthanum hexaboride, an anticancer drug DOX, and a dissolvable PVA- or PVP-supporting array patch. Because of this supporting array, the microneedles are completely inserted into the skin and to the target tissue for regional cancer treatment. Near-infrared light is exposed to the microneedle array, which uniformly heats the target tissue, inducing a large thermal ablation area (melts at 50°C) to release the DOX in a broad area and finally destroy tumors, as shown in Fig. 7.16. The demonstrated microneedle-mediated synergistic therapy completely eradicated 4T1 tumors within 1 week after

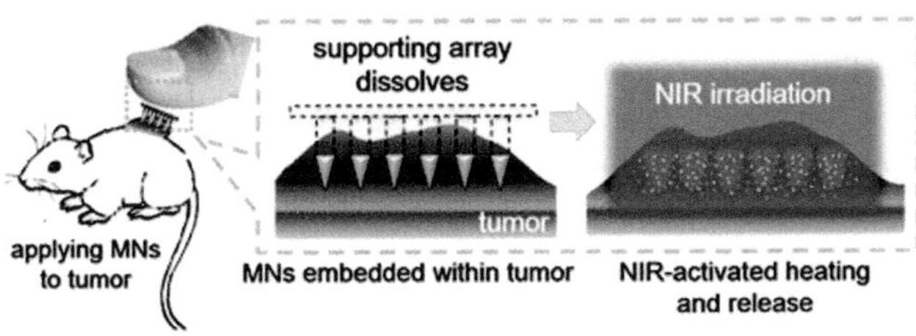

Figure 7.16 Near-infrared optically stimulated microneedle system for the treatment of superficial tumors. Reprinted with permission from Ref. [151]. Copyright (2016) American Chemical Society.

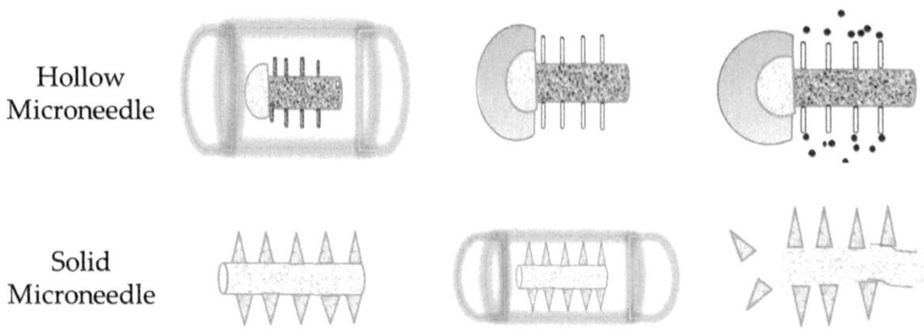

Figure 7.17 Therapeutic concept using a microneedle pill: hollow and solid microneedles. In either case, the pill's needles are first coated with a pH-responsive coating to help in ingestion (left) [152].

a single application of the microneedle and three cycles of laser treatment [151].

The solid, drug-containing polymeric microneedles unfasten from the capsule and move toward GI tissue to release their payload gently. Also the peristaltic motion of the GI tract squeezes the capsule, eventually releasing of drug when the needle and tissue are in close vicinity. The safety evaluation prototype designed with computer-aided design software (Solidworks, Dassault Systèmes, Waltham MA) is shown in Fig. 7.17. The fabrication process involves clear acrylic and 25 G needles protruding 5 mm upwards. The needles were fitted manually into the orifices of the device (2 cm length and 1 cm diameter). A central metallic core was coupled to enhance the radio-opacity for faster radiographic detection [152].

Later, a local delivery system by Zhang et al. showed minimum side effects of the antiobesity therapy. The polymeric microneedle patch is shown in Fig. 7.18 and includes glucose-responsive NPs to encapsulate rosiglitazone as the browning agent for sustained release. The drug-loaded NPs were packed inside a microneedle array for localized delivery in the subcutaneous adipose tissue. The acidic environment gradually degrades pH-sensitive NPs, subsequently releasing the browning agents to adipocytes, transforming them into energy expenditure units from energy storing units. In an in vivo experiment a diet-induced obese mouse model showed 15% less weight gain in comparison to the untreated group. Surprisingly, the microneedle patch showed ~10% difference in the inguinal fat pad size between the treated and untreated sides in the same mouse.

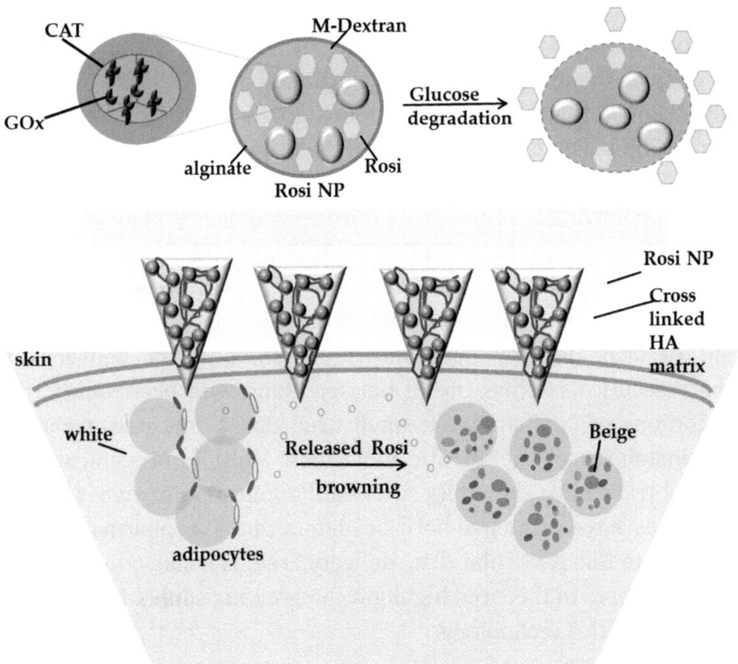

Figure 7.18 Schematic of microneedle-assisted local delivery of the browning agent into adipose tissue for obesity control [82].

Later, the NPs were reinforced in a polymeric microneedle-array patch for the local delivery of browning reagents. Briefly, the NPs were first loaded into the tip region of a silicon microneedle mold by centrifugation, followed by the addition of the base solution. The microneedle array consists of 121 needles in 7×7 mm^2 patches with an intermicroneedle distance of 600 μm. Each microneedle is conical, 300 μm in diameter at the base and 800 μm in height [82].

Leuprorelin acetate (LA) is the clinical drug used for endocrine therapy via hypodermic injection to control the disease progression. But the disadvantage of this technique is the associated pain and that it can lead to infection and well-trained persons are required to administer injections to the patient. Liao et al. developed hydrogel-mediated microneedle patches, readily dissolvable into the skin with painless, self-administered delivery of LA for prostate cancer therapy. In this design, the LA is loaded in the hydrogel, followed by encapsulation within polymer microneedles. After insertion of the microneedles into the skin, the needles dissolve within minutes to release the drug for further sustained drug release. Studies have shown that the activity of the green fluorescent protein–expressing plasmid DNA remains in the original state after loading. The LA encapsulation rate was also measured to be higher by up to 72%, showing a promising future as a delivery platform for a range of biomolecules, including proteins, peptides, vaccines, and nucleotides [153].

Till now the studies conducted on microneedle-assisted oral and GI drug delivery have shown its potential for delivery of pharmaceuticals across the GI barrier, along with overcoming the limitations of comparatively small drug doses, low performance, and instability of therapeutic molecules. Multiple preclinical and clinical trials have shown its successful application in vivo. Clinical successes have attracted both academia and the pharmaceutical industry to this particular drug delivery area. The above-mentioned advancements in this area highlight the growing sophistication and precision of this technology.

7.7.6 Protein and Vaccine Delivery

Dissolvable microneedles are designed for the encapsulation of the drug inside a soluble polymeric matrix and to fully dissolve during insertion into the skin. The associated therapeutic duration lies between hours and days. This variation is dependent on the dissolution kinetics of the selected dissolvable polymer, such as maltose [60], chitosan [40], and PVA [75], and the administration time of the microneedles [47, 154]. For the continuous release of therapeutic proteins for a predetermined period, transdermal microneedle patches have been fabricated from biodegradable polymers. The release profile of the proteins is also affected by the degradation of the polymer and the diffusion of the proteins from the microneedles. For this, the selection of the polymer is crucial and depends on its molecular weight and degradability.

Prausnitz and coworkers developed a layer-by-layer deposition technique for polyelectrolytes to coat stainless-steel microneedles with bovine pancreatic ribonuclease A. The protein release took place by the hydrolytic degradation of the coating layer after microneedle insertion up to a depth of 500 μm in porcine cadaver skin [155]. Recently, conserved human melanoma antigen–reinforced, polyelectrolyte multilayer (PEM)-assembled microneedles were developed. These can act as a platform for assembling cancer vaccine components on microneedle arrays for intradermal delivery from these substrates [156]. Later, Martin et al. assessed a low-temperature vacuum-forming method that demonstrates the fabrication of microneedles from a dehydrated sugar mixture of trehalose anhydrous, trehalose dihydrate, sucrose, and maltose [71]. Rats were used as the model for assessing the feasibility of dextran microneedle preparation by a polypropylene thread. The study shows that the recombinant human growth hormone (rhGH) levels in rat plasma shoot to the peak after 1 h of microneedle administration. The rhGH levels were found to be equivalent to that after intravenously injected rhGH solution [157].

Bouwstra et al. developed HA-based microneedles targeting active monoclonal immunoglobin G (IgG) delivery [158]. HA is a polyanionic disaccharide commonly found in all parts of the human body and employed as a dermal filler [159]. Microneedles with 2% or 10% (w/w) of monoclonal IgG were prepared with the micromolding method. Recovery of 80% of IgG was achieved upon microneedle dissolution, with insignificant conformational changes in the tertiary structure. The tissues surrounding the insertion site started getting HA and IgG within 10 min of needle insertion [158]. PVP microneedles (length 650 μm) also delivered inactivated influenza virus vaccine within a few minutes at a penetration depth of around 200 μm of the porcine cadaver skin [152]. Additionally, Donnelly and coworkers fabricated dissolvable PMVE/MA polymeric microneedle arrays for encapsulating recombinant HIV antigens [160].

Hammond and coworkers reported a highly biodegradable and biocompatible silk fibroin/polyacrylic acid composite microneedle patch [161, 162]. During experiment, they used an aqueous-based micromolding fabrication method at room temperature and atmospheric pressure to maintain the temperature-susceptible therapeutics, like, proteins, antibiotics, and vaccines.

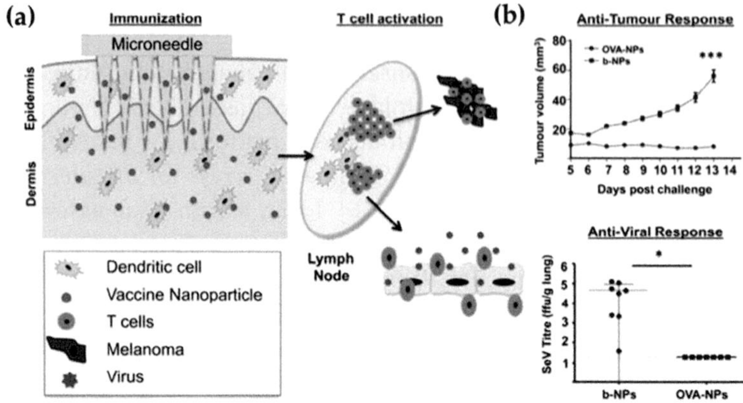

Figure 7.19 (a) Schematic showing dissolvable microneedle arrays loaded with antigen-loaded NPs, increasing vaccine immunogenicity by targeting dendritic cell network–specific antigen in the skin. (b) The mice received robust antigen-specific cellular immune responses against antigen-encapsulated NP vaccination via microneedles. Reprinted with permission from Ref. [163]. Copyright (2013) American Chemical Society (https://pubs.acs.org/doi/abs/10.1021/nn304235j). Further permissions related to the material excerpted should be directed to the ACS.

Zaric et al. reported a microneedle array containing PLGA NPs with chicken OVA for in situ skin delivery. Further, afferent lymphatics transported the NPs with skin antigen-presenting cells (APCs) to cutaneous draining lymph nodes. Antigens show better functionality in conjugation with major histocompatibility complex (MHC) class I molecules to CD8+ T cells in comparison to soluble antigens, thus stimulating OVA-specific interferon-γ positive CD4+ and CD8+ T cells concurrently in the murine model, making the system useful for parainfluenza infection and melanoma vaccination, as shown in Fig. 7.19 [163]. As mentioned in the section on swellable polymer microneedles, Park et al. reported a micromolded swellable PLGA matrix containing 53% (v/v) PNIPAAm microparticle microneedles for long-term topical drug delivery accompanying hydrogel volume expansion in the skin [50].

The transdermal delivery of proteins and peptides is a difficult task owing to unfavorable physicochemical properties (high MW and hydrophilicity), instability, complex structural orientation, and potency. Simpler techniques are needed to overcome the barrier properties of the skin with negligible irreversible damage to the skin. Peptide derivatives, prodrugs, HA derivatives, and other techniques have been reviewed above for promising prospects.

7.7.7 Miscellaneous Applications

Normally, drug delivery for cardiovascular diseases, like atherosclerosis, is less effective due to the poor drug transport through the walls of cardiac muscles. Choi et al. designed microneedles consisting of curved backing layers; these microneedles could wrap around the blood vessel in such a way that the tips of the microneedles were inserted into the tunica adventitia. To fabricate microneedles on a curved surface, discrete thermal drawing and postannealing processes are employed, which help the microneedles to enter the target vascular tissue and release the drug [164].

Microneedles can be used for needle-free delivery of anesthetics. As they are minimally invasive in nature, cause no pain, and are small in size (100 microns), they play a major role in drug delivery. Microneedle administration of anesthetics is identical to one injection of hypodermic needles. During drug delivery the sensitivity and concentration of the anesthetic given to the patient depend on the number and spacing of microneedles in the microneedle patch. Lidocaine is used as a drug for local anesthesia. Baek et al. showed consistent and higher delivery within 2 min by microneedle tips coated with lidocaine. Thus the proposed technique can be employed for painless and faster local anesthesia [165].

Baek et al. developed phenylephrine (PE) microneedles that help to increase PE availability to the anal sphincter, which is an effective treatment for fecal incontinence. The microneedles fabricated by using PE, which is used to increase perianal skin permeability for locally targeted drug delivery, help to increase resistance to anal sphincter pressure to treat fecal incontinence. The microneedle patches were developed by micromolding PLA [166].

Chiu et al. designed a microneedle with a cylindrical roller that allows rapid drug diffusion into the nail. The polymeric NPs act as drug reservoirs for sustained topical drug delivery into microneedle-treated human nail [167].

Thermoresponsive polymers [11] promote in situ gel formation, which helps in the sustainable delivery of a drug without any surgery during its application. Among the poloxamers or pluronics, a polyoxyethylene-polyoxypropylene-polyoxyethylene triblock copolymer, representing a series of central chains of polyoxypropylene and two identical lateral hydrophilic chains of polyoxyethylene, has seen tremendous success in in situ carrier formation. These copolymers have a sol-gel transition behavior in an aqueous medium, reacting to the temperature change at skin (32°C) and body temperature (37°C). At room temperature (<30°C), the aqueous solutions are in a sol state; however, at certain temperatures (>lower critical solution temperature, 32°C) and concentrations, the gel transforms into a liquid form [168].

To treat restenosis, solid microneedles were applied across the internal elastic lamina of rabbit arteries ex vivo. To aid tissue engineering, hollow microneedles were developed to improve cell viability in slices of harvested brain tissue. These systems unlock many pathways for neurobiological and other culture systems [169].

Microneedles are extensively used in drug delivery to the skin, but they can also be used in diagnosis for extraction of analytes from the skin. As in one instance, a few microliters of interstitial fluid were sucked from skin pretreated with microneedles. The concentration of glucose present in the interstitial fluid was coherent with the BG levels in rats and human subjects [170, 171]. For monitoring glucose levels, several integrated diagnostic systems were developed involving microneedles with microactuators, microfluidic controls, and sensors. This system collects blood from the skin [172, 173].

Microneedles also work as bioelectrical interfaces, especially for recording neural activity, stimulation [174–177], electrocardiography, and electroencephalography measurements [178, 179].

Further, microneedle arrays have been upgraded to directly deliver drugs in an intracellular manner. In this approach DNA-coated microneedles transfected nematode cells while the nematode cells moved across the DNA-coated microneedle array. The microneedles pierced the nematode cuticle [180].

Wang et al. found that the vaginal mucosa may be a useful site for vaccination, specifically for targeted delivery of APCs to the vaginal tissue. Their group fabricated microneedle patches with ammonium bicarbonate–loaded multifunctional liposomes for vaginal mucosal vaccine adjuvant-dual delivery system and administered antigen using microneedle patches into the vaginal cavity. It is found to trigger a vigorous antigen-specific immune response systemically in the reproductive tract mucosa [181].

Bhatnagar et al. delivered chemotherapeutic agents such as tamoxifen and gemcitabine through microneedles for breast cancer treatment. Zein microneedles were developed using the micro-molding technique containing 36 microneedles in a 1 cm^2 area. These microneedles were loaded with two anti–breast cancer drugs, tamoxifen and gemcitabine, with different water solubilities. Entrapment or surface coating of chemotherapeutic agents in zein microneedles was optimized to achieve greater loading efficiency. Localized delivery of drugs decreases the adverse side effects [182]. Polymeric microneedles were tested against skin cancer and for localized anticancer drugs delivery. Ye et al. developed a melanin-mediated cancer immunotherapy strategy using a transdermal microneedle patch. B16F10 whole tumor lysate containing melanin is loaded into polymeric microneedles and allows sustained release of the lysate upon insertion of the microneedles into the skin. In combination with the near-infrared light irradiation, melanin produces local heat, which enhances the T cell activities by transdermal vaccines and encourages antitumor immune responses [183].

7.8 Drawbacks of Microneedles

Microneedles have been widely employed for safe and efficient drug and vaccine delivery through microchannels into the skin. However, many studies done related to the safety of microneedles show that the microchannels created by microneedles may infect the injection site by allowing the permeation of pathogens or other toxic materials. Also drug delivery into the skin by using microneedles may not be completely reliable and efficient due to some small deformation of the skin or the presence of a hair in the skin, which can affect microneedle insertion because of its micron-size length. Microneedle applicators increase the reliability of insertion but introduce their own limitations in terms of cost, size, and complexity [23].

Gupta et al. and Gill et al. conducted separate studies targeting the pain associated with the microneedle dimension and quantity. The results showed direct relationship between length or number of microneedles and pain [135, 184]. However, Bal et al. and Kaushik et al. established the safety of microneedles and reported it as a painless, non-invasive drug delivery approach [185, 186]. Although microneedles are safe to use, a few challenges still remain in their development and commercialization for successful transdermal drug delivery. With frequent application of the metallic microneedles, traces of metal may remain beneath the skin, leading to irritation, erythema, swelling, discoloration, etc.

A small amount of drugs can be loaded into microneedles for delivery. However, transfer of large and hydrophilic compounds is still a major challenge. To find a solution for this problem, the design of the system needs to be improved using the proper material with adequate mechanical force to pierce the skin.

After use, microneedle patches require safe disposal since they may have residual drug and account for sharp waste, making them biohazardous. Unless drug delivery is 100% efficient, the use of microneedles will result in residual drug retention in the patch. In some cases, this may not be of significant concern. However, in other cases, this may impose some danger of residual drug exposure to others, including children, animals, and the environment. Used microneedle patches are considered as biohazardous waste due to their contact with body fluids, even though in small amounts (i.e., interstitial fluid of the skin). The used microneedle patches may be considered as sharp waste. Dissolving microneedles are likely not to be considered as sharp waste as no remnant microneedle part is present after usage. Again nondissolving (e.g., coated) microneedles fall in the category of sharp waste, although the hazard they pose is different compared to hypodermic needles or scalpel blades and they may therefore be handled differently. In all of these cases, safe disposal may need to be facilitated through suitable waste streams (e.g., placement in sharps containers or biohazard bags) or through suitable packaging after use (e.g., microneedles contained in a protective housing after use).

7.9 Future Scope

The future of drug delivery is significantly dependent on microfabrication technologies, which will help to realize the development of new and improved devices that will be smaller, cheaper, pain-free, and more convenient, with a wide range of biomedical and other applications. The drug delivery devices made by micro-/nanofabricated technology can enable efficient drug delivery, not attainable with conventional drug delivery techniques, thus enhancing the therapeutic activity of a drug.

However, the ultimate measure of the impact of microneedles is their translation into clinical medicine. This translation requires significant funding from cautious pharmaceutical industries for product development and advanced clinical trials, along with advancement in science and technology. For microneedle patch development research, the greatest challenge is its translation into novel medical products that will improve patient compliance. For research and development, two important outstanding concerns are as follows: (i) the true benefits of microneedles to patients and (ii) the development of low-cost, scalable, and reliable microneedles. Application at different sites every time or variation in skin thickness in individuals may result in variation in bioavailability, which needs to be considered while developing microneedles [187]. Animal studies show that vaccines are generally more immunogenic and can be administered at lower doses. From studies it is proved that administration of drugs using microneedles helps in faster drug uptake by the bloodstream for prolonged duration without any skin irritation. Research is going on for further developments in microneedles that can be potentially valuable in future vaccine delivery. With the help of effective drug delivery, patient compliance, cost saving, and ease of storage, microneedle drug delivery systems will be a milestone in the development and dissemination of vaccines and will potentially contribute to human public health.

7.10 Summary

Microneedles, either as a patch or an array, have demonstrated potential as an effective transdermal carrier for various macromolecular drugs. Studies have found that microneedles have the capability of enhanced permeation into systemic circulation along with the advantages of being a painless, effective, and safe route for drug delivery. Currently, approximately 23 active trials are going on and 39 trials on microneedles have been completed by the National Institutes of Health clinical, and microneedles are present for the treatment of different severe and acute diseases, such as type 1 diabetes and psoriatic plaques, and for topical anesthesia [188]. Continuous advancement in these painless delivery systems make them one of the key devices for controlled drug release of the future. It can be concluded that microneedle systems are efficient and superior carriers in comparison to the other needle-based formulations for transdermal drug delivery.

It is expected that fundamental studies and commercialization programs will jointly accelerate an era of microneedle-mediated drug delivery products to improve human health and quality of life.

Acknowledgments

The authors greatly appreciate the financial support from the Wellcome Trust/DBT India Alliance Fellowship under grant number IA/E/16/1/503062 and the Science and Engineering Research Board (SERB) under grant number ECR/2016/001945, Department of Science and Technology (DST), Government of India. We acknowledge all authors and publishers who provided the copyright permissions.

References

1. T. Santra, F. Tseng (2013). Recent trends on micro/nanofluidic single cell electroporation. *Micromachines*, **4**:333–356; doi:10.3390/mi4030333.
2. T. S. Santra, F.-G. (Kevin) Tseng, T. K. Barik (2014). Biosynthesis of silver and gold nanoparticles for potential biomedical applications: a brief review. *J Nanopharm Drug Deliv*, **2**:249–265; doi:10.1166/jnd.2014.1065.
3. S. Kar, M. Loganathan, K. Dey, P. Shinde, H.-Y. Chang, M. Nagai, T. S. Santra (2018). Single-cell electroporation: current trends, applications and future prospects, *J Micromech Microeng*, **28**:123002; doi:10.1088/1361-6439/aae5ae.
4. D. Sharma (2018). Microneedles: an approach in transdermal drug delivery: a review. *PharmaTutor*, **6**(1):7–15; http://dx.doi.org/10.29161/PT.v6.i1.2018.7
5. K. Ita (2015). Transdermal delivery of drugs with microneedles—potential and challenges. *Pharmaceutics*, **7**:90–105; doi:10.3390/pharmaceutics7030090.
6. H. L. Quinn, M.-C. Kearney, A. J. Courtenay, M. T. C. McCrudden, R. F. Donnelly (2014). The role of microneedles for drug and vaccine delivery. *Expert Opin Drug Deliv*, **11**:1769–1780; doi:10.1517/17425247.2014.938635.
7. H. W. Smith, G. H. A. Clowes, E. K. Marshall (1919). On dichloroethylsulfide (mustard gas) IV. The mechanism of absorption by the skin. *J Pharmacol Exp Ther*, **13**:1–30.
8. R. J. Scheuplein (1967). Mechanism of percutaneous absorption: II. Transient diffusion and the relative importance of various routes of skin penetration. *J Invest Dermatol*, **48**:79–88; doi:10.1038/JID.1967.11.
9. E. Larrañeta, R. E. M. Lutton, A. D. Woolfson, R. F. Donnelly (2016). Microneedle arrays as transdermal and intradermal drug delivery systems: materials science, manufacture and commercial development. *Mater Sci Eng, R*, **104**:1–32; doi:10.1016/J.MSER.2016.03.001.
10. Drug delivery device (1971). https://patents.google.com/patent/US3964482A/en (accessed July 16, 2019).
11. S. Henry, D. V. McAllister, M. G. Allen, M. R. Prausnitz (1998). Micro fabricated microneedles: a novel approach to transdermal drug delivery. *J Pharm Sci*, **87**:922–925; doi:10.1021/js980042+.

12. D. V. McAllister, P. M. Wang, S. P. Davis (2003). Microfabricated needles for transdermal delivery of macromolecules and nanoparticles: fabrication methods and transport studies. *Proc Natl Acad Sci USA*, **100**(24):13755–13760; doi:10.1073/pnas.2331316100.
13. R. F. Donnelly, M. T. C. McCrudden, A. Zaid Alkilani, E. Larrañeta, E. McAlister, A. J. Courtenay, M.-C. Kearney, T. R. R. Singh, H. O. McCarthy, V. L. Kett, E. Caffarel-Salvador, S. Al-Zahrani, A. D. Woolfson (2014). Hydrogel-forming microneedles prepared from "super swelling" polymers combined with lyophilised wafers for transdermal drug delivery. *PLoS One*, **9**:e111547; doi:10.1371/journal.pone.0111547; doi:10.1371/journal.pone.0111547.
14. J. H. Park, M. G. Allen, M. R. Prausnitz (2005). Biodegradable polymer microneedles: fabrication, mechanics and transdermal drug delivery. *J Control Release*, **104**:51–66; doi:10.1016/j.jconrel.2005.02.002.
15. K. Van Der Maaden, W. Jiskoot, J. Bouwstra (2012). Microneedle technologies for transdermal drug and vaccine delivery. *J Control Release*, **161**:645–655; doi: 10.1016/j.jconrel.2012.01.042.
16. S. H. Bariya, M. C. Gohel, T. A. Mehta, O. P. Sharma (2012). Microneedles: an emerging transdermal drug delivery system. *J Pharm Pharmacol*, **64**:11–29; doi:10.1111/j.2042-7158.2011.01369.x.
17. M. R. Prausnitz (2004). Microneedles for transdermal drug delivery. *Adv Drug Deliv Rev*, **56**:581–587; doi:10.1016/J.ADDR.2003.10.023.
18. T. Waghule, G. Singhvi, S. K. Dubey, M. M. Pandey, G. Gupta, M. Singh, K. Dua (2019). Microneedles: a smart approach and increasing potential for transdermal drug delivery system. *Biomed Pharmacother*, **109**:1249–1258; doi:10.1016/j.biopha.2018.10.078.
19. L. W. Wong, W. Q. Sun, N. W. Chan, W. Y. Lai, W. K. Leung, J. C. Tsang, Y. H. Wong, K. L. Yeung (2007). Zeolite microneedles for transdermal drug delivery. *Stud Surf Sci Catal*, **170**:525–530; doi:10.1016/S0167-2991(07)80887-1.
20. J. A. Mikszta, J. B. Alarcon, J. M. Brittingham, D. E. Sutter, R. J. Pettis, N. G. Harvey (2002). Improved genetic immunization via micromechanical disruption of skin-barrier function and targeted epidermal delivery. *Nat Med*, **8**:415–419; doi:10.1038/nm0402-415.
21. D. Prabhakar, J. Sreekanth, K. N. Jayaveera (2016). Transdermal drug delivery patches: a review. *J Drug Deliv Ther*, **3**(4):231–221; doi:10.22270/jddt.v3i4.590.
22. R. Donnelly, D. Douroumis (2015). Microneedles for drug and vaccine delivery and patient monitoring. *Drug Deliv Transl Res*, **5**:311–312; doi:10.1007/s13346-015-0250-2.

23. M. R. Prausnitz (2017). Engineering microneedle patches for vaccination and drug delivery to skin. *Annu Rev Chem Biomol Eng*, **8**:177–200; doi:10.1146/annurev-chembioeng-060816-101514.
24. J. Li, M. Zeng, H. Shan, C. Tong (2017). Microneedle patches as drug and vaccine delivery platform. *Curr Med Chem*, **24**(22):2413–2422; doi:10.2174/0929867324666170526124053.
25. M. S. Ullah, R. Kaw, P. A. Rasool, N. Tripathi, V. Singh (2017). Current trends in MEMS drug delivery techniques. *Int J Eng Technol*, **9**:1355–1364; doi:10.21817/ijet/2017/v9i2/170902284.
26. S. Pradeep Narayanan, S. Raghavan (2017). Solid silicon microneedles for drug delivery applications. *Int J Adv Manuf Technol*, **93**:407–422; doi:10.1007/s00170-016-9698-6.
27. S. P. Narayanan, S. Raghavan (2018). Fabrication and characterization of gold-coated solidsilicon microneedles with improved biocompatibility. *Int J Adv Manuf Technol*, 1–7.e; doi: 10.1007/s00170-018-2596-3.
28. Q. Y. Li, J. N. Zhang, B. Z. Chen, Q. L. Wang, X. D. Guo (2017). A solid polymer microneedle patch pre-treatments enhances the permeation of drug molecules into the skin. *RSC Adv*, **7**:15408–15415; doi: 10.1039/c6ra26759a.
29. L. Daugimont, N. Baron, G. Vandermeulen, N. Pavselj, D. Miklavcic, M. C. Jullien, G. Cabodevila, L. M. Mir, V. Préat (2018). Hollow microneedle arrays for intradermal drug delivery and DNA electroporation. *J Membr Biol*, **236**:117–125; doi:10.1007/s00232-010-9283-0.
30. K. B. Vinayakumar, P. G. Kulkarni, M. M. Nayak, N. S. Dinesh, G. M. Hegde, S. G. Ramachandra, K. Rajanna (2016). A hollow stainless steel microneedle array to deliver insulin to a diabetic rat. *J Micromech Microeng*, **26**; doi:10.1088/0960-1317/26/6/065013.
31. E. García-López, H. R. Siller, C. A. Rodríguez (2018). Study of the fabrication of AISI 316L microneedle arrays. *Procedia Manuf*, **26**:117–124; doi:10.1016/J.PROMFG.2018.07.014.
32. S. Khumpuang, M. Horade, K. Fujioka, S. Sugiyama (2006). Geometrical strengthening and tip-sharpening of a microneedle array fabricated by X-ray lithography. *Microsyst Technol*, **13**:209–214; doi:10.1007/s00542-006-0173-4.
33. S. D. Gittard, A. Ovsianikov, B. N. Chichkov, A. Doraiswamy, R. J. Narayan (2010). Two-photon polymerization of microneedles for transdermal drug delivery. *Expert Opin Drug Deliv*, **7**:513–533; doi:10.1517/17425241003628171.

34. A. Arora, M. R. Prausnitz, S. Mitragotri (2008). Micro-scale devices for transdermal drug delivery. *Int J Pharm*, **364**:227–236; doi:10.1016/j.ijpharm.2008.08.032.
35. R. Mishra, T. K. Maiti, T. K. Bhattacharyya (2018). Development of SU-8 hollow microneedles on a silicon substrate with microfluidic interconnects for transdermal drug delivery. *J Micromech Microeng*, **28**: 105017; doi:10.1088/1361-6439/aad301.
36. K. van der Maaden, J. Heuts, M. Camps, M. Pontier, A. Terwisscha van Scheltinga, W. Jiskoot, F. Ossendorp, J. Bouwstra (2018). Hollow microneedle-mediated micro-injections of a liposomal HPV E743–63 synthetic long peptide vaccine for efficient induction of cytotoxic and T-helper responses. *J Control Release*, **269**:347–354; doi:10.1016/j.jconrel.2017.11.035.
37. M. Wang, L. Hu, C. Xu (2017). Recent advances in the design of polymeric microneedles for transdermal drug delivery and biosensing. *Lab Chip*, **17**:1373–1387; doi:10.1039/C7LC00016B.
38. J. D. Kim, M. Kim, H. Yang, K. Lee, H. Jung (2013). Droplet-born air blowing: novel dissolving microneedle fabrication. *J Control Release*, **170**:430–436; doi:10.1016/j.jconrel.2013.05.026.
39. K. Lee, C. Y. Lee, H. Jung (2011). Dissolving microneedles for transdermal drug administration prepared by stepwise controlled drawing of maltose. *Biomaterials*, **32**:3134–3140; doi:10.1016/j.biomaterials.2011.01.014.
40. M. C. Chen, M. H. Ling, K. Y. Lai, E. Pramudityo (2012). Chitosan microneedle patches for sustained transdermal delivery of macromolecules. *Biomacromolecules*, **13**:4022–4031; doi:10.1021/bm301293d.
41. K. Fukushima, A. Ise, H. Morita, R. Hasegawa, Y. Ito, N. Sugioka, K. Takada (2011). Two-layered dissolving microneedles for percutaneous delivery of peptide/protein drugs in rats. *Pharm Res*, **28**:7–21; doi:10.1007/s11095-010-0097-7.
42. D. D. Zhu, Q. L. Wang, X. B. Liu, X. D. Guo (2016). Rapidly separating microneedles for transdermal drug delivery. *Acta Biomater*, **41**:312–319; doi:10.1016/j.actbio.2016.06.005.
43. Y. H. Park, S. K. Ha, I. Choi, K. S. Kim, J. Park, N. Choi, B. Kim, J. H. Sung (2016). Fabrication of degradable carboxymethyl cellulose (CMC) microneedle with laser writing and replica molding process for enhancement of transdermal drug delivery. *Biotechnol Bioprocess Eng*, **21**:110–118; doi:10.1007/s12257-015-0634-7.

44. Y. K. Demir, Z. Akan, O. Kerimoglu (2013). Characterization of polymeric microneedle arrays for transdermal drug delivery. *PLoS One*, **8**:1–9; doi:10.1371/journal.pone.0077289.

45. J.-H. Park, S.-O. Choi, R. Kamath, Y.-K. Yoon, M. G. Allen, M. R. Prausnitz (2007). Polymer particle-based micromolding to fabricate novel microstructures. *Biomed Microdevices*, **9**:223–234; doi:10.1007/s10544-006-9024-4.

46. W. Yu, G. Jiang, Y. Zhang, D. Liu, B. Xu, J. Zhou (2017). Polymer microneedles fabricated from alginate and hyaluronate for transdermal delivery of insulin. *Mater Sci Eng C*, **80**:187–196; doi:10.1016/j.msec.2017.05.143.

47. X. Hong, L. Wei, F. Wu, Z. Wu, L. Chen, Z. Liu, W. Yuan (2013). Dissolving and biodegradable microneedle technologies for transdermal sustained delivery of drug and vaccine. *Drug Des Dev Ther*, **7**:945–952; doi:10.2147/DDDT.S44401.

48. X. Hong, L. Chen, L. Wei, W. Yuan, Z. Wu, F. Wu (2014). Hydrogel microneedle arrays for transdermal drug delivery. *Nano-Micro Lett*, **6**:191–199; doi:10.5101/nml140021r.

49. J. Jin, V. Reese, R. Coler, D. Carter, M. Rolandi (2014). Chitin microneedles for an easy-to-use tuberculosis skin test. *Adv Healthc Mater*, **3**:349–353; doi:10.1002/adhm.201300185.

50. M. Y. Kim, B. Jung, J. H. Park (2012). Hydrogel swelling as a trigger to release biodegradable polymer microneedles in skin. *Biomaterials*, **33**:668–678; doi:10.1016/j.biomaterials.2011.09.074.

51. B. Jana, A. Wadhwani (2019). Microneedle–future prospect for efficient drug delivery in diabetes management. *Indian J Pharmacol*, **51**:4–10; doi:10.4103/ijp.ijp_16_18.

52. S. Lau, J. Fei, H. Liu, W. Chen, R. Liu (2017). Multilayered pyramidal dissolving microneedle patches with flexible pedestals for improving effective drug delivery. *J Control Release*, **265**:113–119; doi:10.1016/j.jconrel.2016.08.031.

53. R. F. Donnelly, T. R. R. Singh, A. Z. Alkilani, M. T. C. McCrudden, S. O'Neill, C. O'Mahony, K. Armstrong, N. McLoone, P. Kole, A. D. Woolfson (2013). Hydrogel-forming microneedle arrays exhibit antimicrobial properties: potential for enhanced patient safety. *Int J Pharm*, **451**:76–91; doi:10.1016/j.ijpharm.2013.04.045.

54. E. M. Migdadi, A. J. Courtenay, I. A. Tekko, M. T. C. McCrudden, M.-C. Kearney, E. McAlister, H. O. McCarthy, R. F. Donnelly (2018). Hydrogel-forming microneedles enhance transdermal delivery

of metformin hydrochloride. *J Control Release*, **285**:142–151; doi: 10.1016/j.jconrel.2018.07.009.

55. S. Hashmi, P. Ling, G. Hashmi, M. Reed, R. Gaugler, W. Trimmer (1995). Genetic transformation of nematodes using arrays of micromechanical piercing structures. *Biotechniques*, **19**:766–770.
56. Y. Ito, J.-I. Yoshimitsu, K. Shiroyama, N. Sugioka, K. Takada (2006). Self-dissolving microneedles for the percutaneous absorption of EPO in mice. *J Drug Target*, **14**:255–261; doi:10.1080/10611860600785080.
57. J. Wang, J. Lu, Y. L. Suw, V. Maika, T. Baomin, W. K. Adeniyi, R. A. Armendariz (2000). Lab-on-a-cable for electrochemical monitoring of phenolic contaminants. *Anal Chem*, **72**:2659-2663. doi.org/10.1021/ac991054y.
58. A. Ovsianikov, B. Chichkov, P. Mente, N. A. Monteiro-Riviere, A. Doraiswamy, R. J. Narayan (2007). Two photon polymerization of polymer–ceramic hybrid materials for transdermal drug delivery. *Int J Appl Ceram Technol*, **4**:22–29; doi:10.1111/j.1744-7402.2007.02115.x.
59. C. S. Kolli, A. K. Banga (2008). Characterization of solid maltose microneedles and their use for transdermal delivery. *Pharm Res*, **25**:104–113; doi:10.1007/s11095-007-9350-0.
60. T. Miyano, Y. Tobinaga, T. Kanno, Y. Matsuzaki, H. Takeda, M. Wakui, K. Hanada (2005). Sugar micro needles as transdermic drug delivery system. *Biomed Microdevices*, **7**:185–188; doi:10.1007/s10544-005-3024-7.
61. R. F. Donnelly, D. I. J. Morrow, T. R. R. Singh, K. Migalska, P. A. McCarron, C. O'Mahony, A. David Woolfson (2009). Processing difficulties and instability of carbohydrate microneedle arrays. *Drug Dev. Ind Pharm*, 1242–1254; doi: 10.1080/03639040902882280.
62. J.-H. Park, Y.-K. Yoon, S.-O. Choi, M. R. Prausnitz, M. G. Allen (2007). Tapered conical polymer microneedles fabricated using an integrated lens technique for transdermal drug delivery. *IEEE Trans Biomed Eng*, **54**:903–913; doi:10.1109/TBME.2006.889173.
63. J. H. Park, M. G. Allen, M. R. Prausnitz (2006). Polymer microneedles for controlled-release drug delivery. *Pharm Res*, **23**:1008; doi:10.1007/s11095-006-0028-9.
64. M. A. Hopcroft, W. D. Nix, T. W. Kenny (2010). What is the Young's modulus of silicon? *J Microelectromech Syst*, **19**:229–238; doi:10.1109/JMEMS.2009.2039697.
65. R. F. Donnelly, T. R. R. Singh, A. D. Woolfson (2010). Microneedle-based drug delivery systems: microfabrication, drug delivery, and safety. *Drug Deliv*, **17**:187–207; doi:10.3109/10717541003667798.

66. F. J. Verbaan, S. M. Bal, D. J. van den Berg, W. H. H. Groenink, H. Verpoorten, R. Lüttge, J. A. Bouwstra (2007). Assembled microneedle arrays enhance the transport of compounds varying over a large range of molecular weight across human dermatomed skin. *J Control Release*, **117**:238–245; doi:10.1016/J.JCONREL.2006.11.009.

67. M. Niinomi, M. Nakai (2011). Titanium-based biomaterials for preventing stress shielding between implant devices and bone. *Int J Biomater*, **2011**:836587; doi.org/10.1155/2011/836587.

68. A. C. Williams, B. W. Barry (2004). Penetration enhancers. *Adv Drug Deliv Rev*, **56**:603–618; doi:10.1016/J.ADDR.2003.10.025.

69. S. D. Gittard, R. J. Narayan, C. Jin, A. Ovsianikov, B. N. Chichkov, N. A. Monteiro-Riviere, S. Stafslien, B. Chisholm (2009). Pulsed laser deposition of antimicrobial silver coating on Ormocer®microneedles. *Biofabrication*, **1**:041001; doi:10.1088/1758-5082/1/4/041001.

70. J. Gupta, E. I. Felner, M. R. Prausnitz (2009). Minimally invasive insulin delivery in subjects with type 1 diabetes using hollow microneedles. *Diabetes Technol Ther*, **11**; doi:10.1089/dia.2008.0103.

71. C. J. Martin, C. J. Allender, K. R. Brain, A. Morrissey, J. C. Birchall (2012). Low temperature fabrication of biodegradable sugar glass microneedles for transdermal drug delivery applications. *J Control Release*, **158**:93–101; doi:10.1016/J.JCONREL.2011.10.024.

72. F. Pérennès, B. Marmiroli, M. Matteucci, M. Tormen, L. Vaccari, E. Di Fabrizio (2016). Sharp beveled tip hollow microneedle arrays fabricated by LIGA and 3D soft lithography with polyvinyl alcohol. *J Micromech Microeng*, **16**:473–479; doi: 10.1088/0960-1317/16/3/001.

73. S. Aoyagi, H. Izumi, Y. Isono, M. Fukuda, H. Ogawa (2007). Laser fabrication of high aspect ratio thin holes on biodegradable polymer and its application to a microneedle. *Sens Actuators, A*, **139**:293–302; doi:10.1016/J.SNA.2006.11.022.

74. M. Han, D.-H. Hyun, H.-H. Park, S. S. Lee, C.-H. Kim, C. Kim (2007). A novel fabrication process for out-of-plane microneedle sheets of biocompatible polymer. *J Micromech Microeng*, **17**:1184–1191; doi:10.1088/0960-1317/17/6/012.

75. L. Y. Chu, S.-O. Choi, M. R. Prausnitz (2010). Fabrication of dissolving polymer microneedles for controlled drug encapsulation and delivery: bubble and pedestal microneedle designs. *J Pharm Sci*, **99**:4228–4238; doi:10.1002/JPS.22140.

76. Y.-C. Kim, F.-S. Quan, R. W. Compans, S.-M. Kang, M. R. Prausnitz (2010). Formulation of microneedles coated with influenza virus-like particle

vaccine. AAPS *PharmSciTech*, **11**:1193–1201; doi:10.1208/s12249-010-9471-3.

77. R. F. Donnelly, R. Majithiya, T. R. R. Singh, D. I. J. Morrow, M. J. Garland, Y. K. Demir, K. Migalska, E. Ryan, D. Gillen, C. J. Scott, A. D. Woolfson (2011). Design, optimization and characterisation of polymeric microneedle arrays prepared by a novel laser-based micromoulding technique. *Pharm Res*, **28**:41–57; doi:10.1007/s11095-010-0169-8.
78. H. Huang, C. Fu (2007). Different fabrication methods of out-of-plane polymer hollow needle arrays and their variations. *J Micromech Microeng*, **17**:393-402; doi:10.1088/0960-1317/17/2/027.
79. X. Chen, G. J. P. Fernando, M. L. Crichton, C. Flaim, S. R. Yukiko, E. J. Fairmaid, H. J. Corbett, C. A. Primiero, A. B. Ansaldo, I. H. Frazer, L. E. Brown, M. A. F. Kendall (2011). Improving the reach of vaccines to low-resource regions, with a needle-free vaccine delivery device and long-term thermo stabilization. *J Control Release*, **152**:349–55; doi:10.1016/j.jconrel.2011.02.026.
80. A. Vrdoljak, E. A. Allen, F. Ferrara, N. J. Temperton, A. M. Crean, A. C. Moore (2016). Induction of broad immunity by thermo stabilised vaccines incorporated in dissolvable microneedles using novel fabrication methods. *J Control Release*, **225**:192–204; doi: 10.1016/j.jconrel.2016.01.019.
81. C. Uppuluri, A. S. Shaik, T. Han, A. Nayak, K. J. Nair, B. R. Whiteside, B. N. Nalluri, D. B. Das (2017). Effect of microneedle type on transdermal permeation of rizatriptan. *AAPS PharmSciTech*, **18**:1495–1506; doi:10.1208/s12249-016-0702-0.
82. Y. Zhang, Q. Liu, J. Yu, S. Yu, J. Wang, L. Qiang, Z. Gu (2017). Locally induced adipose tissue browning by microneedle patch for obesity treatment. *ACS Nano*, **11**:9223–9230; doi:10.1021/acsnano.7b04348.
83. A. S. Rzhevskiy, T. R. R. Singh, R. F. Donnelly, Y. G. Anissimov (2018). Microneedles as the technique of drug delivery enhancement in diverse organs and tissues. *J Control Release*, **270**:184–202; doi:10.1016/j.jconrel.2017.11.048.
84. L. Wei Ze, H. Mei Rong, Z. Jian Ping, Z. Yong Qiang, H. Bao Hua, L. Ting, Z. Yong (2010). Super-short solid silicon microneedles for transdermal drug delivery applications. *Int J Pharm*, **389**:122–129; doi:10.1016/j.ijpharm.2010.01.024.
85. Y. Qiu, Y. Gao, K. Hu, F. Li (2008). Enhancement of skin permeation of docetaxel: a novel approach combining microneedle and elastic liposomes. *J Control Release*, **129**:144–150; doi:10.1016/j.jconrel.2008.04.019.

86. S. L. Banks, K. S. Paudel, N. K. Brogden, C. D. Loftin, A. L. Stinchcomb (2011). Diclofenac enables prolonged delivery of naltrexone through microneedle-treated skin. *Pharm Res*, **28**:1211–1219; doi:10.1007/s11095-011-0372-2.

87. R. F. Donnelly, D. I. J. Morrow, P. A. McCarron, A. David Woolfson, A. Morrissey, P. Juzenas, A. Juzeniene, V. Iani, H. O. McCarthy, J. Moan (2009). Microneedle arrays permit enhanced intradermal delivery of a preformed photosensitizer. *Photochem Photobiol*, **85**:195–204; doi:10.1111/j.1751-1097.2008.00417.x.

88. M. T. Clementoni, M. B Roscher, G. S. Munavalli (2010). Photodynamic photorejuvenation of the face with a combination of microneedling, red light, and broadband pulsed light. *Lasers Surg Med*, **42**:150–159; doi:10.1002/lsm.20905.

89. J. Ornelas, N. Foolad, V. Shi, W. Burney, R. K. Sivamani (2016). Effect of microneedle pretreatment on topical anesthesia: a randomized clinical trial. *JAMA Dermatol*, **152**:476–477; doi:10.1001/ jamadermatol.2015.5544.

90. M. T. C. Mccrudden, E. Mcalister, A. J. Courtenay, P. Gonzálezn Vázquez, T. R. Raj Singh, R. F. Donnelly (2015). Microneedle applications in improving skin appearance. *Exp Dermatol*, **24**:561–566; doi:10.1111/exd.12723.

91. Y. H. Mohammed, M. Yamada, L. L. Lin, J. E. Grice, M. S. Roberts, A. P. Raphael, H. A. E. Benson, T. W. Prow (2014). Microneedle enhanced delivery of cosmeceutically relevant peptides in human skin. *PLoS One*, **9**:e101956; doi:10.1371/journal.pone.0101956.

92. M. M. Badran, J. Kuntsche, A. Fahr (2009). Skin penetration enhancement by a microneedle device (Dermaroller®) in vitro: dependency on needle size and applied formulation. *Eur J Pharm Sci*, **36**:511–523; doi:10.1016/j.ejps.2008.12.008.

93. D. Bhardwaj (2013). Collagen induction therapy with dermaroller. *Community Based Med J*, **1**:35–37; doi:10.3329/cbmj.v1i1.13854.

94. L. M. Bonati, G. K. Epstein, T. L. Strugar (2017). Microneedling in all skin types: a review. *J Drugs Dermatol*, **16**:308–313.

95. A. Kumar, Y. W. Naguib, Y. C. Shi, Z. Cui (2016). A method to improve the efficacy of topical eflornithine hydrochloride cream. *Drug Deliv*, **23**:1495–1501; doi:10.3109/10717544.2014.951746.

96. R. Dhurat, M. Sukesh, G. Avhad, A. Dandale, A. Pal, P. Pund (2013). A randomized evaluator blinded study of effect of microneedling in androgenetic alopecia: a pilot study. *Int J Trichol*, **5**:6–11; doi:10.4103/0974-7753.114700.

97. Y. Park, K. S. Kim, M. Chung, J. H. Sung, B. Kim (2016). Fabrication and characterization of dissolving microneedle arrays for improving skin permeability of cosmetic ingredients. *J Ind Eng Chem*, **39**:121–126; doi:10.1016/j.jiec.2016.05.022.
98. L. Yan, A. P. Raphael, X. Zhu, B. Wang, W. Chen, T. Tang, Y. Deng, H. J. Sant, G. Zhu, K. W. Choy, B. K. Gale, T. W. Prow, X. Chen (2014). Nanocomposite strengthened dissolving microneedles for improved transdermal delivery to human skin. *Adv Healthc Mater*, **3**:555–564; doi:10.1002/adhm.201300312.
99. Y. Park, J. Park, G. S. Chu, K. S. Kim, J. H. Sung, B. Kim (2015). Transdermal delivery of cosmetic ingredients using dissolving polymer microneedle arrays. *Biotechnol Bioprocess Eng*, **20**:543–549; doi:10.1007/s12257-014-0775-0.
100. L. K. Vora, P. R. Vavia, E. Larrañeta, S. E. J. Bell, R. F. Donnelly (2018). Novel nanosuspension-based dissolving microneedle arrays for transdermal delivery of a hydrophobic drug. *J Interdiscip Nanomed*, **3**:89–101; doi:10.1002/jin2.41.
101. H. Chen, B. Wu, M. Zhang, P. Yang, B. Yang, W. Qin, Q. Wang, X. Wen, M. Chen, G. Quan, X. Pan, C. Wu (2019). A novel scalable fabrication process for the production of dissolving microneedle arrays. *Drug Deliv Transl Res*, **9**:240–248; doi:10.1007/s13346-018-00593-z.
102. At home Clinic: Dermaroller, The Wrinkle Warrior (n.d.). https://fiftyplusfab.wordpress.com/2014/09/19/at-home-clinic-1-dermaroller/ (accessed July 17, 2019).
103. J. Yoon, T. Son, E. Choi, B. Choi, J. S. Nelson, B. Jung (2008). Enhancement of optical skin clearing efficacy using a microneedle roller. *J Biomed Opt*, **13**:021103; doi:10.1117/1.2907483.
104. J. Yoon, D. Park, T. Son, J. Seo, J. S. Nelson, B. Jung (2010). A physical method to enhance transdermal delivery of a tissue optical clearing agent: combination of microneedling and sonophoresis. *Lasers Surg Med*, **42**:412–417; doi:10.1002/lsm.20930.
105. Y. T. Kim, S. I. Ahn, H. J. Park, S. H. Jung, H. K. Hong, H.-K. Lee, D. Lee, J. M. Rhee, K.-H. Lee, G. Khang (2009). Microneedle-mediated transdermal delivery of chondroitin sulphate. *Tissue Eng Regen Med*, **6**:756–761.
106. J. W. Lee, M. R. Prausnitz (2018). Drug delivery using microneedle patches: not just for skin. *Expert Opin Drug Deliv*, **15**:541–543; doi:10.1080/17425247.2018.1471059.
107. R. R. Thakur Singh, I. Tekko, K. McAvoy, H. McMillan, D. Jones, R. F. Donnelly (2017). Minimally invasive microneedles for ocular

drug delivery. *Expert Opin Drug Deliv*, **14**:525–537; doi:10.1080/17425247.2016.1218460.

108. J. Jiang, H. S. Gill, D. Ghate, B. E. McCarey, S. R. Patel, H. F. Edelhauser, M. R. Prausnitz (2007). Coated microneedles for drug delivery to the eye. *Investig Opthalmol Vis Sci*, **48**:4038; doi:10.1167/iovs.07-0066.
109. A. Than, C. Liu, H. Chang, P. K. Duong, C. M. G. Cheung, C. Xu, X. Wang, P. Chen (2018). Self-implantable double-layered micro-drug-reservoirs for efficient and controlled ocular drug delivery. *Nat Commun*, **9**:1–12; doi:10.1038/s41467-018-06981-w.
110. K. S. Yadav, R. Rajpurohit, S. Sharma (2019). Glaucoma: current treatment and impact of advanced drug delivery systems. *Life Sci*, **221**:362–376; doi:10.1016/j.lfs.2019.02.029.
111. S. H. Park, D. H. Jo, C. S. Cho, K. Lee, J. H. Kim, S. Ryu, C. Joo, J. H. Kim, W. Ryu (2018). Depthwise-controlled scleral insertion of microneedles for drug delivery to the back of the eye. *Eur J Pharm Biopharm*, **133**:31–41; doi:10.1016/J.EJPB.2018.09.021.
112. O. Khandan, M. Y. Kahook, M. P. Rao (2016). Fenestrated microneedles for ocular drug delivery. *Sens Actuators, B*, **223**:15–23; doi:10.1016/j.snb.2015.09.071.
113. E. R. Parker, B. J. Thibeault, M. F. Aimi, M. P. Rao, N. C. MacDonald (2005). Inductively coupled plasma etching of bulk titanium for MEMS applications. *J Electrochem Soc*, **152**:C675; doi:10.1149/1.2006647.
114. O. Khandan, A. Famili, M. Y. Kahook, M. P. Rao (2012). Titanium-based, fenestrated, in-plane microneedles for passive ocular drug delivery. In *34th Annual International Conference of the IEEE Engineering in Medicine & Biology Society*, pp. 6572–6575; doi:10.1109/EMBC.2012.6347500.
115. H. B. Song, K. J. Lee, I. H. Seo, J. Y. Lee, S.-M. Lee, J. H. Kim, J. H. Kim, W. Ryu (2015). Impact insertion of transfer molded microneedle for localized and minimally invasive ocular drug delivery. *J Control Release*, **209**:272–279; doi:10.1016/j.jconrel.2015.04.041.
116. H. Chang, M. Zheng, X. Yu, A. Than, R. Z. Seeni, R. Kang, J. Tian, D. P. Khanh, L. Liu, P. Chen, C. Xu (2017). A swellable microneedle patch to rapidly extract skin interstitial fluid for timely metabolic analysis. *Adv Mater*, **29**:1702243; doi:10.1002/adma.201702243.
117. A. Than, K. Liang, S. Xu, L. Sun, H. Duan, F. Xi, C. Xu, P. Chen (2017). Transdermal delivery of anti-obesity compounds to subcutaneous adipose tissue with polymeric microneedle patches. *Small Methods*, **1**:1700269; doi:10.1002/smtd.201700269.

118. Y. Luo, K. R. Kirker, G. D. Prestwich (2000). Cross-linked hyaluronic acid hydrogel films: new biomaterials for drug delivery. *J Control Release,* **69**:169–184; doi: 10.1016/s0168-3659(00)00300-x.
119. Z. Jiang, J. Chen, L. Cui, X. Zhuang, J. Ding, X. Chen (2018). Advances in stimuli-responsive polypeptide nanogels. *Small Methods,* **2**:1700307; doi:10.1002/smtd.201700307.
120. I. Hartman (2008). Insulin analogs: impact on treatment success, satisfaction, quality of life, and adherence. *Clin Med Res,* **6**:54–67; doi:10.3121/cmr.2008.793.
121. Skin perforating device for transdermal medication (1995). https://patents.google.com/patent/US5611806 (accessed July 11, 2019).
122. W. Martanto, S. P. Davis, N. R. Holiday, J. Wang, H. S. Gill, M. R. Prausnitz (2004). Transdermal delivery of insulin using microneedles in vivo. *Pharm Res,* **21**:947–952; doi:10.1023/B:PHAM.0000029282.44140.2e.
123. J. Gupta, E. I. Felner, M. R. Prausnitz (2011). Rapid pharmacokinetics of intradermal insulin administered using microneedles in type 1 diabetes subjects. *Diabetes Technol Ther,* **13**(4); doi:10.1089/dia.2010.0204.
124. R. J. Pettis, B. Ginsberg, L. Hirsch, D. Sutter, S. Keith, E. McVey, N. G. Harvey, M. Hompesch, L. Nosek, C. Kapitza, L. Heinemann (2011). Intradermal microneedle delivery of insulin lispro achieves faster insulin absorption and insulin action than subcutaneous injection. *Diabetes Technol Ther,* **13**(4); doi:10.1089/dia.2010.0184.
125. X. Jin, D. D. Zhu, B. Z. Chen, M. Ashfaq, X. D. Guo (2018). Insulin delivery systems combined with microneedle technology. *Adv Drug Deliv Rev,* **127**:119–137; doi:10.1016/j.addr.2018.03.011.
126. C. J. Rini, E. McVey, D. Sutter, S. Keith, H. J. Kurth, L. Nosek, C. Kapitza, K. Rebrin, L. Hirsch, R. J. Pettis (2015). Intradermal insulin infusion achieves faster insulin action than subcutaneous infusion for 3-day wear. *Drug Deliv Transl Res,* **5**:332–345; doi:10.1007/s13346-015-0239-x.
127. R. J. Pettis, B. Ginsberg, L. Hirsch, D. Sutter, S. Keith, E. McVey, N. G. Harvey, M. Hompesch, L. Nosek, C. Kapitza, L. Heinemann (2011). Intradermal microneedle delivery of insulin lispro achieves faster insulin absorption and insulin action than subcutaneous injection. *Diabetes Technol Ther,* **13**:435-442; doi: 10.1089/dia.2010.0184.
128. X. Jin, D. D. Zhu, B. Z. Chen, M. Ashfaq, X. D. Guo (2018). Insulin delivery systems combined with microneedle technology. *Adv Drug Deliv Rev,* **127**:119–137; doi:10.1016/j.addr.2018.03.011.

129. S. P. Davis, W. Martanto, M. G. Allen, M. R. Prausnitz (2005). Hollow metal microneedles for insulin delivery to diabetic rats. *IEEE Trans Biomed Eng*, **52**:909–915; doi:10.1109/TBME.2005.845240.

130. P. M. Wang, M. Cornwell, J. Hill, M. R. Prausnitz (2006). Precise microinjection into skin using hollow microneedles. *J Invest Dermatol*, **126**:1080–1087; doi:10.1038/sj.jid.5700150.

131. Y. Ito, E. Hagiwara, A. Saeki, N. Sugioka, K. Takada (2006). Feasibility of microneedles for percutaneous absorption of insulin. *Eur J Pharm Sci*, **29**:82–88; doi:10.1016/j.ejps.2006.05.011.

132. K. Migalska, D. I. J. Morrow, M. J. Garland, R. Thakur, A. D. Woolfson, R. F. Donnelly (2011). Laser-engineered dissolving microneedle arrays for transdermal macromolecular drug delivery. *Pharm Res*, **28**:1919–1930; doi:10.1007/s11095-011-0419-4.

133. M. H. Ling, M. C. Chen (2013). Dissolving polymer microneedle patches for rapid and efficient transdermal delivery of insulin to diabetic rats. *Acta Biomater*, **9**:8952–8961; doi:10.1016/j.actbio.2013.06.029.

134. K. Fukushima, T. Yamazaki, R. Hasegawa, Y. Ito, N. Sugioka, K. Takada (2010). Pharmacokinetic and pharmacodynamic evaluation of insulin dissolving microneedles in dogs. *Diabetes Technol Ther*, **12**(6); doi:10.1089/dia.2009.0176.

135. J. Gupta, H. S. Gill, S. N. Andrews, M. R. Prausnitz (2011). Kinetics of skin resealing after insertion of microneedles in human subjects. *J Control Release*, **154**:148–155; doi:10.1016/j.jconrel.2011.05.021.

136. J. J. Norman, M. R. Brown, N. A. Raviele, M. R. Prausnitz, E. I. Felner (2013). Faster pharmacokinetics and increased patient acceptance of intradermal insulin delivery using a single hollow microneedle in children and adolescents with type 1 diabetes. *Pediatr Diabetes*, **14**:459–465; doi:10.1111/pedi.12031.

137. J. Cegla (2015). Microneedle-array patches loaded with hypoxia-sensitive vesicles provide fast glucose-responsive insulin delivery. *Ann Clin Biochem*, **112**:8260–8265; doi:10.1177/0004563215605691.

138. H. Chen, H. Zhu, J. Zheng, D. Mou, J. Wan, J. Zhang, T. Shi, Y. Zhao, H. Xu, X. Yang (2003). Iontophoresis-driven penetration of nanovesicles through microneedle-induced skin micro channels for enhancing transdermal delivery of insulin. *J Control Release*, **139**:63–72; doi:10.1016/j.jconrel.2009.05.031.

139. C. P. P. Pere, S. N. Economidou, G. Lall, C. Ziraud, J. S. Boateng, B. D. Alexander, D. A. Lamprou, D. Douroumis (2018). 3D printed

microneedles for insulin skin delivery. *Int J Pharm*, **544**:425–432; doi:10.1016/j.ijpharm.2018.03.031.

140. E. L. Giudice, J. D. Campbell (2006). Needle-free vaccine delivery. *Adv Drug Deliv Rev*, **58**:68–89; doi:10.1016/j.addr.2005.12.003.
141. J. Di, S. Yao, Y. Ye, Z. Cui, J. Yu, T. K. Ghosh, Y. Zhu, Z. Gu (2015). Stretch-triggered drug delivery from wearable elastomer films containing therapeutic depots. *ACS Nano*, **9**:9407–9415; doi:10.1021/acsnano.5b03975.
142. S. Yang, F. Wu, J. Liu, G. Fan, W. Welsh, H. Zhu, T. Jin (2015). Phase-transition microneedle patches for efficient and accurate transdermal delivery of insulin. *Adv Funct Mater*, **25**:4633–4641; doi:10.1002/adfm.201500554.
143. W. Sun, Q. Hu, W. Ji, G. Wright, Z. Gu (2017). Leveraging physiology for precision drug delivery. *Physiol Rev*, **97**:189–225; doi:10.1152/physrev.00015.2016.
144. J. Yu, Y. Zhang, A. R. Kahkoska, Z. Gu (2017). Bioresponsive transcutaneous patches. *Curr Opin Biotechnol*, **48**:28–32; doi:10.1016/j.copbio.2017.03.001.
145. Y. Lu, A. A. Aimetti, R. Langer, Z. Gu (2016). Bioresponsive materials. *Nat Rev Mater*, 16075; doi:10.1038/natrevmats.2016.75.
146. J. O. You, D. Almeda, G. J. C. Ye, D. T. Auguste (2010). Bioresponsive matrices in drug delivery. *J Biol Eng*, 15; doi:10.1186/1754-1611-4-15.
147. Y. Ye, J. Yu, D. Wen, A. R. Kahkoska, Z. Gu (2018). Polymeric microneedles for transdermal protein delivery. *Adv Drug Deliv Rev*, **127**:106–118; doi:10.1016/j.addr.2018.01.015.
148. Microneedle devices and uses thereof (2012). https://patents.google.com/patent/US20130165772A1/en (accessed August 11, 2019).
149. H.-S. Seon-Woo, H. J. Kim, J. Y. Roh, J.-H. Park (2019). Dissolving microneedle systems for the oral mucosal delivery of triamcinolone acetonide to treat aphthous stomatitis. *Macromol Res*, **27**:282–289; doi:10.1007/s13233-019-7031-6.
150. Y. Ma, S. E. Boese, Z. Luo, N. Nitin, H. S. Gill (2015). Drug coated microneedles for minimally-invasive treatment of oral carcinomas: development and in vitro evaluation. *Biomed Microdevices*, **17**:9944; doi:10.1007/s10544-015-9944-y.
151. M.-C. Chen, Z.-W. Lin, M.-H. Ling (2016). Near-infrared light-activatable microneedle system for treating superficial tumors by combination of chemotherapy and photothermal therapy. *ACS Nano*, **10**:93–101; doi:10.1021/acsnano.5b05043.

152. G. Traverso, C. M. Schoellhammer, A. Schroeder, R. Maa, G. Y. Lauwers, B. E. Polat, D. G. Anderson, D. Blankschtein, R. Langer (2015). Microneedles for drug delivery via the gastrointestinal tract. *J Pharm Sci*, **104**:362–367; doi:10.1002/jps.24182.

153. J.-F. Liao, J.-C. Lee, C.-K. Lin, K.-C. Wei, P.-Y. Chen, H.-W. Yang (2017). Self-assembly DNA polyplex vaccine inside dissolving microneedles for high-potency intradermal vaccination. *Theranostics*, **7**:2593–2605; doi:10.7150/thno.19894.

154. M. An, H. Liu (2017). Dissolving microneedle arrays for transdermal delivery of amphiphilic vaccines. *Small*, **13**:1700164; doi:10.1002/smll.201700164.

155. E. M. Saurer, R. M. Flessner, S. P. Sullivan, M. R. Prausnitz, D. M. Lynn (2010). Layer-by-layer assembly of DNA- and protein-containing films on microneedles for drug delivery to the skin. *Biomacromolecules*, **11**:3136–3143; doi:10.1021/bm1009443.

156. Q. Zeng, J. M. Gammon, L. H. Tostanoski, Y.-C. Chiu, C. M. Jewell (2017). In vivo expansion of melanoma-specific t cells using microneedle arrays coated with immune-polyelectrolyte multilayers. *ACS Biomater Sci Eng*, **3**:195–205; doi:10.1021/acsbiomaterials.6b00414.

157. E. McAlister, M. J. Garland, T. R. R. Singh, R. F. Donnelly (2017). Microporation using microneedle arrays. In *Percutaneous Penetration Enhancers Physical Methods in Penetration Enhancement*, Springer Berlin Heidelberg, Berlin, Heidelberg, pp. 273–303; doi:10.1007/978-3-662-53273-7_18.

158. J. Mönkäre, M. Reza Nejadnik, K. Baccouche, S. Romeijn, W. Jiskoot, J. A. Bouwstra (2015). IgG-loaded hyaluronan-based dissolving microneedles for intradermal protein delivery. *J Control Release*, **218**:53–62; doi:10.1016/j.jconrel.2015.10.002.

159. Y. P. Yekaterina Ostapchuk, Y. Ostapchuk, E. A. Cetin, A. Yilmaz, G. Deniz, S. Talaeva, N. Omarbaeva, I. Oskolchenko, N. Belyaev (2015). Hyaluronan-binding T regulatory cells in peripheral blood of breast cancer patients. *J Clin Cell Immunol*, **06**:1–6; doi:10.4172/2155-9899.1000286.

160. A. Pattani, P. F. McKay, M. J. Garland, R. M. Curran, K. Migalska, C. M. Cassidy, R. K. Malcolm, R. J. Shattock, H. O. McCarthy, R. F. Donnelly (2012). Microneedle mediated intradermal delivery of adjuvanted recombinant HIV-1 CN54gp140 effectively primes mucosal boost inoculations. *J Control Release*, **162**:529–537; doi:10.1016/j.jconrel.2012.07.039.

161. P. C. DeMuth, Y. Min, D. J. Irvine, P. T. Hammond (2014). Implantable silk composite microneedles for programmable vaccine release kinetics

and enhanced immunogenicity in transcutaneous immunization. *Adv Healthc Mater*, **3**:47–58; doi:10.1002/adhm.201300139.

162. K. Tsioris, W. K. Raja, E. M. Pritchard, B. Panilaitis, D. L. Kaplan, F. G. Omenetto (2012). Fabrication of silk microneedles for controlled-release drug delivery. *Adv Funct Mater*, **22**:330–335; doi:10.1002/adfm.201102012.
163. M. Zaric, O. Lyubomska, O. Touzelet, C. Poux, S. Al-Zahrani, F. Fay, L. Wallace, D. Terhorst, B. Malissen, S. Henri, U. F. Power, C. J. Scott, R. F. Donnelly, A. Kissenpfennig (2013). Skin dendritic cell targeting via microneedle arrays laden with antigen-encapsulated poly- d,l-lactide-co-glycolide nanoparticles induces efficient antitumor and antiviral immune responses. *ACS Nano*, **7**:2042–2055; doi:10.1021/nn304235j.
164. C. K. Choi, J. B. Kim, E. H. Jang, Y.-N. Youn, W. H. Ryu (2012). Curved biodegradable microneedles for vascular drug delivery. *Small*, **8**:2483–2488; doi:10.1002/smll.201200441.
165. S.-H. Baek, J.-H. Shin, Y.-C. Kim (2017). Drug-coated microneedles for rapid and painless local anesthesia. *Biomed Microdevices*, **19**:2; doi:10.1007/s10544-016-0144-1.
166. C. Baek, M. Han, J. Min, M. R. Prausnitz, J. H. Park, J. H. Park (2011). Local transdermal delivery of phenylephrine to the anal sphincter muscle using microneedles. *J Control Release*, **154**:138–147; doi:10.1016/j.jconrel.2011.05.004.
167. W. S. Chiu, N. A. Belsey, N. L. Garrett, J. Moger, G. J. Price, M. B. Delgado Charro, R. H. Guy (2015). Drug delivery into microneedle porated nails from nanoparticle reservoirs. *J Control Release*, **220**:98–106; doi:10.1016/j.jconrel.2015.10.026.
168. L. Klouda, A. G. Mikos (2008). Thermo responsive hydrogels in biomedical applications. *Eur J Pharm Biopharm*, **68**:34–45; doi:10.1016/j.ejpb.2007.02.025.
169. Y. Choi, M. A. McClain, M. C. LaPlaca, A. B. Frazier, M. G. Allen (2007). Three dimensional MEMS microfluidic perfusion system for thick brain slice cultures. *Biomed Microdevices*, **9**:7–13; doi:10.1007/s10544-006-9004-8.
170. H. W. Vesper, P. M. Wang, E. Archibold, M. R. Prausnitz, G. L. Myers (2006). Assessment of trueness of a glucose monitor using interstitial fluid and whole blood as specimen matrix. *Diabetes Technol Ther*, **8**:76–80; doi:10.1089/dia.2006.8.76.

171. P. M. Wang, M. Cornwell, M. R. Prausnitz (2005). Minimally invasive extraction of dermal interstitial fluid for glucose monitoring using microneedles. *Diabetes Technol Ther*, **7**:131–141; doi:10.1089/dia.2005.7.131.

172. G. Wang, M. D. Poscente, S. S. Park, C. N. Andrews, O. Yadid-Pecht, M. P. Mintchev (2017). Wearable microsystem for minimally invasive, pseudo-continuous blood glucose monitoring: the e-mosquito. *IEEE Trans Biomed Circuits Syst*, **11**:979–987; doi:10.1109/TBCAS.2017.2669440.

173. W. H. Smart, K. Subramanian (2000). The use of silicon microfabrication technology in painless blood glucose monitoring. *Diabetes Technol Ther*, **2**:549–559; doi:10.1089/15209150050501961.

174. Q. Qing Bai, K. D. Wise, D. J. Anderson (2000). A high-yield microassembly structure for three-dimensional microelectrode arrays. *IEEE Trans Biomed Eng*, **47**:281–289; doi:10.1109/10.827288.

175. Y. C. Kim, J. H. Park, M. R. Prausnitz (2012). Microneedles for drug and vaccine delivery. *Adv Drug Deliv Rev*, **64**:1547–1568; doi:10.1016/j.addr.2012.04.005.

176. S. Rajaraman, J. A. Bragg, J. D. Ross, M. G. Allen (2011). Micromachined three-dimensional electrode arrays for transcutaneous nerve tracking. *J Micromech Microeng*, **21**:085014(13pp); doi:10.1088/0960-1317/21/8/085014.

177. K. D. Wise, A. M. Sodagar, Ying Yao, M. N. Gulari, G. E. Perlin, K. Najafi (2008). Microelectrodes, microelectronics, and implantable neural Microsystems. *Proc IEEE*, **96**:1184–1202; doi: 10.1109/JPROC.2008.922564.

178. L. M. Yu, F. E. H. Tay, D. G. Guo, L. Xu, K. L. Yap (2009). A microfabricated electrode with hollow microneedles for ECG measurement. *Sens Actuators, A*, **151**:17–22; doi:10.1016/j.sna.2009.01.020.

179. P. Griss, H. K. Tolvanen-Laakso, P. Merilainen, G. Stemme (2002). Characterization of micro machined spiked bio potential electrodes. *IEEE Trans Biomed Eng*, **49**:597–604; doi:10.1109/TBME.2002.1001974.

180. S. Hashmi, R. Gaugler (1997). Application of micromechanical piercing structures for genetic transformation of nematodes. In *Recombinant Gene Expression Protocols*, Humana Press, New Jersey, Vol. **62**, pp. 393–398; doi:10.1385/0-89603-480-1:393.

181. N. Wang, Y. Zhen, Y. Jin, X. Wang, N. Li, S. Jiang, T. Wang (2017). Combining different types of multifunctional liposomes loaded with ammonium bicarbonate to fabricate microneedle arrays as a vaginal

mucosal vaccine adjuvant-dual delivery system (VADDS). *J Control Release*, **246**:12–29; doi:10.1016/j.jconrel.2016.12.009.

182. S. Bhatnagar, P. Kumari, S. P. Pattarabhiran, V. V. K. Venuganti (2018). Zein microneedles for localized delivery of chemotherapeutic agents to treat breast cancer: drug loading, release behavior, and skin permeation studies. *AAPS PharmSciTech*, **19**:1818–1826; doi:10.1208/s12249-018-1004-5.
183. Y. Ye, C. Wang, X. Zhang, Q. Hu, Y. Zhang, Q. Liu, D. Wen, J. Milligan, A. Bellotti, L. Huang, G. Dotti, Z. Gu (2017). A melanin-mediated cancer immunotherapy patch. *Sci Immunol*, **2**:eaan5692; doi:10.1126/sciimmunol.aan5692.
184. H. S. Gill, D. D. Denson, B. A. Burris, M. R. Prausnitz (2008). Effect of microneedle design on pain in human volunteers. *Clin J Pain*, **24**:585–594; doi:10.1097/AJP.0b013e31816778f9.
185. S. Kaushik, A. H. Hord, D. D. Denson, D. V. McAllister, S. Smitra, M. G. Allen, M. R. Prausnitz (2001). Lack of pain associated with microfabricated microneedles. *Anesth Analg*, **92**:502–504; doi:10.1097/00000539-200102000-00041.
186. S. M. Bal, J. Caussin, S. Pavel, J. A. Bouwstra (2008). In vivo assessment of safety of microneedle arrays in human skin. *Eur J Pharm Sci*, **35**:193–202; doi:10.1016/j.ejps.2008.06.016.
187. B. Godin, E. Touitou (2007). Transdermal skin delivery: predictions for humans from in vivo, ex vivo and animal models. *Adv Drug Deliv Rev*, **59**:1152–1161; doi:10.1016/j.addr.2007.07.004.
188. Special Diabetes Program: Clinical Trials Recruiting Patients & Families, NIDDK (n.d.). https://www.niddk.nih.gov/about-niddk/research-areas/diabetes/type-1-diabetes-special-statutory-funding-program/special-diabetes-program-clinical-trials-recruiting-patients-families (accessed July 17, 2019).

Chapter 8

Microfluidic Electroporation and Applications

Koyel Dey,[a,*] Srabani Kar,[b,*] Pallavi Shinde,[a] Loganathan Mohan,[a] Saumendra Kumar Bajpai,[c] and Tuhin S. Santra[a]

[a] *Department of Engineering Design, Alumni Ave, Indian Institute of Technology, Chennai, Tamil Nadu 600036, India*
[b] *Department of Electrical Engineering, University of Cambridge, 9 JJ Thomson Ave, Cambridge CB3 0FA, UK*
[c] *Department of Applied Mechanics, Mechanical Sciences Block, Alumni Ave, Indian Institute of Technology, Chennai, Tamil Nadu 600036, India*
santra.tuhin@gail.com, tuhin@iitm.ac.in

The creation of transient, hydrophilic pores on the cell membrane in response to an applied high-amplitude electric field is recognized as electroporation. This method has been extensively applied for delivering different types of cargos into or out of cells for analyzing their interactions. Traditional bulk electroporation (BEP) facilitating the transfection of millions of cells shows significant limitations, such as non-uniform transfection, increase in toxicity, low cell viability, and high voltage requirement. In more recent years, due to a rapid growth in micro-/nanotechnologies, it has been feasible to perform precise single-cell electroporation using

*These authors contributed equally to this chapter.

Microfluidics and Bio-MEMS: Devices and Applications
Edited by Tuhin S. Santra

ISBN 978-981-4800-85-3 (Hardcover), 978-1-003-01493-5 (eBook)
www.jennystanford.com

microelectrodes with high cell viability and high transfection efficiency as compared to bulk electroporation. But with the increasing demand for minimal cell damage and dosage regulatory capabilities, researchers are trying to further miniaturize the electrode structures and perform single-cell nanoelectroporation. In this chapter, we have discussed the fundamental aspects of electroporation and recent cutting-edge technologies to perform bulk- and single-cell micro-/nanoelectroporation and their clinical applications. The advantages, limitations, and future prospects of electroporation are also elaborated.

8.1 Introduction

Molecular delivery into cells with high delivery efficiency and high cell viability is a challenging issue for cell therapy and diagnostic purposes. Over the years, different kinds of physical techniques have been developed in order to generate permeable pores on the cell membrane, such as magnetoporation [1], electroporation [2, 3], mechanoporation [4], thermoporation [5], optoporation [6, 7], and sonoporation [8, 9]. Among these techniques, electroporation has gained remarkable attention due to easily addressable technical requirements [10–13], high transfection efficiency [14–16], and reproducibility [17–19]. Conventional electroporation techniques use strong electric fields (a few KV/cm) [20–23] to transfect millions of cells together, known as bulk electroporation (BEP). The experiment is performed simply by using a cuvette, consisting of two large electrodes inserted in the interior part to produce high electric field distribution around the suspension of millions of cells [24, 25]. Though this approach can transfect a large number of cells simultaneously, it suffers from low dosage controlling capability [26–28] and low cell viability [29–32] and lacks in-depth information of individual targeted cells [33, 34]. Most of the BEP techniques are often bound by low efficiency of plasmid delivery in various cell types because of plasmid degradation, toxicity, and immune response [23].

In the last two decades, the rapid growth in micro-/nanotechnologies has allowed conducting electroporation at the single-cell level for exploring the heterogenetic behavior of individual cell types [35–38]. Further advancements in single-cell electroporation (SCEP) shows nanolocalized transfection [14, 39–41], which can control the dosage of drugs precisely. With the recent developments in nanoscale electroporation techniques, it is possible to deform the cell membrane in a localized area and visualize the transfection [14, 39]. A variety of micro- and nanoscale electroporation techniques have been developed that not only overcome the throughput limitation but also can provide high transfection efficiency and high cell viability [14, 37, 38, 42]. In this chapter, we give an overview of electroporation, including the latest advancements in the micro-/nanoscale. Finally, the challenges and future prospects in electroporation are discussed briefly.

8.2 Brief Overview of Electroporation

The impacts of an externally applied electric field on biological objects have been extensively studied during 1900–1970 [43–49]. In 1936, McKinley explained the effect of an electric field on a cell membrane [23]. In 1951, Hodgkin applied a high electric field across the surface membrane of living cells to study ion-cell interaction [50]. Traditionally, electroporation experiments are carried out within cuvette-shaped chambers. The interelectrode distance usually ranges from centimeters to millimeters. Figure 8.1 shows the schematic setup for BEP [49].

Application of a high voltage across the metal electrodes immersed inside the cuvette results in a potential drop, called the transmembrane potential (TMP), across the cell membrane. As soon as the TMP ($\Delta\psi_E$) reaches a critical value (0.2–1 V [14], lipid molecules inside the cell membrane start getting rearranged in order to generate transient hydrophilic pores from originally created hydrophobic pores to facilitate cargo delivery [51–55].

A model for describing the TMP for a spherical cell of volume V exposed to an external electric field has been discussed here [56, 57]. From the conservation-of-charge principle, the total flow of current through this volume is found to be equivalent to the net flow of charges, which is equal to the rate of decrease of charge within

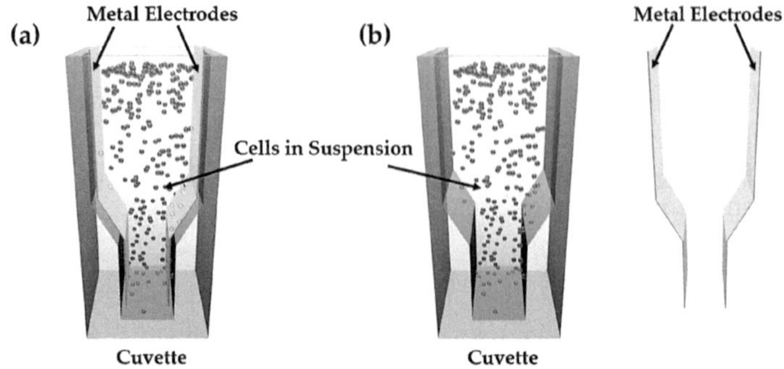

Figure 8.1 Schematic representation of a cuvette with metal electrodes used for bulk electroporation. The gap between two electrodes varies from millimeters to centimeters. (a) Cells are suspended inside the cuvette, and metal electrodes are kept fixed inside the cuvette for the electroporation experiment. (b) Metal electrodes to be incorporated inside the cuvette [49].

volume V. This can be shown as

$$I = \int_{S} J \cdot \mathrm{d}s \tag{8.1}$$

$$= -\frac{\partial}{\partial t}\int_{V} \rho \mathrm{d}V \tag{8.2}$$

Using divergence theorem (shown in Eq. 8.3), Eqs. 8.1 and 8.2 can be modified into Eq. 8.4:

$$\int_{S} \mathbf{A} \cdot \mathbf{n} ds = \int_{V} \nabla \cdot \mathbf{A} dV \tag{8.3}$$

$$\int_{V} \nabla \cdot \mathbf{J} dV = -\int_{V} \frac{\partial \rho}{\partial t} \mathrm{d}V \tag{8.4}$$

Equation 8.4 can be further modified into Eq. 8.5, where $\mathbf{J}$ and ρ represent current density and volume charge density, respectively.

$$\int_{V} \left(\nabla \cdot \mathbf{J} + \frac{\partial \rho}{\partial t} \right) \mathrm{d}V = 0 \tag{8.5}$$

Equation 8.5 remains true without consideration of volume and gives rise to the continuity equation

$$\nabla \cdot \mathbf{J} + \frac{\partial \rho}{\partial t} = 0. \tag{8.6}$$

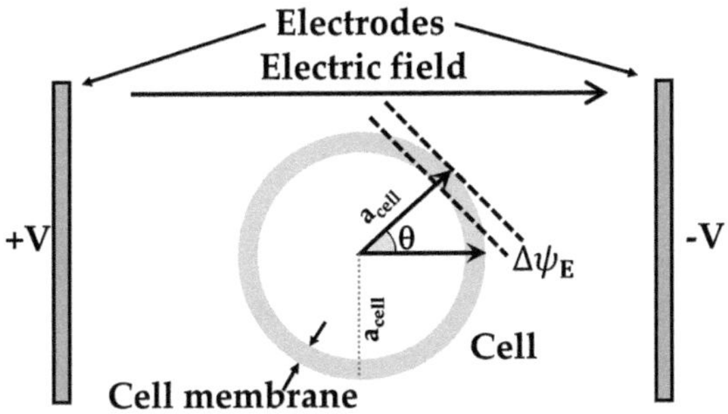

Figure 8.2 Transmembrane potential for a spherical cell of radius a_{cell} placed in an external electric field **E** [57].

This implies the absence of charge density fluctuation over time, that is, $\frac{\partial\rho}{\partial t} = 0$ and $\nabla \cdot \mathbf{J} = 0$. Current density $\mathbf{J} = \sigma\mathbf{E}$, where σ is the conductivity and **E** is the electric field, which is defined as the gradient of electric potential. Maxwell's equation becomes $\nabla^2\psi = 0$, where ψ (electric potential) becomes zero when there is no net charge. If the conductivity of the external medium and cytoplasm becomes much stronger than that of the cell membrane, the field-induced TMP ($\Delta\psi_{\text{E}}$) at a radial position (a_{cell}, θ) across a spherical cell can be expressed as

$$\Delta\psi_{\text{E}} = 1.5a_{\text{cell}}\mathbf{E}_{\text{e}}\cos\theta. \tag{8.7}$$

Here, a_{cell} is the cell radius, $\mathbf{E}_{\text{e}}$ is the applied electric field strength, and θ is the angle between the field line and normal of the point of interest across the cell membrane. Angle θ can be either 0° or 180°. Figure 8.2 demonstrates the model cell for calculating the TMP. Induced TMP is much higher across the poles, whereas it is lower across the equator [56, 57].

After applying an external electric field, the TMP is developed within a very short charging period, which is in the microsecond range. This can be shown as

$$\Delta\psi_{\text{E}}(t) = 1.5a_{\text{cell}}\,\mathbf{E}_{\text{e}}\cos\theta[1 - \exp(-t/\tau)]. \tag{8.8}$$

Here, τ represents the charging time. It relies upon the dielectric properties of the cell membrane. Assuming the cell membrane as a pure dielectric, the charging time (τ) can be expressed as

$$\tau = a_{\text{cell}} C_{\text{memb}} \left(\lambda_{\text{int}} + 2\lambda_{\text{out}}\right) / \left(2\lambda_{\text{int}}\lambda_{\text{out}}\right). \tag{8.9}$$

Here, λ_{int} and λ_{out} are expressed as the conductivity of the intracellular medium and the extracellular medium, respectively, a_{cell} is the outer radius of the cell, and C_{memb} is the membrane capacitance per unit area. The potential difference $\Delta\psi$ across the cell membrane comprises natural $\Delta\psi_0$ and field-induced TMP $\Delta\psi_{\text{E}}$. Hence, the overall potential difference across the cell membrane can be shown as

$$\Delta\psi = \Delta\psi_0 + \Delta\psi_{\text{E}}. \tag{8.10}$$

To supply power to the cell or to transmit a signal, the natural membrane potential is conserved in a live cell membrane. Typically, it varies between −40 to −60 mV, whereas the potential of the outer surface of the membrane is taken to be zero [56, 58].

8.3 Recent Advancement in Electroporation

In traditional electroporation, BEP, millions of cells are transfected together with well-established protocols. Although easy to implement in a lab environment, BEP suffers from significant limitations. Application of a high voltage has a tendency to decrease cell viability due to bubble formation, Joule heating effect, and pH change [11, 32]. Moreover, nonhomogeneous distribution of the electric field over a large number of cells affects the transfection efficiency [30, 31, 44]. In the last two decades, the development of the microfluidic electroporation technique offers SCEP, which has the flexibility of manipulating and targeting single cells with an extremely focused electric field that offers single-cell transfection [37, 59, 60].

With the recent advancements in micro-/nanotechnologies, SCEP is able to overcome adverse limitations of BEP. It provides an opportunity to transfect targeted cell numbers with an electric pulse of a few volts. which further helps in reducing pH variation, Joule heating, electric field distortion, and metal-ion dissolution from metal electrodes [23]. Thus, SCEP provides better delivery efficiency and cell viability as compared to BEP. Moreover, SCEP provides better dosage control facility, which is not possible in the case of BEP. SCEP provides homogeneous field distribution on cells, lower electrolysis effect, low toxicity, precise control for drug delivery, high cell viability, and high transfection efficiency as compared to BEP. Recent research methodology uses the finite element method to study the various aspects of the SCEP process [14, 38, 41, 60, 61]. In this section we will discuss some advanced electroporation techniques.

8.3.1 Gold Nanoparticle–Enhanced Electroporation

One of the foremost shortcomings that traditional electroporation systems encounter is low electrical conductivity of the buffer solution. This results a higher voltage consumption across the electrodes, while the actual voltage drop across the treated cell membrane is much lower than the critical value. Increasing the ion strength of the buffer solution is not allowed due to constraint imposed by cell physiological conditions, such as osmotic balance, structural integrity, and ATP generation [62, 63]. To resolve this issue, Zu et al. [64] demonstrated a simple approach in which highly conductive Au NPs were added to the buffer solution. This reduced the buffer solution resistance and increased the overall conductivity of the solution. With the application of an external pulse, Au NPs near the surface of the cell membrane start behaving as virtual microelectrodes to electroporate the cell locally (Fig. 8.3a). This enhancement of the electroporation methodology solely relies upon the size, concentration, and mixing ratio of Au NPs.

Huang et al. utilized a gold nanoparticle (Au NP)-enhanced electroporation technique for directly delivering therapeutic probes such as siRNA or DNA plasmid into cell cytosol [65]. As shown in Fig. 8.3b, at first, they immobilized the cationic polymer polyethylenimine (PEI) on Au NPs though electrostatic interaction. The PEI, coated on Au NPs, was conjugated with negatively charged therapeutic probes to form a Au NP polyplex. This process helps to minimize the cytotoxicity concerns from the polyplex after delivery. After mixing the Au NP polyplex in the buffer solution, an external electric field was applied. Au NPs sitting close to the cell membrane

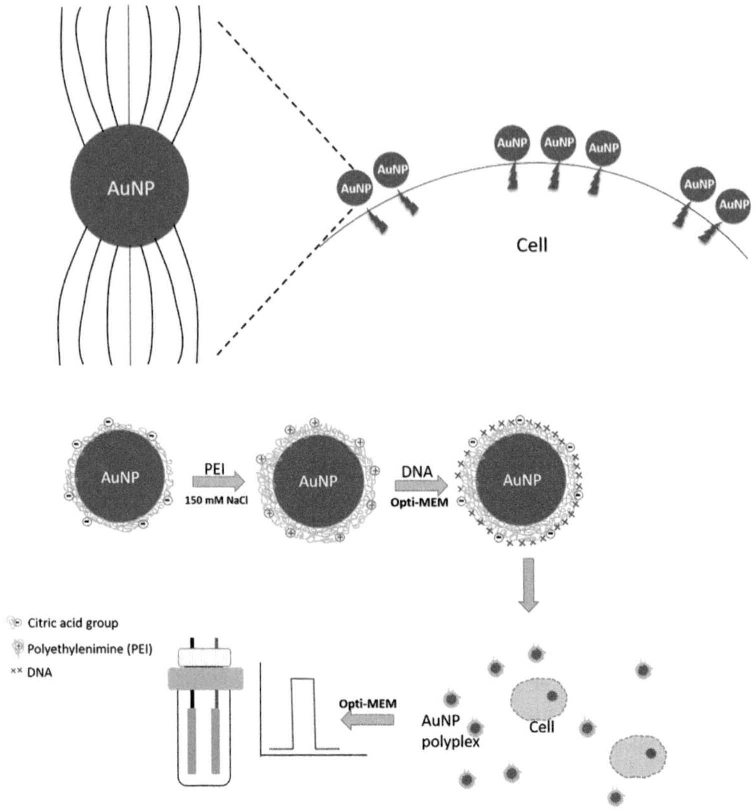

Figure 8.3 (a) Au NPs come closer to the membrane due to the affinity binding with receptors. After applying an external voltage, the electric field is converged on the conductive Au NPs. These Au NPs then start serving as a virtual electrode to polarize the limited area across the membrane [64]. (b) Schematic description of Au NP polyplex synthesis steps and delivery inside a cell [65].

act as a virtual electrode for generating focused electric pulses required for electroporation.

The main reason behind using specifically Au NPs is their biocompatibility and imaging capability [66, 67]. Another advantage is found is the non-invasive tracking of therapeutic probes. These methods have the capability to accelerate the applications of Au NPs in the field of regenerative medicine, diagnosis, and therapeutics.

8.3.2 Carbon Nanotube–Based Electroporation

Ever since the discovery of carbon nanotubes (CNTs) by Iijima in 1991 [68], they have been found to be physically and chemically robust and non-oxidizable under ambient conditions [69]. Timothy et al. conducted an experiment where he showed that a cell can remain viable even after insertion of carbon nanofibers [70]. Thus, CNTs have been regarded as a potential material for drug delivery [71, 72]. Conventional electroporation systems use metallic electrodes that release metal particles a few hundreds of nanometers in size on application of an electric pulse [29]. One more critical concern is the Joule heating effect [30, 73, 74]. Recent research has shown that the use of carbon fibers as electrode materials, particularly in micropipette-based electroporation, produces excellent results [75]. Rojas-Chapana et al. conducted reversible electroporation by employing CNTs [76]. Being an excellent conductor, multiwall CNTs (MWCNTs) enhanced the electrostatic force upon the cell membrane, which in turn formed permeable nanopores on the cell membrane and allowed the biomolecules to enter into the interior of the cells. Giersig et al. applied the same approach to deliver exogenous DNA into the cell [77]. Later on, Suehiro et al. demonstrated a dielectrophoresis (DEP)-based bacteria manipulation and electroporation technique by employing MWCNTs [78]. In their approach, MWCNTs were first mixed with bacteria. Then with the help of DEP, a single *E. coli* bacterium was trapped on the tip or surface of each MWCNT, where the highest DEP force was generated (Fig. 8.4a). Thereafter, a high electric field was applied in order to start the electroporation procedure and deliver drugs into *E. coli* bacteria (Fig. 8.4). This method shows great potential in the area of drug delivery.

8.4 Recent Advancements in Single-Cell Electroporation

Most of the recent advancements in SCEP incorporate single-cell trapping and miniaturized electrodes. Thus, the electric field gets intense in a considerably smaller area of the cell membrane,

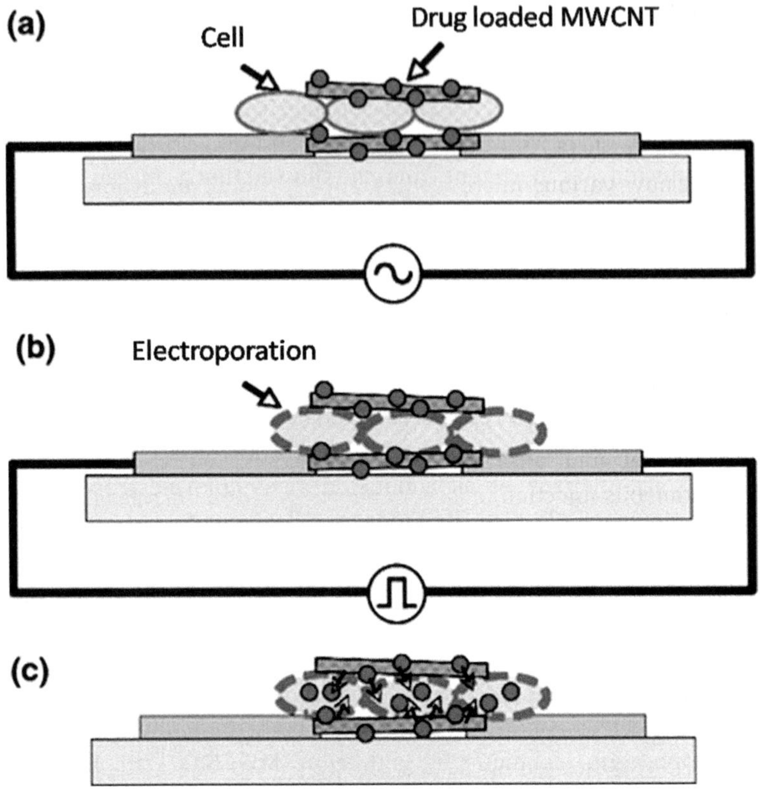

Figure 8.4 (a) Drug-loaded CNTs and *E. coli* are captured in the microelectrode cavity by DEP. (b) The cell membrane is electroporated with the help of a high electric field. (c) Drug molecules are delivered inside the electroporated *E. coli*. Reprinted by permission from Springer Nature Customer Service Centre GmbH: Springer Nature, *Microfluids and Nanofluids*, Ref. [78], copyright (2008).

which conveniently causes negligible cell damage and increases cell viability after delivery. In this section, we discuss different techniques for single cell trapping and recent advances in SCEP.

8.4.1 Single-Cell Trapping and Electroporation

To perform SCEP, individual cells need to be fixed near the electrode's position. Then the cell can be exposed to the electric field to generate pores at the cell membrane for molecular delivery. After resealing of pores, the cells should be released from the electrode's position so that they can be collected for further downstream analysis. In this process, it becomes essential to implement a flexible cell trapping and releasing mechanism. These traps can be contact or noncontact based. In this section, we will briefly discuss several trapping mechanisms subjected to the SCEP technique.

8.4.1.1 Mechanical trap

Traditionally, micropipettes have been used to mechanically hold a cell in a fixed position [79]. With the development of microfluidic devices, patterned structures were developed to constrict individual cells [80]. For example, Khine et al. demonstrated an SCEP chip where cell trapping was performed in a polydimethylsiloxane (PDMS) channel by applying negative pressure (Fig. 8.5a) [81]. Another device was developed by Huang et al. that also used negative pressure to trap a cell in a microhole along with a fluidic channel for electroporation (Fig. 8.5b) [82]. Chen et al. developed an orifice-based trapping mechanism, as shown in Fig. 8.5c [83]. The trap is made inside a PDMS channel, and thus it is biocompatible and transparent for cell observation in the trap region. They also used radio-frequency (RF) biosensors to realize the trapping mechanism. It has also been observed in all trapping techniques mentioned above that the electric field is profoundly enhanced at the contact point between cell and trap. This helps to create pores precisely on the cell membrane during the electroporation process.

8.4.1.2 Optical tweezer

An optical tweezer (OT) uses a tightly focused laser beam to trap and manipulate microscopic dielectric objects by imparting radiation pressure. The simplest kind of OT consists of a Gaussian laser beam and a high-numerical-aperture objective. A schematic of an OT arrangement is demonstrated in Fig. 8.6 [84]. The force exerted on the object can be either attractive or repulsive as per the refractive index of the objects as well as the surrounding medium. The OT has already been established as a potential tool

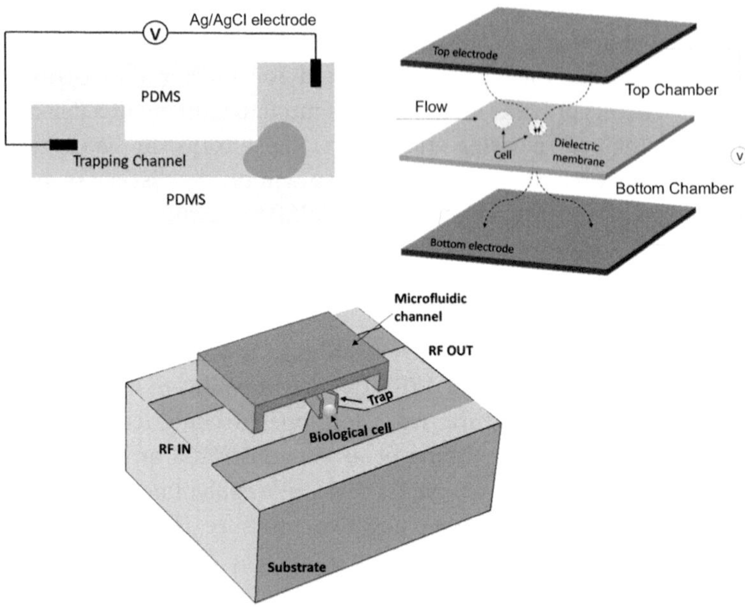

Figure 8.5 (a) A cell is trapped in a small PDMS channel on application of a negative pressure [81]. (b) Trapping of a single cell in a microhole on application of negative pressure. Redrawn from Ref. [82]. (c) RF biosensor–based trapping of a single cell [83].

for manipulating cells and cellular organelles. In electroporation, it has also been implemented for holding K562 cells at particular positions in suspension before applying an electric field to study the dependence of cell size on the minimal electric field required for permeabilization [85]. A detailed analysis has revealed that the minimal permeabilization field for any cell does not rely upon its size; instead it is the average characteristic field for each cell line. Choudhury et al. proposed optical trapping to push targeted cells into a slot for electroporation [86]. Being a non-contact-based technique, OT is safer for cell trapping and manipulation.

8.4.1.3 Droplets microfluidics

The droplet microfluidic technology is used to encapsulate cells as shown in Fig. 8.7. This technique is useful for transfecting cells

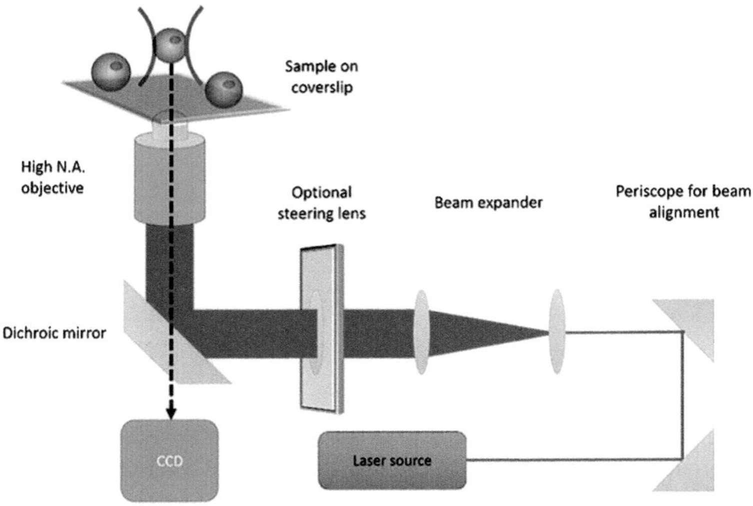

Figure 8.6 Schematic of an optical tweezer setup. Reprinted by permission from Springer Nature Customer Service Centre GmbH: Springer Nature, Ref. [84], copyright (2016).

in suspension. It uses the principle of immiscibility of water in oil to encapsulate individual cells. These cells are then passed through a microfluidic channel embedded with electrodes to carry out electroporation. A prominent advantage of scrutinizing single cells in microdroplets is the fast detection of cell-secreted molecules inside microdroplets. Qu et al. illustrated the use of droplet microfluidics in 2012 to electroporate algal cells [87]. Kurita et al. proposed parallel electroporation by electrostatically manipulating the droplets in a microreactor [88]. Im et al. observed enhanced delivery efficiency using a electroporation needle–based droplet microfluidic chip [89].

8.4.1.4 Magnetic trap

Magnetic traps are another way of arranging cells in specific patterns with high precision. These traps have been used for manipulating and directing nonmagnetic objects, such as different types of cells. An array of magnetic traps generates a high magnetic field at the desired location. Cells coupled with magnetic particles

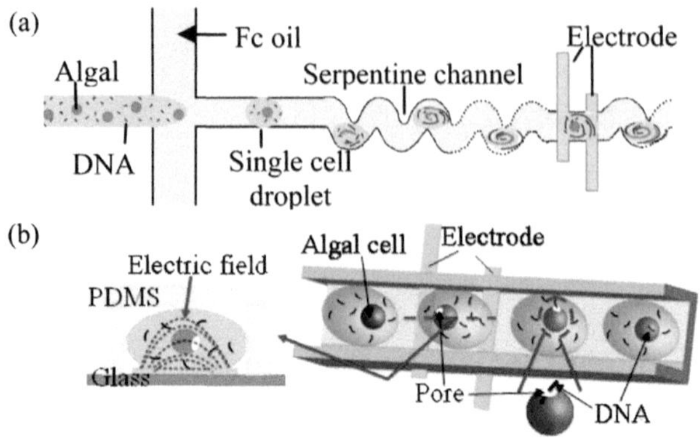

Figure 8.7 (a) Schematic of cell encapsulation in a droplet. (b) Distribution of the electric field near the electrodes followed by electroporation. Reproduced by permission of The Royal Society of Chemistry, from Ref. [87], copyright (2012); permission conveyed through Copyright Clearance Center, Inc.

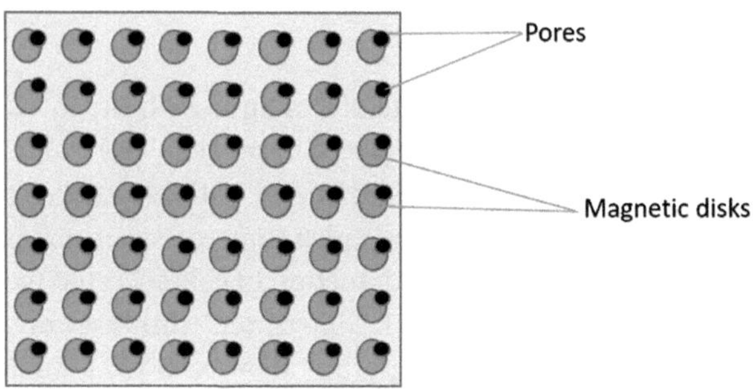

Figure 8.8 Top view of a magnetoelectroporation (MEP) chip containing magnetic traps [92].

can interact with magnetic fields [90]. Vries et al. showed a way of inserting magnetic particles in live cells for controlling and directing trapped cells [91]. Figure 8.8 shows the top view of a magnetoelectroporation chip, which contains magnetic disks working as magnetic traps for cell alignment [92]. However, biological effects of coupling magnetic beads to cells need to be considered for cell-based experiments.

8.4.2 Micro-/Nanochannel-Based Single-Cell Electroporation

Micro-/nanochannel-based SCEP systems have some advantages, such as low voltage requirement, control over pH variation, formation of a uniform electric field, and efficient usage of reagent [11, 93, 94]. The designing of a micro-/nanochannel based SCEP device typically follows some specific format. First, the device must contain a microfluidic channel where the cell mixture is loaded. Subsequently, the channel must contain a single trap or array of traps, which is basically a micro-/nanochannel for isolating a single cell. The third and most important step is introducing the electrodes with dimensions in the range of micrometer or nanometer for creating a strong electric field across the cell membrane [35, 81, 82, 94]. Over the past few years several micro-/nanochannel-based electroporation devices have been reported. One such device, shown Fig. 8.9a, consists of two wide channels, specifically used for cell input and output. One electrode is associated with the input channel and the other electrode is linked to the trapping end. Here, a single cell is isolated in the trap by application of negative pressure. On applying an electric field, the maximum voltage drops across the trap and cell membrane interface. Thus, electroporation takes place [81]. The main advantage of this design is the distance between electrodes and cell which helps to eliminate the adverse effect of electrode reaction.

Valero et al. developed a silicon- and glass-based microfluidic device for SCEP [95]. As demonstrated in Fig. 8.9b, the device consists of two microchannels (20 μm and 50 μm) and nine trapping sites of width 4 μm each. When the sample flows through the upper channel, negative pressure is applied in the lower channel through the trapping sites. Once the cells are trapped, electroporation is performed by applying voltages through the electrodes positioned at each individual trap. The channel, covered with a glass wafer, allows visualization of the trapping mechanism and the electroporation process. Chang et al. [60] introduced a nanochannel electroporation (NEP) system that comprises four parts: a 3D NEP chip, a support platform, a PDMS spacer, and a pair of electrodes, as shown in Fig. 8.9c. The top chamber, made of PDMS, is used for loading

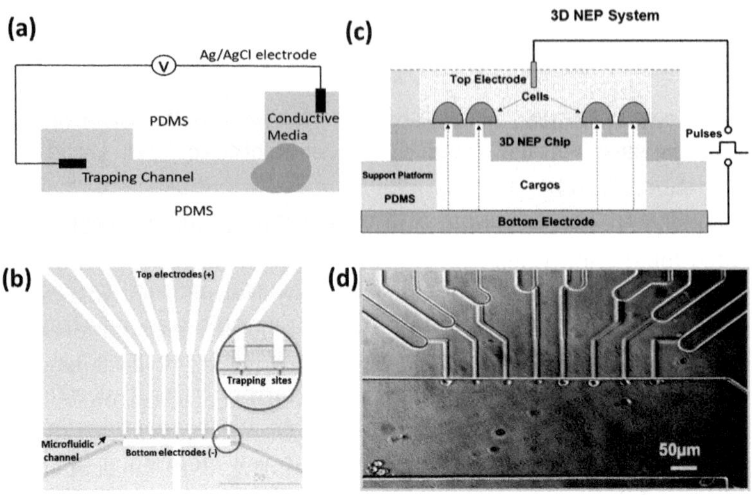

Figure 8.9 (a) Image demonstrating the trapping of a single cell inside the channel. One electrode is linked to the main channel. Another electrode is associated to the trapping channel [81]. (b) Silicon-glass microfluidic device for single-cell electroporation. The image shown is a zoomed-in version of the chip when a single cell is trapped. Republished with permission of The Royal Society of Chemistry, from Ref. [95], copyright (2008); permission conveyed through Copyright Clearance Center, Inc. (c) The 3D model of the NEP system showing all components stacked on a substrate. Republished with permission of The Royal Society of Chemistry, from Ref. [60], copyright (2015); permission conveyed through Copyright Clearance Center, Inc. (d) Bright-field image of a HeLa cell when interfaced with a capillary channel that is electrically and pneumatically individually addressable and connected to the bottom of a single well of a 96-well plate. Reprinted by permission from Springer Nature Customer Service Centre GmbH: Springer Nature, *Biomedical Microdevices*, Ref. [35], copyright (2007).

cells to be transfected, while the bottom chamber is used for loading drug. The top and bottom chambers are separated by the 3D NEP chip. After the cell is loaded, vacuum is applied from the bottom of the chip to pull the cells in the direction of nanopores. A strong electric field is applied across the electrodes. The device showed ~90% transfection efficiency after trapping cells in the nanopores [60]. Ionescu-Zanetti et al. demonstrated a PDMS-based microfluidic chip for performing SCEP [35]. They fabricated a PDMS-based microfluidic chip bonded to a standard bottomless 96-well plate (Nalge Nunc International) with lateral channels. The lateral channels are connected to 3 × 3 μm capillaries that are used for cell trapping and electric field focusing, as shown in Fig. 8.9d. A hollow Ag/AgCl electrode is placed at each of the wells in order to apply pressure and electrical connection to microfluidic channels inside the chip. To position cells at the beginning of the microchannel a slight transient negative pressure is applied. After positioning the cell at the trap, electroporation is performed by applying voltage through a modified patch clamp amplifier. Though the method provides good control over the electroporation area, it requires certain skills and expertise for aligning the cells across the microchannel.

Boukany et al. developed an NEP system by the DNA combining and imprinting (DCI) method [94]. As shown in Fig. 8.10a, the primary part of the NEP chip is two microchannels linked via a nanochannel of diameter ~90 nm. One microchannel is loaded with cells, while the other microchannel is loaded with drugs. An OT is used for positioning the cell at the tip of the nanochannel. After trapping the cell high voltage pulses were applied between two microchannels for performing electroporation. The device has been used for delivering quantum dots and plasmids with high transfection efficiency. In their next work, Boukany et al. employed the same device for injecting a lipoplex into cell cytoplasm [36] (Fig. 8.10b). The device provides precise control over the electroporation area and drug delivery.

For obtaining precise biomolecular delivery during SCEP, Kang et al. proposed the concept of a nanofountain probe (NFP) chip [38]. The chip comprises an array of cantilevers with a fixed microfluidic channel along with dispensing fountains. In the nanofountain probe, the molecular solution is stored in microreservoirs and loaded in microchannels by external pressure. The small probe aperture, as shown in Fig. 8.11a, provides precise control and can transfect the concerned area of a targeted cell [38]. Xie et al. demonstrated a nanostraw-based electroporation system, the schematic of which is shown in Fig. 8.11b [37]. Hollow nanostraws of aluminum oxide with ~250 nm diameter and ~1.5 μm height were fabricated on nanoporous track-etched polycarbonate membranes. The nanostraw membrane was incorporated between the PDMS

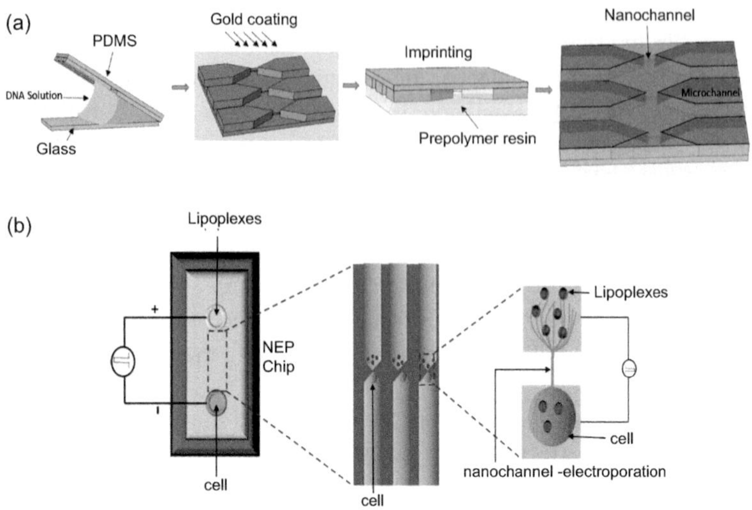

Figure 8.10 (a) Image demonstrating the fabrication process of the NEP chip by DCI [94]. (b) Schematic of nanoelectroporation operation by applying electrical pulses and injecting lipoplex nanoparticles inside the cell [36].

microfluidic channel (at the bottom, where molecules were loaded) and a PDMS well (on top, for cell culture). The indium tin oxide (ITO) substrate was integrated at the bottom of the microfluidic channel. An electric field was applied in between the ITO substrate and the Pt electrode placed in the PDMS well. On application of an external voltage, the electric field gets intense on top of the nanostraw, resulting in cell membrane deformation followed by the creation of transient hydrophilic membrane pores to deliver molecules into cells. The device showed transfection efficiency of ~81% for the delivery of plasmid-encoding red fluorescent protein into Chinese hamster ovary cells and ~67% for plasmid transfection into human embryonic kidney 293T cells, with high cell viability (>98%) [37].

Santra et al. demonstrated nanoelectroporation by using transparent ITO electrodes fabricated on a glass substrate, which is user friendly for microscopic visualization [14, 39–41]. The device, as shown in Fig. 8.12a, comprises an array of ITO electrodes a few microns wide and tenths of a nanometer thick [14]. The gaps between the electrode pairs subjected to selective voltage

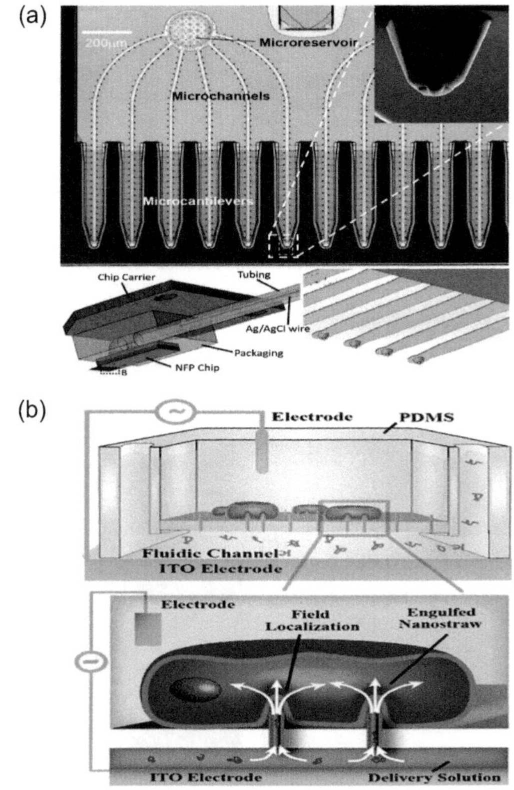

Figure 8.11 (a) Image showing an NFP chip with multiple probes for parallelized single-cell electroporation. Reprinted with permission from Ref. [38]. Copyright (2013) American Chemical Society. (b) Schematic view of a nanostraw-electroporation system. Reprinted with permission from Ref. [37]. Copyright (2013) American Chemical Society.

application varied between 60 nm and 500 nm (Fig. 8.12b) [39]. The surfaces of the ITO electrodes were covered by a silicon dioxide layer in order to passivate the intense electric field that can cause undesired Joule heating, formation of gas bubbles, and turbulent flow of local fluid. When an external voltage is applied, the electric field gets focused in between the nanoelectrode gap, resulting in cell membrane deformation followed by the creation of permeable nanopores that allow the biomolecules to enter gently into targeted cell with 80%–95% cell viability. It was seen that when

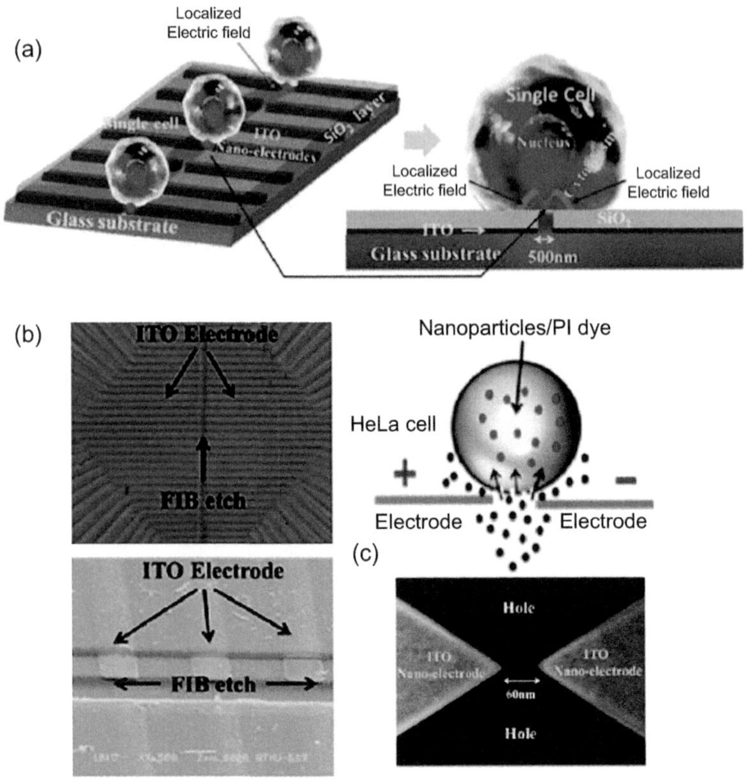

Figure 8.12 (a) Schematic representation of a localized single-cell nanoelectroporation chip. Reprinted from Ref. [14], with the permission of AIP Publishing. (b) Top-left figure: Optical microscope image of a patterned ITO electrode (20×). Bottom-left figure: Scanning electron microscopy (SEM) image of an ITO electrode with a microchannel. Top-right figure: Working principle of localize electroporation. Reprinted by permission from Springer Nature Customer Service Centre GmbH: Springer Nature, *Biomedical Microdevices*, Ref. [39], copyright (2012). (c) SEM image of the nanoelectrode tip with a 60 nm gap between two nanoelectrodes and a 40 nm tip diameter. © 2014 IEEE. Reprinted, with permission, from Ref. [40].

the electrodes are shaped into triangular structures with a 60 nm gap in between and a tip diameter of ~40 nm (Fig. 8.12c) [40], the electric field gets strongly focused only at the nanoelectrode tips, which can promote ultralocalized electroporation with enhanced higher cell viability of 90%–98%) as compared to the previous method [39].

In this section, we have shown different micro-/nanochannel-based electroporation mechanisms. The highly intense localized electric field in the nanochannel accelerates the transfection process. In most of these devices, the affected region of the cell membrane is quite small, resulting in a low percentage of cell death after delivery. Thus, micro-/nanochannel-based SCEP can be applied where safe transfection and dosage control is required.

8.4.3 Magnetoelectroporation

Recent developments in the area of microchannel- and nanochannel-based electroporation techniques showed their capability to handle thousands of cells in parallel. Most nanochannel SCEP systems use negative pressure–based trapping methods to manipulate and trap a single cell. These negative pressure–based methods can only trap a cell and are unable to align cells or move them from one place to another. Thus, in search of a robust ability to optimally align cells and transfect them uniformly, researchers started incorporating the concept of magnetic field in SCEP platforms [96–99]. Chang et al. proposed a 3D microchannel electroporation (MEP) chip integrated with four orthogonal electromagnets and a solenoid (Fig. 8.13) [92]. Here, by controlling an external magnetic field, individual cells are aligned as an array on top of a magnetic disk and each magnetic disk consists of microholes connected with the drug-loaded reservoir. Thus, when an external voltage is applied, the cell membrane deforms and biomolecules get delivered into cells in a parallel fashion.

To avoid cell adhesion at the time of transfection, the chip surface was treated with polyethylene glycolsilane before each trial. All cells were associated with superparamagnetic beads in order to remotely control cells by magnetic fields. The results confirmed that high transfection efficiency can be attained using low voltages and a weak magnetic field. The main advantage of magnetoelectroporation is that cell alignment can be done without any damage to the cells. Thus, the platform offers a promising way for high-throughput gene

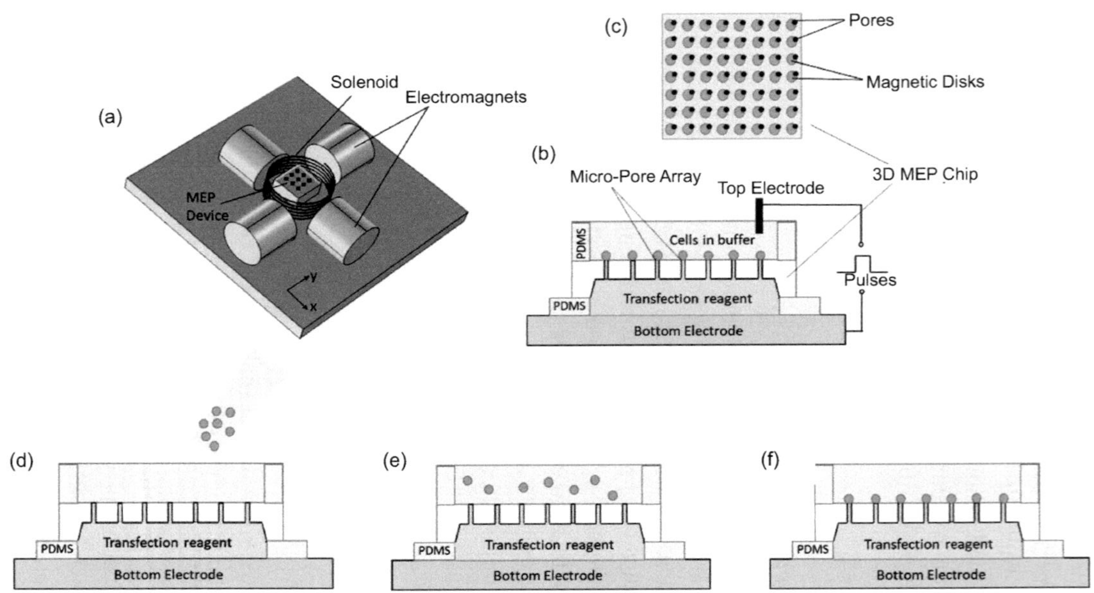

Figure 8.13 Magnetic tweezer–based 3D microchannel electroporation setup (a) The electromagnet system comprises four orthogonal electromagnets that generate the X and Y (in-plane) components of the applied magnetic field and a solenoid that produces a Z (out-of-plane) component. (b) The electroporation device is placed in the middle of the magnets. A gold substrate is used as the bottom electrode, and a PDMS spacer is used as the drug container. The electroporation wafer is placed just above the spacer, and the other PDMS spacer is placed on the upper-most side of the wafer to hold cells in solution. A platinum electrode is placed just above it. (c) Schematic of a 3D MEP chip showing the alignment of magnetic disks with micron-sized pores. (d) Seeding of cells is done by pipetting. (e) The magnetic field is turned on, and cells are falling in an aligned pattern. (f) Cells are aligned [92].

delivery, with the additional advantage of uniform transfection and high cell viability.

8.4.4 Optoelectronic Tweezer for Electroporation

One of the major challenges in the current research on SCEP revolves around the dimensions of the electrodes and their arrangements in microfluidic devices. Smaller electrodes increase the probability of mechanical failure, such as mechanical breakage due to tissue reactions with the electrode material [100]. Furthermore, fixed electrode arrangements can lead to specific experiments. Such limitations can be resolved by using an optoelectronic tweezer. Such a tweezer was first demonstrated by Chiou et al. in 2005 for the parallel manipulation of around 15,000 particle traps on a $1.3 \times 1.0\ mm^2$ area [101]. Valley et al. utilized this technique for the purpose of electroporation in 2009 [102]. The device consists of two electrodes (Fig. 8.14). The transparent electrode, made of ITO coated on a glass substrate, is placed at the upper side. The photosensitive electrode consists of amorphous silicon (a-Si) and a layer of ITO upon a glass substrate and is placed at the bottom. A chamber is created between top and bottom electrodes using SU-8. An aqueous solution containing cells is loaded into the chamber. Then an alternating current (AC) voltage is applied between these two electrodes. In the absence of light, most of the potential drops across the highly resistive a-Si layer. Desired optical patterns are generated by using a PC and are focused on the substrate by using a spatial light modulator/projector. As a consequence, the resistance of the a-Si layer drops drastically by several orders of magnitude in the spatially localized illuminated areas due to the generation of electron-hole pairs. This leads to a spatially patterned conductivity profile throughout the a-Si layer. Therefore, the liquid in the illuminated regions experiences a large electric field due to higher potential drops across the liquid. When the liquid-containing cells are exposed to such light patterns, the illuminated cells encounter a highly concentrated electric field and electroporation can take place above a threshold value of the field. The optimal field for electroporation in the case of HeLa cell should be in the range of 1.4–2.3 kV cm^{-1} [102, 103]. The major advantage of this

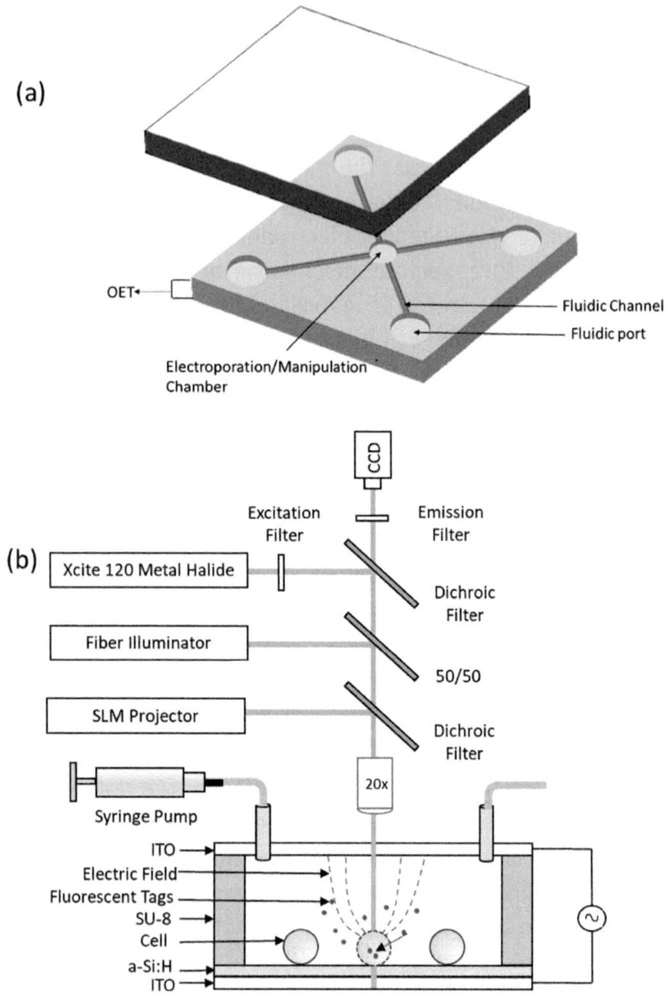

Figure 8.14 (a) Structure of the overall device, showing a microfluidic channel and an electroporation chamber. (b) A cross-sectional view of the device explaining light-induced electroporation. Concentration of the electric field increases across the illuminated cell, which results in selective electroporation [102].

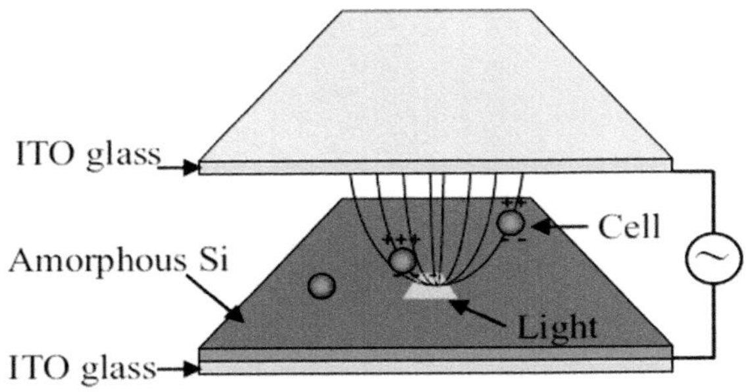

Figure 8.15 Schematic of the main device and the operating principle. The electric field gets concentrated across the cell upon illumination and results in cell lysis on the amorphous silicon layer. Reprinted from Ref. [104], with the permission of AIP Publishing.

device is its dynamic capability of cell alignment to desired patterns through DEP by applying a low bias. The bias is low (~0.2 kV/cm) enough to ensure that the cell membrane does not deform during manipulation. Thereafter, the bias (~1.5 kV/cm) is increased to a value above the critical value to perform electroporation selectively to the patterned cells. Without going into the detailed fabrication process, one can change the electrode geometry just by changing the optical pattern. Thus, an optically induced SCEP process provides greater flexibility in comparison to fixed metal electrode-based electroporation.

In a recent work, Lin et al. employed a similar device for transfecting fibroblast and oral cancer cells [104]. A pictorial representation of the device is given in Fig. 8.15. The effective conditions for lysing the oral cancer cells and fibroblast cells are 9 and 7 V p.p., respectively, at 20 kHz. The illumination power density is 1.17 mW/μm^2, with a beam diameter of 22 μm. One interesting finding is that in the case of optically induced electroporation, the transmembrane potential depends on the illumination power density and the size of the light spot, apart from the magnitude of the electric field. With the above analysis, it can be concluded that light-induced electroporation comes as a promising technique for low-cost, dynamic, and parallel SCEP.

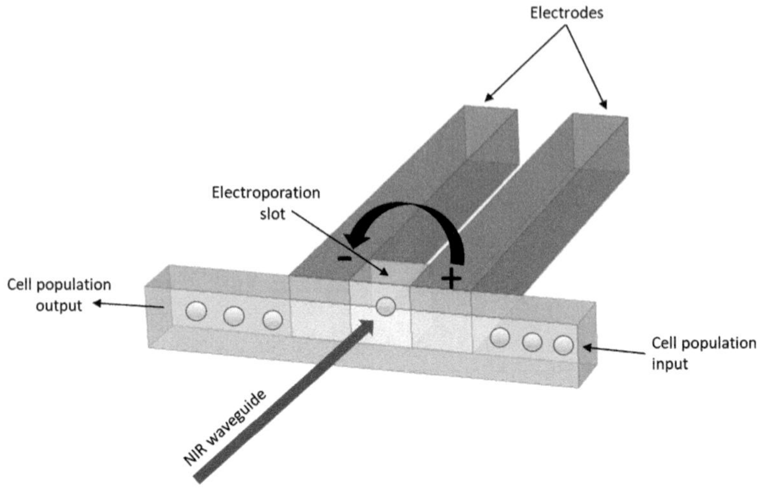

Figure 8.16 Schematic representation of an optofluidic electroporation device [86].

8.4.5 Waveguide-Based Electroporation

High-frequency electromagnetic waves (beyond hundreds of megahertz) can directly interact with the intracellular components when applied across a cell. Several research articles [83, 105–107] have demonstrated microwave sensing at the single-cell level. By using this technique, a single cell can be blocked between two nanoelectrodes and electroporation can be accomplished. For example, Choudhury et al. designed an optofluidic device in which SCEP was performed by employing a monolithically integrated near-infrared (NIR) waveguide [86]. As demonstrated in Fig. 8.16, the device comprises two electrodes and a waveguide setup placed directly opposite to the electroporation slot. Using a waveguide a single cell is captured in the electroporation slot that is placed in between two electrodes. Next, an electric field is applied for performing electroporation. This device is well suited for continuous-flow operation as it allows to target a cell from a population for electroporation through the radiation pressure from an NIR waveguide.

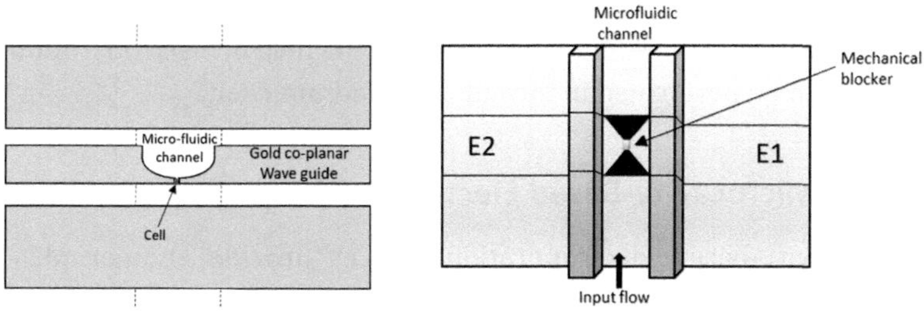

Figure 8.17 (a) Schematic diagram of a coplanar waveguide with a cell trapped by the center electrode [108]. (b) Pictorial representation of a microwave biosensor with a coplanar waveguide and a microfluidic channel on top [110].

In 2015, Ma and his group presented a new design that can block a single cell, using a coplanar waveguide. The device, as shown in Fig. 8.17a, comprises a gold coplanar waveguide inserted at a right angle to the microfluidic channel made by using PDMS. The device can capture a cell and release the same in the series gap by controlling the dielectrophoretic signal as a switch [108]. Using this device, Li et al. performed high-frequency continuous-wave SCEP, where a 2.2 V, 10 MHz signal was used as a dielectrophoretic signal for cell trapping. After the cells were trapped, SCEP was performed by increasing the voltage to a value above 2.2 V and decreasing the frequency to 100 KHz [109]. The viability of Jurkat human lymphoma cells was found to vary between 1 V at 100 kHz and 2.9 V at 10 MHz.

Chen et al. proposed a microwave-based biosensor that enables scrutinization of a single cell within its culture medium [83]. Later, by employing this device, Tamra et al. performed electroporation using microwave dielectric spectroscopy [110]. The microfluidic channel was placed on top of the coplanar waveguide circuit (Fig. 8.17b). The coplanar waveguide circuit consists of a capacitive gap at the center to trap a single cell. A mechanical trap was used to immobilize a single cell. After the cell was trapped, reversible SCEP was accomplished by using 8 pulses of 1 kV/cm with a pulse width of 100 μs and a frequency of 1 Hz with high cell viability (~80%). For conducting irreversible electroporation the electric field was enhanced to 1.5 kV/cm.

The outcomes of aforementioned techniques confirm that the waveguide-based method is promising in the field of electroporation and it can be performed in the cultural medium itself.

8.4.6 Microcavity-Based Electroporation

Microcavity-based electroporation devices provide another platform for performing SCEP. Zhang et al. presented a microfluidic chip design. As depicted in Fig. 8.18a and Fig 8.18b, it consists of

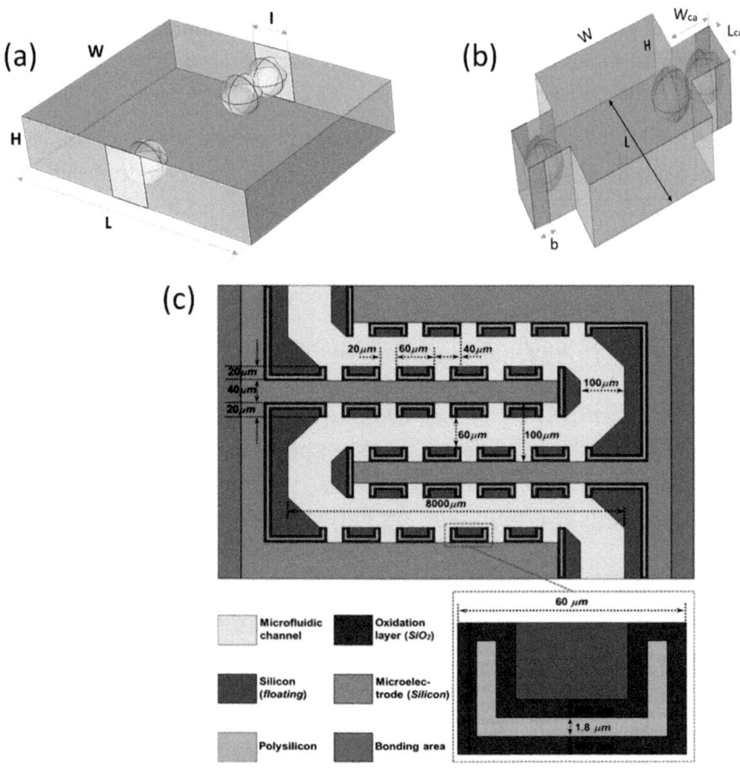

Figure 8.18 (a) Microchannels with a discrete coplanar microelectrode [111]. (b) Microelectrodes inside the microcavity structures [111]. (c) Schematic of a microfluidic device consisting of microelectrode arrays and a microfluidic channel. Reprinted by permission from Springer Nature Customer Service Centre GmbH: Springer Nature, *Microfluids and Nanofluids*, Ref. [112], copyright (2013).

coplanar microelectrodes and microcavity microelectrodes [111]. A drop of cell solution was introduced in the chip. When an AC voltage was applied, cells were attracted to the microelectrodes due to the effect of positive DEP. A high-voltage pulse (amplitude ~14 V, duration ~40 μs, interval between two pulses 1 s, number of pulses 5) was applied for irreversible electroporation. It was found that the required TMP distribution can be achieved by changing the length of the sidewall microelectrode.

Hu et al. [112] designed a microfluidic chip that consists of cavity-based microelectrode arrays fabricated upon a "silicon on insulator" wafer, as demonstrated in Fig. 8.18c. These microelectrode arrays act as electrodes besides acting as the vertical sidewalls of the microfluidic channel. Once cells are trapped at the microcavities, an electric signal (amplitude 9 V, duration 40 μs, interval between two pulses 1 s, number of pulses 5) is applied across the microelectrode arrays. This device can be used for performing both electroporation and electrofusion.

These microcavity-based devices come with several advantages. The TMP can be controlled by changing the length of the microelectrodes. In addition to this, electrofusion can be performed in these cavity structures. However, multiple cell fusion does not take place as the TMP outside the microcavity is very low. Hence, this design is helpful for electroporation and electrofusion-based applications and can be further modified and optimized for performing electroporation in a controlled manner [112].

8.4.7 Electrofusion

Electrofusion is a fusion of two different kinds of cells to generate a completely new kind of cell. This new cell contains the characteristics of both original cells [113, 114]. As an emerging field of SCEP, electrofusion has become immensely popular for its therapeutic application [115–117]. This process is partially based on the concept of reversible electroporation. For electrofusion, initially a mixture of cells is exposed to an external AC field. This AC field, of moderate intensity, helps the targeted cell membranes to gain electric charge, creating a potential difference between the interior and exterior sides of the cell membrane (also known as TMP). When this TMP

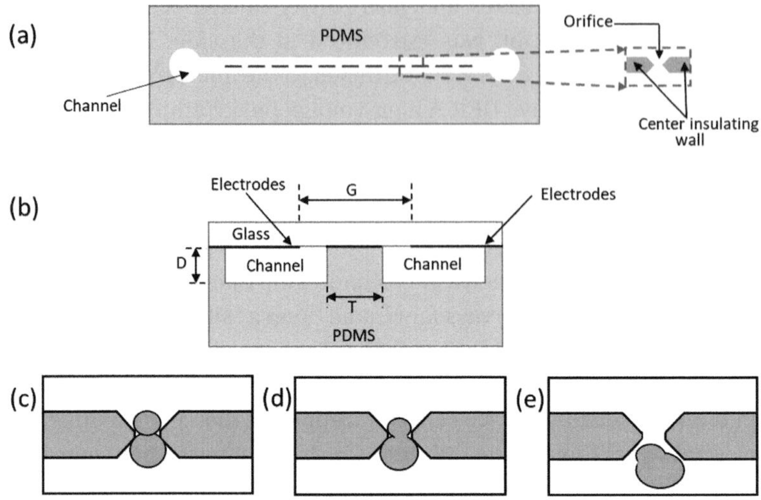

Figure 8.19 Device structure for (a) top view and (b) cross-sectional view. (c) Two distinct cells trapped at the orifice. (d) Merging of cell membranes and cytoplasm on applying an electric pulse. (e) A new kind of cell coming out of the orifice [123].

reaches the breakdown voltage, membrane breakdown occurs and the cell membrane is temporarily destroyed at the place of contact between two cells, which helps in the merging of cell membranes and cytoplasm [117–121]. In conventional electrofusion, high-voltage pulses are needed to develop a strong electric field for electroporation. Also, it is hard to control cell pairing in the case of bulk electrofusion. Thus the fusion efficiency comes out to be low [122]. To improve fusion efficiency, researchers have developed trapping-based microstructures. Techaumnat et al. developed a device that consists of an orifice, as shown in Fig. 8.19, comparable to the cell size [123]. When the high-voltage pulse is applied, most of the voltage drops across the orifice, which leads to the localized breakdown of the membrane at the orifice. The device was found to achieve highly reproducible fusion [123].

Skelley et al. presented a microfluidic control–based approach for electrofusion [124]. Their device consists of a dense array of passive hydrodynamic cell traps. Figure 8.20a demonstrates the flow of cells under the cell trap. This flow directs the cells toward the capture cups. Figure 8.20b shows a schematic of a trap consisting of capture cups on the front and back. Cell capture and pairing includes a three-step cell-loading protocol. First, cells were loaded in the smaller capture cups (Fig. 8.20c). Then, the cells were transferred to the larger capture cups (Fig. 8.20d). Finally, the second type of cells were loaded in the larger capture cups upon the first type of cells (Fig. 8.20e). For applying an electric field, the device was attached to a glass slide that contains metal electrodes, using plasma bonding. After the cells were immobilized, a hypo-osmolar fusion buffer was loaded in the chip at the rate of 0.4 mL/min for 10 min. High-voltage pulses were then applied for electrofusion. The pairing efficiency of this device is 70%. The device is capable of simultaneously pairing thousands of cells.

To fuse mouse myeloma cells with human CD19+ B cells on a chip, Kemna et al. developed a microfluidic PDMS chip [125] comprising one inlet branch divided into four channels for even cell distribution. The chip consisted of 11 rows of PDMS trap arrays with a total of 783 traps. Figure 8.21 shows the trap structure. The distance between adjacent rows was 50 μm, and the distance between the traps (d_t) was varied. A Pyrex glass slide with patterned Pt electrodes was used to seal the PDMS trap array using oxygen plasma. The distance between the electrodes was 40 μm, and each electrode was 25 μm in width. The total depth of the chip was 26 μm. A single trap consisted of a V-shaped large trap 25–29 μm wide and 20–21 μm long and a small trap with varying depths (d_s = 3 or 6 μm) in the middle. Before loading cells into the chip, both types of cells were filtered using a 40 μm filter to ensure a homogenous cell suspension. Subsequently, the cells were loaded by putting suspended types of cells in the inlet reservoir and taking out the cells through the device at a rate of 10 μL/ min using a syringe pump. After loading and pairing the cells, an iso-osmolar electrofusion buffer was introduced. Subsequently, pulses of strength 10 V and duration 50, 100, and 200 μs were applied as per the requirement. The device shows 50%–60% efficiency. The result estimates that the fusion efficiency can be improved by applying multiple pulses. However, it also shows that direct application of pulses for 200 μs can result in cell death [125].

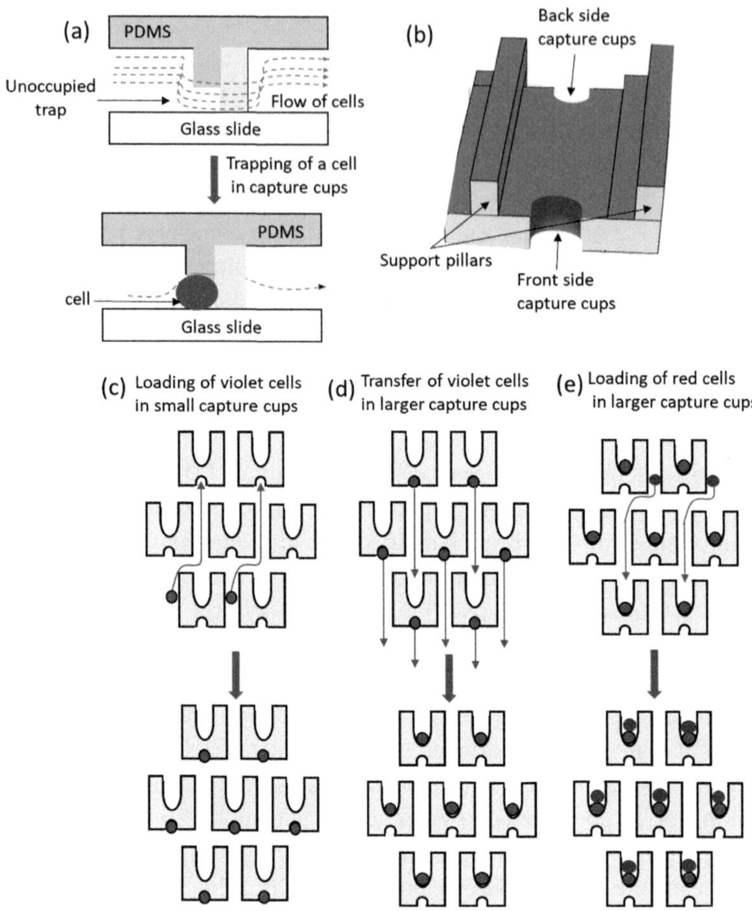

Figure 8.20 (a) Pictorial representation of a microfluidic device explaining the flow of cells under the trap. (b) Trap structure consisting of large capture cups on the front and back. In (c–e) schematics, cell-loading protocols are shown. (c) Cells are first stacked in the smaller capture cup. (d) On reversing the direction of flow, cells are relocated to the larger capture cups. (e) The second cell type is filled in the large capture cups on the first type of cells [124].

The techniques discussed above provide crucial details about electrofusion and directs toward the improvement of current electrofusion protocols. Most of the devices can provide higher

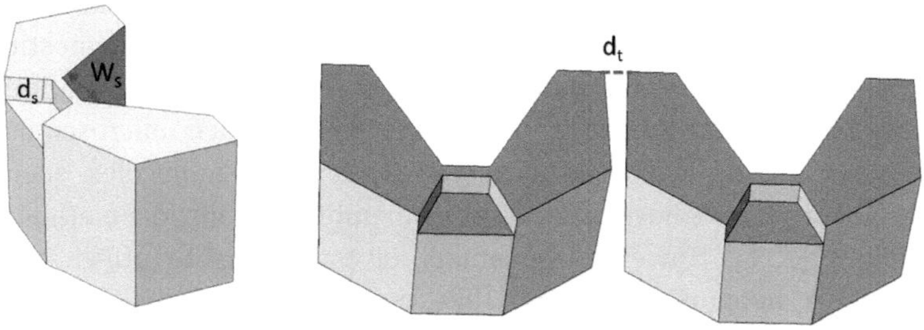

Figure 8.21 (a) Schematic version of a single-trap structure. Here, d_S and W_S represent the depth and width of a single trap. (b) Pictorial representation of trap arrays, where d_t represents the distance between two consecutive traps [125].

efficiency in terms of cell fusion when compared to bulk electrofusion. In the future, electrofusion devices can be optimized to get higher efficiency and cell viability.

8.5 Applications

Over the past few years, electroporation has emerged as a powerful technique for biomedical research. In this section, we will discuss some of the breakthrough applications of electroporation.

8.5.1 Imaging Based on Electrical Properties

Treatment plan assessment is one of the most significant parts for electroporation-based clinical applications. Creating an adequately high electric field in the target tissues is complicated by nontrivial electrode configuration, heterogeneity of cells, optimization of pulse parameters, etc. The basic step of treatment planning is to model the procedure to optimize the position of the electrodes, the electric field amplitude, the frequency, and the pulse duration to perform nonthermal electroporation only of the targeted tissues. Rubinsky et al.'s model stresses on the electrode position that spares nontarget tissues and covers the target ones [126]. Edd and Davalos explained the mathematical modeling of irreversible electroporation for treatment planning using electric field protocol optimization [127]. Zupanic et al. and Garcia et al. [128, 129] derived mathematical treatment planning with the integration of diagnostic imaging [130]. The mechanism of cell membrane electroporation can be understood using phase and fluorescence microscopic studies. During electroporation, various types of molecules, such as DNA and RNA, have different transmembrane transport methodologies [131]. However, the ability of physicians to utilize the available imaging techniques, like magnetic resonance imaging (MRI), computed tomography (CT), ultrasound (US), and positron emission tomography, is a major advantage of electroporation-based techniques in clinics.

During electroporation, US is used in animals as well as in clinical studies for electrode positioning in real time and detecting the instant and post-treatment effects. US outcomes after electroporation are lively and strongly interrelated with histological findings [132–134]. In animals and human, MRI is frequently used to identify the changes in tissues after electrochemotherapy (ECT) [135] and nonthermal irreversible electroporation [136]. Similarly, for electrode positioning, diagnosis, post-treatment examination of tumor regression, and tissue ablation, contrast-enhanced and noncontrast CT is used in patient care and animal experiments [130, 134, 137]. Hewapathirane and Haas studied and optimized the SCEP of fluorescent markers on the brain of *Xenopus laevis* tadpoles, which allowed direct visualization of cell growth, morphology, and intracellular actions over time [138].

In all biological media, ions are naturally present and these ions' transportation through the membrane is worthy evidence of the permeabilization of the membrane. Due to electroporation, tissue impedance decreases instantly, which can be detected and used for analysis [139–141]. Using a galvanic electroporation cell, changes in tissue conductivity can be measured instantaneously [142]. In this method, electroporation is carried out through electrodes with various electrochemical potentials, such as Zn and Cu. Without using other devices, the galvanic current produced by the electrodes and the treated tissue affords a direct measurement of changes in electric impedance. Electrical impedance tomography (EIT), enables real-time observation of tissue electroporation [143]. In addition to EIT, current density imaging and magnetic resonance electrical impedance tomography have been suggested for the measurement of the electric field and current distribution in tissue throughout the experiment [144, 145]. The feasibility of electroporation in living tissue with EIT was theoretically proposed by Davalos et al. [146]. Later, Granot et al. [147] proved it experimentally. EIT generates an image of the electrical properties of the investigated media. Edd et al. developed a range of EIT techniques to observe the minimally invasive approaches like cryosurgery [148] and electroporation. For monitoring irreversible electroporation in living tissues, multifrequency EIT was applied by Granot et al. [147]. Huang and Rubinsky [149], Khine et al. [150], and Lee and Deng [151] observed the change in a cell's electrical properties with electroporation (on a microscopic scale) and utilized it for single-cell microelectroporation technology. Conversely, this technique provided only bulk information, without any particulars on the geometrical structure of the electroporated surface.

Meir and Rubinsky applied EIT technology for imaging electroporation in a single-cell model system [152]. In contrast to former microelectroporation techniques, EIT inherently generates a 3D image with geometrical details on performing SCEP [150, 153]. Depending on the number of electrodes and simulation frequency, an image dataset time can be minimized to milliseconds with the existing techniques. Sun et al. observed the growth of mold cells using EIT at the single-cell level [154]. Dharia et al. developed electrical impedance topography to characterize the cell membrane local capacitance of a single cell [155]. Microelectrode arrays are usually utilized for research on single neurons [156]. Meir and Rubinsky developed mathematical models to simulate the procedures of SCEP and EIT to monitor the electroporation at the single-cell level [152].

8.5.2 Cell Reprogramming

Cell reprogramming is one of the profound emerging areas of nano-electroporation that has many clinical applications. For example, it has become an important method for enhancing the efficiency of DNA vaccination [157]. Many applications also show that the process of delivery of plasmid genes for encoding specific antigens has been improved by employing an NEP chip [11, 158]. Application of a 3D NEP chip in the area of nonviral immune cell engineering has shown significant improvement in antitumoral activity [135]. NEP-based transfections have also the ability to deliver siRNAs at the single-cell level in a time-controlled manner, which has the potential to treat acute myeloid leukemia [159]. Thus nanoelectroporation provides a potential platform for these applications with more deterministic and real-time monitoring capabilities.

8.5.3 Tissue Engineering and Regenerative Medicine

The objective of regenerative medicine is to develop techniques to regrow, repair, or replace damaged cells and tissues. Dosage control and timely delivery of the complex drugs at the single-cell level have taken a leap forward in regenerative medicine. For example, Saijilafu et al. applied electroporation for studying regeneration in adult mouse sensory neurons [160]. Microcapillary-based electroporation was used for analyzing bone diseases by transfecting bone marrow–derived mesenchymal stem cells [161]. The same method was also applied for transfecting siRNAs for analyzing immune homeostasis, inflammation, and liver function [162]. The aforementioned research studies indicate that micro-/nanoscale electroporation has become a powerful tool for different clinical applications.

8.5.4 Intracellular Recording

Recording of intracellular activity in real time can give an insight into critical changes and progress of a disease. Novel technologies, like nanopillar electroporation, have been used for collecting data of HL-1 cells using the intracellular recording method [163]. Hanson and his group demonstrated the same approach for characterizing nuclear biomechanics [164]. Zhan et al. proposed a different technique for recording intracellular activity. They used a 2D NEP chip for the delivery of molecular beacons by intracellular probing. These biosensors also have the potential to extract intracellular molecules selectively, thus moving a step forward in the field of intracellular analysis [165].

8.5.5 Cancer Studies

Electroporation techniques have potential application in cancer treatment [126, 166]. In 2005, Davalos et al. presented a theoretical study that suggests that irreversible electroporation can be used for tissue ablation [167]. Furthermore, Miller et al. showed that the application of 1500 V/cm in three sets of 10 pulses with pulse duration of 300 μs each can completely ablate hepatocarcinoma cells [168]. The study also showed that the application of multiple pulses is more effective than the application of a single pulse of the same energy for cancer cell ablation. Lee et al. presented an image-guided percutaneous irreversible electroporation technique for the treatment of liver cancer [169]. The technique also provides a facility to visualize the effect of irreversible electroporation using real time US. Onik et al. employed irreversible electroporation for prostate ablation [170]. Despite having several clinical applications [168–170], irreversible electroporation comes up with many technical challenges. Bertacchini et al. addressed these technical challenges in their review [171]. With the development of nanoscale electroporation at the single-cell level, it is possible to treat cancer cells without damaging the healthy tissues in a controlled manner. For example, Guduru et al. used the magnetonanoelectroporation method to deliver drugs through the creation of localized nanopores in human ovarian carcinoma cells (SKOV-3) [172]. Furthermore, microcapillary-based electroporation has a huge application in the study of cancers, such as gastric cancer, colorectal cancer, and colon cancer [42].

ECT is a process that uses electroporation for delivering nonpermeant or poorly permeant cytotoxic chemotherapeutic drugs into cells by increasing membrane permeability [173]. In 1991, Mir et al. demonstrated the effectiveness, safety, and feasibility of ECT [174, 175]. From the viewpoint of cancer and tumor treatment, several preclinical studies have been carried out. A recent study shows that the treatment of breast cancer can be accomplished by employing ECT with a high success rate, specifically applicable for metastasis [176]. Byrne et al. treated metastatic melanoma using ECT with a high success rate [177]. According to Gargiulo et al., ECT plays a major role in treating patients with inoperable conditions in head and neck cancer [178]. In 2006, a standard operating procedure (SOP) for electrochemotherapy was established and certified by the European Standard Operating Procedures for ECT [158, 179]. A new version of SOP has been published by a pan-European expert panel for the treatment of tumor and skin metastasis by employing ECT [180]. The adverse effects of ECT include pain after procedure, edema or erythema, and muscle contraction. However, no serious adverse effect or toxicity has been reported regarding this technique [181].

8.5.6 Gene Transfer and DNA Vaccination

Another important application of electroporation is gene electrotransfer (GET) in cells for DNA vaccination and gene therapy. As shown in Fig. 8.22, the basic mechanism of DNA electrotransfer involves several steps: (i) membrane permeabilization, (ii) electrophoretic migration of DNA toward the plasma membrane, (iii) interaction between DNA and membrane, (iv) membrane DNA translocation, (v) intracellular migration of DNA, (vi) entry of DNA into the nucleolus, and (vii) generation of gene expression [3, 182]. In 1982, Neumann et al. demonstrated that the DNA can be introduced inside a living cell with the help of electric pulses [183, 184]. In 1991, Titomirov et al. delivered plasmid DNA to newborn mice by applying exponentially decaying short pulses to skin [185]. For cancer treatment different types of genes have been used, such as IL-12, IL-2, IL-15, and GM-CSF, for performing GET [186, 187]. Daud et al. applied GET by employing IL-12 on patients with metastatic melanoma [188]. The procedure was found to be highly effective as tumor necrosis was found in the treated area.

8.6 Limitations and Future Directions

Over the past few decades, electroporation has emerged as a potential technique for several clinical applications. Under various external influences cells undergo changes. Even if many cells have similar genomes, tissues contain distinct cell types. Hence, it is very important to learn more about the heterogenetic nature of cells

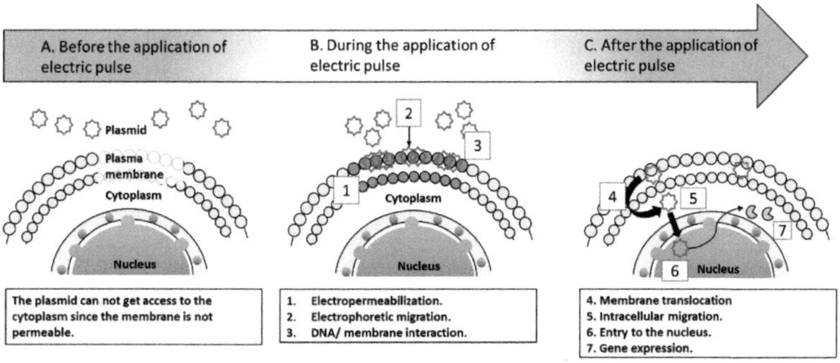

Figure 8.22 Schematic diagram explain the mechanism of electrotransfection [182].

as cell properties directly affect the electroporation parameters. SCEP may be the finest technique for analyzing cell properties, but it is the slowest process as separating a single cell from the bulk is time consuming. Thus massively parallel high-throughput SCEP platform is needed to solve this issue [59]. Additional analysis shows fluorescence-activated cell sorting improves the methodology by automating the whole process. It is also noted that in vivo electroporation needs constant maintenance in a stable environment. Also, further miniaturization of electrodes may provide more selectivity and be less damaging for cells during electroporation. To address real-time clinical and biological applications, it is important to establish a standard protocol for transfecting cells more reliably. Thus, more precise SCEP techniques and methodology are encouraged.

8.7 Conclusions

Electroporation is a powerful technique for cell therapy and analysis. Current micro-/nanofluidic device–based electroporation technologies have the capability to deliver biomolecules of different sizes in different cell types with high cell viability and high transfection efficiency. These devices can perform spatial, temporal, and qualitative dosage control, which can be considered as a major advantage for cell-based research and therapeutic applications. Emerging techniques, such as electrofusion, have helped in enhancing cell fusion, thus helping in eliminating diseases and also helping in cell reconstruction and regeneration. In the future, the focus can be on optimizing current technologies, which may provide unknown insights into cells and provide breakthroughs in cell-based analysis, cell regeneration, and regenerative medicine.

Acknowledgements

The authors greatly appreciate the financial support from the Science and Engineering Research Board (SERB) under grant number ECR/2016/001945, Department of Science and Technology (DST), Government of India, and Wellcome Trust/DBT India Alliance Fellowship under grant number IA/E/16/1/503062. We acknowledge all authors and publishers from whom we got the copyright permissions.

References

1. D. Liu, L. Wang, Z. Wang, A. Cuschieri (2012). Magnetoporation and magnetolysis of cancer cells via carbon nanotubes induced by rotating magnetic fields. *Nano Lett*, **12**:5117–5121; doi:10.1021/nl301928z.
2. G. Chu, H. Hayakawa, P. Berg (1987). Electroporation for the efficient transfection of mammalian cells with DNA. *Nucleic Acids Res*, **15**:1311–1326; doi:10.1093/nar/15.3.1311.
3. J. M. Escoffre, T. Portet, L. Wasungu, J. Teissié, D. Dean, M. P. Rols (2009). What is (still not) known of the mechanism by which electroporation mediates gene transfer and expression in cells and tissues. *Mol Biotechnol*, **41**:286–295; doi:10.1007/s12033-008-9121-0.
4. Y. Zhang, C. B. Ballas, M. P. Rao (2012). Towards ultrahigh throughput microinjection: MEMS-based massively-parallelized mechanoporation. In *Annual International Conference of the IEEE Engineering in Medicine and Biology Society*, pp. 594–597; doi:10.1109/EMBC.2012.6346001.
5. M. Cammalleri, E. Pipitone, T. Rubino, D. M. Geraci, F. Bonura, C. Mammina (2017). Extended validation of dynamic irreversible

thermoporation: a novel thermal process for microbial inactivation. *J Food Process Eng*, **40**:e12300; doi:10.1111/jfpe.12300.

6. P. Soman, W. Zhang, A. Umeda, Z. J. Zhang, S. Chen (2011). Femtosecond laser-assisted optoporation for drug and gene delivery into single mammalian cells. *J Biomed Nanotechnol*, **7**:334–341; doi:10.1166/jbn.2011.1295.
7. L. Gu, S. K. Mohanty (2011). Targeted microinjection into cells and retina using optoporation. *J Biomed Opt*, **16**:128003; doi:10.1117/1.3662887.
8. J. Castle, S. B. Feinstein (2016). Drug and gene delivery using sonoporation for cardiovascular disease. *Adv Exp Med Biol*, **880**:331–338; doi:10.1007/978-3-319-22536-4_18.
9. Y. Zhou (2013). Ultrasound-mediated drug/gene delivery in solid tumor treatment. *J Healthc Eng*, **4**:223–254; doi:10.1260/2040-2295.4.2.223.
10. T. Y. Tsong, T. D. Xie (1992). Mechanisms of cell transfection by electroporation: field induced DNA uptake and transfection efficiency. In *1992 14th Annual International Conference of the IEEE Engineering in Medicine and Biology Society*, pp. 307–308; doi: 10.1109/IEMBS.1992.5760978.
11. L. Chang, L. Li, J. Shi, Y. Sheng, W. Lu, D. Gallego-Perez, L. J. Lee (2016). Micro-/nanoscale electroporation. *Lab Chip*, **16**:4047–4062; doi:10.1039/C6LC00840B.
12. G. Narayanan, M. H. Doshi (2016). Irreversible electroporation (IRE) in renal tumors. *Curr Urol Rep*, **17**(2):15; doi:10.1007/s11934-015-0571-1.
13. C. Månsson, R. Brahmstaedt, A. Nilsson, P. Nygren, B. M. Karlson (2016). Percutaneous irreversible electroporation for treatment of locally advanced pancreatic cancer following chemotherapy or radiochemotherapy. *Eur J Surg Oncol*, **42**(9):1401–1406; doi:10.1016/j.ejso.2016.01.024.
14. T. S. Santra, P.-C. Wang, H.-Y. Chang, F.-G. Tseng (2013). Tuning nano electric field to affect restrictive membrane area on localized single cell nano-electroporation. *Appl Phys Lett*, **103**:233701; doi:10.1063/1.4833535.
15. D. C. Chang, P. Q. Gao, B. L. Maxwell (1991). High efficiency gene transfection by electroporation using a radio-frequency electric field. *Biochim Biophys Acta*, **1092**(2):153–160; doi:10.1016/0167-4889(91)90149-R.

16. W. J. Dower, J. F. Miller, C. W. Ragsdale (1988). High efficiency transformation of E. coli by high voltage electroporation. *Nucleic Acids Res*, **16**(13):6127–6145; doi:10.1093/nar/16.13.6127.
17. Y. Granot, B. Rubinsky (2008). Mass transfer model for drug delivery in tissue cells with reversible electroporation. *Int J Heat Mass Transf*, **51**(23–24):5610–5616; doi:10.1016/j.ijheatmasstransfer.2008.04.041.
18. M. Hjouj, D. Last, D. Guez, D. Daniels, S. Sharabi, J. Lavee, B. Rubinsky, Y. Mardor (2012). MRI study on reversible and irreversible electroporation induced blood brain barrier disruption. *PLoS One*, **7**:e42817; doi:10.1371/journal.pone.0042817.
19. K. T. Powell, A. W. Morgenthaler, J. C. Weaver (1989). Tissue electroporation. Observation of reversible electrical breakdown in viable frog skin. *Biophys J*, **56**(6):1163–1171; doi:10.1016/S0006-3495(89)82763-8.
20. Q. A. Zheng, D. C. Chang (1991). High-efficiency gene transfection by in situ electroporation of cultured cells. *Biochim Biophys Acta*, **1088**:104–110; doi:10.1016/0167-4781(91)90158-i.
21. H. Melkonyan, C. Sorg, M. Klempt (1996). Electroporation efficiency in mammalian cells is increased by dimethyl sulfoxide (DMSO). *Nucleic Acids Res*, **24**:4356–4357; doi:10.1093/nar/24.21.4356.
22. S. Y. Ho, G. S. Mittal (1996). Electroporation of cell membranes: a review. *Crit Rev Biotechnol*, **16**(4):349–362; doi:10.3109/07388559609147426.
23. S. Kar, M. Loganathan, K. Dey, P. Shinde, H. Y. Chang, M. Nagai, T. S. Santra (2018). Single-cell electroporation: current trends, applications and future prospects. *J Micromech Microeng*, **28**(12); doi:10.1088/1361-6439/aae5ae.
24. H. Potter (1988). Electroporation in biology: methods, applications, and instrumentation. *Anal Biochem*, **174**:361–373; doi:10.1016/0003-2697(88)90035-8.
25. D. Zhao, D. Huang, Y. Li, M. Wu, W. Zhong, Q. Cheng, X. Wang, Y. Wu, X. Zhou, Z. Wei, Z. Li, Z. Liang (2016). A flow-through cell electroporation device for rapidly and efficiently transfecting massive amounts of cells in vitro and ex vivo. *Sci Rep*, **6**:18469; doi:10.1038/srep18469.
26. R. C. Lee, L. P. River, F. S. Pan, L. Ji, R. L. Wollmann (1992). Surfactant-induced sealing of electropermeabilized skeletal muscle membranes in vivo. *Proc Natl Acad Sci*, **89**:4524–4528; doi:10.1073/pnas.89.10.4524.

27. A. Fennell, R. Hauptmann (1992). Electroporation and PEG delivery of DNA into maize microspores. *Plant Cell Rep*, **11**(11):567–570; doi:10.1007/BF00233094.

28. J. W. Hooper, J. W. Golden, A. M. Ferro, A. D. King (2007). Smallpox DNA vaccine delivered by novel skin electroporation device protects mice against intranasal poxvirus challenge. *Vaccine*, **25**(10):1814–1823; doi:10.1016/j.vaccine.2006.11.017.

29. J. W. Loomis-Husselbee, P. J. Cullen, R. F. Irvine, A. P. Dawson (2015). Electroporation can cause artefacts due to solubilization of cations from the electrode plates. Aluminum ions enhance conversion of inositol 1,3,4,5-tetrakisphosphate into inositol 1,4,5-trisphosphate in electroporated L1210 cells. *Biochem J*, **277**(Pt 3):883–885; doi: 10.1042/bj2770883.

30. U. Pliquett (2003). Joule heating during solid tissue electroporation. *Med Biol Eng Comput*, **41**(2):215–219; doi:10.1007/BF02344892.

31. M. Jaroszeski, V. Dang, C. Pottinger, J. Hickey, R. Gilbert, R. Heller (2000). Toxicity of anticancer agents mediated by electroporation in vitro. Journals.Lww.Com. (n.d.). https://journals.lww.com/anti-cancerdrugs/fulltext/2000/03000/toxicity_of_anticancer_agents_mediated_by.8.aspx (accessed July 6, 2019).

32. M. P. Rols, C. Delteil, G. Serin, J. Teissié (1994). Temperature effects on electrotransfection of mammalian cells. *Nucleic Acids Res*, **22**(3):540; doi:10.1093/nar/22.3.540.

33. B. M. Medi, S. Hoselton, R. B. Marepalli, J. Singh (2005). Skin targeted DNA vaccine delivery using electroporation in rabbits: I: efficacy. *Int J Pharm*, **294**(1–2):53–63; doi:10.1016/j.ijpharm.2004.12.014.

34. J. Shi, Y. Ma, J. Zhu, Y. Chen, Y. Sun, Y. Yao, Z. Yang, J. Xie (2018). A review on electroporation-based intracellular delivery. *Molecules*, **23**(11):pii.E3044; doi:10.3390/molecules23113044.

35. C. Ionescu-Zanetti, A. Blatz, M. Khine (2008). Electrophoresis-assisted single-cell electroporation for efficient intracellular delivery. *Biomed Microdevices*, **10**:113–116; doi:10.1007/s10544-007-9115-x.

36. P. E. Boukany, Y. Wu, X. Zhao, K. J. Kwak, P. J. Glazer, K. Leong, L. J. Lee (2014). Nonendocytic delivery of lipoplex nanoparticles into living cells using nanochannel electroporation. *Adv Healthc Mater*, **3**:682–689; doi:10.1002/adhm.201300213.

37. X. Xie, A. M. Xu, S. Leal-Ortiz, Y. Cao, C. C. Garner, N. A. Melosh (2013). Nanostraw–electroporation system for highly efficient intracellular

delivery and transfection. *ACS Nano*, **7**:4351–4358; doi:10.1021/nn400874a.

38. W. Kang, F. Yavari, M. Minary-Jolandan, J. P. Giraldo-Vela, A. Safi, R. L. McNaughton, V. Parpoil, H. D. Espinosa (2013). Nanofountain probe electroporation (NFP-E) of single cells. *Nano Lett*, **13**:2448–2457; doi:10.1021/nl400423c.
39. S.-C. Chen, T. S. Santra, C.-J. Chang, T.-J. Chen, P.-C. Wang, F.-G. Tseng (2012). Delivery of molecules into cells using localized single cell electroporation on ITO micro-electrode based transparent chip. *Biomed Microdevices*, **14**:811–817; doi:10.1007/s10544-012-9660-9.
40. T. S. Santra, S. Kar, J. Borana, P.-C. Wang, F.-G. Tseng (2014). Nanolocalized single-cell-membrane nanoelectroporation: for higher efficiency with high cell viability. *IEEE Nanotechnol Mag*, **8**:30–34; doi:10.1109/MNANO.2014.2312031.
41. T. S. Santra, H.-Y. Chang, P.-C. Wang, F.-G. Tseng (2014). Impact of pulse duration on localized single-cell nano-electroporation. *Analyst*, **139**:6249–6258; doi:10.1039/C4AN01050G.
42. W. Lee, U. Demirci, A. Khademhosseini (2009). Microscale electroporation: challenges and perspectives for clinical applications. *Integr Biol (Camb)*, **1**(3):242–251; doi:10.1039/b819201d.
43. S. Weidmann (1951). Effect of current flow on the membrane potential of cardiac muscle. *J Physiol*, **115**(2):227–236; doi:10.1113/jphysiol.1951.sp004667.
44. A. J. H. Sale, W. A. Hamilton (1967). Effects of high electric fields on microorganisms: I. Killing of bacteria and yeasts. *Biochim Biophys Acta*, **148**(3):781–788; doi:10.1016/0304-4165(67)90052-9.
45. A. V. Babakov, L. N. Ermishkin, E. A. Liberman (1966). Influence of electric field on the capacity of phospholipid membranes. *Nature*, **210**:953–955; doi:10.1038/210953b0.
46. S. Weidmann, Electrical characteristics of sepia axons, n.d.
47. B. Katz, O. H. Schmitt (1940). Electric interaction between two adjacent nerve fibres. *J Physiol*, **97**(4):471–488; doi: 10.1113/jphysiol.1940.sp003823.
48. T. Teorell (1948). Membrane electrophoresis in relation to bioelectrical polarization effects. *Nature*, **162**:961; doi:10.1038/162961a0.
49. T. S. Santra, F. G. Tseng (2013). Recent trends on micro/nanofluidic single cell electroporation. *Micromachines*, **4**:333–356; doi:10.3390/mi4030333.

50. A. L. Hodgkin (1951). The ionic basis of electrical activity in nerve and muscle. *Biol Rev*, **26**:339–409; doi:10.1111/j.1469-185X.1951.tb01204.x.

51. J. Gehl (2003). Electroporation: theory and methods, perspectives for drug delivery, gene therapy and research. *Acta Physiol Scand*, **177**:437–447; doi:10.1046/j.1365-201X.2003.01093.x.

52. D. Moldovan, D. Pinisetty, R. V. Devireddy (2007). Molecular dynamics simulation of pore growth in lipid bilayer membranes in the presence of edge-active agents. *Appl Phys Lett*, **91**:204104; doi:10.1063/1.2814876.

53. T. Kotnik, P. Kramar, G. Pucihar, D. Miklavcic, M. Tarek (2012). Cell membrane electroporation: Part 1: the phenomenon. *IEEE Electr Insul Mag*, **28**:14–23; doi:10.1109/MEI.2012.6268438.

54. L. V. Chernomordik, Y. A. Chizmadzhev (1989). Electrical breakdown of lipid bilayer membranes. In *Electroporation Electrofusion in Cell Biology*, Springer, Boston, MA, pp. 83–95; doi:10.1007/978-1-4899-2528-2_5.

55. I. G. Abidor, V. B. Arakelyan, L. V. Chernomordik, Y. A. Chizmadzhev, V. F. Pastushenko, M. P. Tarasevich (1979). Electric breakdown of bilayer lipid membranes: I. The main experimental facts and their qualitative discussion. *J Electroanal Chem Interfacial Electrochem*, **104**:37–52; doi:10.1016/S0022-0728(79)81006-2.

56. T. S. Santra, F.-G. Tseng (2016). Electroporation for Single-Cell Analysis. In *Essentials of Single-Cell Analysis*, Springer, Berlin, Heidelberg, pp. 55–83; doi:10.1007/978-3-662-49118-8_3.

57. T. Kotnik, F. Bobanović, D. Miklavčič (1997). Sensitivity of transmembrane voltage induced by applied electric fields: a theoretical analysis. *Bioelectrochem Bioenerg*, **43**(2):285–291; doi:10.1016/S0302-4598(97)00023-8.

58. J. Teissie, M. Golzio, M. P. Rols (2005). Mechanisms of cell membrane electropermeabilization: a minireview of our present (lack of??) knowledge. *Biochim Biophys Acta*, **1724**(3):270–280; doi:10.1016/j.bbagen.2005.05.006.

59. T. S. Santra, M. A. Teitell, P. E. Chiou (2018). Device for massively parallel high throughput single cell electroporation and uses thereof. http://www.freepatentsonline.com/y2018/0066222.html (accessed June 20, 2019).

60. L. Chang, P. Bertani, D. Gallego-Perez, Z. Yang, F. Chen, C. Chiang, V. Malkoc, T. Kuang, K. Gao, L. J. Lee, W. Lu (2016). 3D nanochannel elec-

troporation for high-throughput cell transfection with high uniformity and dosage control. *Nanoscale*, **8**:243–252; doi:10.1039/c5nr03187g.

61. C. Yao, H. Liu, Y. Zhao, Y. Mi, S. Dong, Y. Lv (2017). Analysis of dynamic processes in single-cell electroporation and their effects on parameter selection based on the finite-element model. *IEEE Trans Plasma Sci*, **45**:889–900; doi:10.1109/TPS.2017.2681433.
62. H. Gissel, R. C. Lee, J. Gehl (2011). Electroporation and cellular physiology. In *Clinical Aspects of Electroporation*, Springer, New York, NY, pp. 9–17; doi:10.1007/978-1-4419-8363-3_2.
63. P. Hojman, H. Gissel, F. M. Andre, C. Cournil-Henrionnet, J. Eriksen, J. Gehl, L. M. Mir (2008). Physiological effects of high- and low-voltage pulse combinations for gene electrotransfer in muscle. *Hum Gene Ther*, **19**:1249–1260; doi:10.1089/hum.2008.059.
64. Y. Zu, S. Huang, W.-C. Liao, Y. Lu, S. Wang (2014). Gold nanoparticles enhanced electroporation for mammalian cell transfection. *J Biomed Nanotechnol*, **10**:982–992; doi:10.1166/jbn.2014.1797.
65. S. Huang, H. Deshmukh, K. K. Rajagopalan, S. Wang (2014). Gold nanoparticles electroporation enhanced polyplex delivery to mammalian cells. *Electrophoresis*, **35**:1837–1845; doi:10.1002/elps.201300617.
66. S. Mukherjee, V. Sushma, S. Patra, A. K. Barui, M. P. Bhadra, B. Sreedhar, C. R. Patra (2012). Green chemistry approach for the synthesis and stabilization of biocompatible gold nanoparticles and their potential applications in cancer therapy. *Nanotechnology*, **23**:455103; doi:10.1088/0957-4484/23/45/455103.
67. C. Shan, H. Yang, D. Han, Q. Zhang, A. Ivaska, L. Niu (2010). Graphene/AuNPs/chitosan nanocomposites film for glucose biosensing. *Biosens Bioelectron*, **25**:1070–1074; doi:10.1016/j.bios.2009.09.024.
68. S. Iijima (1991). Helical microtubules of graphitic carbon. *Nature*, **354**:56–58; doi:10.1038/354056a0.
69. M. Terrones (2003). Science and technology of the twenty-first century: synthesis, properties, and applications of carbon nanotubes. *Annu Rev Mater Res*, **33**:419–501; doi:10.1146/annurev.matsci.33.012802.100255.
70. T. E. McKnight, A. V Melechko, G. D. Griffin, M. A. Guillorn, V. I. Merkulov, F. Serna, D. K. Hensley, M. J. Doktycz, D. H. Lowndes, M. L. Simpson (2003). Intracellular integration of synthetic nanostructures with viable cells for controlled biochemical manipulation. *Nanotechnology*, **14**:551–556; doi:10.1088/0957-4484/14/5/313.

71. A. Bianco, K. Kostarelos, M. Prato (2005). Applications of carbon nanotubes in drug delivery. *Curr Opin Chem Biol*, **9**(6):674–679; doi:10.1016/j.cbpa.2005.10.005.
72. B. S. Harrison, A. Atala (2007). Carbon nanotube applications for tissue engineering. *Biomaterials*, **28**(2):344–353; doi:10.1016/j.biomaterials.2006.07.044.
73. R. V. Davalos, B. Rubinsky, L. M. Mir (2003). Theoretical analysis of the thermal effects during in vivo tissue electroporation. *Bioelectrochemistry*, **61**(1–2):99–107; doi:10.1016/j.bioelechem.2003.07.001.
74. C. B. Arena, R. L. Mahajan, M. N. Rylander, R. V. Davalos (2013). Phase change electrodes for reducing Joule heating during irreversible electroporation. Proceedings Paper, Paper No: SBC2012-80825, pp. 473–474; doi:10.1115/sbc2012-80825.
75. C. C. Chang, M. Mao, Y. Liu, M. Wu, T. Vo-Dinh, F. Yuan (2016). Improvement in electrotransfection of cells using carbon-based electrodes. *Cell Mol Bioeng*, **9**:538–545; doi:10.1007/s12195-016-0452-9.
76. J. A. Rojas-Chapana, M. A. Correa-Duarte, Z. Ren, K. Kempa, M. Giersig (2004). Enhanced introduction of gold nanoparticles into vital *acidothiobacillus ferrooxidans* by carbon nanotube-based microwave electroporation. *Nano Lett*, **4**(5):985–988; doi:10.1021/nl049699n.
77. M. Giersig, I. Firkowska, J. Troszczynska, J. Rojas-Chapana (2005). Cell manipulation and tissue engineering at the nanoscale. *NanoBiotechnology*, **1**(3):290–292; doi:10.1385/Nano:1:3:290.
78. J. Suehiro, N. Ikeda, A. Ohtsubo, K. Imasaka (2008). Fabrication of bio/nano interfaces between biological cells and carbon nanotubes using dielectrophoresis. *Microfluid Nanofluidics*, **5**:741–747; doi:10.1007/s10404-008-0276-6.
79. R. M. Hochmuth (2000). Micropipette aspiration of living cells. *J Biomech*, **33**:15–22; doi:10.1016/S0021-9290(99)00175-X.
80. D. J. Hoelzle, B. A. Varghese, C. K. Chan, A. C. Rowat (2014). A microfluidic technique to probe cell deformability. *J Vis Exp*, e51474; doi:10.3791/51474.
81. M. Khine, A. Lau, C. Ionescu-Zanetti, J. Seo, L. P. Lee (2005). A single cell electroporation chip. *Lab Chip*, **5**:38–43; doi:10.1039/b408352k.
82. Y. Huang, B. Rubinsky (2003). Flow-through micro-electroporation chip for high efficiency single-cell genetic manipulation. *Sens Actuators, A*, **104**:205–212; doi:10.1016/S0924-4247(03)00050-5.
83. T. Chen, F. Artis, D. Dubuc, J.-J. Fournie, M. Poupot, K. Grenier (2013). Microwave biosensor dedicated to the dielectric spectroscopy of

a single alive biological cell in its culture medium. In *2013 IEEE MTT-S International Microwave Symposium Digest (MTT)*, pp. 1–4; doi:10.1109/MWSYM.2013.6697740.

84. D. Casey, J. Dooley (2016). Optical tools for single-cell manipulation and analysis. In *Essentials of Single-Cell Analysis*, Springer, Berlin, Heidelberg, pp. 131–157; doi:10.1007/978-3-662-49118-8_5.
85. B. E. Henslee, A. Morss, X. Hu, G. P. Lafyatis, L. J. Lee (2011). Electroporation dependence on cell size: optical tweezers study. *Anal Chem*, **83**:3998–4003; doi:10.1021/ac1019649.
86. D. Choudhury, L. M. Maestro, W. T. Ramsay, L. Paterson, D. Jaque, A. K. Kar (2012). Continuous flow single cell electroporation in an ultrafast laser inscribed optofluidic device. In *2012 International Conference on Fiber Optics and Photonics (PHOTONICS)*, p. T1A.4; doi:10.1364/PHOTONICS.2012.T1A.4.
87. B. Qu, Y.-J. Eu, W.-J. Jeong, D.-P. Kim (2012). Droplet electroporation in microfluidics for efficient cell transformation with or without cell wall removal. *Lab Chip*, **12**:4483; doi:10.1039/c2lc40360a.
88. H. Kurita, S. Takahashi, A. Asada, M. Matsuo, K. Kishikawa, A. Mizuno, R. Numano (2015). Novel parallelized electroporation by electrostatic manipulation of a water-in-oil droplet as a microreactor. *PLoS One*, **10**:e0144254; doi:10.1371/journal.pone.0144254.
89. D. J. Im, S.-N. Jeong (2017). Transfection of Jurkat T cells by droplet electroporation. *Biochem Eng J*, **122**:133–140; doi:10.1016/J.BEJ.2017.03.010.
90. W. Liu, N. Dechev, I. G. Foulds, R. Burke, A. Parameswaran, E. J. Park (2009). A novel permalloy based magnetic single cell micro array. *Lab Chip*, **9**:2381–2390; doi:10.1039/b821044f.
91. A. H. B. de Vries, B. E. Krenn, R. van Driel, J. S. Kanger (2005). Micro magnetic tweezers for nanomanipulation inside live cells. *Biophys J*, **88**:2137–2144; doi:10.1529/BIOPHYSJ.104.052035.
92. L. Chang, M. Howdyshell, W.-C. Liao, C.-L. Chiang, D. Gallego-Perez, Z. Yang, W. Lu, J. C. Byrd, N. Muthusamy, L. J. Lee, R. Sooryakumar (2015). Magnetic tweezers-based 3D microchannel electroporation for high-throughput gene transfection in living cells. Small, **11**:1818–1828; doi:10.1002/smll.201402564.
93. J. Kim, I. Hwang, D. Britain, T. D. Chung, Y. Sun, D.-H. Kim (2011). Microfluidic approaches for gene delivery and gene therapy. *Lab Chip*, **11**:3941; doi:10.1039/c1lc20766k.

94. P. E. Boukany, A. Morss, W. Liao, B. Henslee, H. Jung, X. Zhang, B. Yu, X. Wang, Y. Wu, L. Li, K. Gao, X. Hu, X. Zhao, O. Hemminger, W. Lu, G. P. Lafyatis, L. J. Lee (2011). Nanochannel electroporation delivers precise amounts of biomolecules into living cells. *Nat Nanotechnol*, **6**:747–754; doi:10.1038/nnano.2011.164.

95. A. Valero, J. N. Post, J. W. van Nieuwkasteele, P. M. ter Braak, W. Kruijer, A. van den Berg (2008). Gene transfer and protein dynamics in stem cells using single cell electroporation in a microfluidic device. *Lab Chip*, **8**:62–67; doi:10.1039/B713420G.

96. M. Li, Y. C. Lin, K. C. Su (2006). Using magnetic nanoparticles to enhance gene transfection on magneto-electroporation microchips. *Mater Sci Forum*, **505–507**:661–666; doi:10.4028/www.scientific.net/MSF.505-507.661.

97. B. Lim, V. Reddy, X. Hu, K. Kim, M. Jadhav, R. Abedini-Nassab, Y.-W. Noh, Y. T. Lim, B. B. Yellen, C. Kim (2014). Magnetophoretic circuits for digital control of single particles and cells. *Nat Commun*, **5**:3846; doi:10.1038/ncomms4846.

98. A. Kaushik, R. D. Jayant, R. Nikkhah-Moshaie, V. Bhardwaj, U. Roy, Z. Huang, A. Ruiz, A. Yndart, V. Atluri, N. El-Hage, K. Khalili, M. Nair (2016). Magnetically guided central nervous system delivery and toxicity evaluation of magneto-electric nanocarriers. *Sci Rep*, **6**:25309; doi:10.1038/srep25309.

99. A. Kaushik, R. Nikkhah-Moshaie, R. Sinha, V. Bhardwaj, V. Atluri, R. D. Jayant, A. Yndart, B. Kateb, N. Pala, M. Nair (2017). Investigation of ac-magnetic field stimulated nanoelectroporation of magneto-electric nano-drug-carrier inside CNS cells. *Sci Rep*, **7**:45663; doi:10.1038/srep45663.

100. T. Kozai, A. L. Vazquez (2017). Wireless micro/nano- stimulation opto-electrode for excitable tissue. US Patent Appl. No. 15/526,135, Pub. No. US 2017/0326381 A1, University of Pittsburgh, PA, USA.

101. P. Y. Chiou, A. T. Ohta, M. C. Wu (2005). Massively parallel manipulation of single cells and microparticles using optical images. *Nature*, **436**:370–372; doi:10.1038/nature03831.

102. J. K. Valley, S. Neale, H.-Y. Hsu, A. T. Ohta, A. Jamshidi, M. C. Wu (2009). Parallel single-cell light-induced electroporation and dielectrophoretic manipulation. *Lab Chip*, **9**:1714; doi:10.1039/b821678a.

103. J. K. Valley, H.-Y. Hsu, S. Neale, A. T. Ohta, A. Jamshidi, M. C. Wu (2009). Assessment of single cell viability following light-induced

electroporation through use of on-chip microfluidics. In *2009 IEEE 22nd International Conference on Micro Electro Mechanical Systems*, pp. 411–414; doi:10.1109/MEMSYS.2009.4805406.

104. Y.-H. Lin, G.-B. Lee (2009). An optically induced cell lysis device using dielectrophoresis. *Appl Phys Lett*, **94**:033901; doi:10.1063/1.3072593.
105. Y. Yang, H. Zhang, J. Zhu, G. Wang, T.-R. Tzeng, X. Xuan, K. Huang, P. Wang (2010). Distinguishing the viability of a single yeast cell with an ultra-sensitive radio frequency sensor. *Lab Chip*, **10**:553; doi:10.1039/b921502f.
106. G. A. Ferrier, S. F. Romanuik, D. J. Thomson, G. E. Bridges, M. R. Freeman (2009). A microwave interferometric system for simultaneous actuation and detection of single biological cells. *Lab Chip*, **9**:3406; doi:10.1039/b908974h.
107. Y. Ning, C. Multari, X. Luo, C. Palego, X. Cheng, J. C. M. Hwang, A. Denzi, C. Merla, F. Apollonio, M. Liberti (2014). Broadband electrical detection of individual biological cells. *IEEE Trans Microw Theory Tech*, **62**:1905–1911; doi:10.1109/TMTT.2014.2342660.
108. X. Ma, X. Du, C. R. Multari, Y. Ning, C. Palego, X. Luo, V. Gholizadeh, X. Cheng, J. C. M. Hwang (2015). Broadband single-cell detection with a coplanar series gap. In *2015 86th ARFTG Microwave Measurement Conference*, pp. 1–3; doi:10.1109/ARFTG.2015.7381459.
109. H. Li, X. Ma, X. Du, X. Cheng, J. C. M. Hwang (2016). High-frequency continuous-wave electroporation of Jurkat human lymphoma cells. In *2016 IEEE MTT-S International Microwave Symposium (IMS)*, pp. 1–4; doi:10.1109/MWSYM.2016.7540431.
110. A. Tamra, D. Dubuc, M.-P. Rols, K. Grenier (2017). Microwave monitoring of single cell monocytes subjected to electroporation. *IEEE Trans Microw Theory Tech*, **65**:3512–3518; doi:10.1109/TMTT.2017.2653776.
111. X. Zhang, N. Hu, X. Chen, T. Fan, Z. Wang, X. Zheng, J. Yang, S. Qian (2017). Controllable cell electroporation using microcavity electrodes. *Sens Actuators, B*, **240**:434–442; doi:10.1016/J.SNB.2016.08.172.
112. N. Hu, X. Zhang, J. Yang, S. W. Joo, S. Qian (2013). A cell electrofusion microfluidic chip with micro-cavity microelectrode array. *Microfluid Nanofluidics*, **15**:151–160; doi:10.1007/s10404-013-1136-6.
113. G. Köhler, C. Milstein (1975). Continuous cultures of fused cells secreting antibody of predefined specificity. *Nature*, **256**:495–497; doi:10.1038/256495a0.

114. J. Vienken, U. Zimmermann, M. Fouchard, D. Zagury (1983). Electrofusion of myeloma cells on the single cell level. *FEBS Lett*, **163**:54–56; doi:10.1016/0014-5793(83)81161-2.

115. M. Lee, S. Kim, H. Lee, G. Im, B. C. Yang, N. H. Kim, D. H. Kim (2011). Trichostatin A promotes the development of bovine somatic cell nuclear transfer embryos. *J Reprod Dev*, **57**(1):34–42.

116. N. H. Mcclenaghan, N. H. Mcclenaghan (2007). Physiological regulation of the pancreatic β-cell: functional insights for understanding and therapy of diabetes. *Author J Compil C*, **92**:481–496; doi:10.1113/expphysiol.2006.034835.

117. J.-Y. Yang, D.-Y. Cao, L.-Y. Ma, W.-C. Liu (2010). Dendritic cells fused with allogeneic hepatocellular carcinoma cell line compared with fused autologous tumor cells as hepatocellular carcinoma vaccines. *Hepatol Res*, **40**:505–513; doi:10.1111/j.1872-034X.2010.00625.x.

118. U. Zimmermann, G. A. Neil (1996). *Electromanipulation of Cells*, CRC Press, Boca Raton, FL.

119. D. C. Chang (1992). *Guide to Electroporation and Electrofusion*, Academic Press, San Diego, CA.

120. U. Pliquett, R. P. Joshi, V. Sridhara, K. H. Schoenbach (2007). High electrical field effects on cell membranes. *Bioelectrochemistry*, **70**:275–282; doi:10.1016/J.BIOELECHEM.2006.10.004.

121. L. Macqueen, M. Thibault, M. Buschmann, M. Wertheimer (2012). Electro-manipulation of biological cells in microdevices. *IEEE Trans Dielectr Electr Insul*, **19**:1261–1268; doi:10.1109/TDEI.2012.6260000.

122. H. Andersson, A. van den Berg (2004). *Lab-on-Chips for Cellomics*, Springer Netherlands, Dordrecht, 2004; doi:10.1007/978-1-4020-2975-2.

123. B. Techaumnat, K. Tsuda, O. Kurosawa, G. Murat, H. Oana, M. Washizu (2008). High-yield electrofusion of biological cells based on field tailoring by microfabricated structures. *IET Nanobiotechnol*, **2**(4):93–99; doi:10.1049/iet-nbt:20080008.

124. A. M. Skelley, O. Kirak, H. Suh, R. Jaenisch, J. Voldman (2009). Microfluidic control of cell pairing and fusion. *Nat Methods*, **6**:147–152; doi:https://doi.org/10.1038/nmeth.1290.

125. E. W. M. Kemna, F. Wolbers, I. Vermes, A. van den Berg (2011). On chip electrofusion of single human B cells and mouse myeloma cells for efficient hybridoma generation. *Electrophoresis*, **32**:3138–3146; doi:10.1002/elps.201100227.

126. B. Rubinsky, G. Onik, P. Mikus (2007). Irreversible electroporation: a new ablation modality—clinical implications. *Technol Cancer Res Treat*, **6**:37–48; doi:10.1177/153303460700600106.

127. J. F. Edd, R. V Davalos (2007). Mathematical modeling of irreversible electroporation for treatment planning. *Technol Cancer Res Treat*, **6**:275–286; doi:10.1177/153303460700600403.

128. P. A. Garcia, J. H. Rossmeisl, R. E. Neal, T. L. Ellis, J. D. Olson, N. Henao-Guerrero, J. Robertson, R. V. Davalos (2010). Intracranial nonthermal irreversible electroporation: in vivo analysis. *J Membr Biol*, **236**:127–136; doi:10.1007/s00232-010-9284-z.

129. A. Zupanic, B. Kos, D. Miklavcic (2012). Treatment planning of electroporation-based medical interventions: electrochemotherapy, gene electrotransfer and irreversible electroporation. *Phys Med Biol*, **57**:5425–5440; doi:10.1088/0031-9155/57/17/5425.

130. I. Edhemovic, E. M. Gadzijev, E. Brecelj, D. Miklavcic, B. Kos, A. Zupanic, B. Mali, T. Jarm, D. Pavliha, M. Marcan, G. Gasljevic, V. Gorjup, M. Music, T. P. Vavpotic, M. Cemazar, M. Snoj, G. Sersa (2011). Electrochemotherapy: a new technological approach in treatment of metastases in the liver. *Technol Cancer Res Treat*, **10**:475–485; doi:10.7785/tcrt.2012.500224.

131. A. Paganin-Gioanni, E. Bellard, J. M. Escoffre, M. P. Rols, J. Teissié, M. Golzio (2011). Direct visualization at the single-cell level of siRNA electrotransfer into cancer cells. *Proc Natl Acad Sci U S A*, **108**(26):10443–10447; doi:10.1073/pnas.1103519108.

132. L. Appelbaum, E. Ben-David, J. Sosna, Y. Nissenbaum, S. N. Goldberg (2012). US findings after irreversible electroporation ablation: radiologic-pathologic correlation. *Radiology*, **262**:117–125; doi:10.1148/radiol.11110475.

133. E. W. Lee, C. Chen, V. E. Prieto, S. M. Dry, C. T. Loh, S. T. Kee (2010). Advanced hepatic ablation technique for creating complete cell death: irreversible electroporation. *Radiology*, **255**:426–433; doi:10.1148/radiol.10090337.

134. R. E. Neal Ii, W. Cheung, H. Kavnoudias, K. R. Thomson (2012). Spectrum of imaging and characteristics for liver tumors treated with irreversible electroporation. *J Biomed Sci Eng*, **5**:813–818; doi:10.4236/jbise.2012.512A102.

135. L. Calmels, B. Al-Sakere, J.-P. Ruaud, A. Leroy-Willig, L. M. Mir (2012). *In vivo* MRI follow-up of murine tumors treated by electrochemotherapy

and other electroporation-based treatments. *Technol Cancer Res Treat*, **11**:561–570; doi:10.7785/tcrt.2012.500270.

136. M. Hjouj, D. Last, D. Guez, D. Daniels, S. Sharabi, J. Lavee, B. Rubinsky, Y. Mardor (2012). MRI study on reversible and irreversible electroporation induced blood brain barrier disruption. *PLoS One*, **7**(8):e42817; doi:10.1371/journal.pone.0042817.
137. L. W. Matthiessen, H. H. Johannesen, H. W. Hendel, T. Moss, C. Kamby, J. Gehl (2012). Electrochemotherapy for large cutaneous recurrence of breast cancer: a phase II clinical trial. *Acta Oncol (Madr)*, **51**:713–721; doi:10.3109/0284186X.2012.685524.
138. D. S. Hewapathirane, K. Haas (2008). Single cell electroporation in vivo within the intact developing brain. *J Vis Exp*, e705; doi:10.3791/705.
139. A. Ivorra, B. Al-Sakere, B. Rubinsky, L. M. Mir (2009). In vivo electrical conductivity measurements during and after tumor electroporation: conductivity changes reflect the treatment outcome. *Phys Med Biol*, **54**:5949–5963; doi:10.1088/0031-9155/54/19/019.
140. D. Cukjati, D. Batiuskaite, F. André, D. Miklavčič, L. M. Mir (2007). Real time electroporation control for accurate and safe in vivo non-viral gene therapy. *Bioelectrochemistry*, **70**:501–507; doi:10.1016/J.BIOELECHEM.2006.11.001.
141. A. Golberg, S. Laufer, H. D. Rabinowitch, B. Rubinsky (2011). In vivo non-thermal irreversible electroporation impact on rat liver galvanic apparent internal resistance. *Phys Med Biol*, **56**:951–963; doi:10.1088/0031-9155/56/4/005.
142. A. Golberg, H. Rabinowitch, B. Rubinsky (2009). Galvanic apparent internal impedance: an intrinsic tissue property. *Biochem Biophys Res Commun*, **389**(1):168–171; doi:10.1016/j.bbrc.2009.08.119.
143. Y. Granot, B. Rubinsky (2007). Methods of optimization of electrical impedance tomography for imaging tissue electroporation. *Physiol Meas*, **28**:1135–1147; doi:10.1088/0967-3334/28/10/001.
144. M. Kranjc, F. Bajd, I. Sersa, E. J. Woo, D. Miklavcic (2012). Ex vivo and in silico feasibility study of monitoring electric field distribution in tissue during electroporation based treatments. *PLoS One*, **7**; doi:10.1371/journal.pone.0045737.
145. M. Kranjc, F. Bajd, I. Sersa, D. Miklavcic (2011). Magnetic resonance electrical impedance tomography for monitoring electric field distribution during tissue electroporation. *IEEE Trans Med Imaging*, **30**:1771–1778; doi:10.1109/TMI.2011.2147328.

146. R. Davalos, B. Rubinsky, D. M. Otten (2002). A feasibility study for electrical impedance tomography as a means to monitor tissue electroporation for molecular medicine. *IEEE Trans Biomed Eng,* **49**(4):400–403; doi:10.1109/10.991168.

147. Y. Granot, A. Ivorra, E. Maor, B. Rubinsky (2009). In vivo imaging of irreversible electroporation by means of electrical impedance tomography. *Phys Med Biol,* **54**:4927–4943; doi:10.1088/0031-9155/54/16/006.

148. J. F. Edd, A. Ivorra, L. Horowitz, B. Rubinsky (2008). Imaging cryosurgery with EIT: tracking the ice front and post-thaw tissue viability. *Physiol Meas,* **29**:899–912; doi:10.1088/0967-3334/29/8/004.

149. Y. Huang, B. Rubinsky (1999). Micro-electroporation: improving the efficiency and understanding of electrical permeabilization of cells. *Biomed Microdevices,* **2**(2):145–150; doi:https://doi.org/ 10.1023/A:1009901821588.

150. M. Khine, C. Ionescu-Zanetti, A. Blatz, L.-P. Wang, L. P. Lee (2007). Single-cell electroporation arrays with real-time monitoring and feedback control. *Lab Chip,* **7**:457; doi:10.1039/b614356c.

151. Y. Lee, P. Deng (2012). Review of micro/nano technologies and theories for electroporation of biological cells. *Sci China Phys Mech Astron,* **55**:996–1003; doi:https://doi.org/10.1007/s11433-012-4747-0.

152. A. Meir, B. Rubinsky (2014). Electrical impedance tomographic imaging of a single cell electroporation. *Biomed Microdevices,* **16**(3):427–437; doi:10.1007/s10544-014-9845-5.

153. Y. Huang, B. Rubinsky (2001). Microfabricated electroporation chip for single cell membrane permeabilization. *Sens Actuators, A,* **89**(3):242–249; doi:https://doi.org/10.1016/S0924-4247(00)00557-4.

154. T. Sun, S. Tsuda, K. P. Zauner, H. Morgan (2010). On-chip electrical impedance tomography for imaging biological cells. *Biosens Bioelectron,* **25**(5):1109–1115; doi:https://doi.org/10.1016/j.bios.2009.09.036.

155. S. Dharia, H. E. Ayliffe, R. D. Rabbitt (2009). Single cell electric impedance topography: mapping membrane capacitance. *Lab Chip,* **9**(23):3370–3377; doi: 10.1039/b912881f.

156. M. Spira, A. Hai (2013). Multi-electrode array technologies for neuroscience and cardiology. *Nat Nanotechnol,* **8**:83–94; doi:https://doi. org/10.1038/nnano.2012.265.

157. M. Rosati, A. Valentin, R. Jalah, V. Patel, A. von Gegerfelt, C. Bergamaschi, C. Alicea, D. Weiss, J. Treece, R. Pal, P. D. Markham, E. T. Marques,

J. T. August, A. Khan, R. Draghia-Akli, B. K. Felber, G. N. Pavlakis (2008). Increased immune responses in rhesus macaques by DNA vaccination combined with electroporation. *Vaccine*, **26**(40):5223–5229; doi:10.1016/j.vaccine.2008.03.090.

158. B. Ferraro, M. Morrow, N. A. Hutnick, T. H. Shin, C. E. Lucke, D. B. Weiner (2011). Clinical applications of DNA vaccines: current progress. *Clin Infect Dis*, **53**(3):296–302; doi:10.1093/cid/cir334.

159. K. Gao, X. Huang, C. L. Chiang, X. Wang, L. Chang, P. Boukany, G. Marcucci, R. Lee, L. J. Lee (2016). Induced apoptosis investigation in wild-type and FLT3-ITD acute myeloid leukemia cells by nanochannel electroporation and single-cell qRT-PCR. *Mol Ther*, **24**(5):956–964; doi:10.1038/mt.2016.6.

160. Saijilafu, E.-M. Hur, F.-Q. Zhou (2011). Genetic dissection of axon regeneration via in vivo electroporation of adult mouse sensory neurons. *Nat Commun*, **2**:543; doi:10.1038/ncomms1568.

161. J.-L. Lin, M. J. Wang, D. Lee, C.-C. Liang, S. Lin (2008). Hypoxia-inducible factor-1alpha regulates matrix metalloproteinase-1 activity in human bone marrow-derived mesenchymal stem cells. *FEBS Lett*, **582**(17):2615–2619; doi:10.1016/j.febslet.2008.06.033.

162. N. Mizutani, T. Sakurai, T. Shibata, K. Uchida, J. Fujita, R. Kawashima, Y. I. Kawamura, N. Toyama-Sorimachi, T. Imai, T. Dohi (2007). Dose-dependent differential regulation of cytokine secretion from macrophages by fractalkine. *J Immunol*, **179**(11):7478–7487; doi:10.4049/jimmunol.179.11.7478.

163. C. Xie, Z. Lin, L. Hanson, Y. Cui, B. Cui (2012). Intracellular recording of action potentials by nanopillar electroporation. *Nat Nanotechnol*, **7**:185–190; doi:https://doi.org/10.1038/nnano.2012.8.

164. L. Hanson, W. Zhao, H. Y. Lou, Z. C. Lin, S. W. Lee, P. Chowdary, Y. Cui, B. Cui (2015). Vertical nanopillars for in situ probing of nuclear mechanics in adherent cells. *Nat Nanotechnol*, **10**:554–562; doi:https://doi.org/10.1038/nnano.2015.88.

165. Y. Zhan, V. A. Martin, R. L. Geahlen, C. Lu (2010). One-step extraction of subcellular proteins from eukaryotic cells. *Lab Chip*, **10**(16):2046–2048; doi:10.1039/c005152g.

166. B. Rubinsky (2007). Irreversible electroporation in medicine. *Technol Cancer Res Treat*, **6**:255–259; doi:10.1177/153303460700600401.

167. R. Davalos, L. Mir, B. Rubinsky (2005). Tissue ablation with irreversible electroporation. *Ann Biomed Eng*, **33**(2):223–231; doi:10.1007/s10439-005-8981-8.

168. L. Miller, J. Leor, B. Rubinsky (2005). Cancer cells ablation with irreversible electroporation. *Technol Cancer Res Treat*, **4**:699–705; doi:10.1177/153303460500400615.

169. E. W. Lee, C. T. Loh, S. T. Kee (2007). Imaging guided percutaneous irreversible electroporation: ultrasound and immunohistological correlation. *Technol Cancer Res Treat*, **6**:287–293; doi:10.1177/153303460700600404.

170. G. Onik, P. Mikus, B. Rubinsky (2007). Irreversible electroporation: implications for prostate ablation. *Technol Cancer Res Treat*, **6**:295–300; doi:10.1177/153303460700600405.

171. C. Bertacchini, P. M. Margotti, E. Bergamini, A. Lodi, M. Ronchetti, R. Cadossi (2007). Design of an irreversible electroporation system for clinical use. *Technol Cancer Res Treat*, **6**:313–320; doi:10.1177/153303460700600408.

172. R. Guduru, P. Liang, C. Runowicz, M. Nair, V. Atluri, S. Khizroev (2013). Magneto-electric nanoparticles to enable field-controlled high-specificity drug delivery to eradicate ovarian cancer cells. *Sci Rep*, **3**:2953; doi:10.1038/srep02953.

173. N. Esmaeili, M. Friebe (2019) Electrochemotherapy: a review of current status, alternative IGP approaches, and future perspectives. *J Healthc Eng*, **2019**:2784516; doi:10.1155/2019/2784516.

174. M. L. Yarmush, A. Golberg, G. Serša, T. Kotnik, D. Miklavčič (2014). Electroporation-based technologies for medicine: principles, applications, and challenges. *Annu Rev Biomed Eng*, **16**:295–320; doi:10.1146/annurev-bioeng-071813-104622.

175. L. M. Mir, M. Belehradek, C. Domenge, S. Orlowski, B. Poddevin, J. Belehradek, G. Schwaab, B. Luboinski, C. Paoletti (1991). Electrochemotherapy, a new antitumor treatment: first clinical trial. *C R Acad Sci III*, **313**:613–618. doi:http://www.ncbi.nlm.nih. gov/pubmed/1723647.

176. L. W. Matthiessen, M. Keshtgar, P. Curatolo, C. Kunte, E. M. Grischke, J. Odili, T. Muir, D. Mowatt, J. P. Clover, S. H. Liew, K. Dahlstroem, J. Newby, V. Letulé, E. Stauss, A. Humphreys, S. Banerjee, A. Klein, R. Rotunno, F. de Terlizzi, J. Gehl (2018). Electrochemotherapy for breast cancer—results from the INSPECT database. *Clin Breast Cancer*, **18**(5):e909–e917; doi:10.1016/j.clbc.2018.03.007.

177. C. M. Byrne, J. F. Thompson, H. Johnston, P. Hersey, M. J. Quinn, T. Michael Hughes, W. H. McCarthy (2005). Treatment of

metastatic melanoma using electroporation therapy with bleomycin (electrochemotherapy). *Melanoma Res*, **15**(1):45–51; doi:10.1097/00008390-200502000-00008.

178. M. Gargiulo, A. Papa, P. Capasso, M. Moio, E. Cubicciotti, S. Parascandolo (2012). Electrochemotherapy for non-melanoma head and neck cancers: clinical outcomes in 25 patients. *Ann Surg*, **255**(6):1158–1164; doi:10.1097/SLA.0b013e31824f68b2.

179. L. M. Mir, J. Gehl, G. Sersa, C. G. Collins, J.-R. Garbay, V. Billard, P. F. Geertsen, Z. Rudolf, G. C. O'Sullivan, M. Marty (2006). Standard operating procedures of the electrochemotherapy: instructions for the use of bleomycin or cisplatin administered either systemically or locally and electric pulses delivered by the Cliniporator^{TM} by means of invasive or non-invasive electrodes. *EJC Suppl*, **4**:14–25; doi:10.1016/j.ejcsup.2006.08.003.

180. J. Gehl, G. Sersa, L. W. Matthiessen, T. Muir, D. Soden, A. Occhini, P. Quaglino, P. Curatolo, L. G. Campana, C. Kunte, A. J. P. Clover, G. Bertino, V. Farricha, J. Odili, K. Dahlstrom, M. Benazzo, L. M. Mir (2018). Updated standard operating procedures for electrochemotherapy of cutaneous tumours and skin metastases. *Acta Oncol (Madr)*, **57**:874–882; doi:10.1080/0284186X.2018.1454602.

181. R. Cadossi, M. Ronchetti, M. Cadossi (2014). Locally enhanced chemotherapy by electroporation: clinical experiences and perspective of use of electrochemotherapy. *Futur Oncol*, **10**:877–890; doi:10.2217/fon.13.235.

182. T. Murakami, Y. Sunada (2011). Plasmid DNA gene therapy by electroporation: principles and recent advances. *Curr Gene Ther*, **11**(6):447–456; doi:10.2174/156652311798192860.

183. F. Andre, L. M. Mir (2004). DNA electrotransfer: its principles and an updated review of its therapeutic applications. *Gene Ther*, **11**(Suppl 1):S33–S42; doi:10.1038/sj.gt.3302367.

184. E. Neumann, M. Schaefer-Ridder, Y. Wang, P. H. Hofschneider (1982). Gene transfer into mouse lyoma cells by electroporation in high electric fields. *EMBO J*, **1**:841–845. http://www.ncbi.nlm.nih.gov/pubmed/6329708.

185. A. V Titomirov, S. Sukharev, E. Kistanova (1991). In vivo electroporation and stable transformation of skin cells of newborn mice by plasmid DNA. *Biochim Biophys Acta*, **1088**(1):131–134; doi:10.1016/0167-4781(91)90162-f.

186. L. C. Heller, R. Heller (2010). Electroporation gene therapy preclinical and clinical trials for melanoma. *Curr Gene Ther*, **10**:312–317; doi:10.2174/156652310791823489.

187. M. Cemazar, T. Jarm, G. Sersa (2010). Cancer electrogene therapy with interleukin-12. *Curr Gene Ther*, **10**:300–311; doi:10.2174/156652310791823425.

188. A. I. Daud, R. C. DeConti, S. Andrews, P. Urbas, A. I. Riker, V. K. Sondak, P. N. Munster, D. M. Sullivan, K. E. Ugen, J. L. Messina, R. Heller (2008). Phase I trial of interleukin-12 plasmid electroporation in patients with metastatic melanoma. *J Clin Oncol*, **26**(36):5896–5903. doi:10.1200/JCO.2007.15.6794.

Chapter 9

Electrical Cell Lysis on Microfluidic Devices

Bashir I. Morshed
Electrical and Computer Engineering Department, University of Memphis, Memphis, TN 38152, USA
bmorshed@memphis.edu

Cell lysis is defined as the destruction of a cell through the rupture of its plasma membrane, often by an external physical or chemical stimulus, and the consequent release of the contents of the cell. Electrical lysis is cell lysis caused by an applied external electrical stimulation. Microfluidic devices, commonly fabricated using etching, polymer deposition, lithography, and other microfabrication processes, can facilitate electrical lysis operation by lowering the required applied voltage for the process. The key advantages of on-chip electrical lysis are rapid operation, low cost, no residue, high throughput, controllability, and suitability for integration with other microfluidic processes. Applications of on-chip electrical lysis are analysis of cell contents (cytology), cancer therapy, genetics, disease diagnosis, oncology, stem cell research, single-cell manipulation, developmental biology, bacterial decontamination, and water treatment system. In this context, this chapter provides a brief overview of the electrical lysis process,

Microfluidics and Bio-MEMS: Devices and Applications
Edited by Tuhin S. Santra

ISBN 978-981-4800-85-3 (Hardcover), 978-1-003-01493-5 (eBook)
www.jennystanford.com

related dynamics and theories, recent reports of electrical lysis performed on microfluidic chips, parameters to consider, associated phenomena, and applications of on-chip electrical lysis.

9.1 Cell Membrane and Electrical Lysis

Biological cells are the building blocks of living animals or plants and provide the basic structural, functional, and biological entities. Animal cells are enclosed by cell membranes. A cell membrane is typically a 4-to-8-nm-thin phospholipid bilayer membrane. It is a complex structure with an intricate mechanism of adhesion to and reaction with the topology and chemistry of the extracellular space and acts as a physical barrier between the intracellular and extracellular environments. A phospholipid has a hydrophilic head containing a phosphate group and two long hydrophobic tails made of fatty acid chains. As both intracellular and extracellular mediums are primarily hydrous fluids (i.e., containing water), the hydrophobic tails will spontaneously arrange to avoid these fluids and lead to this bilayer formation. Generally, animal cell membranes are made of various types of phospholipid molecules, including 1,2-dimyristoyl-*sn*-glycero-3-phosphorylcholine (DMPC) glycerolipid [1], ceramides, sphingolipids, and cholesterol. These are essential to regulate different membrane properties. A schematic representation of the chemical structure of the DMPC phospholipid molecule, a symbolic representation of a phospholipid, and a sketch of a portion of a cell membrane comprising a bilayer membrane are depicted in Fig. 9.1.

To analyze the internal contents (the nucleus, mitochondria, genes, proteins, and other organelles) of these cells, the protective cell membranes need to be broken open by rupturing. This process of cell membrane rupture is known as cell lysis [2]. Various methods can be used for cell lysis, and these can be categorized into two primary stimuli: chemical and physical [3, 4]. In the case of lysis with a chemical stimulus, the common laboratory procedure, lysis agents, such as chemicals or detergents, are mixed with the cell-containing medium so that the chemical agents can bind to the phospholipids and rupture the cell membrane structure [5]. In the case of a

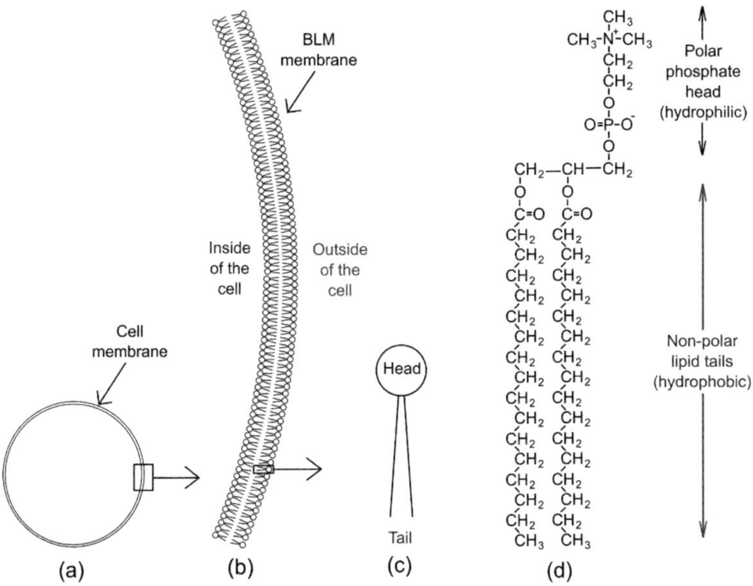

Figure 9.1 Schematic representation of a bilayer lipid membrane (BLM) composition. (a) A cell membrane schematic. (b) A schematic view of the cross section of a portion of BLM membrane composed of two phospholipid layers forming a bilayer. (c) A symbolic representation of one phospholipid molecule depicting a head and two tail sections. (d) An example chemical composition of the phospholipid molecule 1,2-dimyristorl-sn-glycero-3-phosphorylcholine (DMPC) with a polar head and a nonpolar lipid tail [49].

physical stimulus, various physical forces are applied on cells to rupture the membranes, such as mechanical [6, 7], laser [8, 9], thermal [10], magnetic [11], electro-osmotic [12], electrochemical [13], and electrical [14–19]. Cell lysis by a chemical lysis agent, although a commonly used method, is slow, lacks precise control, and requires multiple processing steps, and thus is not optimal for automated operation on microfluidic devices. Among the physical mechanisms, laser and electrical lysis (EL) are the most suitable methods for microfluidic devices. However, laser lysis requires expensive and high-power laser, a precise control mechanism, and human involvement for focusing, which is tedious and slows down the throughput.

EL is the process by which the cell breaks open due to disintegration of the plasma membrane because of a brief, high electric field stimulus [14–19]. This phenomenon is also referred to as an electrical breakdown, an electrodisruption, or irreversible electroporation (EP) [20–24]. The technique of electrical cell lysis, discovered over half a century ago, has recently garnered renewed research interest due to the invention of microfluidic devices that allow on-chip microfluidic operations, process automation, and miniaturization of common laboratory procedures, as well as the discovery of new applications, such as point-of-care diagnosis. The new microfabrication capabilities can leverage key advantages of EL, such as rapid operation, low power consumption, low cost, controllability, automation, and on-chip integration. Advanced microfabrication technologies are used to develop microfluidic devices that include integrated electrodes, microfabricated flow control and actuation mechanisms (such as fluidic valves), and microelectromechanical systems (MEMS). These miniature microfluidic devices have led to new technologies, like lab-on-a-chip (LOC), micro-total-analysis systems (μTAS), and bio-microelectromechanical systems (bio-MEMS). The microfabricated chips miniaturize and integrate macroscale biochemical processes and automate analysis procedures in micro- or nanoscale analytics, which leads to highly compact (often portable), low-cost (both equipment and supplies), and efficient systems.

On-chip EL systems can process microliter cell samples inside microfabricated chips with microchannels (or open cavities) and are excited through integrated or external electrodes. These electrical excitations can be in the form of a direct current (DC), an alternating current (AC), or a pulse stimulation that rapidly lyses the sample cells near the electrodes, and the internal contents of the lysed cells may be analyzed later by collecting the lysate (fluid after the lysis process). In advanced microfabricated chips, this processing can be further integrated downstream on the same chip. EL has shown the potential to be applied in several biomedical fields, such as food and water decontamination, bioanalysis of intracellular contents, single-cell analysis, DNA extraction and analysis, and cancer and tumor treatment [4, 25, 26].

9.2 Mechanism of Pore Formation and Electrical Lysis

9.2.1 Theories

An accurate modeling ability of the effects of an electric field on the bilayer lipid membrane (BLM) is necessary to understand, predict, and optimize various factors, including excitation parameters (such as magnitude, frequency, waveform shape, and number of pulses [27–30]), cell size and shape [31, 32], cell orientation [30], and the molecular update mechanism [33]. The electrical stimulus-related constraints of EL of cells can be related with the following expression [20–22]:

$$V = \frac{1.5\mathbf{E}a}{(1 + I\omega aC(r_{\mathrm{i}}+r_{\mathrm{a}}/2))}, \tag{9.1}$$

where V is the transmembrane potential, **E** is the external electric field strength, a is the radius of the cell, I is the current flow, ω is the frequency of the applied electric field, C is the capacitance of the membranes per unit area, r_{i} is the resistivity of the intracellular electrolyte (i.e., cytoplasm), and r_{a} is the resistivity of the extracellular electrolyte (i.e., medium). In addition, cells interacting with an external electric field can be modeled using an equivalent resistor-capacitor circuit [34, 35].

The phospholipid bilayer membranes of cells are susceptible to external electric fields because of the following two properties [20, 25]:

- **Electric dipole properties of phospholipids:** The phosphate group (head) is electropositive in nature, and the hydrogen chains (tails) are electronegative. Thus, each phospholipid molecule acts as a tiny electric dipole when placed inside an applied external electric field. This leads the phospholipids to undergo reorientation when the electric field is higher than a threshold level (typically in the range of MV/m for many cell lines), which can result in membrane lysis.
- **Electrical permeability of the cell membrane:** The bilayer cell membrane has an electrical permeability that

drastically changes with the formation of pores due to the electric field, a phenomenon called EP. In the presence of a high electric field, ions can flow through these hydrophilic membrane pores, resulting in Joule heating and consequential thermal phase transition, which might lead to EL.

The Joule heat generated during thermal phase transition due to the permeability property of the membrane might damage the contents of the cell. Hence, usually the current flow is kept very low to reduce Joule heating during electrical stimulation. The dipole property of phospholipid molecules, on the other hand, plays the dominant role in an acceptable EL mechanism, achieved typically through the application of a brief but high magnitude electric pulse. In addition, bubble formation during the application of the electric pulse, due to hydrolysis of water molecules, and electrode degradation are some of the concerns that should be carefully considered during microfluidic device design and operation.

9.2.2 Transmembrane Potential and Pore Formation

Transmembrane potential, the potential developed across the cell membrane due to an applied external electric field, is a critical parameter of EL and dictates the onset of pore formation. The exact dynamics involved in pore formation within the cell membrane are still a topic of debate [36]. A commonly accepted hypothesis postulates that a high electric field causes an accumulation of charges on both sides of the cell membrane, causing reorientation of phospholipid molecules [37]. According to this hypothesis, when a cell is suspended in an ionic fluid and is excited by an electric field, ions accumulate along the cell membrane, resulting in induced transmembrane potential across the cell membrane [20]. When this transmembrane potential becomes higher than 1 V, it causes the formation of hydrophilic pores through the EP process. These hydrophilic pores allow any hydrous fluid (e.g., water and its solutes) to pass through the membrane, making the membrane permeable. In contrast, the natural (stable) pores in the cell membrane are hydrophobic or controlled by biosignals such as Na/K protein channels. This transmembrane potential is extrinsic, compared to the intrinsic resting action potential that some cells (e.g., living neuron and muscle cells) exhibit as a characteristic spontaneous electrical discharge of 70 to 90 mV magnitude [3]. The transmembrane potential, a result of the applied external electric field, causes an electrical force on phospholipids to produce the reorientation stress on these phospholipid molecules and initiates the EL process [38, 39].

Another hypothesis postulates that opposite charges accumulated on the inner and outer membranes of the cell, induced by the applied electric field, generate compression pressure on the membrane because of the attraction of opposite charges along the cell membrane, thus thinning the membrane and putting stress on it [40]. As the electric field strength increases, this compression force increases. Eventually, when the transmembrane potential becomes sufficiently large, numerous reversible small pores are formed within the cell membrane due to this compression force, leading to EP and making the membrane permeable to the medium, initiating the lysis process.

The stress related to an applied electric field is the highest where the cell membrane surface is perpendicular to the electric field lines. If the applied electric field is increased, the transmembrane potential will also increase and can reach a critical value of 1 V. The corresponding electric field is called the threshold electric field. This causes the phospholipid molecules to undergo conformational changes within the membrane structure. This process forms a large number of hydrophilic pores throughout the cell membrane, especially at the poles where the cell membrane surface is perpendicular to the electric field lines [20]. If the electric field is smaller than the threshold electric field, intermediate states of pores are created that disappear as these intermediate pores are not stable. If the electric field is higher than the threshold electric field and the pores are stable, they undergo molecular rearrangement to form nanometer-sized hydrophilic pores. The threshold electric field required to initiate pore formation has a typical range of 0.1 to 0.15 MV/m for erythrocytes, protoplasts, and spheroplasts [20]. The time interval between the application of the electric field and the initiation of stable pore formation is known as the "nucleation period" [41].

These pores formed during EP can expand in diameter as time progresses if the external electric field is not withdrawn [42]. Depending on the duration of the applied electric field, the pores formed in the cell membrane can be reversible or become irreversible, leading to EL [43]. EP dynamics consist of a cycle of structural rearrangements of the cell membrane phospholipids for a unidirectional annealing process and a unidirectional pore formation process [42]. The time required for the annealing process to be completed is known as the "relaxation time," and the transition path difference in pore formation and the annealing process is known as EP hysteresis [20, 44]. Experimentally, the critical pore diameter is found to be about 40 nm for irreversible expansion that leads to EL [39, 41, 43, 45].

9.2.3 Transition States and Dynamics

The sequence of events involved in the EL process can be represented by a conceptual state diagram representation, as shown in Fig. 9.2, or a qualitative graph of membrane entropy (ξ) resulting from the reorientation of phospholipids against the elapsed time (t), as shown in Fig. 9.3, where a few representative temporal paths (or traces) of membrane entropies are drawn as electric fields of various strengths when the electrical stimulation is applied (red/lighter traces with line arrows) or withdrawn (blue/darker traces with solid arrows). Membrane entropy, ξ, represents the lack of order (i.e., chaos) in the cellular membrane structure and is primarily responsible for reduced stiffness and nonmonotonic rigidity [46]. To effectively illustrate the dynamics of EL, these two figures are simultaneously discussed here.

- When an electric field beyond the threshold limit is applied on a cell, the stable state of phospholipids within the cell membrane is disrupted and a high transmembrane potential develops across the membrane (Fig. 9.2), which increases the entropy of the membrane structural organization (Fig. 9.3).
- When this transmembrane potential reaches a critical level, many tiny hydrophilic pores begin to form. This entropy

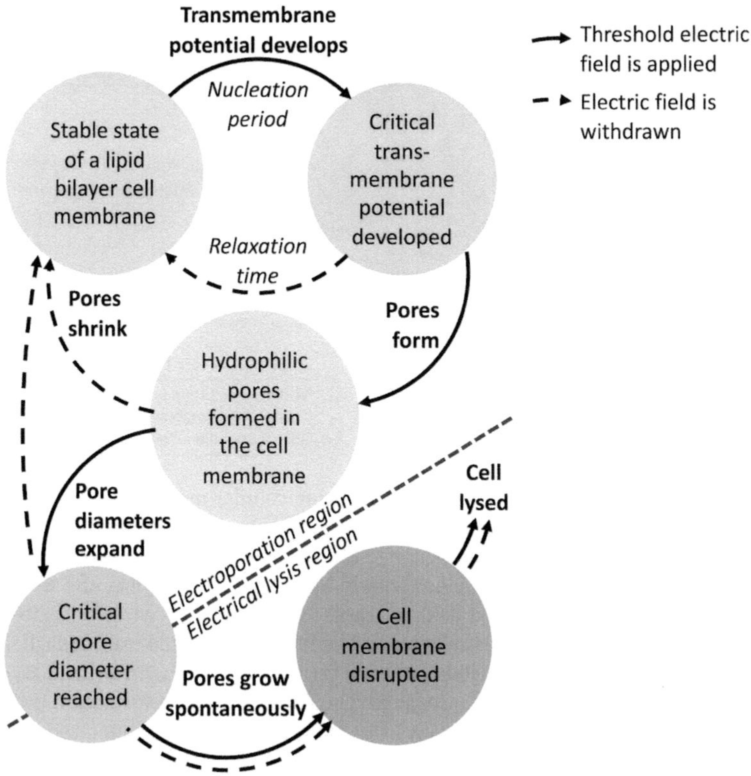

Figure 9.2 Conceptual representation of state transition diagram of the sequence of events during the electroporation (EP) or the electrical lysis (EL) process. The various states are indicated using circles, while transitions are represented using two types of arrows: solid arrows (transitions due to applying an electric field above the threshold limit) and dashed arrows (spontaneous transitions that occurs after the corresponding external electric field is removed). The process remains reversible until the critical pore diameter is reached, after which spontaneously pore growth and irreversible lysis occur. The region from pore formation to reaching the critical diameter is the EP region, while the region after the critical pore diameter is the EL region.

level is represented by ξ_1 in Fig. 9.3. The corresponding times are denoted as nucleation periods. At this entropy, the reorientation force due to the applied electric field is higher than that of the threshold electric field ($\mathbf{E}_{\text{th}}$),

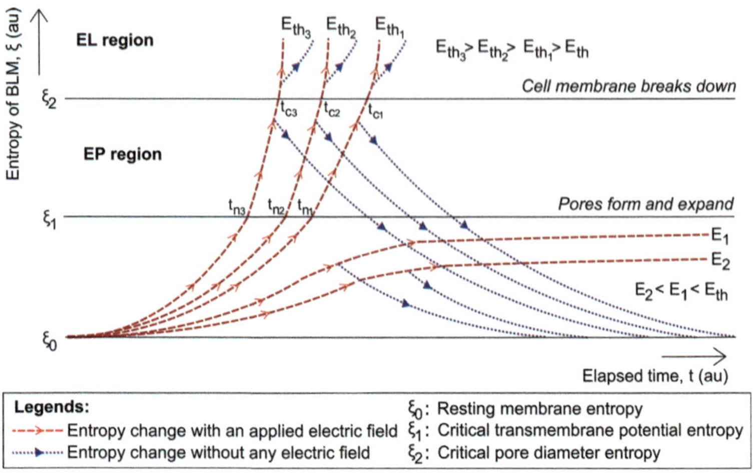

Figure 9.3 Conceptual drawing representing cellular membrane structure entropy as a function of elapsed time (t) after electric stimulation is applied. The traces denote the entropy change for a range of applied electric fields (compared to the threshold electric field $\mathbf{E}_{\text{th}} < \mathbf{E}_{\text{th1}} < \mathbf{E}_{\text{th2}} < \mathbf{E}_{\text{th3}}$ and $\mathbf{E}_{\text{th}} > \mathbf{E}_1 > \mathbf{E}_2$), as denoted by red dashed traces (lighter shade), while that after withdrawal of electric fields as denoted by blue darker dotted traces (darker shade). The nucleation periods for pore formation with $\mathbf{E}_{\text{th1}}$, $\mathbf{E}_{\text{th2}}$, and $\mathbf{E}_{\text{th3}}$ are denoted as t_{n1}, t_{n2}, and t_{n3}, respectively. The critical pore formation time (shown as t_{c1}, t_{c2}, t_{c3}, respectively) for the critical pore diameter reduces as the field strength (i.e., $\mathbf{E}_{\text{th1}}$, $\mathbf{E}_{\text{th2}}$, and $\mathbf{E}_{\text{th3}}$) increases above the threshold electric field ($\mathbf{E}_{\text{th}}$). If the pore diameter does not reach critical entropy (2), the pores reseal (i.e., EP occurs) to the stable state (0), otherwise the cell membrane breaks down (i.e., EL occurs) [49].

which is greater than the orientation force due to the hydrophobicity of phospholipids. The formation of these hydrophilic pores causes EP and makes the cell membrane permeable. However, if the applied electric field is lower than the threshold value, the ξ_1 entropy level would never be reached, irrespective of the stimulation duration.

- If an external electric field higher than the threshold level is sustained, the newly formed hydrophilic pores continue to grow in diameter as the corresponding entropy increases.

When a pore diameter reaches the critical pore diameter for a resealable range, the reversible EP region ends and the irreversible EL region begins (Figs. 9.2 and 9.3). This corresponding entropy is represented by ξ_2 in Fig. 9.3. The process is reversible up to this level of entropy, after which the process becomes irreversible. The critical time duration (e.g., t_{c1}, t_{c2}, t_{c3}), defined as the time required to reach the ξ_2 level of entropy, decreases as the applied electric field (e.g., $\mathbf{E}_{\text{th1}}$, $\mathbf{E}_{\text{th2}}$, and $\mathbf{E}_{\text{th3}}$ beyond the threshold electric field $\mathbf{E}_{\text{th}}$) increases [20].

- If the applied electric field is withdrawn when the entropy is below ξ_2, the membrane entropy decreases and the pores reseal, leading to the EP phenomenon. If the electric field is withdrawn after the entropy exceeds ξ_2, the membrane entropy continues to increase and pores continue to grow spontaneously, leading to complete disruption of the cell membrane structural organization and resulting in cell lysis [41].

Although these dynamics are generally accepted, there are a few experimental observations that are difficult to explain with the above representation of dynamics. For instance, the membrane poles toward the positive electrode side (the hyperpolarized portion of the cell membrane) were reported to show the formation of pores more aggressively compared to the poles toward the negative electrode side (the depolarized membrane) [45].

To explain these aforementioned observations, molecular dynamics representation can be considered as shown in Fig. 9.4 [47–49]. Here, the reorientation of phospholipids within cell membranes due to an applied electric field higher than the threshold electric field is depicted for both hyperpolarized and depolarized portions of the cell membrane.

The phospholipids, acting as tiny electric dipoles, are arranged in a stable bilayer cell membrane formation (Fig. 9.4a) and experience reorientation forces due to the applied electric field, which induces charges (ions) along the cell membrane (Fig. 9.4b) [33, 38]. The resultant electric force–induced stresses attempt to rotate

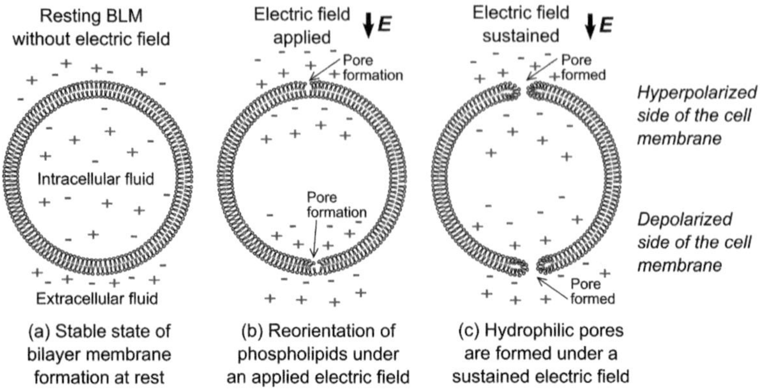

Figure 9.4 Conceptual representation of the reorientation of phospholipids (size significantly exaggerated for clarity) of a BLM on the hyperpolarized (toward the positive electrode) and depolarized (toward the negative electrode) poles. The cell is under an applied electric field beyond the threshold level of the electric field and sustained long enough to cause hydrophilic pore formation due to the accumulation of the opposite charge across the BLM. The figure clearly depicts the reason the pore formation at the hyperpolarized side will be observable, but that of depolarized side is not observable during pore formation [49].

phospholipid molecules and succeed at locations of faults or weak junctions near the poles (Fig. 9.4c).

Notice that only the outer layer of the hyperpolarized membrane and the inner layer of the depolarized membrane would experience the reorientation force (Fig. 9.4b). This might explain why the reorientation of phospholipids is more prominent at the hyperpolarized side of the membrane compared to that at the depolarized side.

9.3 Microfluidic Devices for On-Chip Electrical Lysis

9.3.1 On-Chip Electrical Lysis

Custom-made microfluidic devices using pulse electric field (PEF) excitation to lyse cells with the electric field method were reported as early as in 1998 for yeast and *Escherichia coli* cell lysis [50] and later to lyse cancer cells [40, 43, 51], mammalian cells [45], and leukocytes [38]. A constant electric field to lyse erythrocytes on a chip was reported using a glass-based microfluidic device [52]. Later, polydimethylsiloxane (PDMS)-based microfluidic devices were reported for the lysis with a PEF of red blood cells (RBCs) [15] and *E. coli* and Chinese hamster ovary cells [14, 53]. Alternative electric field stimulation to lyse cultured human colon carcinoma cells was reported with an SU-8-based microfluidic device [34]. In addition, microfluidic devices for lysis were also reported that combine electric field stimulation with other stimuli, such as electromechanical [54] and electrochemical [55].

On-chip EL with integrated electrodes offers many advantages, for example, it is a low-voltage operation, there is low power requirement, a minute amount of sample is needed, precise controllability is possible, and there is ease of integration with other on-chip systems through various fabrication technologies, including bio-MEMS, LOC, and μTAS. With recent advancements in on-chip EL, our understanding of the relevant dynamics is further enhanced and the acquired knowledge can be applied in various scientific areas, including biotechnology, molecular biology, and biomedical engineering [33, 45, 47, 53]. Some applications of these technologies are subcellular analysis [34, 37], gene transfection [38], biofouling treatment [56], integration of multiple biochemical processes [52], high-throughput parallel bioassay preparation [54], cancer treatment [57], and integrated on-chip DNA detection [58]. The key force for the progression of this research field is the advanced microfabrication technologies [26, 59–61].

9.3.2 Microchannel and Electrode Geometry

Microfluidic chips are typically prepared on glass or polymer (PDMS, SU-8, etc.) substrates, while the electrodes are made of inert metals, such as platinum (Pt) and gold (Au). Integrated electrode layers generated with MEMS-based soft lithography techniques allow generation of the high electric fields needed to lyse cells with a low applied potential and low power consumption. This is critical for controlled experimentations with various electrical excitation parameters, to reduce Joule heating, and to broaden the potential of practical applications.

Microchannel geometry can be manipulated to amplify the applied electric field in a certain section of the channel to achieve the threshold electric field required for EL [14, 15, 34, 38, 43]. Such modifications allow the cells to lyse using a lower voltage than otherwise required. Furthermore, such modifications could result in lower Joule heating. However, with the newer fabrication process of microfluidic devices, electrodes can be microfabricated inside the microchannels, which allows the application of an even lower voltage. Some examples of microchannel modifications and electrode designs for on-chip EL are depicted in Fig. 9.5.

Among various reported on-chip EL microfluidic devices with a variety of microchannel configurations and electrode designs, the following four common types are generally noted.

- The electric field inside a microchannel is inversely proportional to the microchannel width for a certain microchannel height [14]. Hence, microchannels can be designed to benefit from this relationship, where a narrow (pinched) section of the microchannel creates a higher electric field compared to a wider section of the microchannel (Fig. 9.5a). Such devices were used to lyse Chinese hamster ovary and *E. coli* cells using electric fields measuring 0.06 MV/m and 0.24 MV/m, respectively [14, 53, 62]. Similar devices lysed RBCs in continuous-flow mode, where a bias voltage of only 50 V created 0.12 MV/m in a narrow section of the microchannel, which was sufficient for EL [15]. As the applied voltages that would have been required without channel modifications are many times higher, the required control circuitry for a PEF are simpler and less expensive with the channel modification. However, such modification allows only the cells that are inside the narrow section to be lysed; hence cells need to flow continuously (preferably in a single profile) through this section of the microchannel.
- Another geometric modification that concentrates the electric field uses a micropore (or microhole) between two

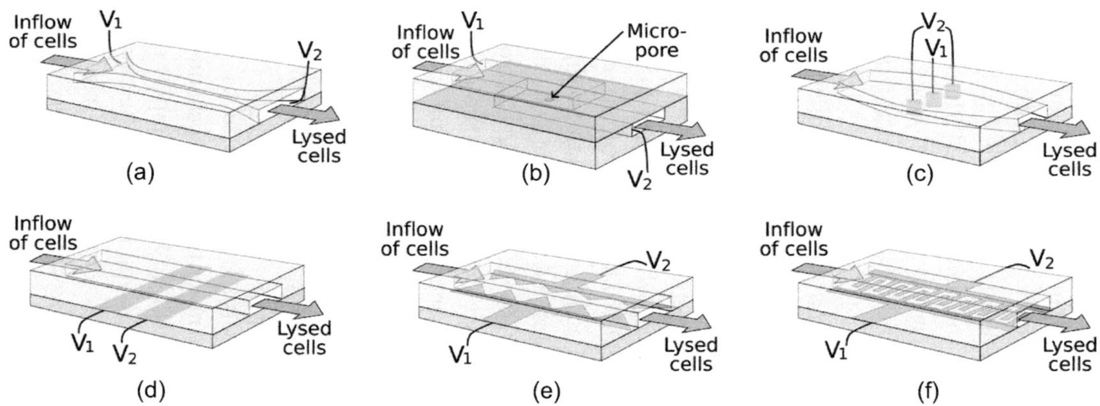

Figure 9.5 Some examples of geometric modification for specialized configurations for on-chip EL experiments. (a) A microchannel configuration with a wide-narrow section, where the cells passing through the narrow section experience a higher electric field compared to the wide section. Integrated electrodes can be placed in the wide section or external electrodes in the fluidic ports. (b) A micropore configuration that captures cells flowing thorough the top microchannel, and the electric field is generated in electrodes in the top and bottom microchannels. (c) Electrodes with a 3D configuration in cylindrical shapes that are placed in the microchannel chamber. (d) A microfabricated parallel electrode pair configuration under the microchannel. (e) A sawtooth configuration of electrodes on the microchannel wall where the cells flow between these electrodes. (f) Integrated electrodes in an interdigitated electrode (IDE) configuration at the bottom of the microchannel.

fluidic chambers, as depicted in Fig. 9.5b [40, 43]. The diameter of the micropore (usually 2 to 4 μm) needs to be smaller compared to the cell diameter (about 20 μm). When a cell settles on top of the micropore, a voltage pulse is applied to lyse (or electroporate) the cell. Human prostate adenocarcinoma (ND-1) cells were lysed with excitation of only 35 V pulse of 100 ms using such a microdevice, much less than without the microhole [43]. However, this modification requires each cell to be positioned on top of the microchannel before it can be lysed. Hence, it is tedious and has low throughput.

- With advanced microfabrication techniques of microfluidic devices, integrated electrodes can be fabricated inside the microchannels to produce a high electric field with a low applied voltage. Electrode designs and configurations are critical to generate an electric field with the preferred distribution at the location where cells need to be lysed. Examples of a planar electrode configuration are a simple electrode pair (Fig. 9.5d), a sawtooth configuration (Fig. 9.5e), and an interdigitated electrode (IDE) configuration (Fig. 9.5f) [15, 34, 37, 63–65]. A simple electrode pair generates a uniform electric field between the electrodes. A sawtooth configuration generates a non-uniform electric field distribution with a localized high electric field at the tip of the electrodes. An IDE configuration creates a relatively uniform electric field distribution in a large area using very small applied voltages with the maximum area coverage for lysis.
- Integrated 3D electrodes (often referred to as 3D electrodes) with a thickness similar to the microchannel depth can be more effective for EL [34, 38]. Examples of these designs are given in Fig. 9.5c and Fig. 9.5e. The advantages of 3D electrodes are smaller applied voltage requirement and more volume in the microchannels energized with a certain amount of energy delivery resulting in higher EL throughput. For example, 30% EL was achieved with 3D electrodes compared to 8% with planar electrodes [38].

9.4 Electrical Lysis Parameters and Considerations

9.4.1 PEF Parameters

PEF excitation parameters have a significant influence on the EL process. These parameters include the pulse magnitude, the pulse duration, the energy per pulse, the number of pulses (in the burst), the repeat rates (frequency), and the pulse shape. The ranges of these pulse parameters are critically cell-line, cell-size, and cell-shape dependent.

It is to be noted that only electric fields above a threshold value induce EL. However, the required electric voltages to create these electric fields rely on the microfluidic device design (i.e., channel and electrode geometry). For example, in the case of a microfabricated device that has top and bottom fluidic chambers connected through a 2 to 4 μm diameter hole (viz. micropore) to concentrate the electric field around it, EL was reported with only 35 V pulses 100 ms in duration as cells were positioned on top of the micropore [43]. However, no EL was observed with 15 V pulses 100 ms in duration in the same device. The micropore configuration that connects the two fluidic chambers separated by a membrane intensifies the electric field above (and below) the micropore. In another study, the release of intracellular content was observed for *E. coli* with an electric field above 0.1 MV/m; however, at a lower level of electric field, minimal content was released [18]. The effectiveness of EL was reported to be increased by using multiple pulses instead of a single pulse [18, 66].

Survivability of the treated cells with pulsed electric fields was reported to depend on pulse widths as well [26, 56]. Above the threshold electric field strength, pulses of longer duration caused a greater death rate to cells as opposed to pulses of shorter duration. For instance, 0.05 MV/m pulses 6 ms in duration were found to be sufficient to lyse Jurkat cells, whereas the cell membranes of the same type of sample were intact with 6 MV/m pulses 60 ns in duration [67]. Nonetheless, cells in the latter case also died because of severe damage to internal organelles. Similar short-duration (nanosecond range) pulses were shown to damage internal organelles that led to cell death without rupturing cell membrane. Leukemia cell membranes were damaged with pulses of 0.14 MV/m 5 μs in duration, but no cell membrane damage was observed with 1.35 MV/m pulses 50 ns in duration, even though over 90% cells died within 3 days after exposure [22]. In addition, increased caspase (enzymes causing programmed cell death or apoptosis) activity was observed for longer pulses (300 ns) compared to shorter pulses (10 ns) [68].

9.4.2 Critical Pore Formation Time

The critical pore formation time is also dependent on the strength of the electric field. Thus, effective EL is a function of the duration as well as the magnitude of the applied electric field. The time between EL initiation and complete disruption, known as lysis time, was reported to be less than 25 ms for erythrocytes with an excitation pulse of 1 s [52]. The time required for lysis was less than 33 ms (within 1 frame of captured video) for rat basophilic leukemia cells with pulses 1 ms in duration, whereas the lysis time increased to more than 1 s with pulses of 100 μs for the same electric field [45]. The lysis time required for acute myeloid leukemia (AML) cells using the EL method is reported to be less than 300 ms after excitation was applied [54]. EL of less than 66 ms was demonstrated for RBL cells through a reduction in fluorescence dye [17].

9.4.3 Excitation Type

Although pulse excitation is the most common excitation source used for EL, other types of excitations were also reported, such as AC and DC. EL of several types of protoplasts was reported with 1 MHz AC excitation [37, 69]. For the case of AC excitations, cells treated with a higher applied voltage and frequency resulted in more cell lysis, whereas cells treated with a lower applied voltage and frequency resulted in more ghost cells (keratinized denucleated cells) [34]. For the case of AC excitations, cells face a dielectrophoretic force in addition to the EL phenomenon. It is thought that this dielectrophoretic force is responsible for the stretching of cells between electrodes, which might additionally contribute to cell lysis [37]. Similarly, EL of *Vaccinia virus* was reported using AC excitation [19]. Most studies for DC excitation consist of microchannels with a high electric field and a steady rate of cell flow, although hydrolysis of water and electrode degradation must be addressed. An example of DC excitation for EL is erythrocytes collected from human blood samples lysed with 1 s excitation of 1.4 kV across a 1 cm long microchannel [52]. Other examples of DC excitation for EL are AML cell lysis with 1.38 kV across a 4.2 cm long microchannel in less than 300 ms [54], Chinese hamster ovary and *E. coli* cells lysis with electric fields measuring 0.06 and 0.24 MV/m, respectively [14, 53], and RBC lysis using an electric field of 0.12 MV/m [15].

9.4.4 Fluidic Medium

As electric field strength is dependent on the electrical parameter of the fluidic medium; EL also depends on the fluidic medium (viz. fluidic buffer) types due to differences in conductivity, permittivity, permeability, osmolarity, and density of ionic molecules that affect the resultant electric fields. For instance, cells were observed to swell due to EP phenomena at different rates with various osmolarities of the buffer medium [53]. The average swelling of cells in a hypotonic buffer was 175%, in an isotonic buffer was 155%, and in a hypertonic buffer was 160% when they were subjected to an electric field of 0.05 MV/m for 300 ms in a microfluidic device.

9.4.5 Cell Line

Various cell lines show different excitation requirements, which might be due to cell structure, shape, strength, and size. Chinese cabbage protoplasts and radish protoplasts were lysed with very small electric fields (~0.15 MV/m) compared to yeast (*S*accharomyces *cerevisiae*) protoplast and *E. coli* (~1 MV/m); however, the former required longer pulses (1 ms) compared to the latter (100 and 500 μs, respectively) [37]. Leukocytes, also known as white blood cells, (1:100 dilution) from blood samples were lysed using 50 MV/m pulses of 10 μs [38]. Chinese hamster ovary, *E. coli*, and RBC were lysed using electric fields of 0.06, 0.24, and 0.12 MV/m, respectively, with DC excitation [14, 15, 53]. These demonstrate that the excitation parameters (electric field, pulse width, etc.) required for cellular breakdown are also dependent on the cell line. In addition, the cell growth phase (as well as differentiated live cells and dead cells) might also have some influence on EL excitation parameters [21].

9.5 EL-Associated Phenomena

9.5.1 Secondary Effects

A high electric field pulse, in addition to EL, might also affect cells in many other ways. These effects might be important for various fields of biochemistry and molecular biology. A variety of morphological changes of cell membranes causing deformation were noted after electric pulse treatment, such as membrane extrusion [34] and vesicle formation [42].

9.5.2 Cellular Spin Resonance

The rotation of cells exposed to external oscillating electric fields can be observed for live as well as dead cells. Spinning often starts at a very sharp frequency of the applied field and ceases in the same manner. This phenomenon is called cellular spin resonance [20]. This technique may be used to separate cells by their different rotational properties, such as characteristic resonance frequency, spinning rate, and spinning sense.

9.5.3 Electrofusion

As PEF excitation is applied, adjacent cell membranes might fuse together, a phenomenon known as electrofusion, observed in several studies [20, 34, 42]. Earlier it was thought to be the result of thermal melting and merging of cell membranes due to Joule heating. However, it is now generally accepted that this phenomenon is due to nonthermal effects related to electric fields below the threshold level [25].

9.5.4 Electroinsertion

Applying a PEF makes the cell membrane permeable, during which target molecules (e.g., proteins) can be inserted within the cell membrane. This process, known as electroinsertion, can be used to correctly incorporate purified cloned membrane proteins into a cell membrane. This electroinsertion technique has potential applications in proteomics [26].

9.5.5 Electroactivation

In the electroactivation process, an oscillating electric field of appropriate amplitude and frequency can synchronize the conformal oscillation of a protein. If the enforced conformal oscillation is coupled to a ligand binding reaction or a substrate-product conversion, the energy transmitted through the electric field may be captured and transferred to drive an endergonic reaction. Electroactivation, or electroconformational coupling, has been shown to interpret how a cell or an organism can recognize very weak electric signals [25].

9.6 On-Chip Electrical Lysis Applications

9.6.1 Advantages of On-Chip Electrical Lysis

EL, particularly performed on a chip by electrically lysing biological cells on microfabricated chips, is applicable for various purposes. In this process, minute amounts (microliter volume) of cell samples are introduced on or inside the chip containing microchannels or open cavities with integrated or external electrodes. For EL, electrical excitation is applied that creates localized electric fields beyond the critical strength around the sample cells, and the electric field must be sustained longer than the critical time duration for lysis. The EL process results in the breakdown of the cell membrane, and the lysate could be collected for analysis or be further processed in assays on downstream chips using LOC or μTAS systems. MEMS- and lithography-based fabrication technologies have allowed researchers to design, develop, and test various types of microfluidic chambers with different configurations with or without integrated electrodes for an optimal EL process.

It has been demonstrated that on-chip EL has many advantages over other lysis methods. EL is faster compared to other lysis techniques except laser-based lysis but offers simple, high-throughput, and low-cost implementation compared to laser-based lysis. Furthermore, controllability of on-chip EL is superior and suitable for automated operation. The operation cost of EL is minimal, and the process does not leave any residue in the lysate. These advantages render the EL method more suitable for the on-chip implementation for LOC, μTAS, and bio-MEMS.

9.6.2 Bioanalysis Device

One application of on-chip EL is to perform bioanalysis of subcellular contents by implementing an integrated EL chamber within a microdevice [34, 37]. Some examples of application of an EL-based microdevice such as this are cytology, genetics and molecular biology, on-chip sample preparation of respiratory bacterial pathogens, chemical cytometry analysis at the single-cell level, and controlled cell growth–interfering agent [23, 62, 70].

9.6.3 Biofouling Treatment

EL can be used to induce apoptosis of bacteria and other microorganisms and can be an effective tool for controlling biofouling (contamination due to the growth of bacteria) by treating food, water, and beverages [21, 56, 66, 67]. This can be economical and effective, with minimal side effects.

9.6.4 Single-Cell Analysis

EL, as well as EP, can be applied for single-cell analysis [43]. This type of analysis is significant for understanding features that cannot be detected in a colony-based approach. Glutathione within individual erythrocytes has been analyzed using the EL method to study the mechanism of various biological processes that occur in a living body [52]. To increase throughput, parallel processing of single-cell analysis using EL followed by capillary electrophoresis technique was also demonstrated through a parallel microchannel–based device [54].

9.6.5 Cancer and Tumor Treatment

EL, as a technique for irreversible EP, is being investigated for tumor ablation in the application of cancer and tumor treatment [22, 57, 59, 71]. There are two main approaches related to EL and EP being investigated. One approach for treating these cancer and tumor cells is to perform selective EL by killing only unwanted cells by zapping them with a high electric field. Another approach is to increase the cytotoxic drug intake of affected cells using EP to increase the permeability of only these cell membranes that have cancer or a tumor.

9.7 Summary

EL is a promising technique for cell lysis as it is fast and inexpensive and can be integrated within microfluidic devices. Furthermore, it does not produce any residue, can obtain high throughput, and is highly controllable. This concept has enabled cell lysis on the micro- and nanoliter scale. With the ability of advanced fabrication technologies, EL has been shown to be an attractive and practical solution with technologies such as LOC, bio-MEMS, and μTAS. EL has many potential usages in various biomedical applications, including bacterial decontamination, microdevices for cytology and genetics, single-cell analysis, as well as cancer and tumor treatment. However, optimized designs and applications of on-chip EL and integration with the preparing, processing, and analyzing protocols of other cell samples in an automated system require further investigation and understanding the dynamics and stratification of the control parameters and resolving other sustained issues.

References

1. J. M. Graham, J. A. Higgins (1997). *Membrane Analysis*, BIOS Scientific, NY, USA.
2. D. Voet, J. G. Voet (2004). *Biochemistry*, 3rd ed., John Wiley & Sons, USA.
3. H. Lodish, A. Berk, P. Matsudaira, et al. (2003). *Molecular Cell Biology*, 5th ed., W. H. Freeman and Co., New York, NY, USA.

4. R. B. Brown, J. Audet (2008). Current techniques for single-cell lysis. *J R Soc Interface*, **5**:S131–S138.
5. S. C. Dixon, J. Horti, Y. Guo, E. Reed, W. D. Figg (1998). Methods for extracting and amplifying genomic DNA isolated from frozen serum. *Nat Biotechnol*, **16**:91–94.
6. H. Kido, M. Micic, D. Smith, J. Zoval, J. Norton, M. Madoua (2007). A novel, compact disk-like centrifugal microfluidics system for cell lysis and sample homogenization. *Colloids Surf,B*, **58**:44–51.
7. R. W. Doebler, B. Erwin, A. Hickerson, et al. (2009). Continuous-flow, rapid lysis devices for biodefense nucleic acid diagnostic systems. *J Assoc Lab Autom*, **14**:119–125.
8. M. D. Dhawan, F. Wise, A. J. Baeumner (2002). Development of a laser-induced cell lysis system. *Anal Bioanal Chem*, **374**:421–426.
9. H. Lai, P. A. Quinto-Su, C. E. Sims, et al. (2008). Characterization and use of laser-based lysis for cell analysis on-chip. *J R Soc Interface*, **5**:S113–S121.
10. Y. P. Bao, M. Huber, T. Wei, S. S. Marla, J. J. Storhoff, U. R. Muller (2005). SNP identification in unamplified human genomic DNA with gold nanoparticle probes. *Nucleic Acids Res*, **33**:e15.
11. M. Babincova, D. Leszczynska, P. Sourivong, P. Babinec (2001). Lysis of photosensitized erythrocytes in an alternating magnetic field. *J Magn. Magn Mater*, **225**:194–196.
12. S. Chen, S. J. Lillard (2001). Continuous cell introduction for the analysis of individual cells by capillary electrophores. *Anal Chem*, **73**:111–118.
13. M. A. McClain, C. T. Culbertson, S. C. Jacobson, N. L. Allbritton, C. E. Sims, J. M. Ramsey (2003). Microfluidic devices for the high-throughput chemical analysis of cells. *Anal Chem*, **75**:5646–5655.
14. H. Wang, A. K. Bhunia, C. Lu (2006). A microfluidic flow-through device for high throughput electrical lysis of bacterial cells based on continuous DC voltage. *Biosens Bioelectron*, **22**:582–588.
15. D. W. Lee, Y. Cho (2007). A continuous electrical cell lysis device using a low DC voltage for a cell transport and rupture. *Sens Actuators, B*, **124**:84–89.
16. Y. Nashimoto, Y. Takahashi, T. Yamakawa, et al. (2007). Measurement of gene expression from single adherent cells and spheroids collected using fast electrical lysis. *Anal Chem*, **79**:6823–6830.
17. P. J. Marc, C. E. Sims, M. Bachman, G. P. Lia, N. L. Allbritton (2008). Fast-lysis cell traps for chemical cytometry. *Lab Chip*, **8**:710–716.

18. N. Bao, C. Lu (2008). A microfluidic device for physical trapping and electrical lysis of bacterial cells. *Appl Phys Lett*, **92**:214103-1-3.
19. K. Park, D. Akin, R. Bashir (2007). Electrical capture and lysis of vaccinia virus particles using silicon nano-scale probe array. *Biomed Microdevices*, **9**:877–883.
20. E. Neumann, A. E. Sowers, C. A. Jordan (1989). *Electroporation and Electrofusion in Cell Biology*, Plenum Press, NY, USA.
21. K. H. Schoenbach, R. P. Joshi, R. H. Stark, F. C. Dobbs, S. J. Beebe (2000). Bacterial decontamination of liquids with pulsed electric fields. *IEEE Trans Dielectr Elect Insul*, **7**:637–645.
22. K. H. Schoenbach, A. Abou-Ghazala, T. Vithoulkas, R. W. Alden, R. Turner, S. Beebe (1997). The effect of pulsed electric fields on biological cells. *IEEE Int Pulsed Power Conf*, **1**:73–78.
23. C. Rosa, P. A. Tilley, J. D. Fox, K. V. Kaler (2008). Microfluidic device for dielectrophoresis manipulation and electrodisruption of respiratory pathogen Bordetella pertussis. *IEEE Trans Biomed Eng*, **55**:2426–2432.
24. R. V. Davalos, B. Rubinsky (2008). Temperature considerations during irreversible electroporation. *Int J Heat Mass Transfer*, **51**:5617–5622.
25. T. Y. Tsong (1991). Electroporation of cell membranes. *Biophys J*, **60**:297–306.
26. S. B. Dev, D. P. Rabussay, G. Widera, G. A. Hofmann (2000). Medical applications of electroporation. *IEEE Trans Plasma Sci*, **28**:206–223.
27. Q. Hu, R. P. Joshi, A. Beskok (2009). Model study of electroporation effects on the dielectrophoretic response of spheroidal cells. *J Appl Phys*, **106**:024701.
28. B. L. Ibey, D. G. Mixon, J. A. Payne, A. Bowman, K. Sickendick, et al. (2010). Plasma membrane permeabilization by trains of ultrashort electric pulses. *Bioelectrochemistry*, **79**:114–21.
29. D. Miklavcic, L. Towhidi (2010). Numerical study of the electroporation pulse shape effect on molecular uptake of biological cells. *Radiol Oncol*, **44**:34–41.
30. K. Maswiwat, D. Wachner, J. Gimsa (2008). Effects of cell orientation and electric field frequency on the transmembrane potential induced in ellipsoidal cells. *Bioelectrochemistry*, **74**:130–141.
31. C. Yao, Y. Mi, C. Li, X. Hu, X. Chen, C. Sun (2008). Study of transmembrane potentials on cellular inner and outer membrane — frequency response model and its filter characteristic simulation. *IEEE Trans Biomed Eng*, **55**:1792–1799.

32. A. Agarwal, I. Zudans, E. A. Weber, O. Orwar, S. G. Weber (2007). Effect of cell size and shape on single-cell electroporation. *Anal Chem*, **79**:3589–3596.
33. Q. Hu, V. Shidhara, R. P. Joshi, J. F. Kolb, K. H. Schoenbach (2006). Molecular dynamics analysis of high electric pulse effects on bilayer membranes containing DPPC and DPPS. *IEEE Trans Plasma Sci*, **34**:1405–1411.
34. H. Lu, M. A. Schmidt, K. F. Jensen (2005). A microfluidic electroporation device for cell lysis. *R Soc Chem*, **5**:23–29.
35. B. I. Morshed, M. Shams, T. Mussivand (2012). Deriving an electric circuit equivalent model of cell membrane pores in electroporation. *Biophys Rev Lett*, **7**:1–12.
36. K. J. Muller, V. L. Sukhorukov, U. Zimmermann (2001). Reversible electropermeabilization of mammalian cells by high-intensity, ultra-short pulses of submicrosecond duration. *J Membr Biol*, **184**:161–170.
37. S. W. Lee, Y. C. Tai (1997). A micro cell lysis device. *Sens Actuators*, **73**:74–79.
38. K. Lu, A. M. Wo, Y. Lo, K. Chen, C. Lin, C. Yang (2006). Three dimensional electrode array for cell lysis via electroporation. *Biosens Bioelectron*, **28**:24–33.
39. D. Moldovan, D. Pinisetty, R. V. Devireddy (2007). Molecular dynamics simulation of pore growth in lipid bilayer membranes in the presence of edge-active agents. *Appl Phys Lett*, **91**:204104-1-3.
40. Y. Huang, B. Rubinsky (2003). Flow-through micro-electroporation chip for high efficiency single-cell manipulation. *Sens Actuators*, **104**:205–212.
41. R. P. Joshi, K. H. Schoenbach (2000). Electroporation dynamics in biological cells subjected to ultrafast electrical pulses: a numerical simulation study. *Phys Rev E*, **62**:1025–1033.
42. E. Neumann (1989). Electric field-induced structural rearrangements in biomembranes. *Studia Biophys*, **130**:139–143.
43. Y. Huang, B. Rubinsky (2001). Microfabricated electroporation chip for single cell membrane permeabilization. *Sens Actuators*, **89**:242–249.
44. E. Neumann (1992). Membrane electroporation and direct gene transfer. *Bioelectrochem Bioenerg*, **28**:37–43.
45. F. Han, Y. Wang, C. E. Sims, et al. (2003). Fast electrical lysis of cells for capillary electrophoresis. *Anal Chem*, **75**:3688–3696.
46. R. Matthews, C. N. Likos (2012). Influence of fluctuating membranes on self-assembly of patchy colloids. *Phys Rev Lett*, **109**:178302-1-5.

47. Q. Hu, S. Viswanadham, R. P. Joshi, K. H. Schoenbach, S. J. Beebe, P. F. Blackmore (2006). Simulations of transient membrane behaviour in cells subjected to a high-intensity ultrashort electric pulse. *Phys Rev E*, **71**:031914-1-9.
48. P. Vulto, G. Dame, U. Maier, S. Makohliso, S. Podszun, P. Zahna, G. A. Urban (2010). A micro fluidic approach for high efficiency extraction of low molecular weight RNA. *R Soc Chem*, **10**:610616.
49. B. I. Morshed, M. Shams, T. Mussivand (2013). Electrical lysis: dynamics revisited and advances in on-chip operation. *Crit Rev Biomed Eng*, **41**:37–50.
50. S. W. Lee, H. Yowanto, Y. C. Tai (1998). A micro cell lysis device. In *Proceedings of the IEEE 11th Annual International Workshop on Micro Electro Mechanical Systems (MEMS 98)*, pp. 443–447.
51. Y. Huang, B. Rubinsky (1999). Micro-electroporation: improving the efficiency and understanding of electrical permeabilization of cells. *Biomed Microdevices*, **2**:145–150.
52. J. Gao, X. Yin, Z. Fang (2004). Integration of single cell injection, cell lysis, separation and detection of intercellular constituents on a microfluidic chip. *Lab Chip*, **4**:47–52.
53. H. Wang, C. Lu (2006). High-throughput and real-time study of single cell electroporation using microfluidic: effects of medium osmolarity. *Biotechnol Bioeng*, **95**:1116–1125.
54. N. R. Munce, J. Li, P. R. Herman, L. Lilge (2004). Microfabricated system for parallel single-cell capillary electrophoresis. *Anal Chem*, **76**:4983–4989.
55. C. T. Culbertson, et al. (2004). Microfluidic systems and methods for transport and lysis of cells and analysis of cell lysate, US Patent 006783647B2.
56. K. H. Schoenbach, F. E. Peterkin, R. W. Alden, S. J. Beebe (1997). The effect of pulsed electric fields on biological cells: experiments and applications. *IEEE Trans Plasma Sci*, **25**:284–292.
57. K. H. Schoenbach, R. Nuccitelli, S. J. Beebe (2006). ZAP: extreme voltage could be a surprisingly delicate tool in the fight against cancer. *IEEE Spectr*, **43**:22–26.
58. S. Li, Q. Yuan, B. I. Morshed, C. Ke, J. Wu, H. Jiang (2013). Dielectrophoretic responses of DNA and fluorophore in physiological solution by impedimetric characterization. *J Biosens Bioelectron*, **41**:649–655.
59. S. S. Saliterman (2006). Fundamentals of bioMEMS and medical microdevices, Wiley-Interscience, WA, USA.

60. F. T. G. Van den Brink, E. Gool, J. P. Frimat, J. Bomer, A. van den Berg, S. Le Gac (2011). Parallel single-cell analysis microfluidic platform. *Electrophoresis*, **32**:3094–3100.
61. G. Mernier, N. Piacentini, T. Braschler, N. Demierre, P. Renaud (2010). Continuous-flow electrical lysis device with integrated control by dielectrophoretic cell sorting. *Lab Chip*, **10**:2077–2084.
62. H. Wang, C. Lu (2006). Electroporation of mammalian cells in a microfluidic channel with geometric variation. *Anal Chem*, **78**:5158–5164.
63. T. Mussivand, B. I. Morshed, M. Shams (2012). Device for electroporation and lysis, US Patent Application 20120219987.
64. B. I. Morshed, M. Shams, T. Mussivand (2011). Identifying severity of electroporation through quantitative image analysis. *Appl Phys Lett*, **98**:143704-3.
65. B. I. Morshed, M. Shams, T. Mussivand (2014). Investigation of low-voltage pulse parameters on electroporation and electrical lysis using a microfluidic device with interdigitated electrodes. *IEEE Trans Biomed Eng*, **61**:871–882.
66. J. R. Beveridge, K. Wall, S. J. Macgregor, J. G. Anderson, N. J. Rowan (2004). Pulsed electric field inactivation of spoilage microorganisms in alcoholic beverages. *Proc IEEE*, **92**:1138–1143.
67. K. H. Schoenbach, R. P. Joshi, J. F. Kolb, et al. (2004). Ultrashort electrical pulses open a new gateway into biological cells. *Proc IEEE*, **92**:1122–1137.
68. S. J. Beebe, P. M. Fox, L. J. Rec, K. Somers, R. H. Stark, K. H. Schoenbach (2002). Nanosecond pulsed electric field (nsPEF) effects on cells and tissues: apoptosis induction and tumor growth inhibition. *IEEE Trans Plasma Sci*, **30**:286–292.
69. N. Ikeda, N. Tanaka, Y. Yangida, T. Hatsuzawa (2007). On-chip single-cell lysis for extracting intracellular material. *Jpn J Appl Phys*, **46**:6410–6414.
70. V. F. Veiga, L. Nimrichter, C. A. Teixeira, et al. (2005). Exposure of human leukemic cells to direct electric current: generation of toxic compounds inducing cell death by different mechanisms. *Cell Biochem Biophys*, **42**:61–74.
71. R. P. Joshi, K. H. Schoenbach (2010). Bioelectric effects of intense ultrashort pulses. *Crit Rev Biomed Eng*, **38**:255–304.

Chapter 10

Microfluidics-Based Metallic Nanoparticle Synthesis and Applications

Kavitha Illath,[a] Ashwin Kumar Narasimahan,[b] Moeto Nagai,[c] Syrpailyne Wankhar,[d] and Tuhin S. Santra[a]

[a] *Department of Engineering Design, Alumni Ave, Indian Institute of Technology, Chennai, Tamil Nadu 600036, India*
[b] *Department of Biomedical Engineering, Rajalakshmi Engineering College, Thandalam, Tamil Nadu 600010, India*
[c] *Department of Mechanical Engineering, Toyohashi University of Technology, 1-1 Hibarigaoka, Tempaku-cho, Toyohashi, Aichi, 441-8580, Japan*
[d] *Department of Bioengineering, Christian Medical College, IDA Scudder Rd, Vellore, Tamil Nadu 632004, India*
tuhin@iitm.ac.in, santra.tuhin@gmail.com

Nanotechnology has shown promise in biomedical applications for diagnosis and treatment purposes. Several materials are used for bioapplications, such as metals, metal oxides, semiconductors, and hybrid nanomaterials of different sizes and shapes. Nanomaterials have various physiochemical properties, such as a large surface-to-volume ratio, excellent optical and magnetic properties, chemical inertness, and less toxicity. Currently, nanomaterials are developed and synthesized by a batch process or in reactors at low volumes to produce efficient homogenous particles. For clinical applications, the requirement of nanomaterials is very high and

Microfluidics and Bio-MEMS: Devices and Applications
Edited by Tuhin S. Santra

ISBN 978-981-4800-85-3 (Hardcover), 978-1-003-01493-5 (eBook)
www.jennystanford.com

thereby developing such nanoparticles on a large scale is mandatory. However, this batch process often involves the use of bench-scale stirred reactors or plug flow reactors, which is inefficient due to poor spatial and temporal control. In recent years, microfluidic platforms have gained attention toward the synthesis and development of nanoparticles in a highly controlled and continuous manner for different biomedical applications. The molecular interaction and formation of metallic nanoparticles using microfluidic devices for the desired size and shape are logical compared to the batch process. In this chapter, we will describe the synthesis of different metal nanoparticles using microfluidic devices and their potential biomedical applications, such as sensing, imaging, and therapeutics and diagnostics. The future challenges of nanoparticle preparation using microfluidic devices and their applications, advantages, and limitations will also be elaborate.

10.1 Introduction

In the early fourth century AD, gold and silver nanoparticles (Au and Ag NPs) were used as coloring agents to coat the Lycurgus cup. The color of these colloidal particles changes when they are excited at a certain wavelength. Later researchers scratched a sample of the powder from the cup to test it and found that it contained silver and gold particles around 70 nm in size [35]. Francis Antonii in 1618 compiled a book about the uses of colloidal gold as a therapeutic agent, along with clinical case studies [9]. In 1857, Au NPs were systematically developed using phosphorous as a reducing agent. The coagulation and the size-dependent optical behavior were explained by Faraday [57]. Moreover, he prepared colloidal gold as the first monodispersed system ever [26]. After this transforming work, several researchers have started to develop particles using a batch process with different reducing agents and studied the mechanism involved in the formation of NPs. The methods of synthesis of NPs can be divided into two: (i) the top-down approach (bulk materials are chiseled down to the nanoscale) and (ii) the bottom-up approach (nanoscale objects are built by using individual atoms or molecules) [109]. Different methods involved in the top-down approach are lithography, laser ablation, ball milling, etc. It produces naked particles that can agglomerate and are unsuitable for bioapplications. The main hindrance of the batch reactor is the usage of expensive equipment and that the process is nonhomogeneous and time consuming, which limits commercialization [109]. However, this approach is able to produce non-uniform particles on a large scale. On the other hand, the bottom-up approach is widely used for developing nanomaterials using chemical precursors. The major driving forces in this technique are nucleation and aggregation when the nanomaterial reaches the desired size and shape [36, 75]. Further, coating with surfactants could reduce the surface energy. The main advantages of using the bottom-up approach are that the instrumentation is flexible, simpler, and inexpensive and produces homogenous particles. The only disadvantage of using this approach is the requirement of a huge amount of precursors, which makes it more expensive, especially in synthesizing Au and Ag NPs. Also, not all the precursor material will convert into NPs [51].

Miniaturization is leading in all the thrust areas of science and technology. Microelectromechanical systems (MEMS) has integrated with microfluidics, chemistry, chemical science, and life science to create labs on chips (LOCs). Thus, LOCs are a new era of science and technology that help to deal with fluids in nano- and picoliters using microsized channels. Microfluidics systems have made a new path in biomedical application for sensing and diagnosing diseases in small volumes [69, 76–78, 81, 87]. Now, microfluidics systems have gained attention in the synthesis and development of nanomaterials. Most importantly, advancements in biomedical research using microfluidics have shown additional benefits in diagnostics, single-cell analysis, screening, cell sorting, and assays for testing and analysis [69, 99–101]. This technique offers high control over biological or chemical individuals, allows conjugation of functional moieties, affords superior diagnostic performance, and is fast and reliable. There is a special focus among researchers on developing metallic nanomaterials due to their various biochemical applications on the clinical side [10, 105]. The physiochemical properties and surface functionalization of nanosized particles can be tuned using microfluidic devices compared to the bulk process.

The most important parameter that helps differentiate between metallic NPs is the surface plasmon resonance, which strongly depends on the composition, shape, and size of the NP [45, 65].

In this book chapter, we review the synthesis of various metallic nanomaterials and the NP formation mechanism using microfluidic devices. Traditional synthesis often involves the use of bench-scale stirred reactors or plug flow reactors, which is ineffective due to the poor spatial and temporal control of parameters. Microfluidic devices are used to overcome these limitations of traditional batch reactors. They allow highly controlled and continuous synthesis of NPs. LOC devices are used to overcome these problems for efficient production of nanomaterials. Metallic nanomaterials exhibit localized surface plasmon resonant oscillations due to a change in morphology. Plasmon resonance can be visually identified by the colors of gold, silver, palladium, and copper materials of different sizes, shapes, and compositions. Optical properties can change either due to individual particles or due to their hybridization with polymer and metal oxide materials. The collective oscillation of surface electrons in noble metals would provide an enhanced output magnitude when compared to the incident light. Properties like maximum peak wavelength, full-width at half maximum of the spectrum, and magnitude play a major role in the healthcare division. These innate properties of metallic NPs are used for various biomedical applications, such as sensing, imaging, therapy, and drug delivery.

10.1.1 Nanoparticle Formation

The mechanism of NP formation can be divided into three stages, as shown in Fig. 10.1: (i) precursors with a reducing agent start nucleation, (ii) there is further growth of nucleation to form particles at different stages by adding more precursors, and (iii) once the particles reach equilibrium, further growth is stopped by an external agent (e.g., surfactant) [51, 54, 55, 71, 98]. During the conventional batch process, the formation of NPs occurs simultaneously rather than in a step-wise process. It is technically challenging to produce consistent particles of definite sizes and structures and with small standard deviations in the nano-/micro-regime by the conventional

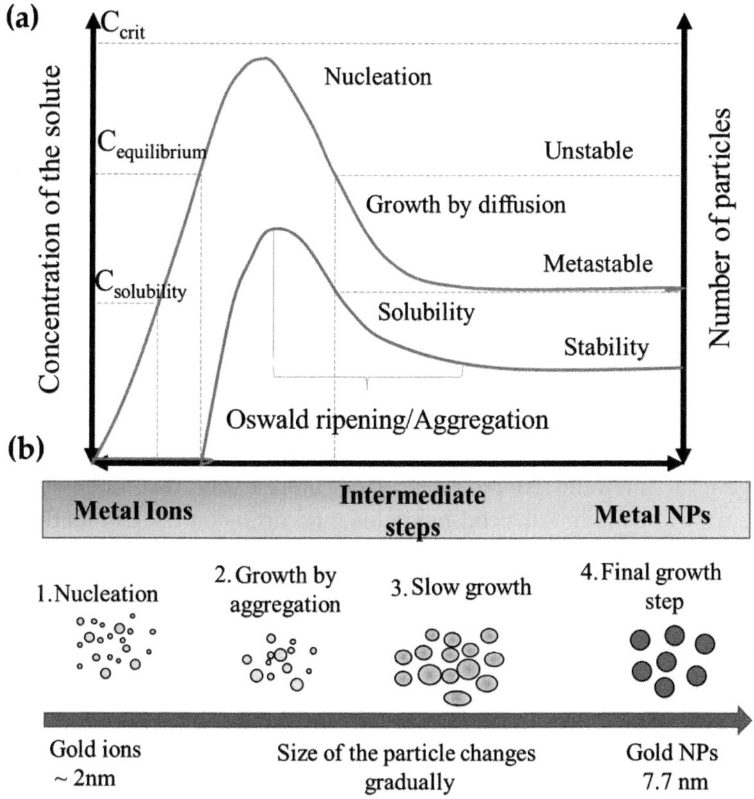

Figure 10.1 (a) Mechanism of nanoparticle formation starting with precursors and going on to nucleation and growth by the LaMer mechanism. Aggregation and Ostwald ripening of the particles occur if the reaction enters the metastable stage. Republished with permission of The Royal Society of Chemistry, from Ref. [114], copyright (2015); permission conveyed through Copyright Clearance Center, Inc. (b) A simple schematic representation of Au NP formation from Au^{3+} to Au^0 using the LaMer mechanism. Reprinted with permission from Ref. [98]. Copyright (2014) American Chemical Society.

batch-by-batch mode. Controlling the mixing of precursors and isolating the particles from the formed solution is difficult. This results in a variation among particles with respect to size and shape, and the batch is inconsistent with the other batches with a high deviation. In large-scale production (upscaling batch process) of NPs also, the same holds. At this level, improper mixing of precursors and the mass transfer problem will worsen the situation even more [3, 61, 63].

As both mechanisms have their own pros and cons, the choice of mechanism depends on the final product quality of metallic NPs. For instance, removal of impurities from the solution, large-scale synthesis, purity of the final product, efficiency of particles, usage of a number of precursors, and a time-consuming process are challenging to obtain the final end product in both the techniques.

10.1.2 Metallic Nanomaterials

Since ancient times, nanosized particles have shown a wide range of applications due to their innate physicochemical properties. In general, metallic NPs can be fabricated using techniques like surfactant-mediated, seed-mediated, and diffusion control method. The diffusion-controlled growth method results in the formation of nanocrystals with facets and edges having regularities and very few irregularities in the crystal lattice and internal grain boundaries [25]. Among nanomaterials, Au NPs have been accepted across the research sectors and have provided new insights, especially in cancer theranostics. These metallic NPs in the size range of 10–100 nm possess variable surface plasmonic resonance in the visible region with unique optical properties based on their size and shape. On the basis of size and shape, the color of the yellow gold precursor turns red, violet, blue, green, and yellow [43]. Similar to Au NPs, Ag NPs also possess shape-dependent optical properties. Further, it has been observed that spherical, triangular, and pentagonal shapes can display the colors blue, red, and green, respectively [14]. The unique feature of NPs is that they possess a high surface-to-volume ratio that enables them to interact with different functional moieties. Due to this property, functionalization of biomolecules like antibodies and drugs on the surface of NPs is much easier and they serve as specific biomarkers. Fluorescent and absorption properties of these metallic NPs serve as a unique marker for the detection of proteins using surface-enhanced Raman spectroscopy (SERS) and fluorescence imaging. Bare Au NPs are highly inert, which enhances their compatibility with tissues, reducing the interaction of proteins in the biological environment. On the other hand, aggregation mechanisms like seed-mediated and surfactant-directed methods lead to the formation of polycrystalline, anisotropic NPs with very large internal grain boundaries and low densities. Also, the mechanism of the growth phase is rapid, leading to the formation of larger-sized particles. To stop further aggregation of particles, the surface of NPs has to be stabilized [43]. The major drawback of synthesizing metallic NPs is the involvement of toxic chemicals. However, developing NPs at high volumes and with specific sizes and shapes is always challenging and requires more trials. Microfluidic devices and how they are I used to synthesize metallic NPs will be described in the following sections.

10.2 Microfluidics

Microfluidics is a branch of MEMS technology, which controls fluids in the range of 10^{-9} to 10^{-18} liters in micron-sized channels for biomedical applications. The essential components of a microfluidic device are micron channels, valves, mixers, and pumps. A fluid flowing through a microfluidic channel shows some interesting characteristics, different from those of macrosystems. Laminar flow is observed in microfluidics, whereas flow through a macrosystem is turbulent. Another interesting characteristic is electro-osmotic flow (EOF), which allows splitting of different charged species within the fluids with high resolution. EOF results when an ion-containing fluid flows through a microfluidic channel that has a fixed charge on its surface and a potential difference across it. A potential characteristic of microfluidic devices is the capability to control or influence water in channels, the dimensions of which match those of the Debye layer [113]. Another key feature of microfluidic reactors is a high surface-area-to-volume ratio, which allows quick heat and mass transfer, allows rapid cooling and heating of reagents, and maintains isothermal conditions. Minimal reagent consumption in microfluidic devices leads to a reduction in cost, making transportation and storage safer. The device also enables enhanced protection in handling toxic and inflammable solvents. The evolution of reactions in the spatial and temporal domain is possible in microfluidics. It can modify the compositions in the reaction by controlling the injection flow rate

Conventional Batch Reactors	Microfluidics Process
Batch by batch synthesis	Continuous Synthesis
Large scale synthesis is highly difficult	Large scale synthesis is possible
Size distribution can vary with batch to batch production	Size distribution will not vary with same design
Reproducibility is lesser	Highly reproducible
Wastage of precursors is high	Wastage of precursors is less
Bench-scale to clinical scale reproducibility is slow	Bench-scale to clinical scale reproducibility is fast

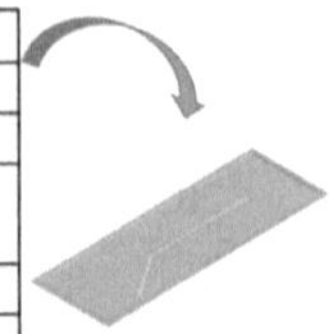

Figure 10.2 Comparison between a conventional batch process and a microfluidic device to fabricate metallic nanoparticles.

at any point. Reactions can be scaled up by conducting the same experiments in multiple reactors in parallel. Further, it is easy to add reagents downstream and quench the reaction once the product is formed [30, 119]. Moreover, nanomaterial with the desired structure and shape can be controlled by using microreactors with smart construction and design. Particles prepared in this manner revealed superior performance in catalytic activity, a low sensing detection limit, enhanced therapeutic efficiency, and targeted cancer imaging [30]. The wide varieties of features mentioned above enhance the synthesis of NPs in microfluidic devices compared to conventional reactors. Figure 10.2 shows the major differences between conventional batch reactors and microfluidic processes.

In general, NPs are synthesized using either a single solvent or multiple miscible solvents. On the basis of the number of phases involved in the reacting species, microfluidics is classified as single phase and multiphase, and the details are given next [37, 113].

10.2.1 Single-Phase Microfluidics

In single-phase microfluidics, reactants are mixed in laminar flow through a diffusion process. Suppose L is the diffusion length and D is the diffusion coefficient. Then the time taken τ for the homogenization of reactants in a microchannel is given by Eq. 10.1 [37].

$$\tau = \frac{L^2}{D} \tag{10.1}$$

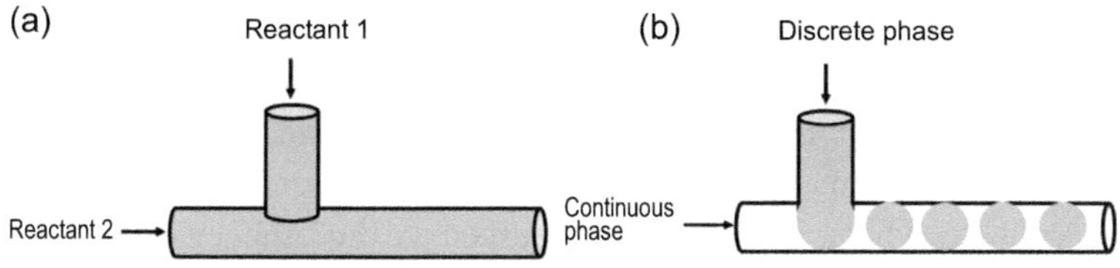

Figure 10.3 Classification of microfluidics based on phase: (a) single-phase microfluidics and (b) multiphase microfluidics [61].

In single-phase microfluidics, variables such as the flow rate of reagents, the reaction temperature, and the reaction time can be controlled and the reaction can be modified continuously to enhance the throughput by conducting multiple reactions in parallel. This ease of operations makes single-phase microfluidics the most commonly used in NP synthesis [120]. Figure 10.3a shows a schematic representation of single-phase microfluidics.

The problem associated with single-phase microfluidics is that it cannot work adequately in connection reactions, which are fast and limited by the diffusion process [63]. Further, the product yield is limited by axial dispersion due to the parabolic velocity profile inside the microchannel [19, 47]. To address these limitations, an additional component, the micromixer, can be incorporated. Micromixers are of two types, active and passive. Active mixing relies on time-dependent disturbance in the fluid flow, whereas passive mixing depends on the geometric properties of the channel through which the fluid flows [19]. Fluid flow in active mixers is controlled by varying either the pressure gradients or the different moving parts [22]. Another major issue associated with single-phase microfluidics is reactor fouling, which might be due to the direct contact of the reactant with the microchannel [11, 19, 47, 59, 107].

10.2.2 Multiphase Microfluidics

Multiphase microfluidics is complex compared to single-phase microfluidics. Its design and function include interaction among various physiochemical phenomena, such as chemical reactions, heat and mass transport, and the hydrodynamics of fluid and the surface phenomenon [30]. It enhances and extends the performance of single-phase microfluidics. The limitations of single-phase microfluidics, such as Taylor dispersion and long diffusion times, can be overcome by multiphase microfluidics design. Here, reaction occurs when two or more partially or fully immiscible fluids come in contact, where one fluid is dispersed in the other, which leads to a segmented flow of immiscible fluids. The reactive fluid can be continuous as shown in Fig. 10.3b, dispersed, or both, and mixing is through recirculation in fluid segments, which helps in keeping narrow residence time distribution [28, 30, 91]. Low residence time can be achieved by increasing the dispersed phase velocity and reducing the channel width. Some advantages of multiphase microfluidics are fast mixing, large interfacial areas, and lowered mass transfer limitations [29].

Depending on the nature of phases of the reaction, multiphase microfluidics is divided into two: liquid-liquid flow and gas-liquid flow microfluidics. In the liquid-liquid flow system, droplets of liquid reagents are formed [28, 90], whereas, in the gas-liquid flow, gas is dispersed in liquid in the form of bubbles [46, 116]. Reactor fouling is prevented in liquid-liquid flow due to the formation of reagents as droplets, while it is present in gas-flow since the reagents are in direct contact with the microchannel [28, 46, 90, 116]. The droplet method has the advantages of low dispersion and rapid mixing. Monodisperse semiconductor nanocrystals were successfully prepared with this approach. Dependence of flow patterns on fluid properties limits this method in general applications [20]. Single-and multiphase microfluidics with specific reagents and reaction conditions are used for successful synthesis of metallic NPs. Depending on the applications and requirements, microfluidic devices can be fabricated using materials such as polymers, silicon, and metals [40].

10.3 Microfluidic Devices for Metallic Nanoparticle Synthesis

Continuous- (laminar) and segmented-flow types of microfluidic processes have been used to synthesize metallic NPs, with their unique physical and chemical characteristics and controlled nanostructure.

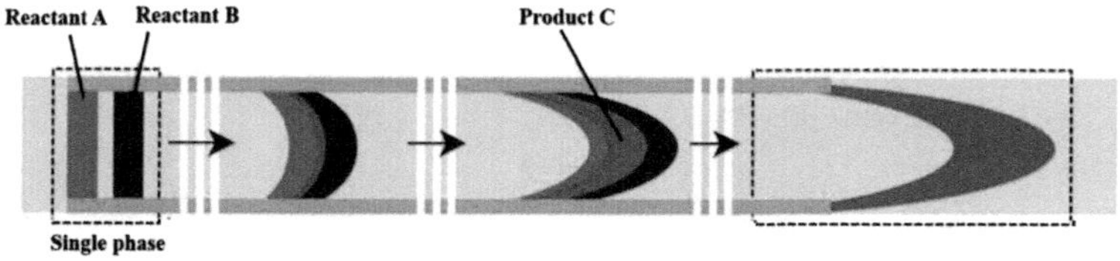

Figure 10.4 Schematic representation of continuous flow in microfluidics: reactants A and B combine due to diffusion in a single continuous flow to form product C. Republished with permission of The Royal Society of Chemistry, from Ref. [89], copyright (2010); permission conveyed through Copyright Clearance Center, Inc.

10.3.1 Continuous-Flow Microfluidics

Continuous-flow microfluidics uses diffusion mixing due to shear forces and surface tension at a low Reynolds number [92]. A schematic representation of continuous-flow microfluidics is shown in Fig. 10.4. Reactants A and B are injected through the inlets at different concentrations and with different flow rates. They diffuse in a single phase and form product C after covering long distances [31, 89, 93]. A laminar-based microfluidic device is easy to design, manipulate, and scale up. It is easier to manipulate and control the characteristics of NPs with continuous flow–based microfluidic devices [2]. The key parameter to achieve the desired shape and size in continuous-flow microfluidics is effective diffusive mixing [31]. These devices are suitable for large-scale integration, perfusion, and centrifuge platforms. The problems associated with a laminar flow are cross contamination, Taylor dispersion, surface interaction, lengthy microchannels, and the requirement of a large volume of reagents. To overcome the issues associated with laminar flow, segmented flow–based microfluidics evolved.

10.3.2 Segmented-Flow Microfluidics

Segmented-flow microfluidic devices have two phases, gas and liquid. The gas evolved in the reaction or inlet air forms the gas phase, whereas the liquid phase is due to immiscible liquids such as oil. Segmented-flow microfluidics is depicted in Fig. 10.5 [89]. The different reactant segments are separated by the gas phase by

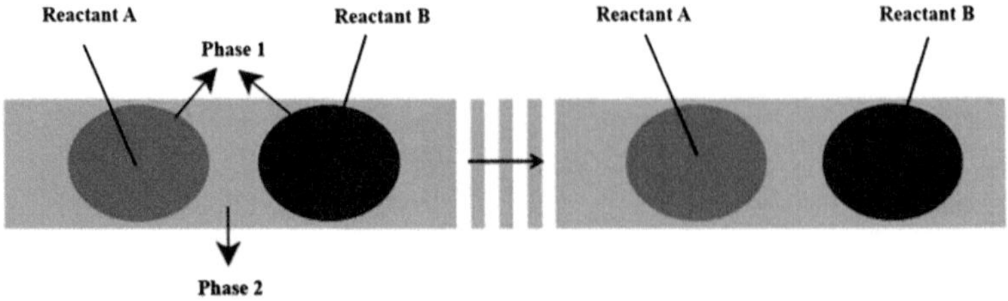

Figure 10.5 Schematic representation of segmented flow in microfluidics: reactants A and B in the form of droplets form one phase, and another phase is the fluid in which drops are formed. Republished with permission of The Royal Society of Chemistry, from Ref. [89], copyright (2010); permission conveyed through Copyright Clearance Center, Inc.

forming gas "bubbles," evolved as a result of the difference between surface tensions of the liquid and gas phases.

The formation of bubbles helps to restrain the reaction in the liquid segment. The main problem associated with a segmented-flow reactor is the physical contact between reagents and the channel wall, which leads to contamination issues. Generally, this is avoided by considering the hydrophobicity of the channel walls. Droplet-based microreactors have compartmentalization, which allows rapid mixing, a contamination-free environment, and good control of the residence time and reagents. The capillary number must be considered during droplet formation, and above a particular value of the capillary, the droplets break off. Electrowetting on dielectric methods and dielectrophoresis-driven methods are important electrohydrodynamic methods for droplet formation. The effective formation of droplet depends on fluid viscosities, relative flow rates, interfacial tension, and inlet geometry. The difficulty to achieve stable droplets is the main drawback of segmented flow, and the flow rate cannot be increased too much compared with the laminar flow [12, 20, 31, 90, 92, 93].

10.3.3 Mixing Strategies in Microfluidics

Various mixing strategies in continuous-laminar-flow mixing and droplet-based segmented mixing are shown in Figs. 10.6a and 10.6b, respectively. The winding microchannels and distributed mixing

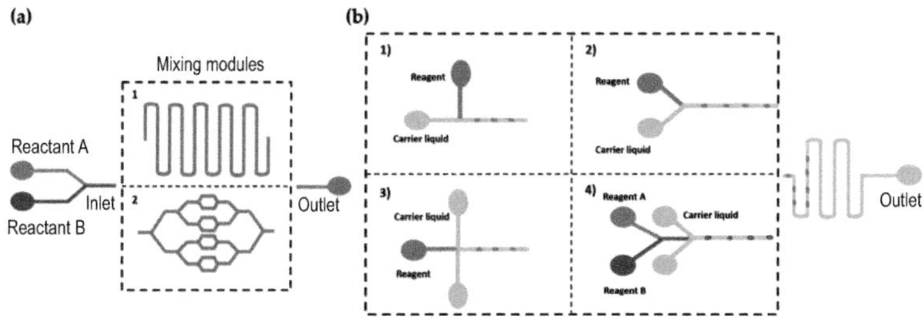

Figure 10.6 (a) Different mixing modules in continuous-laminar-flow microreactors: (1) winding microchannels and (2) distributed mixing microreactors. (b) Various mixing strategies for the formation of droplets in microreactors: (1) T-junction, (2) Y-junction, (3) cross channel, and (4) combination [31].

form continuous-flow micromixing. The mixing geometries to form droplets are T-junction, Y-junction, cross channel, and a combination of any of these. The design of T- and Y-junction geometries is relatively simple, and the size of the synthesized particle depends on the channel width as well as the flow rate. In cross-channel flow focusing, the optimization of synthesis depends on the geometry as well as flow conditions [31]. Another efficient micromixing strategy is slug-based segmented flow, which was first introduced by Günther et al. in 2005 [29]. For this purpose, he first added two liquid phases and removed the inert gas afterward in a planar capillary separator.

Continuous microfluidics with winding microchannels for the synthesis of spherical Au NPs is shown in Fig. 10.7. The reactor has two residence zones, three inlet ports, and one outlet port. All the reagents were injected through polypropylene syringe pumps, and inside the reactor channel, reagents mix continuously through diffusion. The reactor has a volume capacity of 2.3 μL, and components in the reactor were connected through a flexible polytetrafluoroethylene (PTFE) tube of inner diameter 0.3 mm [107]. Distributed micromixing was utilized by Wagner and Köhler for the synthesis of Au NPs of diameter 5–50 nm [106]. The microreactor has eight splits, and they were connected as shown in Fig. 10.8. The splitting arrangement promotes effective mixing through interdiffusion of the fluid as it flows along the horizontal and vertical directions and changes direction at the branches.

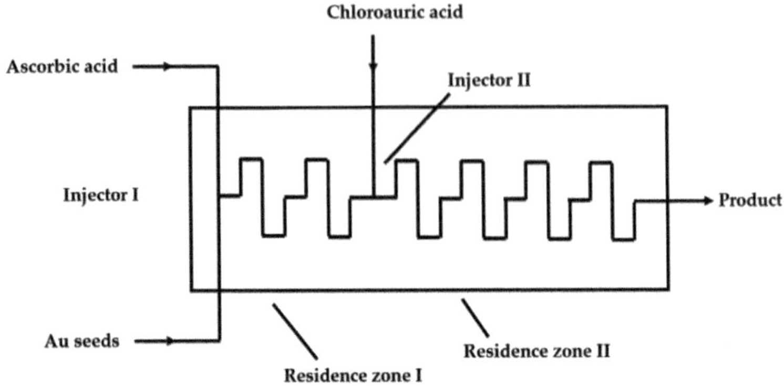

Figure 10.7 Schematic image of winding micromixing for the synthesis of spherical Au NPs in microfluidics. It has two reaction zones and three inlet ports. The reagents were allowed to pass through inlet ports at specific flow rates, and the product was taken from the outlet [107].

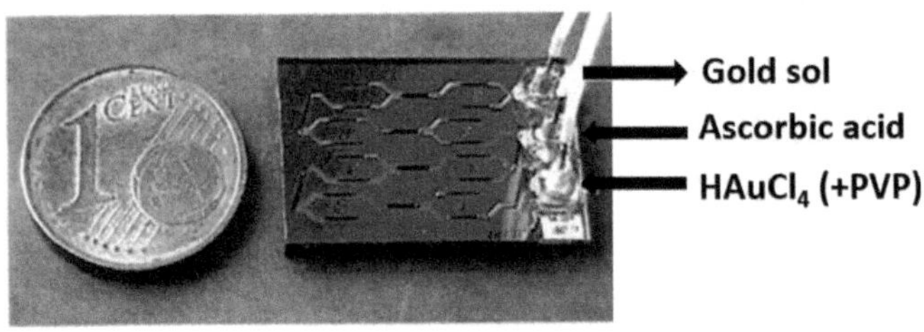

Figure 10.8 Photographic image of the distributed micromixing-based chip. The chip is 22 mm in length and 14 mm in width; the microchannel depth is 160 μm, and the width ranges from 178 μm to 700 μm. Reprinted with permission from Ref. [106]. Copyright (2005) American Chemical Society.

Two issues reported with this approach were reactor fouling and unsuitability for continuous synthesis. Hence, the microchannel surface was modified to work at an augmented pH or the surface was made hydrophobic via silanization.

Duraiswamy and Khan [20] compared T-junction and flow-focusing geometries. They found that the performance of the T-junction geometry is better than that of the flow-focusing geometry for the synthesis of Au nanocrystals since rapid and severe clogging was reported at the small orifices of the flow-focusing geometry. The

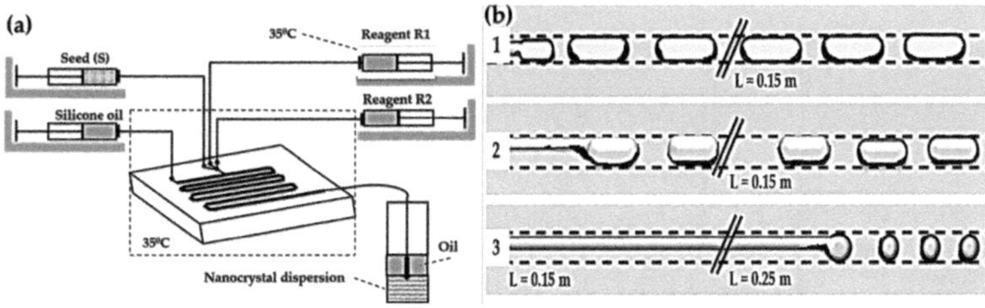

Figure 10.9 Synthesis of gold nanocrystals using the droplet method (a) Schematic representation of the experimental setup. The microchannel was fabricated in the PDMS material on a silicon substrate through the photolithography process on an SU-8 2050 negative photoresist. The seed solution (S) and growth reagents (R1 and R2) were injected continuously, and at the T-junction, they mixed with silicone oil to form the crystals. (b) Breakup location of the droplet in different flow ratios. (1) At a flow ratio of 1.2, the breakup location is close to the junction; (2) at a flow ratio of 0.88 breakup happens after the droplet covers a small distance, and (3) at a flow ratio of 4.16, breakup happens after the droplet travels a long distance [20].

experimental setup for T-junction geometry–based Au NP synthesis is shown in Fig. 10.9a. Silicone oil was used as the immiscible fluid since it has the ability to wet the polydimethylsiloxane (PDMS) channel and because of its immiscible nature with aqueous solutions. At the T-junction, it mixes with reagents and seed through chaotic advection. Droplets are continuously generated, and the oil acts as a lubricating layer, which helps to prevent contact with the channel wall. The breakup of droplets in the channel depends on flow ratios (oil and aqueous). As the ratio increases, the breakup of the aqueous thread location also increases. Here, two modes are important: dripping mode and jetting mode. In the dripping mode (flow ratio below 2.5), the breakup occurs near the downstream of the T-junction, whereas in the jetting mode (flow ratio above 2.5), it occurs far from the downstream of the T-junction. A combination of the T-junction and the spiral reaction tube was used for the synthesis of functionalized Ag NPs. When fluid flows through the spiral reaction tube, the fluid particles possess a higher drag force and the flow direction changes due to a lifting force [53]. Figure 10.10 shows a schematic of the T-junction geometry, along with the spiral micromixer arrangement used to synthesize Ag NPs.

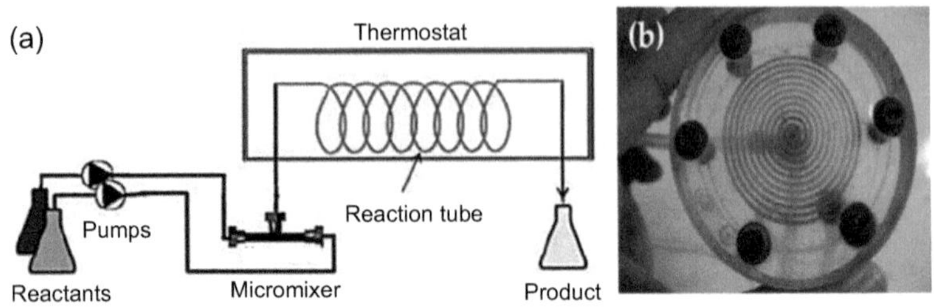

Figure 10.10 Ag NP synthesis in a T-junction micromixer connected to a spiral microreactor. (a) Schematic diagram of a microreactor. Reactants are continuously injected through the pumps into the two arms of the T-junction, and at the junction, they mix together and enter into the spiral reaction tube. The temperature is maintained at 90°C inside the reaction tube. (b) Photographic image of a spiral millireactor. Republished with permission of The Royal Society of Chemistry, from Ref. [53], copyright (2010); permission conveyed through Copyright Clearance Center, Inc.

From the T-mixer, the fluid enters the reaction tube, the entire setup is immersed in a thermostat, and the temperature is maintained at 90°C [53].

A Y-shaped microchannel geometry is used to generate pulsed mixing of reagents. Pulsed mixing is an adequate method for mixing reagents, and it helps control the mixing time. Pumping the reagents one by one through the Y-channel can create a pulsed layer, and the thickness of the layer can be varied as the reagent moves toward the expandable microchannel. The experimental setup of pulsed mixing to synthesize Au NPs is shown in Fig. 10.11a. It consists of a Y-shaped microchannel, two piezoelectric micropumps, and an expandable microchannel. The valveless micropump is connected to an inlet channel, and it consists of a chamber with a diaphragm and a pair of outlet channels. The back of the diaphragm is equipped with piezoelectric zirconate titanate (PZT). Prior to the experiments, flow rates in the two pumps should be calibrated. These micropumps can precisely control the flow rate as well as the switching frequency, which can essentially have an impact on the layer thickness in the expanded channel. It has been observed that if the layer thickness decreases in the expanded channel, it can help in the rapid mixing of two solutions since the diffusion time for mixing is directly proportional to the square of the layer thickness. By varying the

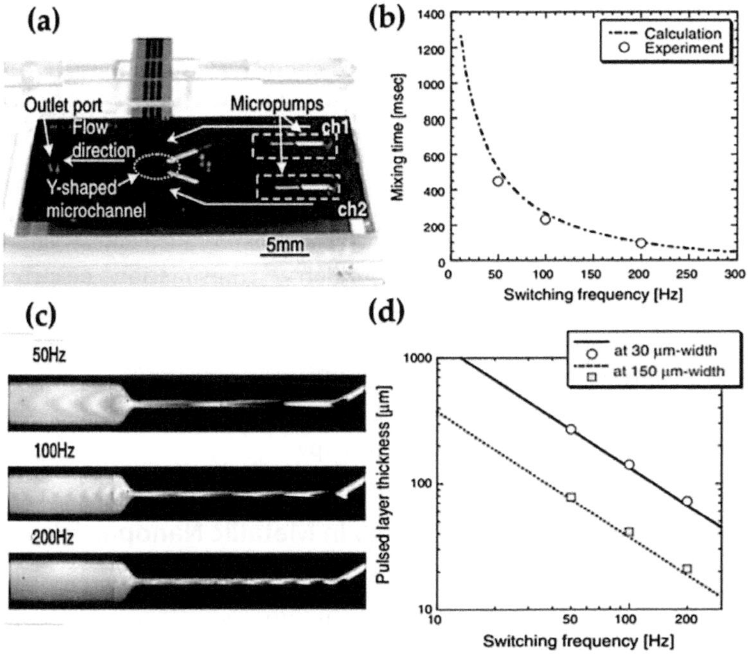

Figure 10.11 Synthesis of Au NPs using a pulsed mixing microfluidic system. (a) Photographic image of the Y-channel microchip. The chip consists of a Y-shaped microchannel connected to an expanded channel and two micropumps. Reagents are supplied through the micropumps one by one to produce a pulsed layer of each reagent, and once it reaches the expanded layer, it diffuses at the common interface. (b) A comparison of the mixing time at different switching frequencies. (c) Fluorescence image of a microchannel at various pulse frequencies. As the frequency increases, the width of the pulse decreases. (d) Variation in the thickness of the pulsed layer at various switching frequencies. Reprinted by permission from Springer Nature Customer Service Centre GmbH: Springer Nature, *Microfluids and Nanofluids*, Ref. [95], copyright (2010).

switching frequency of micropumps, the mixing speed of gold salt and reagents can be controlled and hence the size of the particles can be controlled. An inverse relation exists between switching frequency and mixing time, as given in Fig. 10.11b. For higher frequencies, the pumped volume in each channel is not sufficient to fill the channel to form pulses. It has been found that as the switching frequency increases, the coefficient of variation of the particle diameter decreases. Alternately, it can be said that a shorter mixing time leads to a smaller coefficient of variation and it has an effect on the number of nuclei generated. Figures 10.11c and 10.11d show the fluorescence image of a microchannel at various pulsed frequencies and the variation in the thickness of the pulsed layer at different frequencies, respectively [95].

Synthesized NPs can be characterized using ultraviolet-visible (UV-vis) absorption spectrophotometry, transmission electron microscopy (TEM), high-resolution transmission electron microscopy, high-resolution scanning transmission electron microscopy (HR-STEM), selected-area electron diffraction, atomic force microscopy, energy-dispersive X-ray spectroscopy (EDS), fiber spectrometry, and scanning electron microscopy.

10.3.4 Controlling Parameters in Metallic Nanoparticle Synthesis

The intrinsic characteristics of metallic NPs vary with shape, size, composition, and crystal structure, and they can be achieved by controlling the parameters of the microfluidic device or the reaction entities. Shape and size have significant roles in controlling the physical and chemical properties of metallic NPs. As the size in the nanoscale decreases, the surface becomes more prominent and starts to show interesting properties, as described in the introduction. By tuning the reaction entities at each stage (mixing, reaction, nucleation, growth, aggregation, and termination stages) of particle formation, the shape and size of the particle can be precisely controlled [51, 92, 97]. The species in the early stages of a redox reaction in metallic NP formation are highly reactive and have short lifetimes. Their redox reaction depends on the cluster size. Due to this size-dependent nature of a redox reaction, metal salt reduction greatly depends on the geometry of the microchannel as well as the reaction conditions.

Generally, the nucleation and growth steps of metallic NP synthesis are difficult to control in order to achieve monodispersed particles [41]. Successful dissolution between the nucleation and growth phases is required to control the shape, size, and composition of the material. Since the nucleation and growth phases happen almost simultaneously, reducing the concentration of reagents to a level below the critical supersaturation level can limit the nucleation. Otherwise, adopting different temperature kinetics for nucleation and growth phases can favor the growth process by limiting the nucleation. By using weaker reducing agents, this issue can be addressed [104]. Another solution is the seed-mediated growth approach mentioned in the introduction. In this method, NPs prepared in advance serve as the seed for the synthesis of another set of particles, which can be added along with reagents. The ratio of concentration of the seed to that of the metal salt can be varied to get the desired size [41]. Both seeded and unseeded methods are used to synthesize metallic NPs. The unseeded method produces low-aspect-ratio NPs since the approach is very sensitive to the synthesis parameters [104]. The confined growth of metallic NPs is possible by the addition of surfactants and polymers. LaMer nucleation followed by the Ostwald ripening process helps to reshape polydispersed NPs into monodispersed ones [66]. These can form a lamellar structure and get adsorbed on the surface of the crystal plane of the metal. Then in between the layers, metal starts growing into a 2D structure [102, 104].

Parameters such as flow rate, concentration ratio of reactants, residence time, seed quantity, seed aging time, temperature, heating time, and microchannel wall thickness can change the size of metallic NPs. It has been found experimentally that the flow rate of reagents has an effect on the reaction kinetics and synthesized NPs [24, 85, 106–108]. As the flow rate increases, the particles' mean diameter decreases [106, 107]. A higher flow rate favors the nucleation phase in shorter time compared to the growth time, which supports the narrow size distribution of particles. A higher flow rate means a higher gradient between the lamellae of the laminar flow stream and a shorter interdiffusion period among them. It favors more effective and rapid mixing, which in turn helps quick availability of the reducing agent to gold ion lamellae. Hence, the process of nucleation is favored and growth is suppressed. The influence of the flow rate on the size of the spherical Au NPs is shown in Fig. 10.12a [106].

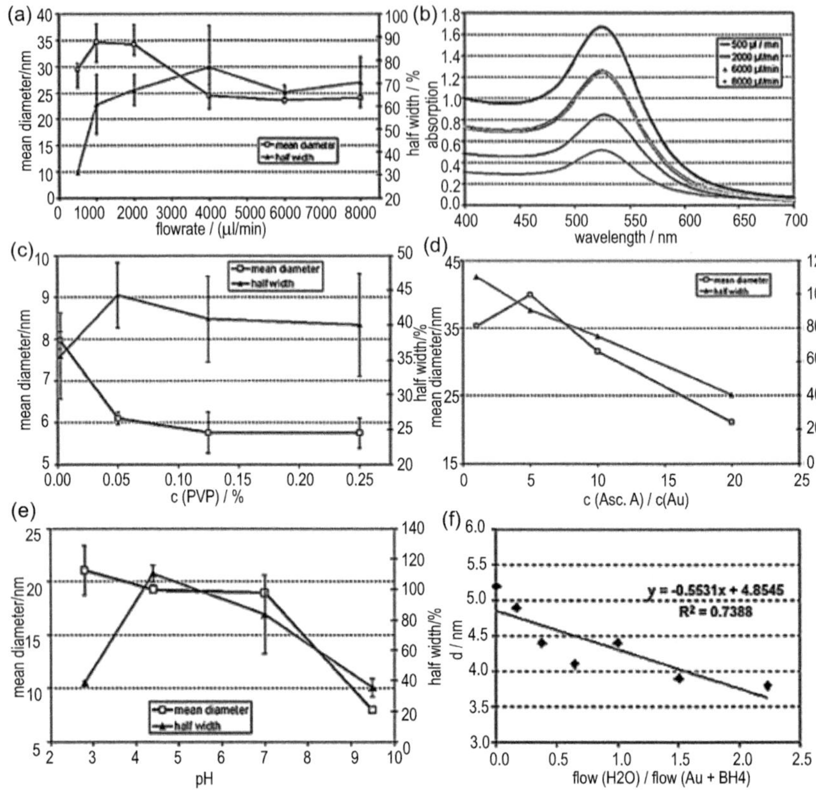

Figure 10.12 Effect of parameters on the size of Au NPs. (a) Influence of flow rate on mean diameter and (b) absorption band, and (c) dependence of mean diameter on addition of PVP; ratio of concentration of AA to (d) Au and (e) pH [108]. (f) Correlation of diameter on the ratio of the flow rate of reagents. Reprinted with permission from Ref. [106]. Copyright (2005) American Chemical Society.

Plasmonic properties of metallic NPs depend indirectly on the absorption spectra. Consider the case of a spherical Au NP, which has a single absorption peak at 520 nm. The intensity of the plasmon absorption band increases with an increased flow rate in an untreated microreactor (i.e., the presence of reactor fouling), as shown in Fig. 10.12b. In other words, the adsorption of NPs on the inner surface of the channel can be treated as a function that varies with flow rate. At a low flow rate, Au ions reduce to elemental

Au on the surface of the reactor and the surface itself acts as a nucleation starter. This stage is identified by the formation of violet color inside the reactor. A high flow rate enhances higher surface shear energy, which reduces the surface interaction and leads to less reactor fouling [85, 106].

The addition of polymer polyvinylpyrrolidone (PVP) helps to reduce the diameter of particles as illustrated in Fig. 10.12c. The actual mechanism behind this is not clear, but one explanation is as follows: The addition of PVP causes Au particles to get covered by the bulky PVP chains. This results in changes in particle properties, such as a reduction in the particle density as well as a change in the flow resistance of the particle moving through the liquid during sedimentation. These changes lead to reduced sedimentation velocity, and hence the average diameter of the particles with PVP is reduced [107]. Further, the ratio of concentration of reducing agent to that of gold ions is inversely proportional to the size of the particle, as depicted in Fig. 10.12d. This might be due to the availability of a higher number of nuclei formed through reduction with an available reducing agent. The growth period is suppressed as more nuclei are available to grow, and once the Au is used, the reaction is quenched. If a weak capping agent is present, such as ascorbic acid (AA), it does not inhibit the growth, unlike a strong capping agent, like thiol. Hence, a narrow size distribution of particles is obtained [106, 107]. Liu and coworkers observed that if glucose is used as a reducing agent, then the particle size is not affected much. But it can affect the yield, monodispersity, and morphology of Ag NPs [60].

The correlation between the size of the NP and pH in AA as a reducing agent is shown in Fig. 10.12e. It is obvious that the mean diameter decreases as the pH increases, which might be due to the dependence of the redox potential of AA on pH. Higher pH means a higher negative redox potential and hence more reducing power; this leads to a greater rate of production of metal nuclei and hence less growth time. Accordingly, the mean diameter reduces with a narrow distribution [106]. The size of the particle also depends on the addition of a reductant in thiol-capped particles. Rapid addition leads to small-sized particles in solution since they are exposed to an abundance of thiol. Further, thiol-capped particles show less monodispersity since the concentration of thiol reduces as time progresses, allowing the particles to grow in size [85].

The concentration of a metal precursor solution has an influence on the synthesis of NPs. Usually, the precursor solution is formed by the precursor in PVP. For example, by varying the concentration of PVP, high-yield Ag NPs can be synthesized. An increase in PVP concentration leads to an increase in the particle size and after a particular concentration, the change is negligible, as shown in Fig. 10.13. PVP plays the role of surface protectant in Ag NP synthesis. In the absence of PVP, the formed Ag nuclei cannot be encapsulated, leading to the formation of nonspherical NPs. Also, the incomplete encapsulation leads to a small electrostatic force of repulsion between nanoparticles and results in partial agglomeration. As the PVP becomes sufficient for enough electrostatic force and encapsulation, proper dispersibility and spherical shape can be achieved. The absorbance spectra given in Fig. 10.13f indicate that samples 9 and 10 have a blue shift. Hence, if the PVP concentration is below the precursor concentration, it shows a blue shift. The maximum absorbance of other samples doesn't show any notable variations [60].

The reducing agent also has a significant role in controlling the size of NPs. If the synthesis is based on borohydride reduction, the flow rate does not have an impact on the Au NPs' size and distribution. It indicates that rate-independent nucleation is possible with borohydride reduction in Au NPs [108]. In contrast to this result, Fu et al. showed that a higher flow rate results in ultrathin rolled Au nanoplates, while rigid flat-surfaced nanoplates were obtained with low flow rates. At a low flow rate, the growth time of the NPs is longer and they grow thicker, whereas a high flow rate provides shorter growth time. It results in very thin NPs that tend to roll up so as to reduce the surface energy [24]. The contradicting results might be due to the different mixing procedures in both cases. Also, the diameter tends to decrease as the ratio of flow rates of water to $Au^{+}BH_4$ increases. It is due to the fast growth of metal atoms at a higher concentration under the nucleation threshold. Addition of thiolated ligands leads to the aggregation of particles and hence a reduction in their size [108]. The added ligands change the plasmonic absorption of Ag NPs as follows: Spherical Ag NPs have

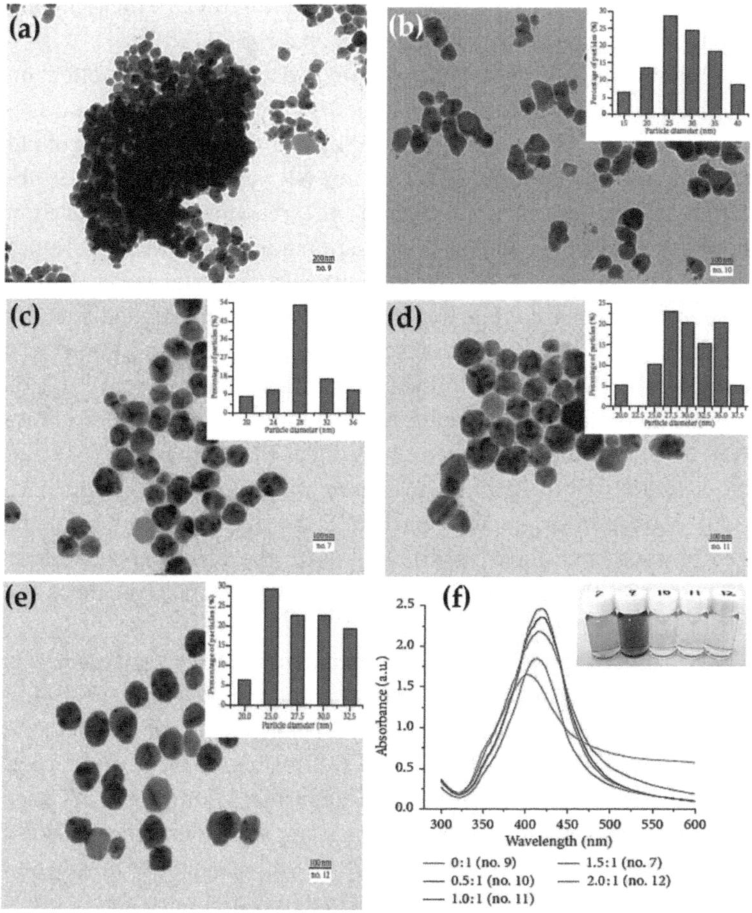

Figure 10.13 Characterization of Ag NPs with PVP in different concentrations. TEM images at the following PVP concentrations: (a) no PVP (no. 9), (b) 0.5 times Ag (no. 10), (c) 1.5 times Ag (no. 7), (d) equal to Ag (no. 11), and (e) twice Ag (no. 12). (f) Absorbance spectra of corresponding particles. Reproduced from Ref. [60], under a Creative Commons Attribution 4.0 International License (https://creativecommons.org/licenses/by/4.0/).

an absorption peak at 398 nm, and their shape is similar to that of Au NPs. Addition of the ligand mercaptoundecanoic acid shifts the absorption peak (422 nm) toward the right along with higher absorption, and the band becomes broader. Mercaptoundecanol ligands also shift the absorption peak (454 nm) toward the right and broaden it further, with a reduction in absorption [108].

A stabilizing agent such as sodium hydroxide has little effect on the size and distribution of metallic NPs [60]. Bifunctional surfactants stearic acid sophorolipid (SASL) and oleic acid sophorolipid (OASL) have a different effect on Ag NP synthesis. Both require different amounts of time to finish the reaction. A slow reaction happens with OASL, while a fast reaction happens with SASL. Hence as time progresses, OASL-based synthesis yields large-sized polydispersed particles, while small-sized particle and monodispersity are observed in SASL-based synthesize [53]. The addition of a complexant (NH_4OH and ethylenediaminetetraacetic acid [EDTA]) has a role on longitudinal surface plasmon resonance (LSPR) of Ag NPs. The addition of a complexant reduces the absorbance peak and changes the band. Particles prepared without a complexant have two absorption peaks (393 nm and 554 nm). The addition of the complexant ammonium hydroxide shifts the absorbance peak to 421 nm and the band becomes narrower. EDTA also redshifts the absorbance peak (407 nm) [115].

Two more tuning parameters for controlling the size of the NPs in seed-mediated growth are the seed size and the seed age. As the seed size increases, the synthesized NP yield decreases. It has been found that seed age has an influence on the formation of Au nanorods. Immediately after the synthesis of nanorods, diluting the Au seed solution can bring down the rate of aging and hence good yield of Au nanorods can be achieved. Dilution will essentially make their structure come closer to the natural state [110]. It also has an effect on the intensity of the absorption band and the size of the particle synthesized. Figure 10.14 depicts the influence of seed aging time on the particle size, its distribution, the corresponding absorption spectrum, and the extinction spectrum. It illustrates that longer aging time of seeds leads to the formation of larger seed particles and a corresponding absorption peak is observed after 520 nm. If the aging time is below 125 min, there is no absorption observed for the seeds. It might be owing to the impact of the quantum size on the conduction band due to low electron density. At a higher aging

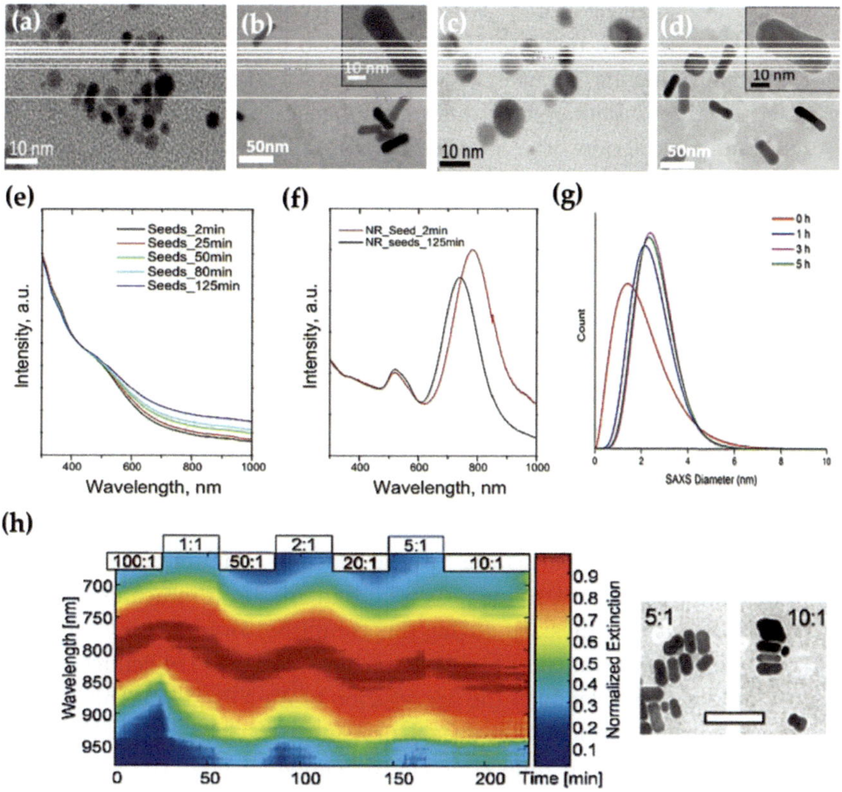

Figure 10.14 Correlation between synthesized Au nanorods and seed age. TEM images obtained after 2 min for (a) seeds and (b) Au nanorods; TEM images obtained after 125 min for (c) seeds and (d) Au nanorods; UV-vis extinction spectra of (e) Au seeds and (f) Au nanorods for seeds of different ages. Reprinted from Ref. [104], copyright (2016), with permission from Elsevier. (g) Size distribution of Au nanorods for seeds of different ages, taken from SAXS analysis. Reprinted with permission from Ref. [110]. Copyright (2015) American Chemical Society (https://pubs.acs.org/doi/10.1021/acs.chemmater.5b02675). Further permissions related to the material excerpted should be directed to the ACS. (h) Measured color-coded extinction with respect to time with various growth-to-seed-solution ratios and corresponding TEM images. Republished with permission of The Royal Society of Chemistry, from Ref. [11], copyright (2006); permission conveyed through Copyright Clearance Center, Inc.

time, large-sized nanorods with a lower aspect ratio are prepared [104]. Apart from these findings, it can also be observed that two resonance maxima occur in the absorption spectra of Au nanorods. The first peak implies the transverse plasmon resonance and the second peak denotes the longitudinal plasmon resonance (LPR) of anisotropic particles. The information from the spectrum, such as the breadth of the peak and a comparison of the peaks, indicates the polydispersity nature and aspect ratio of the synthesized particles. As the aspect ratio increases, the LPR peak is observed to show a redshift [20, 104]. The extinction spectrum analysis reveals that the ratio of growth to seed can vary with the aspect ratio of NPs. As the ratio decreases, the aspect ratio increases. Fewer ratios mean fewer growth solutions available for seeds, which enables slow growth of nanorods [11].

The heating time of a microfluidic reaction has an impact on the UV-vis absorption peak as well as the size of the synthesized NP. It is observed that as the heating reaction time increases, the absorption peak corresponds to 537 nm increases. It indicates that more NPs are synthesized during this condition. More heating time favors the rapid transfer of reagents, which results in the reduction of ionic gold and leads to the aggregation of Au NPs [112]. Thicker Au nanorods can be synthesized by raising the temperature while keeping the heating time constant, as shown in Fig. 10.15. This might be due to the changes in the reaction kinetics at higher temperatures, and the surface mobility increases and favors higher

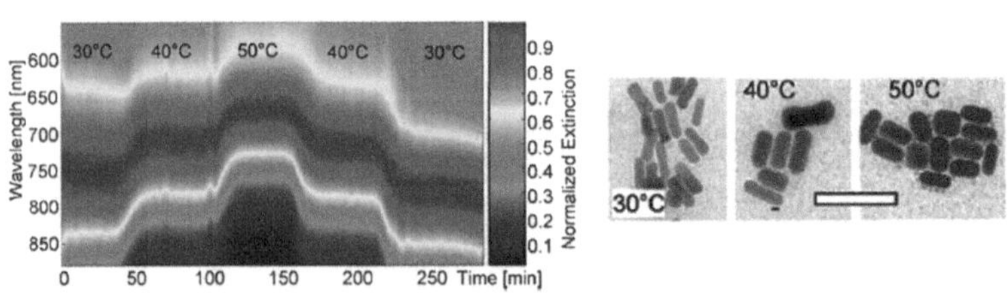

Figure 10.15 Measured extinction color-coded as a function of time with various temperatures in Au nanorod synthesis and corresponding TEM image. Republished with permission of The Royal Society of Chemistry, from Ref. [11], copyright (2006); permission conveyed through Copyright Clearance Center, Inc.

permeability of cetyltrimethylammonium bromide (CTAB) into Au atoms. The maximum extinction also shifts toward the left of the band, as shown in Fig. 10.15.

The size distribution of nanocrystals in segmented-flow micromixing depends on axial dispersion and mixing conditions triggered by internal recirculation. The residence time in segmented flow has an effect on the slug length of the continuous phase. Shorter residence time produces small slugs, as shown in Fig. 10.16a. Along with the slip velocity between two fluids, the slug length also plays a

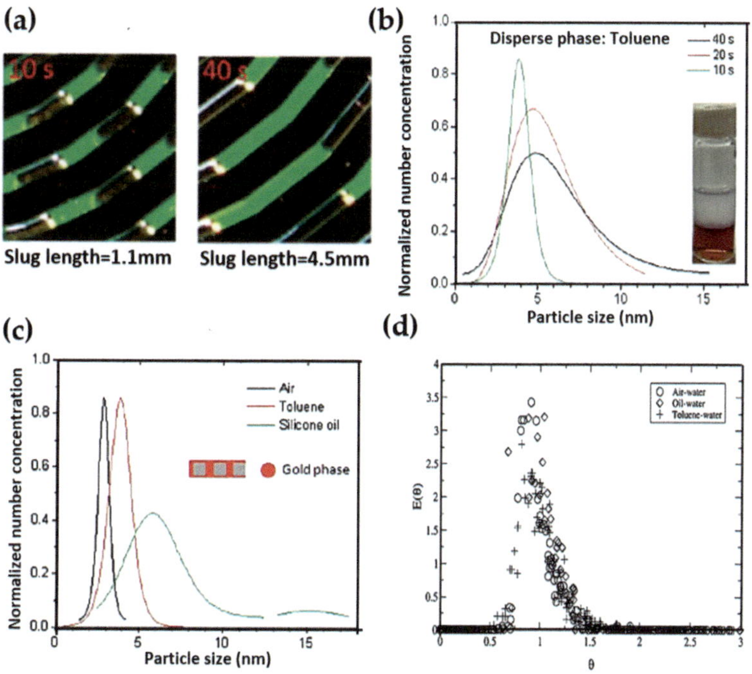

Figure 10.16 Characterization of gold nanocrystals synthesized using segmented-flow mixing. (a) Segmented slugs formed with residence times of 10 s and 40 s with the toluene-aqueous phase; (b) particle size distribution at various residence times with the toluene-aqueous phase; (c) particle size distribution with different segmenting fluids; (d) distribution of residence time in various dispersed phases; θ is the time divided by the mean residence time, and $E(\theta)$ is the measured axial dispersion. Reprinted with permission from Ref. [12]. Copyright (2012) American Chemical Society.

crucial role in the slug mass transfer. Hence, good control over mass transfer is possible, thereby decreasing the particle polydispersity [12]. Moreover, it is found that particle size distribution broadens as the residence time increases, as shown in Fig. 10.16b, which might be due to the availability of internal mixing with shorter residence time, favoring narrow particle distribution. Figure 10.16c presents the consequences of various dispersed phases on the Au nanocrystal particle size distribution. Silicone oil as the dispersed phase gives broadened bimodal particle distribution, toluene gives monodispersed, and air gives narrow monodispersed distribution. In the case of silicone oil, due to an increase in the viscosity, the slip velocity is reduced and the magnitude of recirculation decreases. Hence the particle size distribution is broader. Mobility may increase in the phase interface of air as the dispersed phase, which leads to a greater circulation rate and hence smaller-sized and narrowed particle forms. Among all the dispersed phases, toluene has the lowest interfacial tension and the lowest viscosity. Hence it has moderate internal circulation and moderate particle distribution. Experimental results show that mixing within the continuous phase is easier with air compared to silicone oil and toluene. Further, air is more convenient to handle. Figure 10.16d shows that there are no significant differences among axial dispersion in various dispersed phases. In other words, it can be said that the efficiency of NP formation in segmented-flow mixing solely depends on internal mixing and that the contribution of axial dispersion is less [12].

Figure 10.17 shows the comparison of the effect of the carrier phase on synthesized Ag NPs. For synthesizing Ag nanocubes and octahedral, air is found to be better compared to silicone oil. Silicone oil gives irregularly shaped Ag nanocrystals at a flow rate of 100 μL/min since O_2 cannot diffuse into the droplets and hence the Ag nanocrystals are blocked in the tubes. When air is used as the carrier phase, O_2 can diffuse easily and as the flow rate increases, more O_2 is available for the droplets. Hence nitric oxide (NO) can be accommodated easily [118].

The way they vary with size, optical absorbance spectra can vary according to the shape of NPs. Unlike spherical Au NPs, spheroids possess a 600 nm peak apart from a 520 nm one. These particles are bigger and have multiple crystalline faces. The analysis of UV-

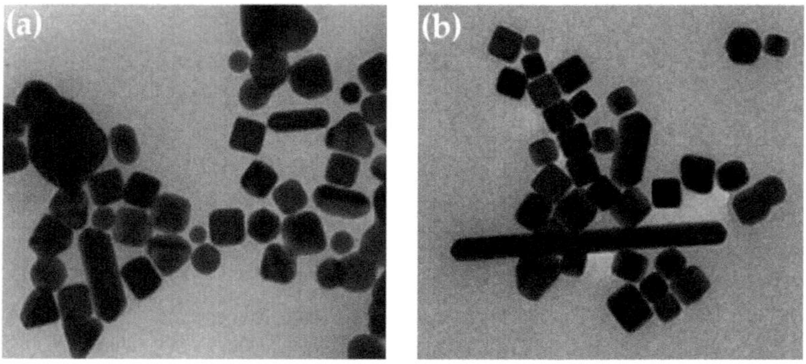

Figure 10.17 Comparison of the carrier phase on the shape of a metallic nanoparticle. TEM images of synthesized Ag nanocubes with (a) air and (b) silicone oil as a carrier phase. Reprinted with permission from Ref. [118]. Copyright (2013) American Chemical Society.

vis absorption spectra of sharp-edged nanocrystals for various reaction conditions reveals that high concentrations of Ag and AA give a sharp-tipped dog bone structure (1), while relatively low concentrations give tetrapods, cubes, and four-edged stars (2). Both structures possess two peaks, and a higher concentration of AA as a reducing agent can give broad spectra. The intensity is also higher with a higher concentration of AA [20].

Higher nanostructures can be synthesized by fusing the smaller structures or by varying the concentrations of reactants. Ag nano-octahedra structures can be synthesized from Ag nanocubes of {100} facets by varying the concentration of PVP in an ethylene glycol (EG) solution. Ag octahedra covered by {111} facets can be prepared below a critical value for the concentration of PVP. Figure 10.18 shows the characterization and formation of Ag octahedra from 40 nm Ag nanocubes. If the reaction time increases by more than 20 min, self-termination of {111} facets further suppresses the growth process. The UV-vis absorption spectra analysis shows that a major LSPR peak is red-shifted from 438 nm to 466 nm, while the other two peaks remain at the same wavelength with more intensity [118].

Generally, low-dimensional nanostructures, such as nanodisks and nanospheres, possess face-centered cubic (fcc) and hexagonal close-packed (hcp) lattices. It is difficult to control the parameters

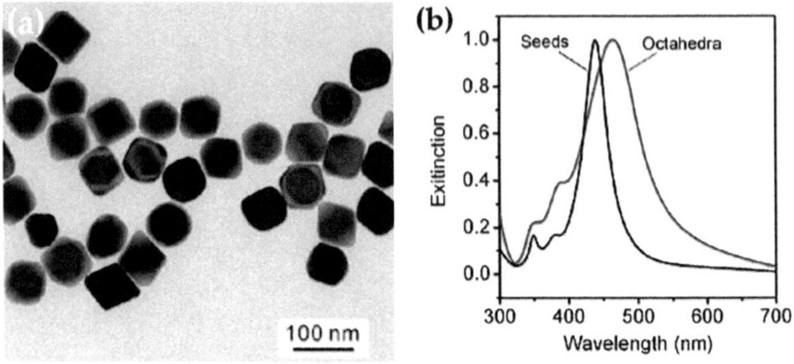

Figure 10.18 Characterization of Ag octahedra synthesized using droplet-based microfluidics; (a) TEM image of the octahedra formed after 20 min of reaction time from 40 nm Ag seeds and (b) the corresponding UV-vis absorption spectra. Reprinted with permission from Ref. [118]. Copyright (2013) American Chemical Society.

to obtain the desired face in a conventional batch process because the energy difference between two phases is very small. But in microfluidic devices, it is possible to tune the kinetic parameters, such as reaction time, growth period, flow rates, reaction temperature, and quenching procedures, precisely and sensitively [88].

Cobalt (Co) is a transition metal that possesses magnetic properties, and its NPs have three different phases: fcc, hcp, and epsilon (ε). Magnetization depends on the temperature of the material and varies according to the blocking temperature (T_b). T_b corresponds to the temperature below which particles are magnetized spontaneously. For fcc and hcp structures, magnetization decreases as the temperature increases beyond T_b, and for ε the value of T_b is higher than 330 K. fcc possesses soft magnetic properties, whereas hcp has permanent magnetic properties with high magnetic coercivity. The net anisotropy in each phase has less variation, and the kinetic energy difference is very small. Thus, phase-controlled synthesis of Co NPs is very difficult to achieve. The formation of different phases of Co NPs is depicted in Fig. 10.19. It is observed that without the quenching step and along with a long Ostwald ripening period (OR) (around 8 h), the synthesis of ε NPs is possible, while the formation of fcc and hcp demands the quenching

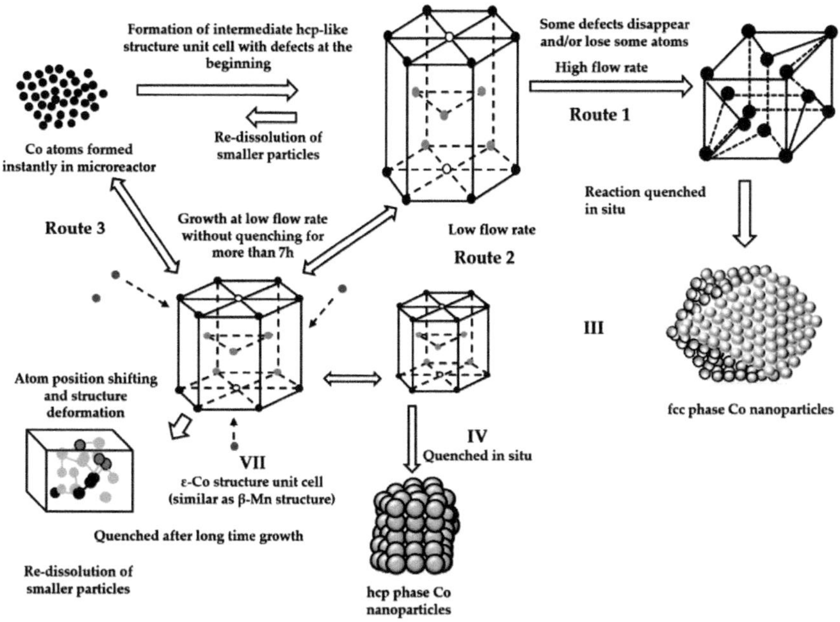

Figure 10.19 Formation of three phases of Co NPs with different flow rates in a microfluidic device [92].

step. A high flow rate (high kinetic energy) of 0.9 mL/min is required for fcc, whereas hcp requires a low flow rate of 0.08 mL/min. It can also be seen that the crystal structure of route 2 is similar to that of route 3, implying that the hcp structure formed with a low flow rate can eventually shift to a metastable ε structure. All these analyses show that the three phases possess a nearly isoenergetic crystal structure among themselves [84, 92].

Similar to Co, four phases (from fcc to hcp) of nickel nanoparticles (Ni NPs) can also be synthesized in a microfluidic device by tuning the flow velocity. Nickel is a ferromagnetic transition metal, and its crystalline structure shows different magnetic properties. Several wet chemistry approaches are there to synthesize fcc- and hcp-based Ni NPs. Many studies have been reported about fcc Ni nanocrystals, whereas fewer studies synthesized and examined hcp nanocrystals [67]. Magnetic properties of nickel nanocubes are considerably improved compared to bulk [88]. Chaudret and coworkers showed that an amine-capped structure possesses magnetization properties [17]. They developed spherical Ni NPs and nanorods with an fcc structure with magnetic properties. Contradictory results were reported regarding the magnetic property of hcp Ni nanostructures. Apart from fcc and hcp, a hexagonal structure also shows magnetic properties, but results are contradictory [70]. The studies revealed that magnetic properties depend not only on the crystalline structure but also on the average size and polydispersity. Net ferromagnetic behavior was observed in fcc nanocrystals, whereas hcp showed weak magnetic features. Further, fcc has large magnetic saturation, which can be slightly enhanced at a lower temperature.

X-ray powder diffraction pattern and TEM image analysis of different crystal phases of Ni NPs indicate the effect of temperature information of different phases. The structures separated from higher temperatures turned into hcp and possessed a low-intensity profile. Mild temperature ($>290°C$) treatment results in the formation of fcc. Some structures have both fcc and hcp characteristics due to the presence of a different solvent. TEM analysis reveals that the particle size depends on the addition of a surfactant but not on the temperature in hcp. Similar results are observed in fcc also. Structures prepared at a higher temperature are bulky and large in size [67]. Shviro and Zitoun synthesized Ni NPs in different facets (from fcc to hcp) by controlling the flow kinetic energy level in the microreactor. It is observed that by carefully adjusting the flow rates, fluid kinetic energy can be changed and the desired facets can be grown inside the microreactor. With a low kinetic energy level or a low flow rate (25 μL/min), metastable hcp-based Ni NPs can be formed, whereas a high flow rate (around 100 μL/min) yields fcc crystalline structured NPs. It is very challenging to control the kinetic energy level precisely to get the required crystalline structure [88].

The higher demand for the synthesis of NPs promotes the development of various types of microreactors. The design varies from the simplest tubular to complicated modular and automated reactors. Further, the application of microreactors in the biomedical field propels its use in the industry [31]. Depending on the requirement of the synthesis of specific NPs, different microreactors with various flow and mixing strategies can be used. Next, an overview is provided of the synthesis and biological applications of important metallic NPs with their specific shapes, sizes, and composition using different microfluidic devices.

10.3.5 Synthesis of Gold Nanoparticles

Gold is a material with unique characteristics such as tunable optical properties, ease of functionalization, nontoxic behavior, and catalytic, physiochemical, and biological properties. These features enable us to use it in a variety of applications like therapeutics, analytical science, catalysis, sensing, medical diagnosis, biomedical imaging, and drug delivery [1, 9, 38, 77, 79, 80, 110]. Among the metallic NPs, gold possesses shape- and size-dependent properties, which makes it useful in chemical catalysis and in bionanotechnology applications [12]. In 1951, Turkevich produced 15 nm Au NPs by reducing tetrachloroauric ($HAuCl_4$) acid at a high temperature using trisodium citrate [103]. For this method, the citrate acts as a reducing agent, a pH mediator, and a capping agent. The synthesized particles can be used as a seed for producing large-sized particles.

Fast nucleation can result in small monodisperse Au NPs. Since trisodium citrate is a weak reducing agent and the nucleation reaction is not fast, a strong reducing agent like sodium borohydride ($NaBH_4$) is used. The synthesis of metal NPs by employing borohydride as a reducing agent has great importance in improving the quality of particles in microfluidic devices. Borohydride reduction is also used to synthesize folded and rigid ultrathin Au NPs of polycrystalline facets. But it dominates over $NaBH_4$ because of low toxic behavior and ease of functionalization with trisodium citrate. Another approach is the Brust–Schiffrin method, which includes a capping agent and a strong reducing agent along with thiols. This method is generally used to synthesize Au NPs less than 5 nm in size [38]. Synthesis of thiol-protected Au NPs involves nucleation, growth, and passivation processes. Passivation is the formation of a monolayer of thiol surrounding Au NPs [85]. AA and iron (Fe(II)) can also be used as reducing agents to synthesize clustered and isolated Au NPs [50]. With a microfluidics approach, Au NPs of ultrasmall sizes (up to 1.8 nm) have been synthesized using

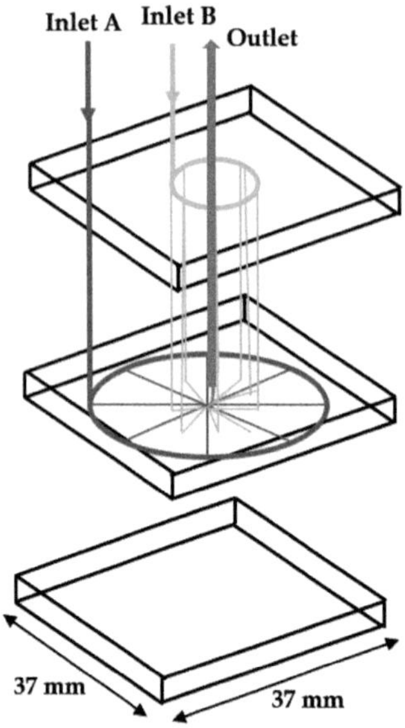

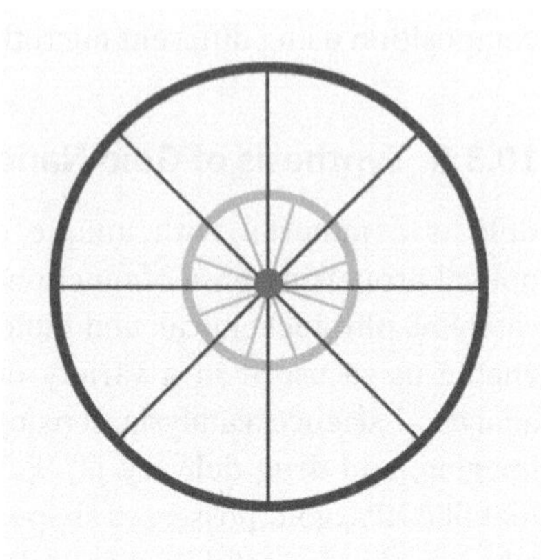

Figure 10.20 Schematic representation of the 3D view of a radial interdigitated mixer and the top view of the mixer. The microchannels are visible in the photograph with the addition of a dye [85].

the citrate reduction method [38]. Electrodeposition is the other approach for synthesizing Au NPs of various shapes, such as spheres, rods, hexagons, plates, hollow shells, and cubes. By changing the condition of electrochemical deposition on the conducting surfaces, Au NPs with various facets can be synthesized [24].

Three important microfluidic devices for the synthesis of Au NPs are described here. Figure 10.20 shows a 3D view and schematic illustration of the top view of a radial interdigitated mixer for the synthesis of thiol-functionalized Au NPs (<10 nm). The reaction mechanism is through diffusive mixing of reagents between 16 input laminae streams. The whole 16-stream channel has a width of 150 μm and an average depth of 60 μm. Mixing is carried out in three substrate layers; the first two layers allow the reagents to flow to the two similar circular channels, where they split into eight fluid laminae and deliver the reagents to a central mixing chamber. The third layer protects the channels and acts as a guide for input and output capillaries [85]. Figure 10.21 gives

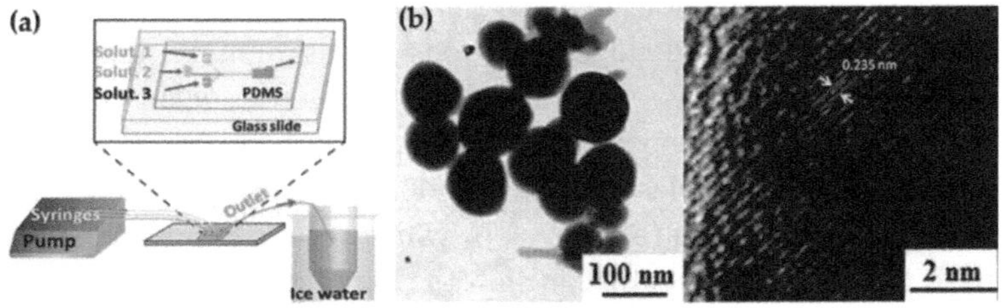

Figure 10.21 Flow-based microfluidics for the synthesis of folded and rigid ultrathin gold nanoplates; (a) schematic diagram of the experimental setup; (b) high-resolution TEM image of flat-surface Au nanoplates synthesized at a flow rate of 5 μL/min. The chip is 3 cm in length and has a 400 μm × 400 μm cross section and consists of three inlets, one microchannel, and one outlet. Republished with permission of The Royal Society of Chemistry, from Ref. [24], copyright (2015); permission conveyed through Copyright Clearance Center, Inc.

a diagrammatic representation of the experimental setup for Au nanoplate synthesis and a TEM image of flat nanoplates produced at a lower flow rate. To remove the excess surfactant, centrifuge was performed at 12,000 rpm for less than 15 min. After this, the precipitates were washed with water twice [24]. Another device, adequate for mixing of reagents and their transportation and reaction, was used for the synthesis of hexagonal Au NPs using the citrate reduction method [111, 112]. It assimilates micropumps, a micromixer, microheaters, microvalves, and a micro-temperature sensor. A rotary micromixer helps in the transportation and mixing of microfluidics, while sample transportation was achieved through PDMS-based microvalves and pumps. PDMS was chosen because it is able to deform during pneumatic actions. Moreover, it has two control modules, a micro-temperature control, and a microfluidic control module. Micro-temperature control was achieved through a heater and micro-temperature sensor assembly. A microfluidic control module consists of a micromixer, a microvalve, and a peristaltic pneumatic valve. Peristaltic movement of the micromixer helps compressed air to move into and out of the chip. The electron diffraction pattern for a single NP synthesized in this way is a polycrystalline and uniform distribution of particles achieved without using any surfactants and polymers [111, 112].

Plasmonic oscillations of Au NPs make them suitable for application in the biomedical field. Au nanorods possess multiple plasmonic absorption bands and transverse surface plasmon resonance (TSPR) and LSPR [110]. The aspect ratio can change the band frequency and hence can vary the plasmonic oscillations. Initially, particles grow along the direction of the long axis, which results in a high aspect ratio. A higher aspect ratio can shift the longitudinal band from the visible to the near-infrared (NIR) region. Once it reaches a maximum after 10–15 min, it grows along the short axis [11]. Hydrophobic cosurfactants, which act as stabilizers like CTAB, can improve the aspect ratio. Nanorod synthesis with CTAB addition involves four steps: (i) Au precursor activation—Au(III) to Au(I) by AA, (ii) Au seed formation—Au(III) to Au(0), (iii) Au nanorod formation; and (iv) Au nanorod PEGylation—addition of poly(ethylene glycol) (PEG). Au nanorods have been synthesized in microfluidic devices, which have the advantage of separating the seed formation and the nanorod formation steps clearly by controlling the residence time in each stage. The stages of synthesis are Au precursor formation, Au seed formation, Au nanorod growth, and PEGylation. Surface PEGylation is performed inside the reactor, which helps to reduce the cytotoxicity introduced by the CTAB nanorod [104]. Since Au nanorods are prepared from seeds, the size, shape, and aging of seeds have an impact on particle formation. Watt and coworkers demonstrated that freshly prepared seeds influence the synthesis of Au nanorods [110] due to the presence of high energy, which results in ultrasmall catalytically active seed NP formation and hence anisotropy. As the seed ages, it grows and increases in size and nanorod yield decreases. This can be overcome by diluting the seed solution. Figure 10.22 shows the schematic of a polyether ether ketone (PEEK)-based microfluidic device for Au nanorod formation and real-time analysis.

Compared to other nanostructures, hollow gold nanoparticles (HAu NPs) synthesis is very difficult due to reproducibility issues and the demand for strict dimensional control. Gomez et al. successfully synthesized HAu NPs from Co NPs [27] in four steps: (i) the synthesis of Co NPs, (ii) Au shell growth, (iii) shell-PEG functionalization, (iv) and downstream sterilization. Co NPs are synthesized from sodium borohydride reduction and mixed with

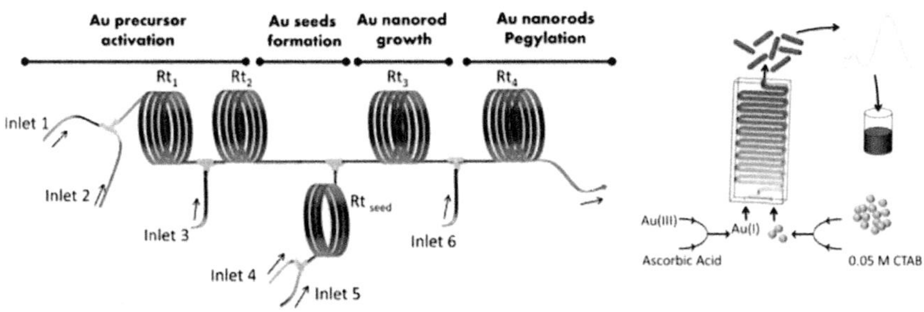

Figure 10.22 Schematic diagram of the microfluidic chip used to investigate the seed age in real-time Au nanorod synthesis. The microreactor has six inlets and six residence zones. Reagents were fed through syringe pumps and connected to the Y-shaped mixer of PEEK through PTFE tubes. Reprinted with permission from Ref. [110]. Copyright (2015) American Chemical Society (https://pubs.acs.org/doi/10.1021/acs.chemmater.5b02675). Further permissions related to the material excerpted should be directed to the ACS.

$HAuCl_4$. Stabilization of Co NPs was observed with the addition of PVP, due to the presence of an amide group. Further, a higher concentration of PVP promotes dispersed HAu NPs. The residence time for the complete synthesis was 22 min, and for high-throughput synthesis, it requires 23 min. Only 7 min was needed to prepare HAu NPs. A solid Au shell was observed with a low Co/Au molar ratio. This might be due to the secondary growth of a Au shell and evolved hydrogen gas. With a higher ratio, the collapse of thin shells results in undefined nanostructures. The formed Au shell contains a Co inner core and hence a galvanic reaction happens between the Au outer electrode and the Co inner electrode. This leads to the flow of anions and cations, resulting in porosity. PEGylated NPs were sterilized by coupling the outlet with the glass microreactor [27].

For in situ synthesis of Au NPs, 3D microchannels were introduced by Agrawal et al. [4]. The 3D circular microchannels were prepared by using PDMS material with the help of microwires. Circular microchannels provide an easy way to deliver the reagent (capillary force) without using any micropumps. Isolated and clustered Au NPs can be synthesized using a three-step static micromixer.

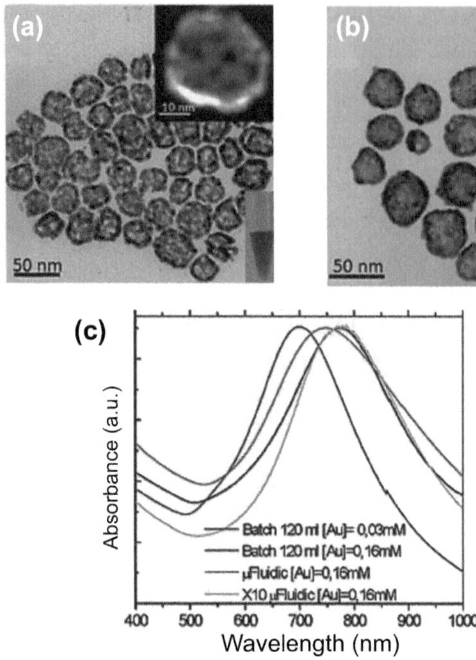

Figure 10.23 Hollow gold nanoparticle characterization: (a) TEM image of HAu NPs synthesized using Co NPs, (b) TEM image of synthesized of HAu NPs scaled up 10 times, and (c) UV-vis absorption spectra of HAu NPs synthesized for different batches. Republished with permission of The Royal Society of Chemistry, from Ref. [27], copyright (2014); permission conveyed through Copyright Clearance Center, Inc.

The hydroxyl group present in PDMS after oxygen plasma treatment can help reduce $HAuCl_4$. The PDMS surface was modified in order to get a stable hydroxyl group using chitosan. After $HAuCl_4$ was incubated for 48 h inside the microchannel, stabilized Au NPs were prepared inside the channel as well as on the walls. A comparison of the UV-vis absorption spectra in the chitosan-modified surface and the oxygen plasma–treated surface shows no visible shift of the absorption band. Instead, the peak value of absorption is present in the chitosan-modified surface. This might be due to the adsorption of Au particles by the chitosan to prevent their aggregation [4].

CTAB and silver are used for the synthesis of Au nanorods with a controlled aspect ratio by separating the seed formation and growth stages within the microreactor. With the help of CTAB and silver, seeds develop into nanorods during the heterogeneous crystallization stage. They help to adsorb on the specific crystal face with high surface energies and render Au NPs with a specific aspect ratio. Even though CTAB helps to improve the aspect ratio, over time it results in morphological changes in the particle due to the loss of CTAB coating; hence PEGylation is performed after CTAB coating. The TEM analysis of PEGylated nanorods reveals that it preserves their aspect ratio for at least 1 month. The effect of bromide ion (Br^-) at each stage of synthesis of Au nanorods was investigated, and it was found that Br^- results in a high-aspect-ratio structure [104].

The HR-STEM analysis of Au nanorods indicates that they are single crystalline in nature within the entire length. The analysis of a high-angle annular dark-field STEM image of PEGylated Au nanorods prepared in a $HAuCl_4$ and $HAuBr_4$ solution reveals that nanorods are single crystalline in nature along a single rod. For modifying the aspect ratio of nanorods, the concentrations of halide anions such as Cl^- and Br^- can be varied. An EDS analysis of a single Au nanorod shows a reduced amount of Ag atoms and no traces of Cl^- and Br^- ions (below the detection limit) [104]. TEM images of HAu NPs synthesized from Co NPs and the corresponding absorption spectrum are shown in Fig. 10.23a–c [27], in where the SPR peak varies from 700 nm to 800 nm and the spectrum shifts toward the right after the synthesis is scaled up. An analysis of Au nanocrystals shows that tuning the reaction parameters can help shift the optical resonance.

Table 10.1 summarizes the microfluidic devices used for the synthesis of Au NPs. It tabulates the tuning parameters, details of reagents, flow rates of reagents, and the shapes and sizes of the synthesized NPs.

10.3.6 Synthesis of Silver Nanoparticles

Ag NPs have tremendous applications in biological, agricultural, and medical fields. They can act as catalysts, chemical sensors, and chemical probes due to properties such as the tunneling, quantum

Table 10.1 Overview of a microfluidic device for the synthesis of Au NPs

Reagents	Flow rates	Shape	Size	Tuning parameters	Refs.
Trisodium citrate dehydrate, chloroauric acid trihydrate, AA, polyvinyl pyrrolidone	10 to 50 μL/min	Sphere	15–24 nm	Flow rate, reagent concentration ratio, product fraction	[107]
Chloroauric acid trihydrate, AA, PVP	500 to 8000 μL/min	Sphere	5–50 nm	Flow rate, concentration ratio, pH variation	[106]
Tetrachloroaurate, toluene, tetraoctylammonium bromide, 1-dodecanethiol, sodium borohydride	400 mL/min to bulk	Sphere	<10 nm	Flow rate, thiol gold ratio	[85]
Tetrachloroaurate, potassium carbonate, water	200 to 4000 μL/min	Sphere	4–7 nm	Flow rate, ratio of flow rate, ratio of concentration of reagents, effect of ligands and pH	[108]
Hydrogen tetrachloroaurate trihydrate, cetyltrimethylammonium chloride, sodium borohydride, AA	5 to 70 μL/min	Folded and rigid nanoplate	<1–5.5 nm	Flow rate, electrocatalytic activity	[24]
Tetrachloroaurate, trisodium citrate	Not mentioned	Hexagon	35 nm (mean diameter)	Volume of reducing agent, heating time	[112]
Hexadecyltrimethylammonium bromide, chloride solution, silver nitrate, AA, hydrogen tetrabromoaurate (III) hydrate, sodium borohydride, poly(ethylene glycol) methyl ether thiol	0.62 to 5.85 mL/h	Nanorod		Concentration of halide anions, aging time	[104]

(*Contd.*)

Table 10.1 (*Continued*)

Reagents	Flow rates	Shape	Size	Tuning parameters	Refs.
Gold (III) chloride hydrate, hexadecyltrimethylammonium bromide, $AgNO_3$, HCl, $NaBH_4$, *N*-octyldiisopropylchlorosilane	5 to 7 μL/min	Nanorod	<2 nm to higher	Flow rate, seed aging, reaction time	[110]
CTAB + silver nitrate solution, AA, Au seed	2.6 to 20 μL/min	Sphere, rod and extended, sharp-edged structure	20 nm	Ratio of flow rates of seed to reagents, concentration of reducing agent	[20]
Tetrachloroauric acid, sodium citrate		Not specified	40 nm	Flow rate, concentration of reagents, switching frequency	[95]
Cobalt chloride hexahydrate, sodium citrate tribasic dihydrate, PVP, sodium borohydride, chloroauric acid trihydrate, poly(ethylene glycol) methyl ether thiol	15 to 25 mL/h and 150 to 250 mL/h for production scaled up 10 times	Hollow gold nanoparticle	32.4 ± 4.7 nm, 40.5 ± 3.2 nm	Residence time, PVP concentration, Co/Au molar ratio	[27]
Sodium borohydride, chloroauric acid, tetradecyltrimethylammonium bromide, segmenting fluid	27 to 300 μL/min	Nanocrystal		Inlet flow rates, segmented fluid	[12]
Chitosan, tetrachloroauric (III) acid	Not mentioned	Not mentioned	Not mentioned	Effect of chitosan coating	[4]
Ferric chloride hexahydrate, ferrous chloride tetrahydrate, sodium hydroxide, hydrochloric acid, gold (III) chloride trihydrate acid, sodium borohydride	100 μL/min (decoration of Au on magnetite)	Sphere	10.4 ± 0.3 nm ($Fe_3O_4^-$ NPs), 3.9 ± 0.2 nm (Au NPs)	Concentration of magnetite, carrier phase	[13]

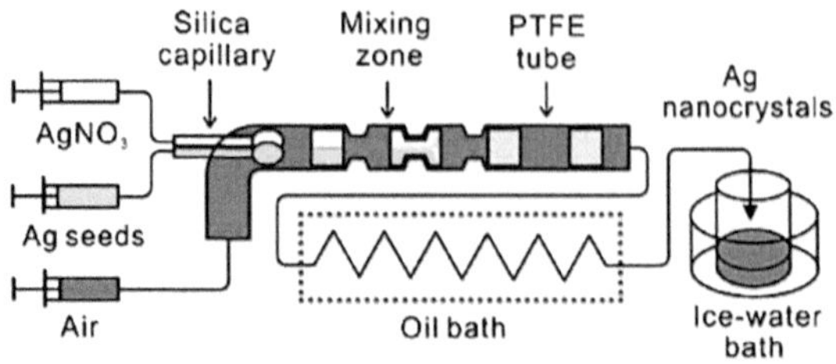

Figure 10.24 A schematic representation of seed-mediated growth of Ag nanocrystals in a droplet-based microreactor. $AgNO_3$ and seed solution form the aqueous phase, and air forms the carrier phase. Seeds were produced by reacting diethylene glycol, NaSH (sodium hydrosulfide), HCl (hydrochloric acid), and CF_3COOAg (silver trifluoroacetate) at 150°C under vigorous stirring. Air was introduced sequentially after the aqueous phase. Reprinted with permission from Ref. [118]. Copyright (2013) American Chemical Society.

size, and surface and volume effects. These properties depend on particle size, shape, and distribution [60, 80]. The general method for the synthesis of Ag NPs is the reduction of aqueous metal salt using borohydride solution. The following chemical reaction describes the production of Ag NPs [60, 115]. Ag NPs synthesized from sophorolipid possess antibacterial activity. Sophorolipids such as OASL and SASL can be used for this purpose, and they serve as both capping agent and reducing agent [53].

$$AgNO_3 + NaBH_4 \rightarrow Ag + \frac{1}{2}H_2 + \frac{1}{2}B_2H_6 + NaNO_3 \qquad (10.2)$$

Droplet-based microfluidic devices are most commonly used for Ag NP synthesis. Figure 10.24 provides a schematic representation of the synthesis of Ag nanocubes and octahedral structures using air as a carrier phase. Mineral oil, air, and silicone oil were used as a continuous phase. Further, EDTA was used as a complexant, which played a crucial role in forming spherical crystals [115, 118].

Microreactors with a coaxial flow can be used to synthesize Ag NPs as shown in Fig. 10.25. Here, trisodium citrate acts as a surfactant. Stages of nucleation, reaction, and stabilization of Ag NPs depend on the concentration of the reducing agent and the surfactant with respect to the precursor [5]. A liquid-phase reduction mechanism was utilized in piezoelectrically actuated three-phase flow pulsating mixing in a microfluidic channel to

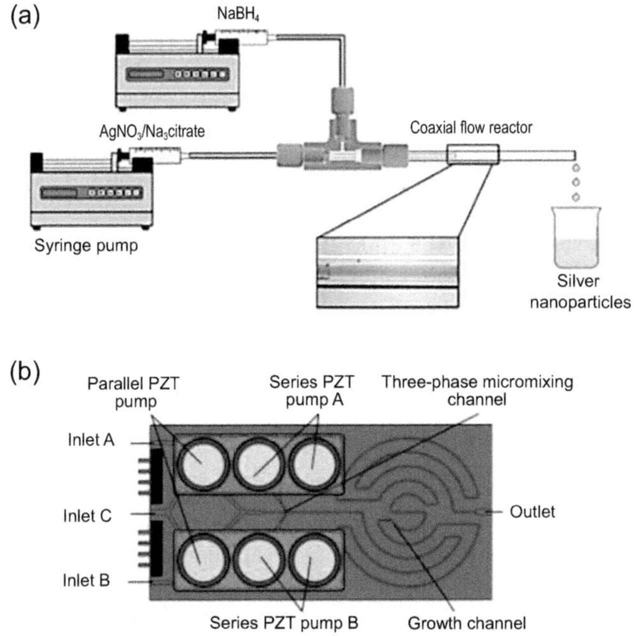

Figure 10.25 Schematic representation of (a) coaxial-flow microreactor. Reproduced from Ref. [5], under a Creative Commons Attribution 3.0 Unported license (https://creativecommons.org/licenses/by/3.0/). (b) A three-phase micromixer for Ag NPs synthesize. The three-phase micromixer consists of three inlets, three series of PZT pumps in parallel with another set, a microchannel in between the parallel PZT, a growth channel, and an outlet. A precursor, a reducing agent, and accelerating agents were injected through inlets A, B, and C respectively. Reproduced from Ref. [60], under a Creative Commons Attribution 4.0 International License (https://creativecommons.org/licenses/by/4.0/).

prepare Ag NPs. Glucose was used as a reducing agent, silver nitrate in PVP as the precursor solution, and sodium hydroxide as an accelerating agent, as shown in Fig. 10.25b. PZT micropumps were driven by a sine wave signal having a phase difference of 180 degrees, and each pump was working at a frequency of 200 Hz. The reducing agent and the stabilizing agent were pumped alternately in pulsed form, whereas the precursor solution was given in continuous form. By effectively tuning the parameters, the contact area of the microchannel can be improved and the micromixing time can be shortened, which can effectively improve the mixing efficiency [60].

At nanoscale, surface modification of Ag microflowers within the microchannel leads to the formation of uniform SERS enhancement [64]. Apart from these, the characterization of synthesized NPs can be carried out in the microfluidic channel itself. The device enables resonant light scattering analysis quantitatively on different particles, such as Au NPs and Ag NPs [68]. Table 10.2 shows how Ag NPs of different shapes and sizes are synthesized by tuning the reagent flow rate in microfluidic devices.

10.3.7 Synthesis of Other Metallic Nanoparticles for Biomedical Applications

Apart from Au and Ag, metals such as palladium (Pd), nickel (Ni), cobalt (Co), platinum (Pt), ruthenium (Ru), iron (Fe), and rhodium (Rh) are also important in biological applications. The important features and their microfluidic-based synthesis are discussed here.

Pd nanocrystals are important for catalytic activity in organic coupling reactions and hydrogen reactions. This catalytic activity can be decided by the shape of the NPs, which depends on the arrangement of the edges, facets, and corners. It has been experimentally demonstrated that Pd nanorods and nanobars were synthesized by using PVP with a water/polyol system. Further, the experiment showed that Br^- anions are responsible for $\{100\}$ and $\{110\}$ facet formation. To improve preferential growth, oxidative etching is useful [82].

Pd NPs were synthesized using either laminar or segmented flow-based microfluidic devices. An example of a continuous segmented flow-based microfluidic device is a spiral microreactor with two zones—the cold zone, where the reactants mix, and the hot zone, where the actual reaction happens. A microchannel 400 μL wide and deep and with a capacity of 100 μL was used to synthesize nanorods with a mean diameter of 4 nm. For this NP preparation, the precursor Na_2PdCl_4+KBr+H_2O was injected through one inlet. PVP in EG (polyol) was administered through another inlet, and air as a segmented phase was also injected through a separate inlet. The injection of gas and reagents was controlled in order to control

Table 10.2 Overview of a microfluidic device for the synthesis of Ag NPs

Reagents	Flow rates	Shape	Size	Tuning parameters	Refs.
$AgNO_3$ (NH_3), H_2O, $NaBH_4$, mercaptoundecanoic acid, mercaptoundecanol	200–8000 μL/min	Sphere	15 nm (mean)	Concentration ratio of metal salt to the reducing agent, flow rate	[108]
$AgNO_3$, KOH, oleic acid, oleic acid sophorolipid, stearic acid sophorolipid	1 mL/min, 100 μL/min, 35 μL/min	Sphere	5.5–20 nm	Residence time, flow rate, sophorolipid	[53]
Silver nitrate, sodium borohydride, polyvinylpyrrolidone, ammonium hydroxide, ethylenediaminetetraacetic acid, sodium hydroxide	10–30 mL/h	Sphere	Not specified	Flow rate, concentration ratio of metal salt to the reducing agent	[115]
Diethylene glycol, NaSH, HCl, CF_3COOAg, PVP, $AgNO_3$	100–300 μL/min (air), 25 μL/min (reaction solution)	Nanocube; octahedra	30–100 nm (cubes)	Carrier phase, reaction time, flow rate of the carrier phase	[118]
Silver nitrate, sodium borohydride, trisodium citrate	1–14 mL/min	Sphere	Not mentioned	Flow rate, concentration of ligand, and precursor	[5]
$AgNO_3$, $C_6H_{12}O_6$, NaOH, PVP	3.5 mL/min	Sphere	29 nm (average)	Inlet flow rate, synthesis temperature, concentration of reagents, PVP dosage	[60]
$AgNO_3$, AA, cetyltrimethylammonium bromide, sodium borohydride		Nanorod			[11]

the residence time as well as the flow velocity. The same device was used to achieve a laminar flow. It has been shown that temperature has an influence on the shape of the NPs formed, where a higher temperature yields nanorods at a residential time of 120 s. Oxidative etching through dissolved oxygen produced from water and Cl^- ions released from a Pd precursor promotes the anisotropic growth of NPs. Br^- ions inhibit the growth of Pd, while oxidative etching inhibits further growth of Br^- ions, enabling the formation of $\{100\}$ facets of an fcc structure. In segmented flow, the ratio of the flow of polyol to the precursor controls anisotropic growth. As the ratio increases, the aspect ratio also increases. But it is observed that the aspect ratio decreases after the polyol-to-precursor ratio crosses 4. Oxidative etching was controlled by changing the ratio of the flow rate of air to the reagents. Decrease in the ratio indicates less availability of oxygen and hence lowered oxidative etching, which results in poor anisotropic growth and low-aspect-ratio nanorods are prepared [82, 84].

The general method of synthesis of Ni NPs involves surfactant-assisted solvothermal reactions in the presence of polymers, ligands, or ionic surfactants. In the presence of a hydride, polyol, long-chain amine, or amino-borane reducing agent, Ni salt turns into spherical NPs. It has been reported that by alloying of Ni with metals such as Pt, catalytic addition of $Fe(CO)_5$, and seeding of Ni on manganese powder, triangular and hexagonal Ni nanoplates with heterogeneous nucleation are produced. Organometallic precursors play a significant role in the production of Ni NPs with a clean metal surface. Precursors such as $Ni(cyclooctadiene)_2$ and $Ni(\eta^4\text{-}C_8H_{12})_2$ can produce NPs of different sizes. Ni-faceted nanorods were synthesized by decomposing $Ni(\eta^4\text{-}C_8H_{12})_2)$ in the presence of hydrogen and hexadecylamine ligand. Ni nanocubes were synthesized by the hydrogen-assisted decomposition of metal-organic $Ni(acac)_2$ [88].

In a microfluidic device, organometallic precursors were used to synthesize Ni NPs with distinct shapes, such as spheres, trigonal pyramids, tetrapods, and cubes, by tuning the kinetic control of crystal growth. The shape change of Ni NPs is achieved by tuning the parameters to change the crystallographic plane. It has been observed that by controlling the ligands, the size of NPs can be controlled. Addition of ligands like alkylamines and triphenylphosphine produces spherical NPs with a mean diameter of 3–6 nm. Using pyramidal Ni NPs, single-crystalline fcc tetrapods were formed by controlling the shape kinetically. In the case of exposed facets of pyramidal NPs, growth was observed for tetrapods in the $\{111\}$ direction. At room temperature, tetrapods are stable but turn spherical after annealing at 150°C [88].

Pt is an important metal for biomedical application, which has catalytic activity, and it has been found that the size of the NPs has an impact on the catalytic action. Small-sized NPs, of approximately 10 nm, are found to have greater catalytic action. The common method to synthesize Pt NPs is reductive crystallization with the addition of a protective reagent. Pt^{4+} or Pt^{2+} reduces ions in the presence of sodium polyacrylate, PVP, or H_2 gas. Reducing agents such as $NaBH_4$ and polyethyleneimine are used for this purpose [31].

10.4 Biomedical Applications

Over the decade, microfluidics has evolved and been used widely in biomedical applications, such as sensing, drug delivery, disease diagnostics, imaging, and therapeutics. For instance, plasmonic properties of gold and silver; magnetic properties of iron; fluorescent properties of metal clusters for sensing; and therapeutic properties of materials like graphene, cobalt, nickel, platinum, and copper are used. It is becoming easier to develop chips to fabricate monodispersed nanoparticles with better-engineered surface functional groups. This combination has helped scientists to enhance the investigations by improving the detection limits in relevant fields. Microfluidic devices utilize a wide range of NPs to improve labs on chips (LOCs). In this context, some metal NPs are synthesized using microfluidic devices for different biomedical purposes.

10.4.1 Bioimaging

Because of their innate properties, metallic NPs are used as contrast agents for bioimaging applications. Using microfluidic technologies, NPs can be developed clinically at a faster pace. Metallic NPs individually or in combination have the potential to be used as fluorescent markers and contrast agents for imaging. Computed tomography (CT), magnetic resonance imaging (MRI), optical imaging, ultrasound, and positron emission tomography require contrast agents to improve the signal-to-noise ratio. Materials like gold, silver, copper, tungsten, graphene, iron, cobalt, and bismuth at nanoscale showed better properties in the imaging sector. In general, a stream of precursor materials and a reducing agent are mixed in a microfluidic channel to produce nanomaterials over time. For example, colloidal gold in different shapes, like nanorod, nanostar, nanosphere, nanoflower, and NP, is promising in point-of-care biochemical assay detection and bioimaging. In general, a large amount of contrast agents is required for bioimaging, and producing monodispersed particles with a specific size and shape is limited in a batch reactor compared to microfluidics devices.

Optical coherence tomography (OCT) is a non-invasive tool, analogous to an ultrasound system, and provides anatomical tissue information. The optical scattering phenomenon of OCT provides distinct tissue layer segmentation [39]. Sebastian et al., in 2012, demonstrated that the use of Au nanorods, synthesized by using a microfluidic device, can enhance the imaging of breast tissues at micron-scale resolution, as shown in Fig. 10.26a–c [83]. Au nanorods exhibit high, intense longitudinal plasmonic resonance peaks in the far-red and NIR regions. Photoacoustic signals are produced by nonradiative relaxation of the absorbed light of Au nanorods. These signals release heat, and change in the temperature in the environment produces a strong signal for detection. Appropriate concentration, toxicity, specificity, bioavailability, effect of light interaction with Au nanorods, and in vivo fate must be evaluated before using Au nanorods as contrast agents [62]. Au nanorods can be synthesized using bioavailable proteins, without any conventional toxic material, that is, CTAB. Au nanorods functionalized with lysine (acting as a reducing agent) makes them suitable for targeting specific parts of the tissues. Au nanorods with a concentration of 3 pM were injected into breast tissues and excited using a laser source running at 240 kHz. By applying photothermal modulation, the axial resolution of the accumulated Au nanorod in the phantom was improved by an order of 5 compared to control phantom [62].

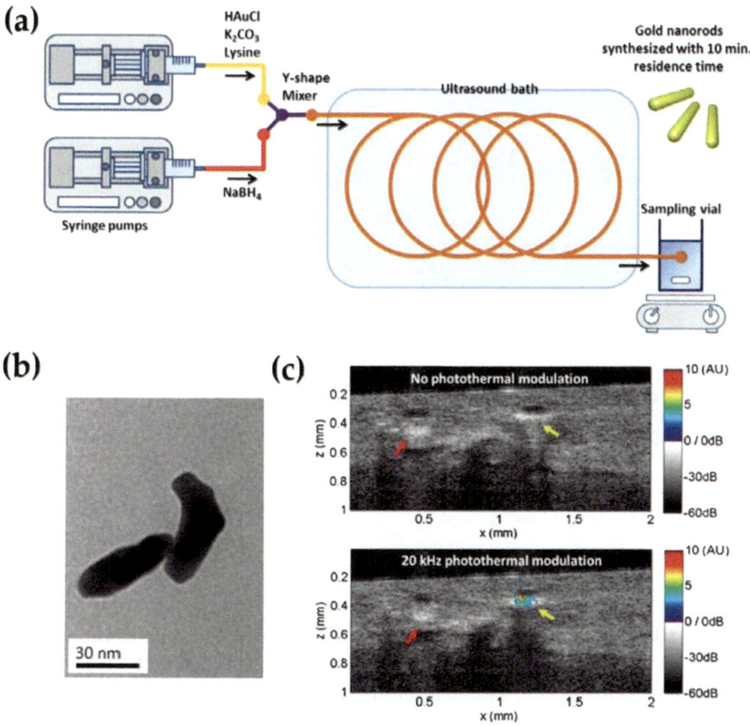

Figure 10.26 Optical coherence tomography images of freshly excised human breast tissue using gold nanorods as contrast agents (a) without photothermal modulation and (b) with photothermal modulation (photothermal OCT signal with SNR >5 is superimposed as a pseudocolor on the gray-scale image). (c) Red arrows denote fibrostromal regions, whereas yellow arrows indicate the administration site of Au nanorods. Republished with permission of The Royal Society of Chemistry, from Ref. [83], copyright (2012); permission conveyed through Copyright Clearance Center, Inc.

Metallic NPs combined with other metal oxide nanomaterials are potentially applicable for multimodal imaging functionality and SERS detection. In this context, Hassan et al., in 2009, developed three interconnected microreactors to study and understand the optimization and chemical reactions of composite nanomaterials. These help reduce the steps in conventional microfluidics-based synthesis and can help tailor the required chemical properties for bioimaging applications [32]. In continuation with the previous

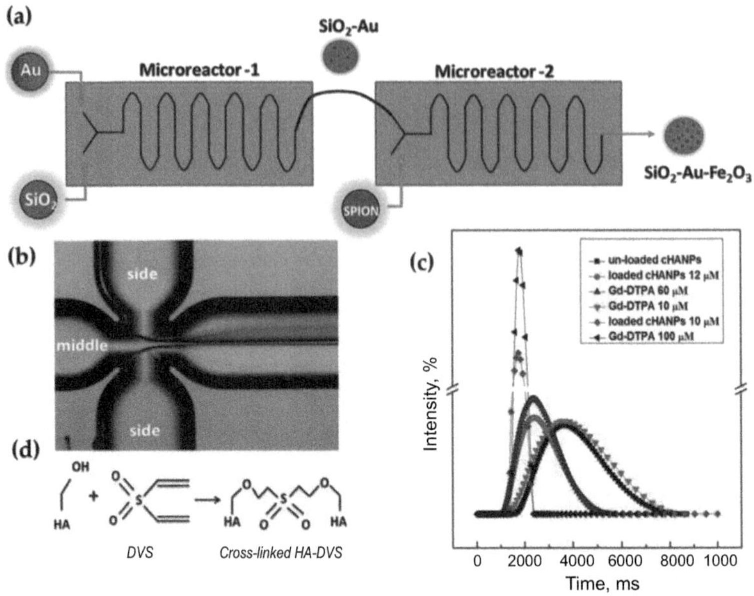

Figure 10.27 (a) Illustration of a multistep device for continuous-flow synthesis of magnetic and fluorescent nanoparticles [32]. (b) Hyaluronic acid linked with DVS to load gadolinium-DTPA at different concentrations. (c) On entrapping conventional MR contrast agent in hyaluronic acid polymers, the polymers showed enhanced relaxivity at 10 μM of Gd-DTPA-loaded cHANPs compared to 100 μM of Gd-DTPA. (d) The linking mechanism of the cross-linking DVS and HA. Reproduced from Ref. [73], under a Creative Commons Attribution 4.0 International License (https://creativecommons.org/licenses/by/4.0/).

work, Hassan developed self-assembled SiO_2*-Au-γ-Fe_2O_3 nanostructures using a microfluidic device, as shown in Fig. 10.27a–d [32]. Developing composite nanomaterials in large concentrations aids in trimodal imaging applications. Gold, Fe_2O_3, and fluorescent-tagged molecules can be used as contrast agents in CT, MRI, and optical imaging.

Other than iron oxide nanomaterials, rare earth metal ions are used as the conventional contrast agent for MRI. MRI is a non-invasive technique approved clinically as a diagnostic imaging modality to visualize the anatomical structure and precise location of disease. MRI uses a high magnetic field (1.5 tesla and above) that provides 3D images with a high signal-to-noise ratio. MRI works on the basic principle of relaxation of water molecules in both transverse and longitudinal modes of paramagnetic protons (^{1}H) to provide images. But the major concern is the tissue contrast between normal and abnormal tissues [21]. To obtain a better contrast, superparamagnetic molecules, like gadolinium complexes and iron oxide NPs, are used. Since this chapter is based on metal-based NPs, we are focusing only on gadolinium-based metal-ion complexes and not discussing metal oxide NPs. Several companies in the US and Europe have developed Food Drug Administration (FDA)-approved Gd^{3+} complexes wherein the fate of these complexes is instable and nontargeted. Thus, conjugating a Gd^{3+} metal-ion complex in a polymer functionalized with the targeted moiety would enhance the imaging properties [8]. Recently, incorporating these conventional agents into specific nanocarriers has helped to enhance imaging and improve the bioavailability.

FDA-approved gadolinium diethylenetriamine penta-acetic acid (Gd^{3+}-DTPA) has gadolinium (a rare earth element), which is used as a magnetic resonance (MR) contrast agent. This metallic element was loaded into cross-linked hyaluronic acid nanoparticles (cHANPs) using a microfluidic device. Developing such clinically relevant NPs continuously facilitates a controlled reaction with highly monodispersed particles. For the first time, Russo et al. developed Gd-DTPA-linked cHANPs hydrogels, which enhance MR relaxometric property compared to the bare contrast agent [73]. The loading encapsulation capability of Gd-DTPA in cHANPs was 89%, and Gd-DTPA was able to achieve lower relaxation in T_1 time when compared to control. Furthermore, the report suggested that under physiological conditions, a high payload of gadolinium complexes showed controlled release behavior up to 96 h. Thus, entrapping MR contrast agents in Gd-cHANPs enhanced relaxivity at minimal dose, reduced toxicity, enhanced the scan time, was inexpensive, showed rapid renal clearance, and allowed easy functionalization due to the presence of the HA polymer. Kim et al. developed seamless microgel particles via a single-step microfluidic device by incorporating nanomaterials like Au, FeO, and quantum dots as shown in Fig. 10.28a–d [48]. High-density lipid (HDL)-derived nanomaterials were easy to visualize using CT, MRI, and optical

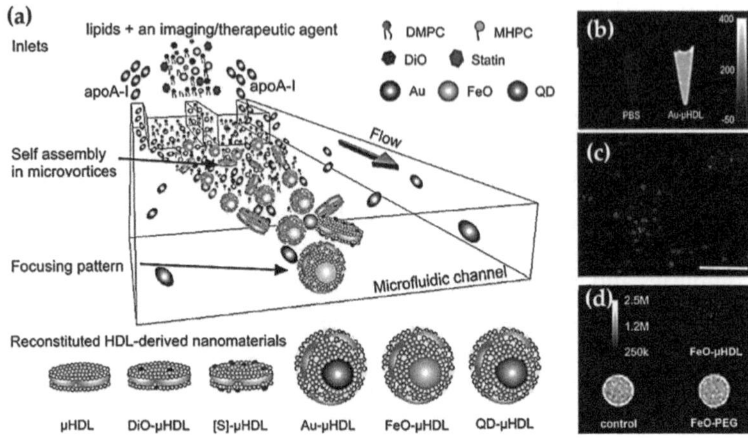

Figure 10.28 (a) One-step microfluidic synthesis of μHDL loaded with gold, iron oxide, and quantum dots produced on a large scale. (b) CT image of the phantom with Au-μHDL compared with control PBS. (c) Confocal microscope of QD tagged with macrophages. Scale bar: 100 μm. (d) T2-weighted MR image of macrophages incubated with PEG-FeO and FeO-μHDL. Reprinted with permission from Ref. [48]. Copyright (2013) American Chemical Society.

imaging modalities depending on the presence of nanomaterials. Biological properties of μHDL particles were similar to the properties of conventional HDL and maintained their parameters by manipulating the precursors in microfluidic channels. This μHDL-loaded-nanomaterial approach can be effectively applied for the delivery of various drugs, identification of biomarkers, and discovery of drugs [48].

Song et al. synthesized three phases of Co NPs via a microfluidic device by controlling the design pattern and flow rates [94]. As described earlier, metallic Co possesses ferromagnetic property when it turns into NPs. When large-sized Co NPs turn ultrasmall, they become superparamagnetic, which is required for MRI. Like iron oxide NPs, Co NPs play a major role in biological systems as cobalamin, a basic constituent of vitamin B. Like Au NPs, Cu is another metal used as a nanoprobe after it is doped with a sulfur group to form copper sulfide (CuS) nanomaterials. Ortiz et al. developed CuS NPs using a microfluidics-based technique that reduced the synthesis time about fourfold compared to the batch process. CuS degrades into copper sulfates at certain pH conditions, and this property of CuS would help it replace plasmonic materials like Au- and carbon-based nanomaterials [18, 52].

10.4.2 Biosensing

Biosensing is an important tool for early detection and sensing of medical parameters for healthcare applications. Optical properties of metallic NPs provide many opportunities in biosensing. For example, metallic NPs exhibit LSPR phenomena, where conduction electrons at the surface of the NPs are induced to oscillate at the resonant frequency by the incident photons, as a result of which, metal NPs can absorb light at that frequency. This resonant frequency of NPs depends on the size, shape, composition, and refractive index of the surrounding medium of the NPs, thereby enabling metal NPs to act as a sensitive label for biosensing. In the last two decades, several FDA-approved microfluidic devices have found an important niche in point-of-care diagnostics in real clinical settings. A microfluidic chip coupled with nanotechnology provides accurate sensing, monitoring, and quantitative analysis in real time with minimal sample volume [96]. For instance, circulating tumor cells (CTCs) have the ability to spread cancer via the bloodstream and detecting these cells is challenging. In this context, Sheng et al. reported the first use of a microfluidic device for capturing and isolating CTCs at a high efficiency using aptamer-conjugated Au NPs [86]. Au NPs with a size of 13 nm have the ability to bind 95 aptamers, showing high efficiency and capability to recognize CTCs. The results of flow cytometry reveal that gold-aptamer conjugates show 39-fold binding with CTCs compared to the control aptamer, as shown in Fig. 10.29a. The frequency of capturing cells using a microfluidic device with a herringbone structure was similar at higher flow rates (2.4 μL/s). Later, Park et al. isolated mesenchymal and metastatic breast cancer cells by interacting blood with thiolated Au NPs on a CTC chip, as shown in Fig. 10.29b. Due to their unique 3D design, the cell and developed chip showed negligible stress between them compared to other developed microfluidic devices. The capture-and-release approach

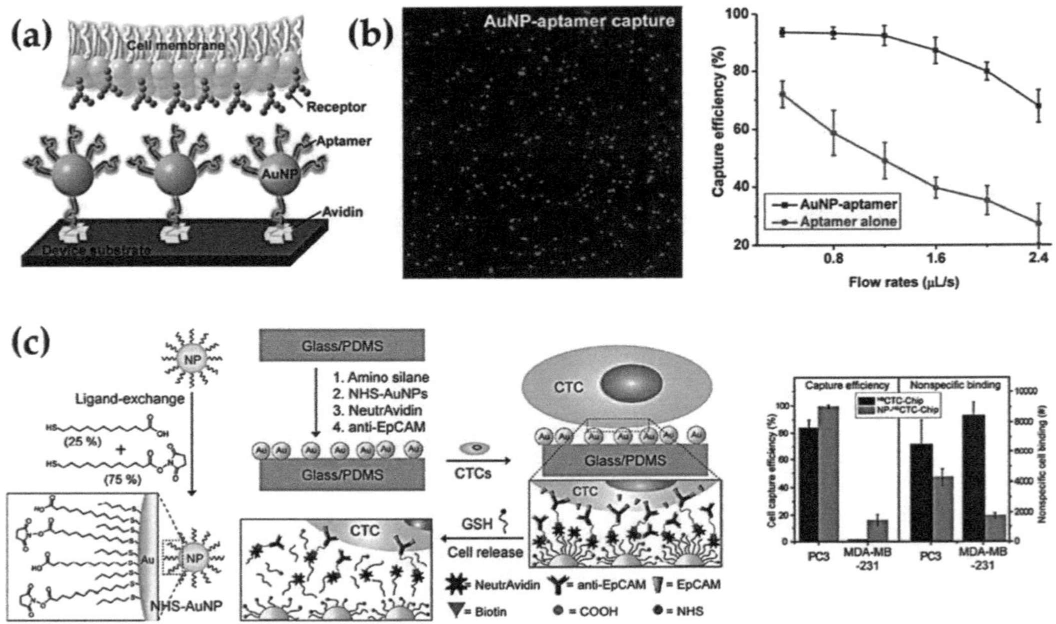

Figure 10.29 (a) Detection of CTCs using a microfluidic chip coupled with Au NPs. Reproduced from Ref. [96], under a Creative Commons Attribution 4.0 International License (https://creativecommons.org/licenses/by/4.0/). (b) A microfluidic chip coated with aptamer-conjugated gold nanoparticles for the detection of CTCs. At 2.4 μL/s, the detection rate was 95% in aptamer-GNPs compared to the control aptamer. Reprinted with permission from Ref. [86]. Copyright (2013) American Chemical Society. (c) Thiol-modified Au NPs captured CTCs by a ligand exchange mechanism. Reproduced from Ref. [96], under a Creative Commons Attribution 4.0 International License (https://creativecommons.org/licenses/by/4.0/).

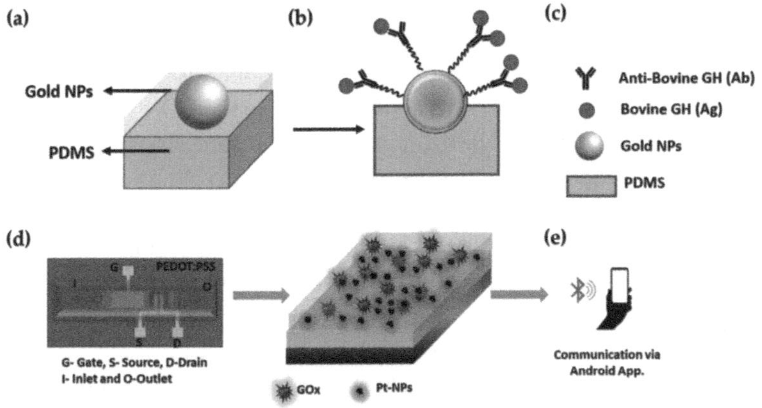

Figure 10.30 Biosensing experiments with an annealed microfluidics biosensor prepared by incubating microfluidics cell with a 2% gold solution for 48 h. (a) Cross-sectional view of a microfluidic cell with Au NPs and PDMS. (b) Steps of interaction of bovine GH (Ag) biosensing. (c) Legend of schematics [74]. (d) Schematic diagram of a glucose transmitter sensor consisting of glucose oxidase and platinum nanoparticles embedded with chitosan. (e) An electrochemical transistor with a glucose sensor communicating with a mobile phone via an android app [42].

of this CTC chip shows low stress on the cells, which can be further used to perform ribose nucleic acid sequencing.

Sadabdi et al. proposed an alternate protein detection mechanism and developed a nanointegrated microfluidic biosensor device to detect hormone growth (recombinant bovine somatotropin [rbST]) [74, 84]. In this device, 125 nm immobilized Au NPs were produced via an in situ reaction between a cross-linking polymer and a gold precursor, as shown in Fig. 10.30a–e. This in situ biosensor is able to detect rbST in the surrounding medium with a sensitivity of 74 nm/RIU and a detection limit as low as 185 pM, or 3.25 ng/mL.

Organic electrochemical transistor (OET) is one type of organic bioelectronics-based sensor to detect biomolecules, which monitors cellular activities, and acquires electrocardiograph signals. But detection of biomolecules like glucose and lactose non-invasively using the OET method at low concentrations is difficult. Implementing NPs in an OET would enhance the detection limit with an extremely low consumption volume. Ji et al. reported a highly sensitive glucose detection kit that interfaced via an android application to present the parameters on a mobile phone, as shown in Fig. 10.30d,e [42]. To perform this activity, PVP-capped Pt NPs along with glucose oxidase (GOx) were deposited on the gate electrode by the dip-coating technique. Pt NPs play a major role in catalyzing hydrogen peroxide, thereby making glucose detection highly sensitive. Further treating Pt NPs with UV ozone, catalytic activity of nanomaterials was enhanced to achieve the glucose detection limit to around 10^{-7} M. The overall time of detecting glucose using this kit was around 1 min with a minimal volume of 30 μL.

Apart from other techniques, Raman scattering is a powerful phenomenon used for sensing biological materials. It is a two-photon event, involving absorption of photons and scattering them inelastically. The shift in the wavelength of the scattered photons corresponds to specific vibrational modes present in the molecule, giving rise to a specific Raman signature for each molecule. Raman scattering is a weak process, where only 1 in every 10^6–10^8 photons is scattered inelastically. For sensitive detection of this scattered light, different enhancement mechanisms are used, such as resonant Raman scattering and SERS [34, 56]. Between these two techniques, SERS shows efficient enhancement for metallic nanomaterials. When an analyte molecule is attached on a metallic nanomaterial surface, it results in changes in the oscillation of plasmons, increasing the strength of the Raman signals. A scattered perpendicular signal is detected using a detector, which shows an enhancement as large as 9×10^7, reported by Haynes et al. in 2002 [33]. Several similar works have been carried out with different metallic NPs, such as gold, silver, and composite metallic nanomaterials. In 2014, a magnetic iron oxide/silver nanocomposite (Fe_3O_4@Ag@streptavidin@anti-IgG) was fabricated that can selectively detect human immunoglobulin G (IgG) in blood samples via SERS, as shown in Fig. 10.33a,b [7]. The magnetic property of the nanocomposite enables separation of the composite from a blood sample after anti-IgG antibodies have bound with the target. The spectral peaks at amide I (1650 cm^{-1} shift) and amide II (1539 cm^{-1} shift) bands observed show the interaction between antibody and nanocomposite that enables them to capture IgG. The ratio of these peaks was used for the quantification of the IgG target. This method enabled better sensitivity at 0.6 ng L^{-1} when compared to other similar non-SERS immunoassays.

Droplet-based microfluidics was developed to synthesize PNIPAM@Au nanoparticles for detection of Cu ions in vivo, as shown in Fig. 10.31a–d. Highly monodispersed particles were synthesized in less than 8 min from 24 h in a conventional batch process. Copper ions are sensed in rat serum by an aggregation-induced change in the intensity in an UV-visible absorption spectrometer from 522 to 574 nm [72]. Choi et al. developed multifunctional six-compartment-based microfluidic channels fused with SERS to determine immunoreactions automatically, as shown in Fig. 10.32c,d [16]. The chip works as follows: (i) droplets are formed with a specific antigen, (ii) SERS nanomaterials and antigen interact, (iii) the main droplet splits into daughter droplets (i.e., magnetic immunocomplexes and supernatant SERS nanotags), and (iv) unbound SERS nanotags are detected in the supernatant solution present in the droplets. The Raman signal of the unbound F1 antigen was detected in nanoliters of the sample, with a detection limit of 59.6 pgm L^{-1}, in less than 10 min. This work motivates the development of different automatic chips for easy and fast detection of hazardous chemicals in the biological or chemical environment.

Recent studies have shown that bimetallic NPs are more stable compared to monometallic particles. For instance, an electrochemical biosensor showed a higher, stable, and linear reduction of current versus concentration. Ko et al. in 2017 showed a composite nanomaterial fabricated inside a microfluidic chip for the continuous detection of H_2O_2 using cyclic voltammetry [49]. In this work, the sensor contains Au-Ag bimetallic nanomaterials modified electrochemically on a single-wall carbon nanotube (CNT) integrated with a PDMS channel, as shown in Fig. 10.33a,b. Cyclic voltammetry results suggested that the reduction sensitivity of H_2O_2 was around 13.1 μA cm^{-2} mM^{-1} with a detection limit of 26.8 μM. Similarly, Bai et al., in 2018, developed a novel 2D metal nanostructure on a 3D glass microfluidic channel by using femtosecond laser for real-time SERS sensing applications [6]. Here a Cu-Ag layer was fabricated by a two-step process: femtosecond laser ablation for the etching process followed by electroless metal plating. Further, laser was irradiated to

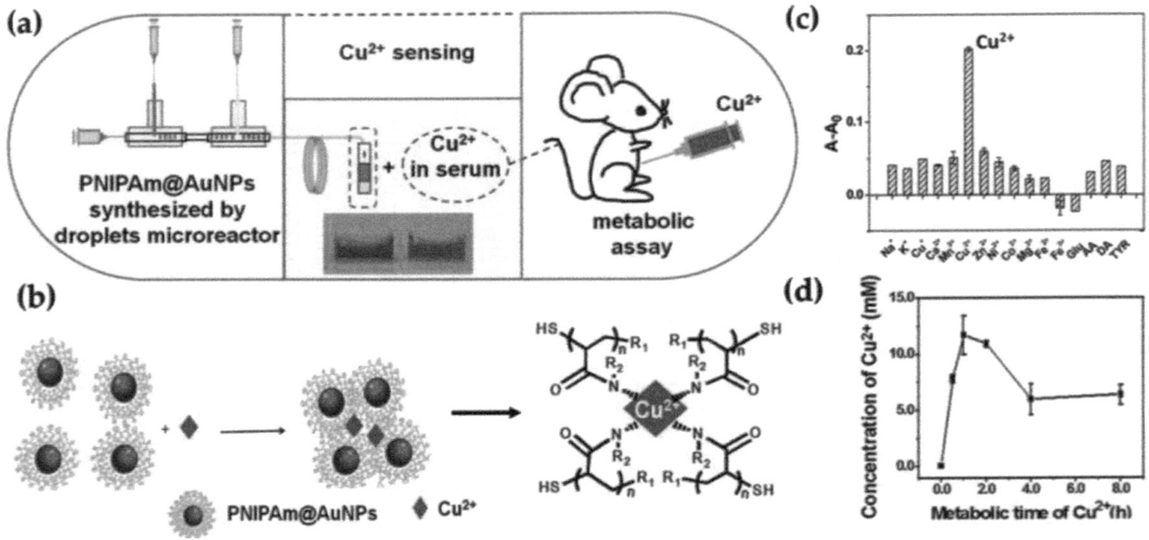

Figure 10.31 Detection of serum copper ions using PNIPAm-protected gold nanoparticles. (a) Schematic diagram of the microfluidic reactor, a photograph of a PNIPAm@Au NPs solution before and after the addition of copper ions, and a schematic of the metabolic assay of copper ions in rat serum. (b) Schematic representation of the copper ion reaction by the colorimetric aggregation of PNIPAm@Au NPs. (c) UV-visible spectrum of copper ions showed enhanced effect compared to other metal ions and proteins. (d) Selective sensitivity of Cu^{2+} ions at different times in rat serum injected via the abdomen. Reprinted with permission from Ref. [72]. Copyright (2017) American Chemical Society.

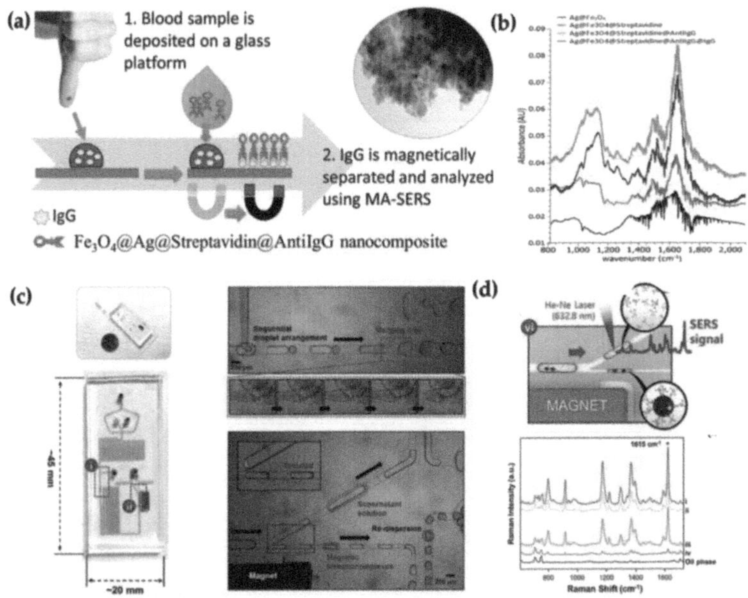

Figure 10.32 Metal-magnetic nanocomposite materials used as a biosensor. (a) Schematic illustration of a magnetic nanocomposite to determine the IgG in whole blood using SERS. (b) The IR spectra at 1650 cm^{-1} of the modified nanocomposites showed an enhanced signal compared to bare particles. Reprinted with permission from Ref. [7]. Copyright (2014) American Chemical Society. (c) Schematic representation of a wash-free immunoreaction assay detection chip injected with red ink with different phases of channels; formation, reaction, isolation, and detection of the supernatant of the immunoreaction via SERS. (d) The detection of IgG via SERS by magnetic separation. Reprinted with permission from Ref. [16]. Copyright (2017) American Chemical Society.

form periodic nanostructures of Cu-Ag nanodots on the 3D chip. Cu-Ag nanodots could detect rhodamine 6G (as a model analyte) and toxic Cd^{2+} ions in real time (at a concentration of 10 ppb, or 0.01 ppm) with the presence of crystal violet, as shown in Fig. 10.33a,b.

10.4.3 Photothermal Therapy

In clinical translation, the use of metallic nanomaterials for cancer photothermal therapy is slow compared to small-molecule

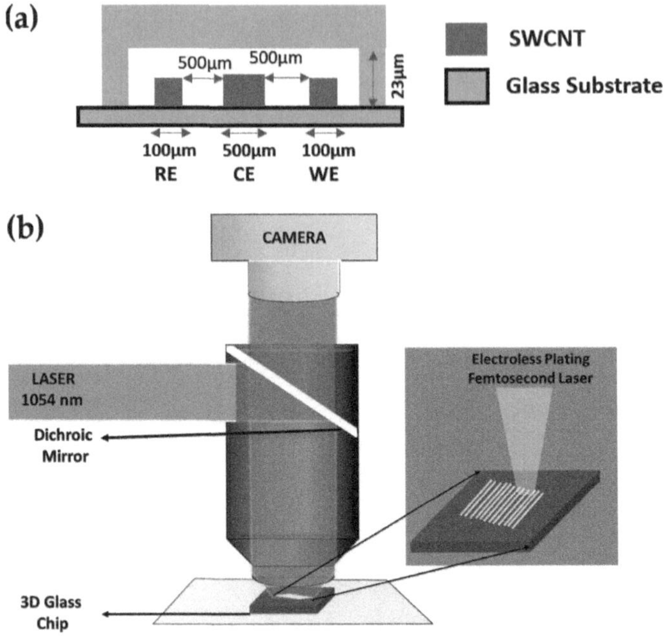

Figure 10.33 Cu-Ag bimetallic nanoparticles for biosensing applications. (a) Schematic illustration of different electrode configurations shown with a 100 µm reference electrode (RE) and a 500 µm of continuous electrode (CE) [49]. (b) Schematic representation of a 3D microfluidic biochip coated with Cu-Ag nanodisks under a SERS microscope [6].

inhibitors and liposome-loaded drugs for therapeutic applications. This is due to the fact that the reproducibility of nanomaterial synthesis is not uniform and the conventional process cannot be tuned and functionalized easily. Cancer treatment uses various techniques, like surgery, radiotherapy, chemotherapy, photothermal therapy (PTT), and hyperthermia, alone or in combination, depending on the tumor stage [1]. PTT uses a laser to illuminate plasmonic nanomaterials, creating resonance that produces heat locally. The mechanistic effect of PTT is clearly demonstrated both in experimental and theoretical studies. The main limitation of this method is the high tissue absorption, the penetration power, and the fast diffusion of the plasmonic particles, resulting in penetration of a few millimeters [117]. Clinically, hyperthermia is a nonradiative

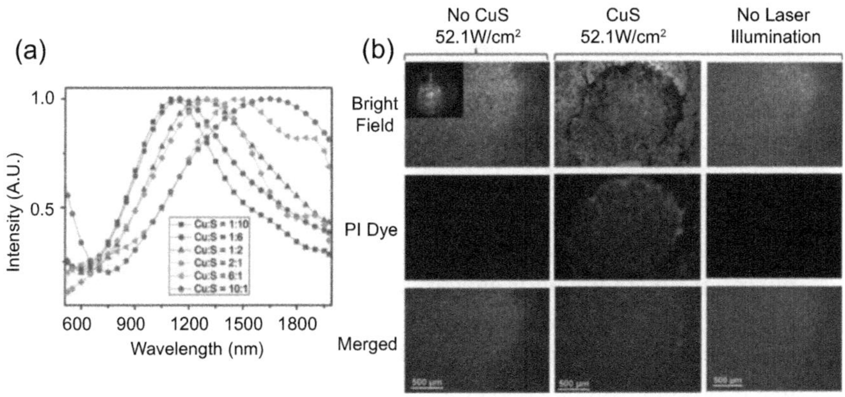

Figure 10.34 Photothermal therapy of $Cu_{2-x}S$ nanoparticles using laser irradiation. (a) The plasmon of CuS nanoparticles was tuned according to different ratios of Cu and S precursors. (b) RAW264.7 cells were exposed at 52.1 W/cm^2 along with CuS NPs, showing better hyperthermia applications compared to control (laser-exposed without CuS and no laser illumination). Republished with permission of The Royal Society of Chemistry, from Ref. [15], copyright (2016); permission conveyed through Copyright Clearance Center, Inc.

and harmless technique used as an adjuvant method to chemo- and radiotherapy. Hyperthermia involves heating of abnormal tissues non-invasively with specific nanomaterials by means of radio frequency (graphene and CNTs, laser ablation [copper and gold], and magnetic field [iron oxide nanomaterials]). Next, we will look at different NPs produced by microfluidic chips for therapeutic applications.

Several works have shown that plasmonic gold in the red and NIR regions shows better cellular death when exposed to laser parameters. But a few recent reports have shown a controversial result of Au NP heating (different sizes and concentrations) when Au NPs are irradiated with radio frequency for cancer treatment [44]. $Cu_{2-x}S$ plasmonic NPs have been developed using continuous flow via a millifluidic system, as shown in Fig. 10.35 [15]. These nanomaterials are widely used in laser-based hyperthermia and sensing applications. Developed $Cu_{2-x}S$ nanocrystals of different sizes showed plasmonic resonance in the far-infrared region (1115–

1644 nm) when the ratio of copper and sulfur precursors was modified [15]. Photothermal effects from $Cu_{2-x}S$ NPs irradiated at 52.1 W cm^{-2} with infrared laser induced high cell death in RAW264.7 exposed at 27 μM particle concentration. Lee et al. fabricated a hydrogel microfluidic device to test the efficacy of Au NP–mediated laser-based hyperthermia and cell migration studies. In this device, MCF-7 (breast cancer cells) and U87MG (human glioblastoma cells) were co-cultured to study and visualize migration and therapeutic effects. Also, Fang et al. reported a simple heating technique in microfluidics PDMS channels by embedding gold nanocrystals to enable better analysis of biological samples and diagnostic applications [23, 58].

10.5 Future Prospects

Microfluidics is now the state-of-the-art technology to produce NPs on a larger scale with minimal investment time, making it ideal for developing countries that require to monitor health and environment. Among the wide varieties of NPs, metal NPs (especially gold and silver) have shown immense potential in disease detection and sensing applications. In the future, microfluidic devices can help answer a lot of questions on the interaction chemistry in the formation of NPs. Also, biomedical assays can be performed in a more convenient way at less cost. Detecting chronic diseases in the early stages would help in effective prognosis and treatment planning. Nanomaterial synthesis on microfluidic chips with specific protocols will help biologists to quickly detect various biomolecules. Several studies have shown that utilizing NPs improves the capability to image single cells to study the interaction between a cell and biomolecules. Detecting single cells would aid clinicians to personalize cancer treatment and monitor disease prognosis. The use of nanotechnology and microfluidics could provide a simple and cost-effective way for point-of-care diagnostics in many rural areas, where the medical facilities are usually inadequate.

10.6 Conclusions

In this chapter, an overview of metallic NPs developed using microfluidic chips is provided, along with their biomedical applications. The experimental and theoretical information on developing nanomaterials for potential application in science and technology is also provided. Metallic NPs exhibit interesting characteristics compared to their macrosized particles. Microfluidic devices with a laminar or segmented-flow pattern are used to synthesize different metallic NPs. Effective mixing is the key parameter to achieve desired NPs. Winding and distributed mixing provides laminar flow–based microfluidic, while T-junction, flow focusing, Y-junction, and cross channel and their combinations are used for segmented-flow micromixing. By tuning the reaction parameters and microchannel dimensions, NPs of different shapes, sizes, compositions, and crystalline structures can be synthesized. The size of the particle depends on the flow rate of reactants, the concentration ratio, the seed age, the seed size, the temperature, the heating time, etc. Shape varies with the reducing agent and its concentration and flow rate and the carrier phase. Both shape and size can change the absorbance spectra of metallic NPs. NPs with different crystalline phases can be obtained by controlling the reaction kinetic energy or the flow rate. Metallic NPs possess a wide variety of properties that make them useful in biomedical applications such as bioimaging, biosensing, therapeutics, and diagnostics. In conclusion, there are many fields that require nanomaterials in bulk, and this bulk production with the highest performance at a very low cost can be achieved using LOC devices.

Acknowledgments

The authors greatly appreciate the financial support from the Science and Engineering Research Board (SERB) under grant number ECR/2016/001945, Department of Science and Technology (DST), Government of India, and Wellcome Trust/DBT India

Alliance Fellowship under grant number IA/E/16/1/503062. We acknowledge all authors and publishers who provided the copyright permissions.

References

1. N. S. Abadeer, C. J. Murphy (2016). Recent progress in cancer thermal therapy using gold nanoparticles. *J Phys Chem C*, **120**:4691–4716.
2. A. Abou-Hassan, R. Bazzi, V. Cabuil (2009). Multistep continuous-flow microsynthesis of magnetic and fluorescent gamma-Fe_2O_3@SiO_2 core/shell nanoparticles. *Angew Chem Int Ed Engl*, **48**:7180–7183.
3. A. Abou-Hassan, O. Sandre, S. Neveu, V. Cabuil (2009). Synthesis of goethite by separation of the nucleation and growth processes of ferrihydrite nanoparticles using microfluidics. *Angew Chem Int Ed Engl*, **48**:2342–2345.
4. S. Agrawal, A. Morarka, K. M. Paknikar, D. Bodas (2012). Microelectronic engineering in situ synthesis of Au nanoparticles in 3D circular microchannels in PDMS using a simple and reliable molding method. *Microelectron Eng*, **90**:104–107.
5. R. Baber, L. Mazzei, N. T. K. Thanh, A. Gavriilidis (2015). Synthesis of silver nanoparticles in a microfluidic coaxial flow reactor. *RSC Adv*, **5**:95585–95591.
6. S. Bai, D. Serien, A. Hu, K. Sugioka (2018). 3D microfluidic surface-enhanced Raman spectroscopy (SERS) chips fabricated by all-femtosecond-laser-processing for real-time sensing of toxic substances. *Adv Funct Mater*, **28**:1706262.
7. A. Balzerova, A. Fargasova, Z. Markova, V. Ranc, R. Zboril (2014). Magnetically-assisted surface enhanced Raman spectroscopy (MA-SERS) for label-free determination of human immunoglobulin G (IgG) in blood using Fe_3O_4@Ag nanocomposite. *Anal Chem*, **86**:11107–11114.
8. E. Blumfield, D. W. Swenson, R. S. Iyer, A. L. Stanescu (2019). Gadolinium-based contrast agents-review of recent literature on magnetic resonance imaging signal intensity changes and tissue deposits, with emphasis on pediatric patients. *Pediatr Radiol*, **49**:448–457.

9. A. M. Boies, P. Lei, S. Calder, W. G. Shin, S. L. Girshick (2011). Hot-wire synthesis of gold nanoparticles. *Aerosol Sci Technol*, **45**:654–663.
10. J. Boken, S. K. Soni, D. Kumar (2016). Microfluidic synthesis of nanoparticles and their biosensing applications. *Crit Rev Anal Chem [Internet]*, **46**:538–561.
11. J. Boleininger, A. Kurz, V. Reuss, C. Sönnichsen (2006). Microfluidic continuous flow synthesis of rod-shaped gold and silver nanocrystals. *Phys Chem Chem Phys*, **8**:3824–3827.
12. V. S. Cabeza, S. Kuhn, A. A. Kulkarni, K. F. Jensen (2012). Size-controlled flow synthesis of gold nanoparticles using a segmented flow microfluidic platform. *Langmuir*, **28**:7007–7013.
13. F. C. Cabrera, A. F. A. A. Melo, J. C. P. de Souza, A. E. Job, F. N. Crespilho (2015). A flexible lab-on-a-Chip magnetic separation of magnetite decorated with gold nanoparticles. *Lab Chip*, **15**:1835–1841.
14. S. Chen, D. L. Carroll (2002). Synthesis and characterization of truncated triangular silver nanoplates. *Nano Lett*, **2**:1003–1007.
15. T. L. Cheung, L. Hong, N. Rao, C. Yang, L. Wang, W. J. Lai, P. H. J. Chong, W.-C. Law, K.-T. Yong (2016). The non-aqueous synthesis of shape controllable Cu(2-x)S plasmonic nanostructures in a continuous-flow millifluidic chip for the generation of photo-induced heating. *Nanoscale*, **8**:6609–6622.
16. N. Choi, J. Lee, J. Ko, J. H. Jeon, G. E. Rhie, A. J. deMello, J. Choo (2017). Integrated SERS-based microdroplet platform for the automated immunoassay of F1 antigens in Yersinia pestis. *Anal Chem*, **89**:8413–8420.
17. N. Cordente, C. Amiens, B. Chaudret, M. Respaud, F. Senocq, M.-J. Casanove (2003). Chemisorption on nickel nanoparticles of various shapes: influence on magnetism. *J Appl Phys*, **94**:6358–6365.
18. I. O. de Solorzano, M. Prieto, G. Mendoza, T. Alejo, S. Irusta, V. Sebastian, M. Arruebo (2016). Microfluidic synthesis and biological evaluation of photothermal biodegradable copper sulfide nanoparticles. *ACS Appl Mater Interfaces*, **8**:21545–21554.
19. A. J. deMello (2006). Control and detection of chemical reactions in microfluidic systems. *Nature*, **442**:394–402.
20. S. Duraiswamy, S. A. Khan (2009). Droplet-based microfluidic synthesis of anisotropic metal nanocrystals. *Small*, **5**:2828–2834.
21. J. Estelrich, M. J. Sánchez-Martín, M. A. Busquets (2015). Nanoparticles in magnetic resonance imaging: from simple to dual contrast agents. *Int J Nanomed*, **10**:1727–1741.

22. H. T. Evensen, D. R. Meldrum, D. L. Cunningham (1998). Automated fluid mixing in glass capillaries. *Rev Sci Instrum*, **69**:519–526.
23. C. Fang, L. Shao, Y. Zhao, J. Wang, H. Wu (2012). A gold nanocrystal/poly (dimethylsiloxane) composite for plasmonic heating on microfluidic chips. *Adv Mater*, **24**:94–98.
24. Q. Fu, G. Ran, W. A. Xu (2015). Microfluidic-based controllable synthesis of rolled or rigid ultrathin gold nanoplates. *RSC Adv*, **5**:37512–37516.
25. R. Genç, M. Ortiz, C. K. O'Sullivan (2014). Diffusion-controlled synthesis of gold nanoparticles: nano-liposomes as mass transfer barrier. *J Nanopart Res*, **16**:2329.
26. D. V. Goia, E. Matijevic (1998). Preparation of monodispersed metal particles. *New J Chem*, **22**:1203–1215.
27. L. Gomez, V. Sebastian, S. Irusta, A. Ibarra, M. Arruebo, J. Santamaria (2014). Scaled-up production of plasmonic nanoparticles using microfluidics: from metal precursors to functionalized and sterilized nanoparticles. *Lab Chip*, **14**:325–332.
28. A. Günther, K. F. Jensen (2006). Multiphase microfluidics: from flow characteristics to chemical and materials synthesis. *Lab Chip*, **6**:1487–1503.
29. A. Günther, M. Jhunjhunwala, M. Thalmann, M. A. Schmidt, K. F. Jensen (2005). Micromixing of miscible liquids in segmented gas-liquid flow. *Langmuir*, **21**:1547–1555.
30. A. Günther, S. A. Khan, M. Thalmann, F. Trachsel, K. F. Jensen (2004). Transport and reaction in microscale segmented gas-liquid flow. *Lab Chip*, **4**:278–286.
31. N. Hao, Y. Nie, J. X. J. Zhang (2018). Microfluidic synthesis of functional inorganic micro-/nanoparticles and applications in biomedical engineering. *Int Mater Rev*, **63**:461–487.
32. N. Hassan, V. Cabuil, A. Abou-Hassan (2013). Continuous multistep microfluidic assisted assembly of fluorescent, plasmonic, and magnetic nanostructures. *Angew Chem Int Ed Engl*, **52**:1994–1997.
33. C. Haynes, R. P. V. Duyne (2002). Plasmon scanned surface-enhanced raman scattering excitation profiles. *Material Research Society Symposium Proceedings*, pp. S10.7.1–S10.7.6.
34. C. L. Haynes, A. D. McFarland, R. P. V. Duyne (2002). Surface-enhanced Raman spectroscopy. *Anal Chem*, **77**:338 A–346 A.
35. F. J. Heiligtag, M. Niederberger (2013). The fascinating world of nanoparticle research. *Mater Today [Internet]*, **16**:262–271.

36. H. Heinz, C. Pramanik, O. Heinz, Y. Ding, R. K. Mishra, D. Marchon, R. J. Flatt, I. Estrela-Lopis, J. Llop, S. Moya, R. F. Ziolo (2017). Nanoparticle decoration with surfactants: molecular interactions, assembly, and applications. *Surf Sci Rep [Internet]*, **22**:1–58.

37. V. Hessel, H. Löwe, F. Schönfeld (2005). Micromixers - a review on passive and active mixing principles. *Chem Eng Sci*, **60**:2479–2501.

38. H. Huang, H. du Toit, M. O. Besenhard, S. Ben-Jaber, P. Dobson, I. Parkin, A. Gavriilidis (2018). Continuous flow synthesis of ultrasmall gold nanoparticles in a microreactor using trisodium citrate and their SERS performance. *Chem Eng Sci*, **189**:422–430.

39. D. Huang, E. Swanson, C. Lin, J. Schuman, W. Stinson, W. Chang (1991). Optical coherence tomography. *Science*, **254**:1178–1181.

40. L. Hung, A. P. Lee (2006). Microfluidic devices for the synthesis of nanoparticles and biomaterials. *J Med Biol Eng*, **27**:1–6.

41. N. R. Jana, L. Gearheart, C. J. Murphy (2001). Seed-mediated growth approach for shape-controlled synthesis of spheroidal and rod-like gold nanoparticles using a surfactant template. *Chem Mater*, **13**:2313–2322.

42. X. Ji, H. Y. Lau, X. Ren, B. Peng, P. Zhai, S.-P. Feng, P. K. L. Chan (2016). Highly sensitive metabolite biosensor based on organic electrochemical transistor integrated with microfluidic channel and poly(N-vinyl-2-pyrrolidone)-capped platinum nanoparticles. *Adv Mat Technol*, **1**:1600042.

43. C. J. Johnson, E. Dujardin, S. A. Davis, C. J. Murphy, S. Mann (2002). Growth and form of gold nanorods prepared by seed-mediated, surfactant-directed synthesis. *J Mater Chem*, **12**:1765–1770.

44. P. Kaur, M. L. Aliru, A. S. Chadha, A. Asea, S. Krishnan (2016). Hyperthermia using nanoparticles-promises and pitfalls. *Int J Hyperthermia*, **32**:76–88.

45. K. L. Kelly, E. Coronado, L. L. Zhao, G. C. Schatz (2003). The optical properties of metal nanoparticles: the influence of size, shape, and dielectric environment. *J Phys Chem B*, **107**:668–677.

46. S. A. Khan, K. F. Jensen (2007). Microfluidic synthesis of titania shells on colloidal silica. *Adv Mater*, **19**:2556–2560.

47. S. A. Khan, A. Gunther, M. A. Schmidt, K. F. Jensen (2004). Microfluidic synthesis of colloidal silica. *Langmuir*, **20**:8604–8611.

48. Y. Kim, F. Fay, D. P. Cormode, B. L. Sanchez-Gaytan, J. Tang, E. J. Hennessy (2013). Single step reconstitution of multifunctional high-

density lipoprotein-derived nanomaterials using microfluidics. *ACS Nano*, **7**:9975–9983.

49. E. Ko, V.-K. Tran, Y. Geng, W. S. Chung, C. H. Park, M. K. Kim, G. H. Jin, G. H. Seong (2017). Continuous electrochemical detection of hydrogen peroxide by Au-Ag bimetallic nanoparticles in microfluidic devices. *J Electroanal Chem*, **792**:72–78.
50. J. M. Köhler, J. Wagner, J. Albert (2005). Formation of isolated and clustered Au nanoparticles in the presence of polyelectrolyte molecules using a flow-through Si chip reactor. *J Mater Chem*, **15**:1924–1930.
51. K. S. Krishna, Y. Li, S. Li, C. S. Kumar (2013). Lab-on-a-chip synthesis of inorganic nanomaterials and quantum dots for biomedical applications. *Adv Drug Deliv Rev*, **65**:1470–1495.
52. G. Ku, M. Zhou, S. Song, Q. Huang, J. Hazle, C. Li (2012). Copper sulfide nanoparticles as a new class of photoacoustic contrast agent for deep tissue imaging at 1064 nm. *ACS Nano*, **6**:7489–7496.
53. D. V. R. Kumar, M. Kasture, A. A. Prabhune, C. V. Ramana, B. L. V. Prasad, A. A. Kulkarni (2010). Continuous flow synthesis of functionalized silver nanoparticles using bifunctional biosurfactants. *Green Chem*, **12**:609–615.
54. V. K. LaMer (1952). Nucleation in phase transitions. *Ind Eng Chem Res*, **44**:1270–1277.
55. V. K. LaMer, R. H. Dinegar (1950). Theory, production and mechanism of formation of monodispersed hydrosols. *J Am Chem Soc*, **72**:4847–4854.
56. C. W. Lee, F. G. Tseng (2018). Surface enhanced Raman scattering (SERS) based biomicrofluidics systems for trace protein analysis. *Biomicrofluidics*, **12**:011502.
57. J. H. Lee, S. U. S. Choi, S. P. Jang, S. Y. Lee (2012). Production of aqueous spherical gold nanoparticles using conventional ultrasonic bath. *Nanoscale Res Lett*, **7**:1–7.
58. J. M. Lee, H. I. Seo, J. H. Bae, B. G. Chung (2017). Hydrogel microfluidic co-culture device for photothermal therapy and cancer migration. *Electrophoresis*, **38**:1318–1324.
59. X. Z. Lin, A. D. Terepka, H. Yang (2004). Synthesis of silver nanoparticles in a continuous flow tubular microreactor. *Nano Lett*, **4**:2227–2232.
60. G. Liu, L. Ma, X. Sun, Y. Jia, T. Wang (2018). Controllable synthesis of silver nanoparticles using three-phase flow pulsating mixing microfluidic chip. *Adv Mater Sci Eng*, **2018**:1–14.

61. J. Ma, S. M. Lee, C. Yi, C.-W. Li (2017). Controllable synthesis of functional nanoparticles by microfluidic platforms for biomedical applications: a review. *Lab Chip*, **17**:209–226.

62. S. Manohar, C. Ungureanu, T. G. V. Leeuwen (2011). Gold nanorods as molecular contrast agents in photoacoustic imaging: the promises and the caveats. *Contrast Media Mol Imaging*, **6**:389–400.

63. S. Marre, K. F. Jensen (2010). Synthesis of micro and nanostructures in microfluidic systems. *Chem Soc Rev*, **39**:1183–1202.

64. G. Mettela, S. Siddhanta, C. Narayana, G. U. Kulkarni (2014). Nanocrystalline Ag microflowers as a versatile SERS platform. *Nanoscale*, **6**:7480–7488.

65. V. Mody, R. Siwale, A. Singh, H. Mody (2010). Introduction to metallic nanoparticles. *J Pharm Bioallied Sci [Internet]*, **2**:282.

66. M. P. Morales, T. Gonzalez-Carreno, C. J. Serna (1992). The formation of a–Fe_2O_3 monodispersed particles in solution. *J Mater Res Technol*, **7**:2538–2545.

67. S. Mourdikoudis, K. Simeonidis, A. Vilalta-Clemente, F. Tuna, I. Tsiaoussis, M. Angelakeris, C. Dendrinou-Samara, O. Kalogirou (2009). Controlling the crystal structure of Ni nanoparticles by the use of alkylamines. *J Magn Magn Mater*, **321**:2723–2728.

68. J. R. G. Navarro, M. H. V. Werts (2013). Resonant light scattering spectroscopy of gold, silver and gold–silver alloy nanoparticles and optical detection in microfluidic channels. *Analyst*, **138**:583–592.

69. K. W. Oh (2012). Lab-on-chip (LOC) devices and microfluidics for biomedical applications. In *MEMS for Biomedical Applications*, Woodhead Publishing Series in Biomaterials, pp. 150–171.

70. D. A. Papaconstantopoulos (1989). Ferromagnetism in hexagonal-close-packed elements. *Phys Rev B*, **39**:2526–2528.

71. J. Polte (2015). Fundamental growth principles of colloidal metal nanoparticles: a new perspective. *CrystEngComm*, **17**:6809–6830.

72. J. Qiao, H. Ding, Q. Liu, R. Zhang, L. Qi (2017). Preparation of polymer@AuNPs with droplets approach for sensing serum copper ions. *Anal Chem*, **89**:2080–2085.

73. M. Russo, P. Bevilacqua, P. A. Netti, E. Torino (2016). A microfluidic platform to design crosslinked hyaluronic acid nanoparticles (cHANPs) for enhanced MRI. *Sci Rep*, **6**:37906.

74. H. SadAbadi, S. Badilescu, M. Packirisamy, R. Wuthrich (2013). Integration of gold nanoparticles in PDMS microfluidics for lab-on-a-

chip plasmonic biosensing of growth hormones. *Biosens Bioelectron*, **44**:77–84.

75. P. R. Sajanlal, T. S. Sreeprasad, A. K. Samal, T. Pradeep (2011). Anisotropic nanomaterials: structure, growth, assembly, and functions. *Nano Rev*, **2**:5883.
76. T. S. Santra, F.-G. Tseng (2014). Micro/nanofluidic devices for single cell analysis. *J Micromach*, **5**:154–157.
77. T. S. Santra, F.-G. Tseng (2016). Pulse laser activated photoporation for high efficient intracellular delivery using nano-corrugated mushroom shape gold nanoparticles. In *The 20th International Conference on Miniaturized Systems for Chemistry and Life Sciences* (μ-*TAS-2016*), Dublin, Ireland, pp. 1097–1098.
78. T. S. Santra, F.-G. Tseng (2018). *Handbook of Single Cell Technology*, Springer Nature.
79. T. S. Santra, F.-G. Tseng, T. K. Barik (2014). Green biosynthesis of gold nanoparticles and biomedical applications. *Am J Nanores Appl*, **2**:5–12.
80. T. S. Santra, F.-G. Tseng, T. K. Barik (2014). Biosynthesis of silver and gold nanoparticles for potential biomedical applications-a brief review. *J Nanopharm Drug Deliv*, **2**:249–265.
81. T, S. Santra, T.-H. Wu, P. Y. Chiou (2016). Photothermal microfluidics. In *Optical MEMS for Chemical Analysis and Biomedicine*, The Institute of Engineering and Technology (IET), NJ, USA.
82. V. Sebastian, S. Basak, K. F. Jensen (2016). Continuous synthesis of palladium nanorods in oxidative segmented flow. *AIChE J*, **62**, 373–380.
83. V. Sebastian, S. K. Lee, C. Zhou, M. F. Kraus, J. G. Fujimoto, K. F. Jensen (2012). One-step continuous synthesis of biocompatible gold nanorods for optical coherence tomography. *Chem Commun (Camb)*, **48**:6654–6656.
84. V. Sebastian, C. D. Smith, K. F. Jensen (2016). Shape-controlled continuous synthesis of metal nanostructures. *Nanoscale*, **8**:7534–7543.
85. D. Shalom, R. C. R. Wootton, R. F. Winkle, B. F. Cottam, R. Vilar, A. J. DeMello, C. P. Wilde (2007). Synthesis of thiol functionalized gold nanoparticles using a continuous flow microfluidic reactor. *Mater Lett*, **61**:1146–1150.
86. W. Sheng, T. Chen, W. Tan, Z. H. Fan (2013). Multivalent DNA nanospheres for enhanced capture of cancer cells in microfluidic devices. *ACS Nano*, **7**:7067–7076.

87. P. Shinde, L. Mohan, A. Kumar, K. Dey, A. Maddi, A. N. Patananan, F.-G. Tseng, H.-Y. Chang, M. Nagai, T. S. Santra (2018). Current trends of microfluidic single cell technologies. *IJMS*, **19**:3143.
88. M. Shviro, D. Zitoun (2013). Nickel nanocrystals: fast synthesis of cubes, pyramids and tetrapods. *RSC Adv*, **3**:1380–1387.
89. X. C. I. Solvas, A. Demello (2011). Droplet microfluidics: recent developments and future applications. *Chem Commun*, **47**:1936–1942.
90. H. Song, D. L. Chen, R. F. Ismagilov (2007). Reactions in droplets in microfluidic channels. *Angew Chem Int*, **45**:7336–7356.
91. H. Song, J. D. Tice, R. F. Ismagilov (2003). A microfluidic system for controlling reaction networks in time. *Angew Chem Int Ed*, **42**:768–772.
92. Y. Song, D. Cheng, L. Zhao (2018). Microfluidic synthesis of metallic nanomaterials. In *Microfluidics Fundamentals, Devices, and Applications*, Wiley-VCH, Weinheim, pp. 419–444.
93. Y. Song, S. Ji, Y. J. Song, R. Li, J. Ding, X. Shen, R. Wang, R. Xu, X. Gu (2013). In situ redox microfluidic synthesis of core-shell nanoparticles and their long-term stability. *J Phys Chem C*, **117**:17274–17284.
94. Y. Song, H, Modrow, L. L. Henry, C. K. Saw, E. E. Doomes, V. Palshin, J. Hormes, C. S. S. R. Kumar (2006). Microfluidic synthesis of cobalt nanoparticles. *Chem Mater*, **18**:2817–2827.
95. K. Sugano, Y. Uchida, O. Ichihashi, H. Yamada, T. Tsuchiya, O. Tabata (2010). Mixing speed-controlled gold nanoparticle synthesis with pulsed mixing microfluidic system. *Microfluid Nanofluid*, **9**:165–174.
96. D. Sun, Z. Chen, M. Wu, Y. Zhang (2017). Nanomaterial-based microfluidic chips for the capture and detection of circulating tumor cells. *Nanotheranostics*, **1**:389–402.
97. Y. Sun, B. Mayers, Y. Xia (2003). Transformation of silver nanospheres into nanobelts and triangular nanoplates through a thermal process. *Nano Lett*, **3**:675–679.
98. N. T. Thanh, N. Maclean, S. Mahiddine (2014). Mechanisms of nucleation and growth of nanoparticles in solution. *Chem Rev*, **114**:7610–7630.
99. F.-G. Tseng, T. S. Santra (2015). *Micro/Nanofluidic Devices for Single Cell Analysis*, MDPI AG Basel, Switzerland, ISBN 978-3-03842-090-3.
100. F.-G. Tseng, T. S. Santra (2016). *Essentials of Single-Cell Analysis*, Springer-Verlag, Berlin Heidelberg, Germany, ISBN 978-3-662-49116-4.

101. F.-G. Tseng, T. S. Santra (2016). *Single Cell Analysis in Biotechnology and System Biology*, MDPI AG Basel, Switzerland, ISBN 978-3-03842-194-8.

102. M. Tsuji, M. Hashimoto, Y. Nishizawa, M. Kubokawa, T. Tsuji (2005). Microwave-assisted synthesis of metallic nanostructures in solution. *Chem: Eur J*, **11**:440–452.

103. J. Turkevich, P. C. Stevenson, J. Hillier (1951). A study of the nucleation and growth processes in the synthesis of colloidal gold. *Discuss Faraday Soc*, **11**:55–75.

104. L. Uson, V. Sebastian, M. Arruebo, J. Santamaria (2016). Continuous microfluidic synthesis and functionalization of gold nanorods. *Chem Eng J*, **285**:286–292.

105. P. M. Valencia, O. C. Farokhzad, R. Karnik, R. Langer (2012). Microfluidic technologies for accelerating the clinical translation of nanoparticles. *Nat Nanotechnol [Internet]*, **7**:623–629.

106. J. Wagner, J. M. Köhler (2005). Continuous synthesis of gold nanoparticles in a microreactor. *Nano Lett*, **5**:685–691.

107. J. Wagner, T. Kirner, G. Mayer, J. Albert, J. M. Köhler (2004). Generation of metal nanoparticles in a microchannel reactor. *Chem Eng J*, **101**:251–260.

108. J. Wagner, T. R. Tshikhudo, J. M. Köhler (2007). Microfluidic generation of metal nanoparticles by borohydride reduction. *Chem Eng J*, **135**:104–109.

109. Y. Wang, Y. Xia (2004). Bottom-up and top-down approaches to the synthesis of monodispersed spherical colloids of low melting-point metals. *Nano Lett*, **4**:2047–2050.

110. J. Watt, B. G. Hance, R. S. Anderson, D. L. Huber (2015). Effect of seed age on gold nanorod formation: a microfluidic, real-time investigation. *Chem Mater*, (27):6442–6449.

111. C. H. Weng, C. C. Huang, C. S. Yeh, G. B. Lee (2007). Synthesis of gold nanoparticles using microfluidic reaction systems. In *2007 7th IEEE Conference on Nanotechnology (IEEE NANO)*, pp. 462–466.

112. C. H. Weng, C. C. Huang, C. S. Yeh, G.-B. Lee (2007). Synthesis of hexagonal gold nanoparticles using microfluidic reaction systems. *J Micromech Microeng*, **18**:462–466.

113. G. M. Whitesides (2006). The origins and the future of microfluidics. *Nature*, **442**:368–373.

114. Z. Wu, S. Yang, W. Wu (2016). Shape control of inorganic nanoparticles from solution. *Nanoscale*, **8**(3):1237–1259.

115. L. Xu, J. H. Peng, M. Yan, D. Zhang, A. Q. Shen (2016). Droplet synthesis of silver nanoparticles by a microfluidic device. *Chem Eng Proc*, **102**:186–193.

116. B. K. H. Yen, A. Günther, M. A. Schmidt, K. F. Jensen, M. G. Bawendi (2005). A microfabricated gas-liquid segmented flow reactor for high-temperature synthesis: the case of CdSe quantum dots. *Angew Chem Int Ed*, **44**:5447–5451.

117. C. Zeng, W. Shang, X. Liang, X. Lieng, Q. Chen, C. Chi, Y. Du, C. Fang, J. Tian (2016). Cancer diagnosis and imaging-guided photothermal therapy using a dual-modality nanoparticle. *ACS Appl Mater Interfaces*, **8**:29232–2941.

118. L. Zhang, Y. Wang, L. Tong, Y. Xia (2013). Seed-mediated synthesis of silver nanocrystals with controlled sizes and shapes in droplet microreactors separated by air. *Langmuir*, **29**:15719–15725.

119. X. Zhang, C. Wiles, S. L. Painter, P. Watts, S. J. Haswell (2006). Microreactors as tools for chemical research. *Chim Oggi*, **24**:43–45.

120. C. X. Zhao, L. He, S. Z. Qiao, A. P. J. Middelberg (2011). Nanoparticle synthesis in microreactors. *Chem Eng Sci*, **66**:1463–1479.

Chapter 11

Microfluidic Particle Separation and Biomedical Applications

Ida Laila Ahmad[a] **and Mohd Ridzuan Ahmad**[b]
[a] *Department of Electronic Engineering, Faculty of Electrical and Electronic Engineering, Universiti Tun Hussein Onn Malaysia, 86400 Batu Pahat, Johor, Malaysia*
[b] *Division of Control and Mechatronics Engineering, School of Electrical Engineering, Faculty of Engineering, Universiti Teknologi Malaysia, 81310 Skudai, Johor, Malaysia*
ida@uthm.edu.my, mdridzuan@utm.my

Microfluidic developments have made recent progress in seeking a simple and effective separation for a variety of applications such as food industries, biodiagnosis, and medical diagnostics. In this chapter, by using hydrodynamic separation, we describe a tapered microfluidic system for passive continuous separation of microparticles. Successful size-based separation is demonstrated by leveraging the hydrodynamic properties of fluid flow and the physical characteristics of microparticles. The tapered microfluidic device has expanding geometries with regard to a specific taper angle that amplifies the sedimentation effect of particles from various sizes. Using taper angles of 20° and 25°, a mixture of 3 μm and 10 μm polystyrene (PS) microbeads is successfully separated. These results agree with the finite element simulation performed

Microfluidics and Bio-MEMS: Devices and Applications
Edited by Tuhin S. Santra

ISBN 978-981-4800-85-3 (Hardcover), 978-1-003-01493-5 (eBook)
www.jennystanford.com

in 3D form using Abaqus 6.12. In addition, the effectiveness of this biological separation process is demonstrated through the use of polydisperse samples consisting of 3 μm PS microbeads and human epithelial cervical carcinoma (HeLa) cells. At outlet 1 and outlet 3, with a fluid rate of 0.5–3.0 μL/min, 98% of sample purity is regenerated. Despite taking a passive approach to separation, our system is interesting. This approach allows multiparticle separation in a stand-alone system to be simple, label-free, and continuous without the need for bulky instruments. This platform can, therefore, become an enabling technology for point-of-care diagnostic tools and hold potential for applications for a micro–total analysis system (μTAS).

11.1 Introduction

For industries like the pharmaceutical industries, food processing, and biological parts preparations for medical analysis, particles and cell separation are essential [1–2]. For example, it is important to isolate circulating tumor cells (CTCs) to screen the progression of cancer in patients [3–5]. With regard to immune deficiency and genetic abnormalities, blood components are essential for the diagnosis of blood [6]. Microfabrication developments have allowed microfluidic work to flourish and eventually become one of the most effective separation solutions. Microfluidic advantages such as lower reagent use, reduced power requirements, and lower sample volume at lower cost have been shown to be greater than traditional bench-top systems [7–10]. Sensing/detection [11, 12], micromanipulation, micromixer, and microseparation are examples of microfluidic applications [13]. It is possible to classify microfluidic separation system into two major groups known as active and passive.

Studies such as microfabricated particle sorters based on acoustic waves [14–16], dielectrophoresis (DEP) [17, 18], optical patterns [19–22], fluorescence-activated cell sorter (FACS) [23], and magnet-activated cell sorting (MACS) [24–27] have previously explored various active separation techniques. Although these methods can separate the particles successfully, due to external forces, they are subject to several drawbacks. To explain, the unwanted Joule heating effect and the experience of electrolysis during separation from DEP will damage biological cells and alter their properties. While costly biochemical markers used for both FACS and MACS may not be appropriate for all cell types, cell rheology may also be interfering. In addition, the difficulty of preparation procedures and integration with other modules would impede chip integration in order to create a full lab-on-a-chip (LOC) system.

The separation process of microparticles, including cells, was demonstrated as a result of passive separation techniques such as microfiltration [28, 29], microvortex manipulation (MVM) [21], and pinching flow fractionation (PFF) [33, 34]. Such methods are safe and very attractive since biological cells are treated in a gentle manner. Hydrodynamic separation is among these techniques, which has a simple nondestructive nature and is one of the most frequently used. The theory of hydrodynamic separation lies in the manipulation of the stream and the interaction between particles and fluid or walls. Instead, it uses no external force or stimulus that induces the unique behavior of separation because of hydrodynamic spreading. Separation is accomplished by contact between particle and fluid by regulating the flow rate ratio or the active forces acting on particles, for example, inertial force. In addition, hydrodynamic separation through the contact between the particle and the wall can use either a continuous wall or a separate wall. The technique has been used for the segregation of size, shape, and even density of the blood, plasma, and microparticles [35].

For hydrodynamic devices, separation efficiency is usually limited. Kuo et al. engineered a curved microchannel to produce centrifugal force pushing the CTCs through the 90% efficiency filters in order to increase separation efficiency. The design, however, has limited ability to deplete small particles [36]. Sethu et al. designed a diffuse microchannel that created a greater overbound pressure drop, and 90% white blood cells (WBCs) had been effectively sorted [37]. Meanwhile, Chiu et al. developed a filtration system with a high aspect ratio (HAR) using new pressure bonding and buffer flow to sort particles of different sizes [38]. The inertial force was used to further enrich the samples, as Di Carlo et al. have shown [39, 40]. In comparison, prior dilution of samples is necessary, and the cellular portion is directly affected by viscosity, thereby increasing the length of the microchannel [48–49]. Huh et al. have shown the effect of gravity on the microchannel, which improves separation efficiencies by >99%. However, the process is very slow and the volumetric flux rate is limited to approximately 1 mL/h [41]. Through a 3D microchannel, work by Li et al. improvised the separation device. Amid improved separation performance, the device requires complex manufacturing and repetitive bonding alignment procedure [42].

In this case, an alternative separation method is needed to produce high-purity samples using a low-complexity system. We suggest a passive microfluidic separation device using hydrodynamic principles and a sedimenting effect to overcome this problem.

11.1.1 Theory and Mechanism

The microfluidic separator consists of a single inlet, concentrating field, main channel, and three outlets with a gradually growing width called the tapered width. The width of the inlet and outlet was limited to 2 mm, and the main channel length was 4 mm. The concentrating field is 1 mm long with 100 μm width. As shown in Fig. 11.1, four different angles have been designed, which were 6°, 12°, 20°, and 25°. On the basis of the taper angle, the gradually increasing width can affect the overall hydraulic diameter. A small angle (6°, 12°) corresponds to a tapered microfluidic device in a more rectangular shape. The larger 20° and 25° angles were used to demonstrate the taper angle effect.

The hydrodynamical separation of micro-objects required manipulation of hydrodynamic spreading by the tapered device. The tapered microchannel is known as a microchannel whose channel width increases. This unique design allows faster speed distribution so that the pressure gradient is created between the main channel and the three outlets. In addition, increasing the tapered microchannel width creates expansion streamlines that are divided into the carrier flow and the initial sample flow. Higher speeds mean lower hydrodynamic resistance that is essential to ensure that particles are effectively trapped in the designated outlets. Since the short focus region used to converge particles

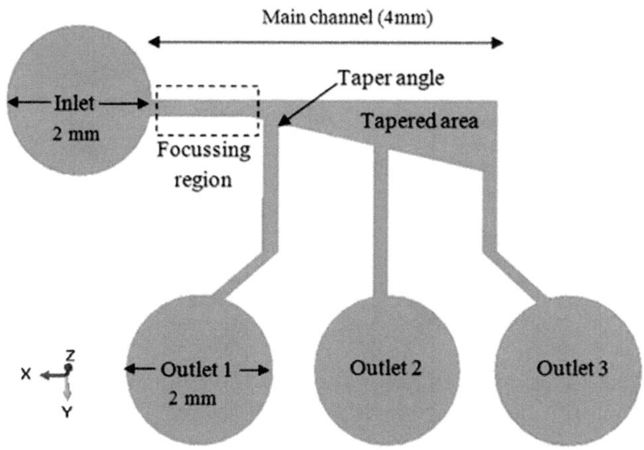

Figure 11.1 Schematic of the proposed separation mechanism. The tapered area will change accordingly on the basis of the taper angle size. The main channel includes both a focusing region and a tapered area. © 2017 IEEE. Reprinted, with permission, from Ref. [45].

together, no buffer or sheath flow is required. Regardless of the size, form, and density of particles, they will be compacted together and migrate around the narrow area. When the flow of samples is introduced to the inlet continually, the particles converge in the focusing region.

At the end of the focusing area, particles began moving on their own trajectory, resulting in a differential vertical position between particles of different sizes. This spreading activity resulted from an abrupt channel expansion or tapered field. As velocity becomes very slow at the tapered area (see velocity colored contour in Fig. 11.2), the gravity is strong and allows the sedimentation to occur. Larger particles were forced into trajectory carrier flows depending on the size. Owing to higher diffusion and faster elution, small particles follow the original sample flow. Due to very limited dimensions, particularly the height of the channel, sedimentation is pronounced in the tapered microchannel. The 10-μm-narrow channel height causes rapid sedimentation. In addition, the rate of sedimentation depends on the inclination angle (taper angle). As a result, the closest outlet (outlet 1) collects larger particles, while the smallest particles are collected from the furthest outlet (outlet

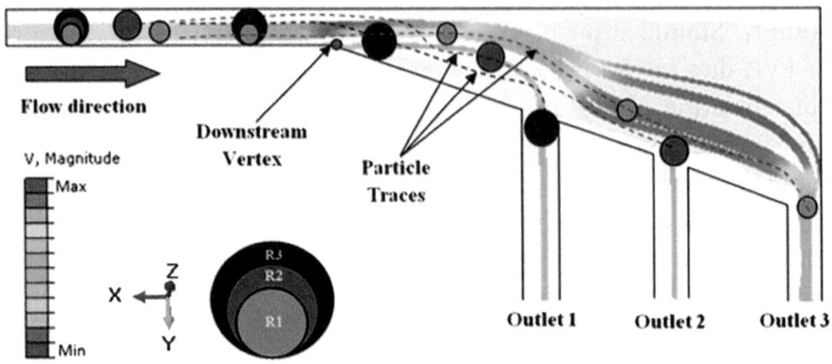

Figure 11.2 Working principle of a tapered microfluidic device used for separation. The flow direction is from left to right. Three particles of different radii ($R1$, $R2$, $R3$) are separated into three different target outlets. © 2017 IEEE. Reprinted, with permission, from Ref. [45].

3). In principle, midsize particles are taken out of the middle outlet (outlet 2), as shown in Fig. 11.2.

11.1.2 Flow Resistance

The sample flow through a separation system can be modeled using an analogous fluid circuit consisting of discrete R_{sense} resistive elements (for the rectangular section) and R_{h1}, R_{h2}, and R_{h3} for the tapered field and three collection outlets, as shown in Fig. 11.3.

Since the tapered microchannel incorporated both configurations, the approach to segmentation was followed to derive the

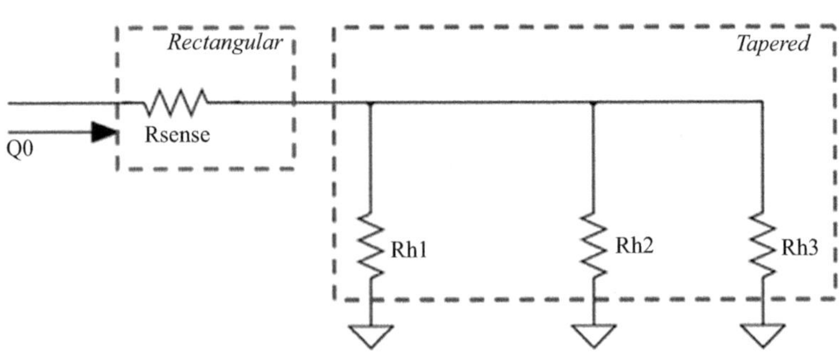

Figure 11.3 Equivalent fluidic circuit to represent a tapered microfluidic device. © 2017 IEEE. Reprinted, with permission, from Ref. [45].

respective total flow resistance within the microchannel, as shown in Fig. 11.3. The hydrodynamic flow resistance is given in Eq. 11.1 based on the Navier–Stokes equation for Newton's incompressible fluid.

$$\Delta P = QR_{\mathrm{h}}, \tag{11.1}$$

where Q is a volumetric flow rate and ΔP is the difference in pressure between the outlet and the channel inlet and the channel hydrodynamic resistance R_{h}. The lower R_{h} outlet is distinguished by the outlet with the highest speed, thereby affecting the movement of suspended particles to escape via this outlet inside the tapered microfluidic. R_{h} computation is extremely geometry dependent, so rectangular and tapered field equations are as follows [43]:

$$R_{\mathrm{h,rect}} = \frac{32\,\mu\mathrm{L}}{D^2 \omega d} \tag{11.2}$$

and

$$R_{\mathrm{h,tapered}} = \frac{12\,\mu\mathrm{L}}{h^3} \frac{\ln(\frac{\omega_1}{\omega_0})}{\omega_1 - \omega_0} \tag{11.3}$$

where h is the height; ω, ω_0 and ω_1 are the widths ($h \leq \omega$); the linear dimensions are L; and the hydraulic diameter is D. It can be seen that for determining R_{h} within the microchannel, geometric parameters are significant. To measure the resistance values of the tapered microchannel consisting of both configurations (rectangular and tapered), the segmentation method was used. Previously, region segmentation has been used by Ref. [44] to evaluate R_{h}. Figure 11.4 shows the segmentation areas, and the overall R_{h} evaluation was performed on the basis of the summation of individual R_{h} from the respective areas, as shown in Table 11.1.

11.2 Materials and Methods

Advanced numerical simulation offers a solution for multiapplied physics and the phenomenon of engineering. The simulation tool typically known as the predictive command allows for cross-disciplinary product development. Liu et al. suggested a rapid and continuous separation of rare tumor cells on the basis

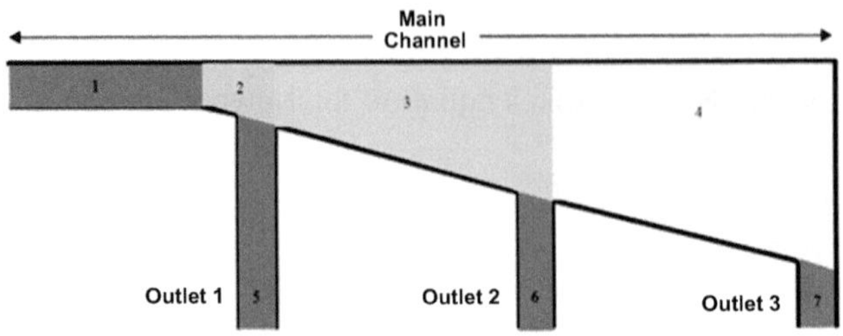

Figure 11.4 Segmentation of areas inside a tapered microfluidic device for hydrodynamic resistance (R_h) evaluation. © 2017 IEEE. Reprinted, with permission, from Ref. [45].

Table 11.1 Hydrodynamic resistance evaluation

Area	Evaluation parameter (Summation of areas)
Main channel	1 + 2 + 3 + 4
Outlet 1	2 + 5
Outlet 2	3 + 6
Outlet 3	4 + 7

of hydrodynamic effects without labels [45]. In the numerical simulation analysis using commercial CFD software FLUENT 6.4 (Ansys Inc), hydrodynamic size-based cell separation of spiked HeLa cells in diluted blood cells in the curved double spiral microchannel was analyzed. The binary blend of 5 μm and 15 μm cells could be separated with a double spiral channel based on simulation results and showed 95% of separation purity at 30 mL/h flow rate. The double spiral channel also achieved high throughput (2.5×10^8 cells/min) and effective cell separation (more than 90%) of spiked HeLa cells and 20 × nearly diluted blood cells. In another study, Sun et al. researched an immersed boundary (IB)–lattice Boltzmann system (LBM)-coupled model based on hydrodynamic simulation to separate microchannel particles [46]. Over a normal Eulerian grid fluid–particle interaction, the incompressible fluid flow was solved using LBM and the IB model was used to analyze particle interaction. Under a pressure gradient and forces caused by self-rotation, the multisize microparticle was successfully divided into an equilibrium location. The size-dependent cell division and sorting channel proposed by Jing-Tao et al. was based on an asymmetrical controllable pinching flow using an IB-LBM method [47]. By controlling the cell size, inlet flow ratio, and flux ratio, LBM solved the simulation and numerical model with a D2Q9 lattice model and an AsPFF cell sorter model. The channel is constructed with two upstream branches, one transitional channel and four downstream branches (D-branches). On the basis of the simulation, the releasing cells of different dimensions were aligned with the hydrodynamic forces in a transitional channel line and fed into target outlets by regulating the channel flow. Samla et al. have conducted a recent study using multiphysics simulation software COMSOL [48] to isolate DNA and superparamagnetic bead microparticles into a microfluidic channel. The developed microfluidic platform is able to separate 83.26% of the DNA output and superparamagnet bead particles on the basis of different particle sizes using four inlets and two outlets of different sizes. In computational fluid dynamics, the particle tracing module was employed to visualize and calculate the direction of a particle in an electromagnetic force–induced mixing scenario.

11.2.1 Flow Streamlines Simulation

There are two investigative simulations in this research work. The simulation of flow streamlines was performed to test microparticle flow patterns for various flow rates. Subsequently, simulation of microparticle separation was studied to distinguish inserted particles on the basis of their size. Using Abaqus 6.12 (Dassaults Systemes Simulia Corp, Providence, RI, USA), the flow fields or streamlines produced in the microfluidic device were obtained from finite element simulation. The tapered microfluidic channel used the “Rigid” component, while the use of the “Eulerian” fluid was specified with water characteristics [49]. It consists of 2622 nodes, and 1765 EC3D8R elements form in a rectangular mesh with hexahedral elements. Due to the interaction between fluid and structure, the simulation was performed using Couple Eulerian Lagrangian (CEL) analysis. The “Rough” limits of fluid flow inside the microfluid, along with in-flow and out-flow limits, have been

Figure 11.5 Flow streamlines inside a tapered microchannel when the flow rate is 2 μL/min. © 2017 IEEE. Reprinted, with permission, from Ref. [45].

set accordingly. The fluid rate varies according to the channel configuration when the flow rate is applied at the inlet, and this influences the internal flow. Laminar streamlines were observed, as shown in Fig. 11.5, when flow stability was reached. Laminar streamlines within microfluidics are important since excessive flow turbulences are avoided. The image's velocity contour reveals that outlet 3 provided the highest velocity, which means the lowest hydrodynamic resistance relative to other outlets (outlet 1 and outlet 2). This velocity property can be further manipulated to establish successful particle separation or particle trapping in accordance with the hydrodynamic principle.

11.2.2 Microparticle Separation Simulation

Finite element analysis (FEA) was performed for better anticipation and visualization in order to realize the proposed separation process. The finite element model has been developed to reflect the microbeads of polystyrene (PS) and human cervical epithelial carcinoma (HeLa) cells. Abaqus 6.12 FEA's ability to simulate solid–fluidic interaction in a 3D environment has been shown to be advantageous, particularly in solving complex engineering issues. The material properties obtained from [50–52] are described in Table 11.2 below.

The flow field for incompressible fluid was governed by the Navier–Stokes equation. At all outlets, boundary conditions were set to reflect a fixed environmental pressure state, while at the inlet flow, velocity varying from 0.5 μL/s to 5 μL/s is described. The 3D simulation assembly consists of both the 3D model (PS microbeads

Table 11.2 Physical properties for the finite element model

Properties (unit)	Polystyrene microbeads	HeLa cell
Radius (μm)	10	10
Density (kg/m^3)	1050	2290
Young's modulus (Pa)	125×10^9	35×10^3
Poisson ratio	0.275	0.39
Mesh type	C3D10	C3D10
No. of element	1246	1246
Part	3D deformable	3D deformable

Table 11.3 Separation simulation results

Device angle (°)	Flow rate (μL/min)	Polydisperse sample
6		No separation
12	0.5–3	No separation
20		Outlet 1: HeLa cells
		Outlet 2: None
		Outlet 3: PS microbeads
25		Outlet 1: HeLa cells
		Outlet 2: None
		Outlet 3: PS microbeads

and HeLa cells) within the tapered microfluidic-containing buffered fluid. The successful separation was shown using higher taper angles of 20° and 25°, collecting HeLa cell from outlet 1 and collecting PS microbeads from outlet 3. Table 11.3 summarizes the results of the simulation.

11.2.3 Microparticle Separation Technique

Separation is possible by electrical force, magnetic field force, acoustic force, or optical deflection in the microfluidic channel. Depending on the force applied, the inserted microparticle can respond differently and the deflected output can be collected at the outlet channel. Differential mechanical forces can further clarify the separation of the sample on-chip. Microfiltration is one of the common microfluidic methods used to isolate microparticles. To capture targeted molecules, the microfilter can serve as a membrane [48]. Kim et al. conducted a study of a capillary-driven microfluidic device made of polydimethylsiloxane (PDMS) and cross-filter architecture for plasma separation using human cells. A PDMS substrate was developed by adding hydrophilic surface superfactants and improving capillary action to reduce external drive force utilization of fluid flow [53]. While the autonomous plasma extraction chip's capillary filling behavior is regarded as simple to use, the analysis of the results is somewhat constrained. The output was restricted by passive capillary transport and cell clogging as obstacles to the filtration process. The method of cell trapping was used to separate cells in a channel geometry design analysis by Wilding et al. [54]. In the channel, the larger cells are trapped, while the smaller cells are washed out. Then it is possible to collect the trapped cells for further study. Nevertheless, cell isolation was partially successful as the larger cells could eventually squeeze through the narrow space and reduce the overall efficiency, in addition to the microfluidic chip's clogging problem. Therefore, the microfilter was adjusted in the follow-up analysis using added micropost weir-type filters primarily to catch large cells rather than to flow through the filter surface. According to the study, WBC separation was an achievement with a higher yield and no obstruction was noted [55]. A microchannel was produced with an integrated microcolumn and packed silica microbeads in a recent study by Shim et al. [56]. Some researchers tried to make a biomarker separation microfluid chip based on the capillary-driven microchannel flow and cross-flow array [57–59]. The results showed improved separation and developed microanalysis systems for the diagnosis of points of care.

From the filtration approach to the separation of cells, Zweifach–Fung's hydrodynamic effect, known as the bifurcation rule, was used to distinguish biomarkers by lateral-laminar-flow migration [60]. Through controlling flow rate streamlines for separation and extraction, targeted biomarkers can also be induced to flow in a specific direction. For the preparation of the sample of nucleic acid separation and subsequent reverse transcription polymerase chain reaction (RT-PCR) for molecular amplification, Cao et al. produced a hybrid microfluidic chip with solid-phase extraction (SPE) [61]. The main objective of this chip was to isolate the influenza virus

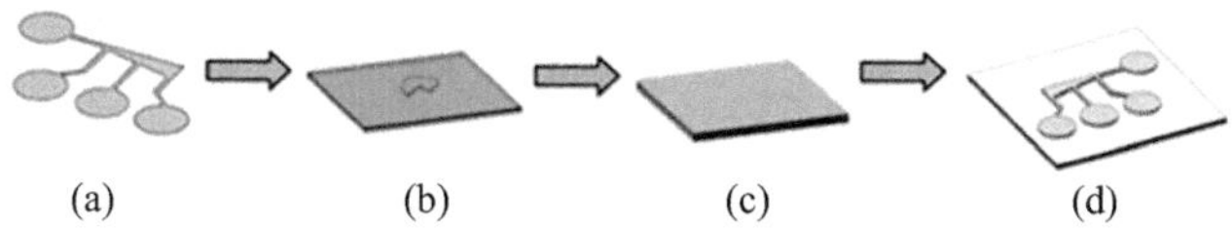

Figure 11.6 Soft lithography techniques. Illustrations of silicone master mold fabrication procedures: (a) tapered microchannel design on a chrome mask; (b) SU-8 photoresist on a silicone substrate; (c) a silicone substrate coated with SU-8; (d) the positive relief of the master mold after development. © 2017 IEEE. Reprinted, with permission, from Ref. [45].

in human nasopharyngeal aspirate (NPA) and nasopharyngeal swab (NPS) specimens. The entire sample separation process was carried out on the basis of capillary electrophoresis. Oblath et al. [62] developed SPE-based filtration with a monolithic aluminum oxide membrane (AOM) for the affected saliva cell and real-time polymerase chain amplification reaction. They reported effective *Streptococcus mutans* identification in a saliva sample with 300 fg (100–125 copies) amplification of both methicillin-susceptible *Staphylococcus aureus* (MSSA) and methicillin-resistant *S. aureus* (MRSA) genomic DNA (gDNA) in less than 2.5 h.

In this study, hydrodynamic separation was used to produce a tapered microfluidic device for passive continuous separation of microparticles. The techniques include producing a microfluidic tapered channel, sample preparation, and the entire testing cycle for the cell separation process. As shown in Fig. 11.6, the tapered microfluidic was produced using regular soft lithography and replica molding techniques. The tapered microchannel design was drawn using the program of the Layout Editor software. Using a direct laser writing technique on a μPG system (Heidelberg Instruments, Germany), a 2.5 inch glass chrome mask filled with AZP 1350 positive photoresist was used to transfer the pattern. A silicone wafer for the substrate was used for the preparation of the master mold. During master mold production, SU8-3005 negative photoresist (Kayaku Microchem, Japan) was used. To produce the PDMS unit, the prepolymer and curing agent liquid PDMS (SILPOT 184, SILPOT CAT Dow Corning, USA) were mixed in a 9:1 ratio.

Air bubble removal and degassing were subsequently performed and the mixture was poured over a mold master. The mixture was

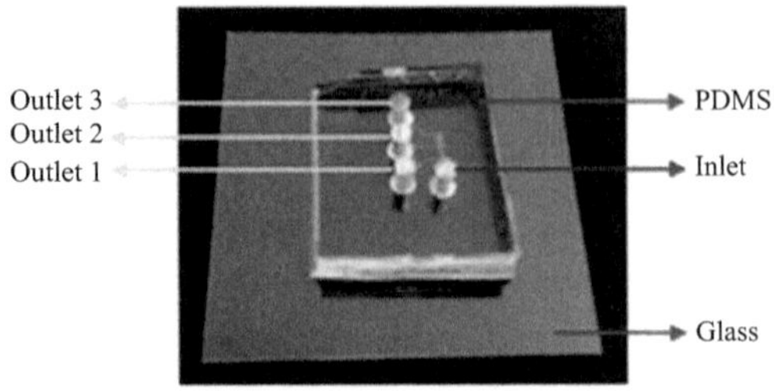

Figure 11.7 Tapered microfluidic separation device. © 2017 IEEE. Reprinted, with permission, from Ref. [45].

cured for 2 h at 65°C in a convection oven for a good cross-linking effect. The cured PDMS was removed from the master, leaving the microchannel pattern's negative cast. Using a 2 mm puncher (Harris, Redding, CA), inlets and outlets were punched manually. At 20 mA for 3 min, plasma oxidation treatment (PR500, Yamato Scientific Corp., Japan) was performed to seal the PDMS with a glass slide. Tygon tubing was attached to the outlets and fixed to plastic vials that were secured for output collection by an adjustable X–Y manipulator. The completed microfluidic chip is shown in Fig. 11.7.

Nonfluorescent PS microbeads of 3 μm and 10 μm diameter (Polysciences. Inc.) and HeLa cells were used as samples to demonstrate the function of the fabricated device. During preparations, by using Eq. 11.4, the number of particles per milliliter, N, before handling was determined.

$$N = \frac{6x \times 10^{12}}{y \times \pi \times z^3}, \tag{11.4}$$

where the suspended microbeads concentration (g/mL) is given as x, y is the density (g/mL), and z is the diameter in micrometers [52]. HeLa cells were then prepared following protocols for cell harvesting. Using a water bath, a cell tube containing 5×10^5 of cells/mL was thawed and the cell culture was prepared on a clean bench. The remaining cell was kept in an incubator at 37°C with CO_2 level 5 after the cultivation process was completed. Before use, cell viability was verified under a confocal microscope. A mixture of PS microbeads and HeLa cells was prepared and further diluted before centrifuging in the culture medium. An increased dilution factor helped distribute the particle concentration and thus later increased

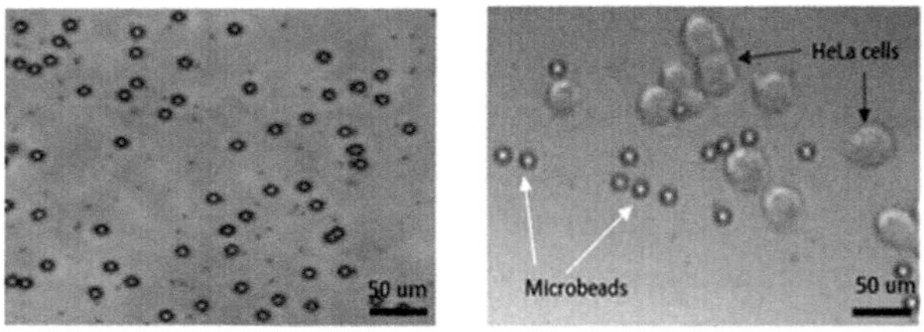

(a) (b)

Figure 11.8 Samples used. (a) Polystyrene microbeads only and (b) a mixture of polystyrene microbeads and HeLa cells. © 2017 IEEE. Reprinted, with permission, from Ref. [45].

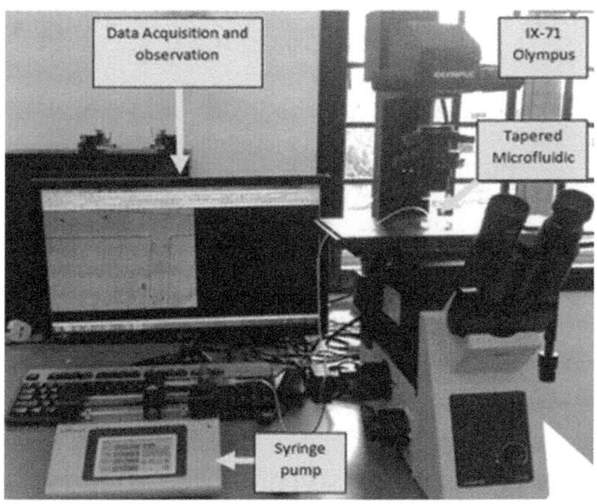

Figure 11.9 Experimental setup. © 2017 IEEE. Reprinted, with permission, from Ref. [45].

observability at collection outlets sites. Figure 11.8a,b shows the two samples.

The experimental environment used to test the proposed device is shown in Fig. 11.9. A KDS Scientific syringe pump was attached to the inlet throughout the experiment and was given a steady flow rate.

To remove excessive bubbles, phosphate-buffered saline (PBS) or culture medium solutions were first introduced into the microchannel. The sample was then inserted into the inlet. Separation behavior was documented using a high-speed charge-coupled device (CCD) camera (ORCA ER, Hamamatsu, Japan) and an inverted microscope (IX-71 Olympus Corp., Japan). This high-speed imagery was a requirement for achieving higher resolution and improved separation process observation. Particles were counted and tabulated at all outlets. The rate and purity of recovery are standard criteria used to test the efficiency of separation. The rate of recovery is defined as the ratio of the number of particles recovered from the target outlet to the number of particles entered in the inlet, while purity is defined as the ratio between the number of target particles from the outlet and the number of total outlet particles, given by Eqs. 11.5 and 11.6, respectively [38].

$$\text{Recovery rate} = \frac{\left(N_{\text{Large+Small}}\right)\text{ Outlet}}{\left(N_{\text{Large+Small}}\right)\text{ Inlet}} \times 100\% \tag{11.5}$$

and

$$\text{Purity} = \frac{\left(N_{\text{Target}}\right)\text{ Outlet}}{\left(N_{\text{Total}}\right)\text{ Outlet}} \times 100\%, \tag{11.6}$$

where N_{Target} is a particle of interest and N_{Total} represents the whole samples.

11.3 Results and Discussions

11.3.1 Separation of Polystyrene Microbeads

To assess the proposed device's separation ability, samples of 3 μm and 10 μm were tested. The same sample mixture concentrations were used at 1×10^5 μL^{-1}. A constant flow rate ranging from 0.5 to 3.0 μL/min was selected as a flow rate higher than this may cause the system to have leakage problems. Nonetheless, this range is within appropriate values according to Refs. [35, 38, 63, 64] for microfluidic-based separation devices. The collected samples are illustrated in Fig. 11.10a,b, in which separation between 10 μm and 3 μm samples occurred by using a 25° taper angle device. Outlet 1 contained a majority of 10 μm samples, while outlet 3 was occupied with 3 μm samples. In outlet 2, no sample was obtained.

This was in accordance with the proposed mechanism by which the larger particle was sorted at the closest outlet (outlet 1) and

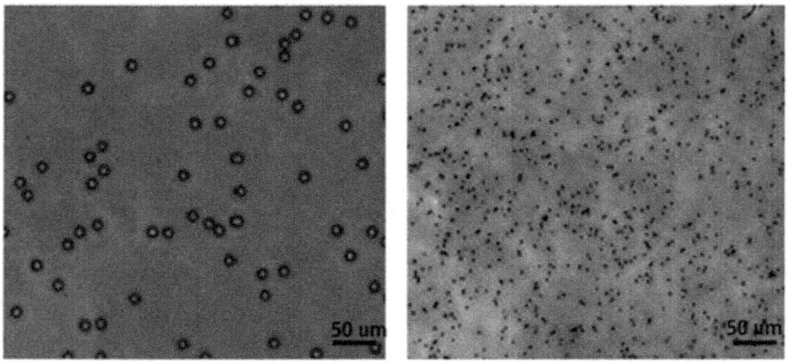

Figure 11.10 (a) 10 μm microbeads at outlet 1. (b) 3 μm microbeads at outlet 3. © 2017 IEEE. Reprinted, with permission, from Ref. [45].

the smallest particle moved to the farthest outlet (outlet 3). The recovery rate and purity of samples are shown in Fig. 11.11a,b at different flow rates. Separation performance was measured and evaluated at various flow rates. Both recovery rate and purity of 3 μm and 10 μm samples at a low flow rate were lower and gradually increased up to 3.0 μL/min at the highest values. At outlet 1 and outlet 3, all of the targeted samples could be found. Due to the relatively significant size difference between the samples, the effect of expanding streamlines due to the tapered microchannel increased the separation process.

In addition, a moving suspended particle within the microfluidic would experience different effects of lifting force/buoyancy that affected the particle sedimentation rate [41]. Within the microfluidic, sedimentation is determined by the Stokes number and is highly dependent on mass. The larger particle thus appears to sediment more rapidly and continues to deviate from the initial flow trajectory, and was obtained from the closest outlet (outlet 1). On the other hand, much less sedimentation would result in smaller particles moving along the flow route and thus escaping through the furthest outlet (outlet 3) [65]. These results showed that by using a tapered microfluidic with higher taper angles (20° and 25°), a successful separation mechanism with high separation efficiency can be achieved.

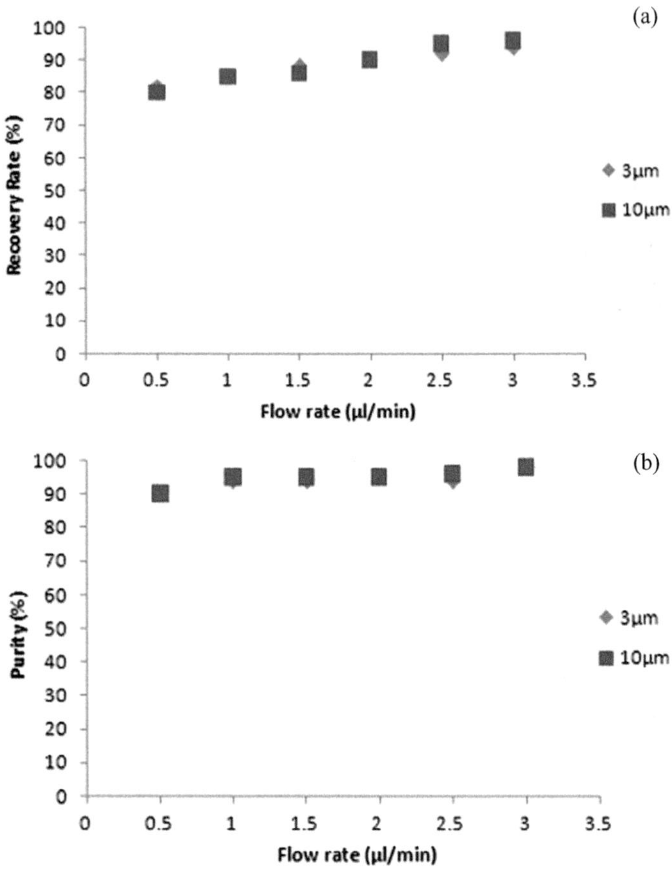

Figure 11.11 (a) Recovery rate at outlets. (b) Purity rate at the outlets. © 2017 IEEE. Reprinted, with permission, from Ref. [45].

11.3.2 Separation of Polydisperse Samples

The system was later tested using samples of polydisperse (PS microbeads and HeLa cells) to further demonstrate its ability to separate. A 1:1 combination of 1×10^5 of cells μl^{-1} and 1×10^5 μl^{-1} microbeads was used in the mixture. The device was then tested at various flow rates and sample purity for its recovery rates. The larger and denser HeLa cells were collected in outlet 1 in accordance with our hypothesis, whereas smaller microbeads

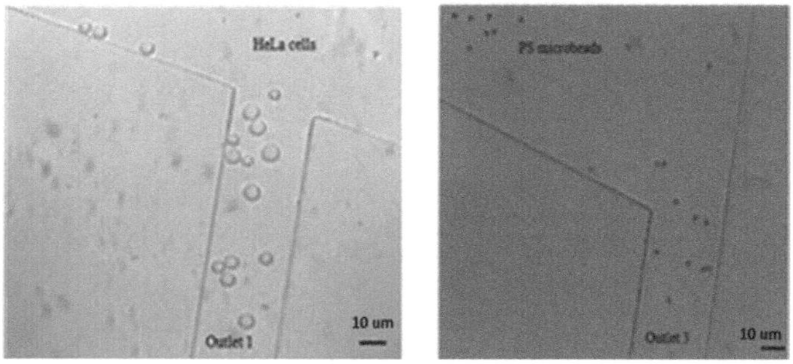

(a) (b)

Figure 11.12 (a) HeLa cells collected at outlet 1; (b) PS microbeads collected at outlet 3. © 2017 IEEE. Reprinted, with permission, from Ref. [45].

were collected through outlet 3 by a 20° and 25° tapered angle. The collected samples in the outlets are shown in Fig. 11.12a,b. Although the physical properties of biological cells and PS microbeads vary greatly (biological cells are softer and deformable), this device was able to separate by size. In addition, variations in density between biological cells and PS microbeads illustrate the sensitivity of the system up to 10^3 power of magnitude in resolution.

Figure 11.13a,b indicates the recovery rate and purity. In addition, on the basis of our observation, a higher flow rate results in a faster separation time but does not necessarily produce the best rate of recovery or purity. Furthermore, separation time was recorded and can be seen in Fig. 11.14a,b. Separation time is the amount of time a particle needs to travel from the inlet to the outlet.

This chapter addressed a novel, passive tapered microfluidic device that uses hydrodynamic theory coupled with a sedimentation effect to continuously perform micro-object separation. The tapered configuration includes a variation of 6°, 12°, 20°, and 25° taper angles. To eradicate the use of a sheath buffer, a focusing region stretched for 1 mm with 100 μm width and 10 μm height was added to the design. In addition, particle self-alignment, irrespective of size, type, and density, was generated after sample migration from inlet to outlet. Due to narrow dimensions, particles will move together in one stream; thus a further reduction in reagent and complexity

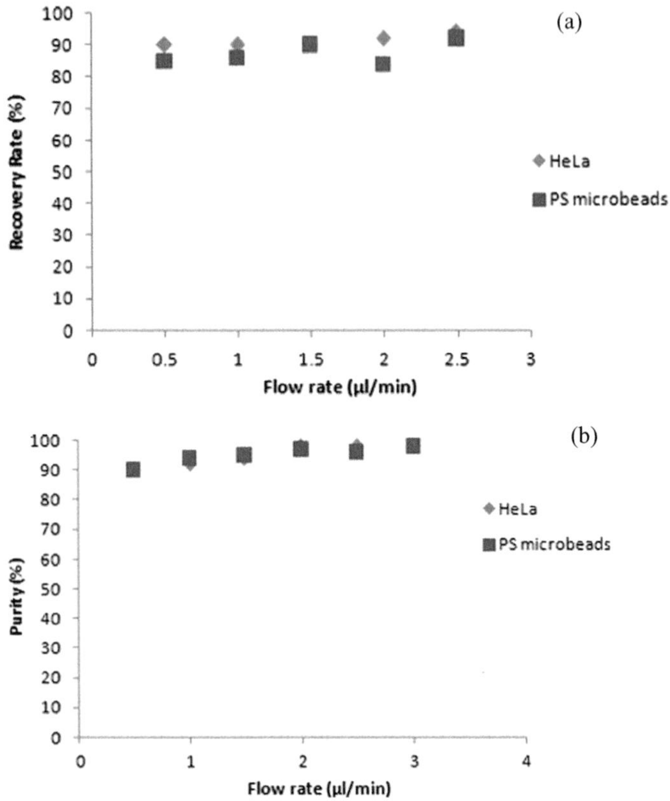

Figure 11.13 (a) Recovery rate; (b) purity rate of polydisperse samples. © 2017 IEEE. Reprinted, with permission, from Ref. [45].

of handling can be achieved. To improve the purity of samples, the sedimentation principle was implemented as a coupling effect of hydrodynamic separation.

The taper angle plays an important role in our tapered microfluidic since the velocity of sedimentation depends on this angle of inclination. Hence, faster sedimentation could be observed from the larger taper angle. In addition, our device has a very low channel height that causes rapid sedimentation compared to the rectangular microchannel. Consequently, as shown in Fig. 11.11b, better sample purity can be achieved by separation. Turbulence

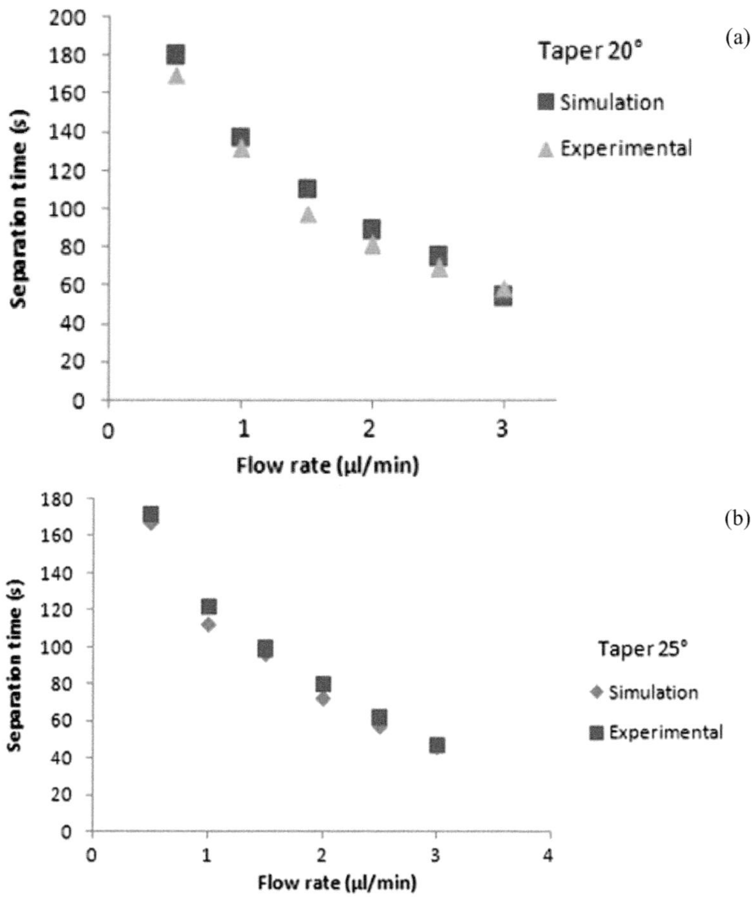

Figure 11.14 Separation time recorded for various flow rates used; (a) taper 20°; (b) taper 25°. © 2017 IEEE. Reprinted, with permission, from Ref. [45].

is a major microfluidic device issue. Verification of the laminarity properties of flow within the condensed microfluidic was performed through simulations of streamlines (see Fig. 11.5). Because of the small dimensions of the real device, experiments found that a flow rate higher than 3.5 μL/min was considered inappropriate as this could lead to higher turbulence and device leakage possibilities.

The optimal distance between the outlets was calculated to be >30 μm from a series of simulations in order to successfully isolate the particles. Before fabrication and testing, it was determined that the study would require only two particles of different sizes (monodisperse) and two particles of different types (polydisperse). Outlet 1 and outlet 3 were configured to be >30 μm apart, while outlet 2 was located at <30 μm to test this positional criterion for efficient separation of samples.

The device was produced using the soft lithography process and tested with the combination of PS microbeads and HeLa cell samples. With the use of larger taper angles (20° and 25°) compared to smaller angles (6° and 12°), positive separation was observed. The largest and densest particles were always collected at outlet 1 compared, with the smaller or lighter ones at outlet 3. During the separation in outlet 2, no particles were collected. Such results were confirmed by a simulation of finite elements and reproduced by experimental works.

11.4 Conclusions

The use of a new passive microfluidic separation device was successfully discussed in this chapter. By using wider taper angles of 20° and 25°, the proposed device demonstrated its ability to separate monodisperse and polydisperse samples. It has been shown that the device works by combining the hydrodynamics with sedimentation, which increases the sample purity up to 98%. A high sample purity and recovery rate can be obtained, demonstrating that despite working on a passive platform, our system is capable of delivering simple and efficient separation. It has been proved useful for applications of both particles and cell separation to address previously associated hydrodynamic separation limitations. This device offers a new perspective on a label-free passive separation device that can eventually be a catalyst for the development of a point-of-care (POC) device. In addition, the flexibility and ease of design add to this advantage when designing a separator that is applicable to the μTAS innovation.

Acknowledgments

The research was supported by the Ministry of Higher Education of Malaysia (grant number 4L640) and Universiti Teknologi Malaysia (grant numbers 03E11, 03G47, 4J148, 02G46, and 03H80); we thank them for funding this project and for their endless support.

References

1. P. Sajeesh, A. K. Sen (2014). Particle separation and sorting in microfluidic devices: a review. *Microfluid Nanofluid*, **17**(1):1–52.
2. D. R. Link, E. Grasland-Mongrain, A. Duri, F. Sarrazin, Z. Cheng, G. Cristobal, M. Marquez, D. A. Weitz (2006). Electric control of droplets in microfluidic devices. *Angew Chem Int Ed*, **45**(16):2556–2560.
3. R. Rosenberg, R. Gertler, J. Friederichs, K. Fuehrer, M. Dahm, R. Phelps, S. Thorban, H. Nekarda, J. R. Siewert (2002). Comparison of two density gradient centrifugation systems for the enrichment of disseminated tumor cells in blood. *Cytometry*, **49**(4):150–158.
4. H. W. Hou, Q. S. Li, G. Y. H. Lee, A. P. Kumar, C. N. Ong, C. T. Lim (2009). Deformability study of breast cancer cells using microfluidics. *Biomed Microdevices*, **11**:557–564.
5. J. G. Moreno, M. C. Miller, S. Gross, W. J. Allard, L. G. Gomella, L. W. Terstappen (2005). Circulating tumor cells predict survival in patients with metastatic prostate cancer. *Urology*, **65**(4):713–718.
6. W. Azim, S. Azim, K. Ahmed, H. Shafi, T. Rafi, M. Luqman (2004). Diagnostic significance of serum protein electrophoresis. *Biomed Biochim Acta*, **20**.
7. C. Rivet, H. Lee, A. Hirsch, S. Hamilton, H. Lu (2011). Microfluidics for medical diagnostics and biosensors. *Chem Eng Sci*, **66**(7):1490–1507.
8. A. M. Streets, Y. Huang (2013). Chip in a lab: microfluidics for next generation life science research. *Biomicrofluidics*, **7**(1):11302.
9. G. M. Whitesides (2006). The origins and the future of microfluidics. *Nature*, **442**(7101):368–373.
10. B. Mosadegh, T. Bersano-Begey, J. Y. Park, M. A. Burns, S. Takayama (2011). Next-generation integrated microfluidic circuits. *Lab Chip*, **11**(17):2813–2818.

11. E. Ghafar-Zadeh, M. Sawan (2007). A hybrid microfluidic/CMOS capacitive sensor dedicated to lab-on-chip applications. *IEEE Trans Biomed Circuits Syst*, **1**(4):270–277.
12. M. A. Miled, S. Member, M. Sawan (2013). Correction to 'dielectrophoresis-based integrated lab-on-chip for nano and micro-particles manipulation and capacitive detection.' *IEEE Trans Biomed Circuits Syst*, **7**(4):557.
13. D. Mark, S. Haeberle, G. Roth, F. von Stetten, R. Zengerle (2010). Microfluidic lab-on-a-chip platforms: requirements, characteristics and applications. *Chem Soc Rev*, **39**(3):1153–1182.
14. Y. Liu, K.-M. Lim (2011). Particle separation in microfluidics using a switching ultrasonic field. *Lab Chip*, **11**(18):3167–3173.
15. B. Hammarström, M. Evander, H. Barbeau, M. Bruzelius, J. Larsson, T. Laurell, J. Nilsson (2010). Non-contact acoustic cell trapping in disposable glass capillaries. *Lab Chip*, **10**(17):2251–2257.
16. J. Dykes, A. Lenshof, I. B. Åstrand-Grundström, T. Laurell, S. Scheding (2011). Efficient removal of platelets from peripheral blood progenitor cell products using a novel micro-chip based acoustophoretic platform. *PLoS One*, **6**(8):e23074.
17. H. Shafiee, M. B. Sano, E. A. Henslee, J. L. Caldwell, R. V. Davalos (2010). Selective isolation of live/dead cells using contactless dielectrophoresis (cDEP). *Lab Chip*, **10**(4):438–445.
18. D. F. Chen, W. H. Li, H. Du, M. Li (2012). Continuous sorting of microparticles using dielectrophoresis. *J Nanosci Nanotechnol*, **12**(4):3035–3039.
19. S. K. Hoi, C. Udalagama, C. H. Sow, F. Watt, A. A. Bettiol (2009). Microfluidic sorting system based on optical force switching. *Appl Phys B*, **97**(4):859–865.
20. A. Y. Lau, L. P. Lee, J. W. Chan (2008). An integrated optofluidic platform for Raman-activated cell sorting. *Lab Chip*, **8**(7):1116–1120.
21. Y. Song, H. Yin, W. E. Huang (2016). Raman activated cell sorting. *Curr Opin Chem Biol*, **33**:1–8.
22. W. Liang, S. Wang, Z. Dong, G. Bin Lee, W. J. Li (2012). Optical spectrum and electric field waveform dependent optically-induced dielectrophoretic (ODEP) micro-manipulation. *Micromachines*, **3**(2):492–508.
23. G. J. Shah, A. T. Ohta, E. P.-Y. Chiou, M. C. Wu, C.-J. Kim (2009). EWOD-driven droplet microfluidic device integrated with optoelectronic tweezers as an automated platform for cellular isolation and analysis. *Lab Chip*, **9**(12):1732–1739.

24. M. Zborowski, J. J. Chalmers (2011). Rare cell separation and analysis by magnetic sorting. *Anal Chem*, **83**(21):8050–8056.
25. N. Pamme, J. C. T. Eijkel, A. Manz (2006). On-chip free-flow magnetophoresis: separation and detection of mixtures of magnetic particles in continuous flow. *J Magn Magn Mater*, **307**(2):237–244.
26. N. Pamme (2007). Continuous flow separations in microfluidic devices. *Lab Chip*, **7**(12):1644–1659.
27. N. Pamme, A. Manz (2004). On-chip free-flow magnetophoresis: continuous flow separation of magnetic particles and agglomerates. *Anal Chem*, **76**(24):7250–7256.
28. H. M. Ji, V. Samper, Y. Chen, C. K. Heng, T. M. Lim, L. Yobas (2008). Silicon-based microfilters for whole blood cell separation. *Biomed Microdevices*, **10**(2):251–257.
29. V. VanDelinder, A. Groisman (2006). Separation of plasma from whole human blood in a continuous cross-flow in a molded microfluidic device. *Anal Chem*, **78**(11):3765–3771.
30. S. K. Murthy, P. Sethu, G. Vunjak-Novakovic, M. Toner, M. Radisic (2006). Size-based microfluidic enrichment of neonatal rat cardiac cell populations. *Biomed Microdevices*, **8**(3):231–237.
31. X. Chen, D. F. Cui, L. L. Zhang (2009). Isolation of plasma from whole blood using a microfludic chip in a continuous cross-flow. *Chin Sci Bull*, **54**(2):324–327.
32. J. McGrath, M. Jimenez, H. Bridle (2014). Deterministic lateral displacement for particle separation: a review. *Lab Chip*, **14**(21):4139–4158.
33. M. Yamada, K. Kano, Y. Tsuda, J. Kobayashi, M. Yamato, M. Seki, T. Okano (2007). Microfluidic devices for size-dependent separation of liver cells. *Biomed Microdevices*, **9**(5):637–645.
34. T. Morijiri, S. Sunahiro, M. Senaha, M. Yamada, M. Seki (2011). Sedimentation pinched-flow fractionation for size- and density-based particle sorting in microchannels. *Microfluid Nanofluid*, **11**(1):105–110.
35. J. Alvankarian, A. Bahadorimehr, B. Yeop Majlis (2013). A pillar-based microfilter for isolation of white blood cells on elastomeric substrate. *Biomicrofluidics*, **7**(1):014102(1)–014102(16).
36. J. S. Kuo, Y. Zhao, P. G. Schiro, L. Ng, D. S. W. Lim, J. P. Shelby, D. T. Chiu (2010). Deformability considerations in filtration of biological cells. *Lab Chip*, **10**(7):837.
37. P. Sethu, A. Sin, M. Toner (2006). Microfluidic diffusive filter for apheresis (leukapheresis). *Lab Chip*, **6**(1):83–89.

38. Y. Y. Chiu, C. K. Huang, Y. W. Lu (2016). Enhancement of microfluidic particle separation using cross-flow filters with hydrodynamic focusing. *Biomicrofluidics*, **10**(1):011906(1)–011906(13).
39. D. Di Carlo, J. F. Edd, D. Irimia, R. G. Tompkins, M. Toner (2008). Equilibrium separation and filtration of particles using differential inertial focusing. *Anal Chem*, **80**(6):2204–2211.
40. D. Di Carlo, D. Irimia, R. G. Tompkins, M. Toner (2007). Continuous inertial focusing, ordering, and separation of particles in microchannels. *Proc Natl Acad Sci USA*, **104**(48):18892–18897.
41. D. Huh, J. H. Bahng, Y. Ling, H. H. Wei, O. D. Kripfgans, J. B. Fowlkes, J. B. Grotberg, S. Takayama (2007). Gravity-driven microfluidic particle sorting device with hydrodynamic separation amplification. *Anal Chem*, **79**(4):1369–1376.
42. X. Li, W. Chen, G. Liu, W. Lu, J. Fu (2014). Continuous-flow microfluidic blood cell sorting for unprocessed whole blood using surface-micromachined microfiltration membranes. *Lab Chip*, **14**(14):2565–2575.
43. J. Takagi, M. Yamada, M. Yasuda, M. Seki (2005). Continuous particle separation in a microchannel having asymmetrically arranged multiple branches. *Lab Chip*, **5**(7):778–784.
44. P. Rupprecht, L. Golé, J. P. Rieu, C. Vézy, R. Ferrigno, H. C. Mertani, C. Rivière (2012). A tapered channel microfluidic device for comprehensive cell adhesion analysis, using measurements of detachment kinetics and shear stress-dependent motion. *Biomicrofluidics*, **6**(1):1–12.
45. C. Liu, J. Sun, M. Li, et al. (2013). Size-based hydrodynamic rare tumor cell separation in curved microfluidic channels. *Biomicrofluidics*, **7**(1):011802.
46. D.-K. Sun, D. Jiang, N. Xiang, K. Chen, Z.-H. Ni (2013). An immersed boundary-lattice boltzmann simulation of particle hydrodynamic focusing in a straight microchannel. *Chin Phys Lett*, **30**(7).
47. J.-T. Ma, Y.-Q. Xu, X.-Y. Tang (2016). A numerical simulation of cell separation by simplified asymmetric pinched flow fractionation. *Comput Math Methods Med*, **2016**:2564584.
48. G. Samla, K. B. Gan, S. M. Then (2017). Modeling microfluidic DNA extraction using superparamagnetic bead particles in COMSOL multiphysics simulation. *Microsyst Technol*, **23**(10):4435–4440.
49. Dassaults Systemes (2013). Abaqus 6.12/CAE user's manual.
50. K. Tomankova, P. Kolar, J. Malohlava, H. Kolarova (2012). Mechanical characterisation of HeLa cells using atomic force microscopy. In *Current*

Microscopy Contributions to Advances in Science and Technology, 1st ed., A. Méndez-Vilas, pp. 549–554.

51. R. G. Breuls, C. V Bouten, C. W. Oomens, D. L. Bader, F. P. Baaijens (2003). Compression induced cell damage in engineered muscle tissue: an in vitro model to study pressure ulcer aetiology. *Ann Biomed Eng*, **31**(11):1357–1364.
52. P. H. Guide (2015). Microspheres & particles handling guide.
53. P. Y. Lee, J. Costumbrado, C. Y. Hsu, Y. H. Kim (2012). Agarose gel electrophoresis for the separation of DNA fragments. *J Visualized Exp*, **62**:3923.
54. P. Wilding, L. J. Kricka, J. Cheng, G. Hvichia, M. A. Shoffner, P. Fortina (1998). Integrated cell isolation and polymerase chain reaction analysis using silicon microfilter chambers. *Anal Biochem*, **257**(2):95–100.
55. J. A. Davis, D. W. Inglis, K. J. Morton, D. A. Lawrence, L. R. Huang, S. Y. Chou, J. C. Sturm, R. H. Austin (2006). Deterministic hydrodynamics: taking blood apart. *Natl Acad Sci USA*, **103**(40):1479–1884.
56. S. T. Weiss, M. S. Shin (2016). Infrastructure for personalized medicine at partners health care. *J Pers Med*, **6**(1):13.
57. T. A. Crowley, V. Pizziconi (2005). Isolation of plasma from whole blood using planar microfilters for lab-on-a-chip applications. *Lab Chip*, **5**(9):922–929.
58. J. Moorthy, D. J. Beebe (2003). *In situ* fabricated porous filters for microsystems. *Lab Chip*, **3**:62–66.
59. V. VanDelinder, A. Groisman (2006). Separation of plasma from whole human blood in a continuous cross-flow in a molded microfluidic device. *Anal Chem*, **78**(11):3765–3771.
60. N. Xia, T. P. Hunt, B. T. Mayers, E. Alsberg, G. M. Whitesides, R. M. Westervelt, D. E. Ingber (2006). Combined microfluidic-micromagnetic separation of living cells in continuous flow. *Biomed Microdevice*, **8**:299–308.
61. J. Wu, W. Wen, R. Kodzius, W. Cao (2013). Extraction, amplification and detection of DNA in microfluidic chip-based assays. *Microchim Acta*, **181**(13–14):1611–1631.
62. W. Oblath, A. Emily, H. Hampton, P. A. Jean, R. Michael (2013). A microfluidic chip integrating DNA extraction and real-time PCR for the detection of bacteria in saliva. *Lab Chip*, **13**(7):1325–1332.
63. M. Seki, M. Yamada, M. Yasuda, Y. Sai (2006). Continuous separation of particles using a microfluidic device equipped with flow rate control valves. *J Chromatogr A*, **1127**(1–2):214–220.

64. R. Aoki, M. Yamada, M. Yasuda, M. Seki (2009). In-channel focusing of flowing microparticles utilizing hydrodynamic filtration. *Microfluid Nanofluid*, **6**(4):571–576.

65. Z. Wu, K. Hjort (2009). Microfluidic hydrodynamic cell separation: a review. *Micro Nanosyst*, **1**(3):181–192.

Index

ABF, *see* artificial bacterial flagella
absorption, 58, 114, 128, 130, 280, 434, 448, 451–52, 454, 457–58, 466, 484
  direct, 278
  high tissue, 488
  plasmonic, 450
absorption spectrophotometry, 446
AC, *see* alternating current
ACEO, *see* alternating current electro-osmosis
acoustophoresis, 50
acute myeloid leukemia (AML), 378, 418–19
adhesion, 9, 14, 65, 156–57, 172, 300, 313, 402
  nonspecific, 72
  particle-electrode, 80
Ag NPs, 430–31, 434, 444, 449–52, 456, 467, 470–72
  functionalized, 443
  high-yield, 450
ALD, *see* atomic layer deposition
alternating current (AC), 51–55, 70, 75, 80–81, 365, 371, 404, 418–19
alternating current electro-osmosis (ACEO), 51, 53, 68, 78–80
aluminum oxide membrane (AOM), 515
Alzheimer's disease, 162, 192
AML, *see* acute myeloid leukemia
anastomosis, 222–25
  vascular, 223
angiogenesis, 216, 221–25, 227–29, 231, 237, 239, 241
  physiological, 237
angiogenic sprouting, 222–24, 228–29, 235–36
antigen-presenting cells (APCs), 277, 318, 321
AOM, *see* aluminum oxide membrane
APCs, *see* antigen-presenting cells
arrays, 9, 26, 28, 61–62, 158, 276, 278, 284–85, 306, 308–9, 313, 355, 357, 359–60, 363
  cross-flow, 514
  dense, 372
  dynamic, 37
  interconnected microvascular, 218
  inverted floating droplet, 28
  microfluidic gradient, 177
  microhole, 168
  micropipette, 38
  radial, 154
artificial bacterial flagella (ABF), 103, 119–20, 129
astrocytes, 164, 190, 193, 233
atomic layer deposition (ALD), 116, 118
axons, 153–57, 162, 164–65, 170, 172–77, 182–83, 190–91, 193, 196
  extended, 175

hippocampal, 181
motoneuron, 156
single dopaminergic, 155
axotomy, 183

barrier functions, 217, 226, 232, 234
endothelial, 218, 237
ocular, 305
selective, 214, 233
vascular, 241
barriers, 128, 165, 213, 215, 230, 233–35, 276, 278, 281, 301–2, 305, 307, 311, 319
biochemical, 163
blood–brain, 232
immunological, 277
microgroove-embedded, 183
ocular, 305
physical, 9, 173, 402
physiological, 131
primary, 301
semipermeable, 278
vascular, 232
BBB, *see* blood–brain barrier
BDNF, *see* brain-derived neurotrophic factor
BEP, see bulk electroporation
Bessel beam, 60
Bessel intensity profile, 60
bilayer lipid membrane (BLM), 403, 405, 412
bioavailability, 283, 288, 294, 305, 310–11, 323, 476, 479
biocatalyst, 108, 118
biochips, 255–56, 258–59, 267
biocompatibility, 83, 125, 138, 242, 295, 310, 350
biodegradability, 138, 310
biohybrid micro-robots, 110–13
bioimaging, 475–76, 491
biological cells, 49, 59–62, 65, 73, 83, 402, 421, 505, 521
biomarkers, 16, 32, 76, 293, 434, 480, 514
biomaterials, 158–59, 172–73, 175–77, 214–16, 243
bio-MEMS, *see* bio-microelectromechanical systems
bio-microelectromechanical systems (bio-MEMS), 404, 413, 422–23
biomicroswimmers, 109–13, 138
biomolecules, 18, 66, 173, 276, 316, 351, 361, 363, 381, 434, 483, 490
diffusible, 163
biosensors, 110, 132–33, 189, 353, 378, 487
microfluidics, 483
electrochemical, 485
microwave, 369
BLM, *see* bilayer lipid membrane
blood–brain barrier (BBB), 163–64, 175, 190–91, 193, 232–33
BMP, *see* bone morphogenetic protein
bone morphogenetic protein (BMP), 186
brain-derived neurotrophic factor (BDNF), 152, 183
Brownian motion, 97
Brust–Schiffrin method, 461
bubble-propelled micro-robots, 105–9, 129, 137
bulk electroporation (BEP), 343–45, 348–49

carbon nanotubes (CNTs), 25, 81, 172, 351, 485, 489
cargo, 17–18, 22, 24, 107–8, 128–29, 277–78, 343, 345
adhered, 107
embedded, 126

foreign, 17
inert, 107
large-sized, 18
microscale, 111
catalytic micro-robots, 106, 137–38
CEL, *see* Couple Eulerian Lagrangian
cell–cell interactions, 162–63, 165, 193
cell lysis, 31, 35, 73, 367, 401–3, 411, 418–19, 423
cell manipulation, 1–2, 4, 6, 8, 10, 12, 14, 16, 18, 20, 22, 38, 50–51, 67–68, 77–78
cell trapping, 3–4, 6, 9–13, 34, 36–37, 62, 65, 74, 158–59, 161, 352–54, 358–59, 369, 514
cell viability, 18–19, 21, 24, 33, 39, 320, 343–44, 348–49, 352, 361, 363, 375, 516
central nervous system (CNS), 115, 128, 136, 185, 188, 232
CFNCs, *see* circulating fetal nucleated cells
chemotactic micro-robots, 111
circulating fetal nucleated cells (CFNCs), 26
circulating tumor cells (CTCs), 25, 27–28, 38, 76, 134, 237, 481, 483, 504–5
CMDs, *see* compartmentalized microfluidic devices
CNA, *see* compartmentalized neuron array
CNS, *see* central nervous system
CNTs, *see* carbon nanotubes
co-culture, 154, 156–57, 162–63, 165–66, 227, 230, 233, 240
co-culturing, 152, 156–57, 162–65, 188, 193, 223, 235
compartmentalized microfluidic devices (CMDs), 154, 165
compartmentalized neuron array (CNA), 158–59
computed tomography (CT), 376, 476, 478–80
Co NPs, 458–59, 464–67, 480
continuous flow, 33, 65, 154, 175, 196, 368, 414, 439, 441, 489
Coulomb force, 52, 55
Couple Eulerian Lagrangian (CEL), 511
cross-linking, 125, 217, 293, 478, 483
CT, *see* computed tomography
CTCs, *see* circulating tumor cells

DAB, *see* droplet-borne air blowing
DBOTs, *see* dual-beam optical traps
DC, *see* direct current
Dean vortex, 19–20
Debye layer, 435
DEP, *see* dielectrophoresis
DHM, *see* digital holographic microscopy
dielectrophoresis (DEP), 4, 15, 50–52, 54, 68, 70, 72, 74, 76, 79–80, 160–61, 168, 351–52, 367, 371, 504–5
diffusion, 70, 163, 170, 175–76, 184, 271, 276–77, 279, 283, 300, 305, 434, 436–39, 441, 444
digital holographic microscopy (DHM), 195–96
dipole–dipole interaction, 80
direct current (DC), 52, 404, 418–20
dissolvable microneedles, 303, 310–11, 316
DNA, 16, 31–32, 35–37, 123, 128, 134, 153–54, 255–61, 266, 268–72, 276, 320, 376–77, 380, 511
double-stranded, 37, 124

exogenous, 351
foreign, 17
genomic, 515
droplet-borne air blowing (DAB), 288
drug delivery, 128–29, 190, 240–41, 275–79, 281, 283, 285–87, 302, 312, 316, 319–24, 349, 351, 359, 432
invasive, 305
ocular, 306–7, 309
oral, 276
passive, 129
sustained topical, 319
targeted, 95, 305–6, 319
dual-beam optical traps (DBOTs), 60–62

ECM, *see* extracellular matrix
ECs, *see* endothelial cells
ECT, *see* electrochemotherapy
EDL, *see* electric double layer
EDS, *see* energy-dispersive X-ray spectroscopy
EHD, *see* electrohydrodynamic
EIT, *see* electrical impedance tomography
EL, *see* electrical lysis
electrical impedance tomography (EIT), 376–77
electrical lysis (EL), 401–11, 414, 416–23
electric double layer (EDL), 53–54, 80
electric fields, 13–15, 50–56, 68–70, 77–78, 80–81, 345, 347–48, 350–54, 356–57, 359–62, 366–69, 373, 375, 405–14, 416–21
electrochemotherapy (ECT), 376, 379–80
electrofusion, 371–75, 382, 420
electrohydrodynamic (EHD), 79–82
electrolysis, 18–19, 39, 349, 505
electro-osmotic flow (EOF), 4, 35, 53–55, 258, 267, 435
electroporation (EP), 3, 17–21, 35, 39, 74–75, 343–57, 359–60, 365–72, 375–81, 404, 406–11, 419, 422–23
bulk, 344, 346
light-induced, 366–67
microcapillary-based, 378–79
micropipette-based, 351
nanochannel, 357
nanopillar, 378
nanoscale, 378–79
nonthermal, 375
selective, 366
tissue, 376
ultralocalized, 362
electrothermal flow (ETF), 51, 55, 68, 78, 80–81
encapsulation, 6–8, 13, 31–32, 287, 290, 310, 316, 450
endothelial cells (ECs), 163, 185, 193, 215–28, 230–35, 237–38, 241–42
endothelial progenitor cells (EPCs), 221, 230
energy-dispersive X-ray spectroscopy (EDS), 446
entropy, 408–11
EOF, *see* electro-osmotic flow
EP, *see* electroporation
EPCs, *see* endothelial progenitor cells
ETF, *see* electrothermal flow
extracellular matrix (ECM), 65, 169, 173, 188, 219–23, 226–28, 232–33, 235, 237–39

face-centered cubic (fcc), 457–60, 474

FACS, *see* fluorescence-activated cell sorter
fcc, *see* face-centered cubic
FDA, *see* Food and Drug Administration
FEA, *see* finite element analysis
FESEM, *see* field emission scanning electron microscopy
FGF, *see* fibroblast growth factor
fibroblast growth factor (FGF), 186–87, 189, 221
Fickian diffusion, 185, 278
field emission scanning electron microscopy (FESEM), 120–21
finite element analysis (FEA), 512
flow cytometry, 15–16, 25, 481
fluorescence-activated cell sorter (FACS), 504–5
Food and Drug Administration (FDA), 96, 241, 304, 479, 481

Gaussian beams, 57–58, 60, 64
gDNA, *see* genomic DNA
gene expression, 25, 256, 272, 380
  altered, 158
  drug-resistant, 76
genome, 31, 380
genomic DNA (gDNA), 515
GFP, *see* green fluorescent protein
GLAD, *see* glancing angle deposition
glancing angle deposition (GLAD), 119–21
green fluorescent protein (GFP), 24, 316

HAu NPs, 464–67
hcp, hexagonal close-packed
HD, *see* Huntington's disease
HDL, *see* high-density lipid
helical micro-robots, 103–4, 119–21, 129
hexagonal close-packed (hcp), 457–60
high-density lipid (HDL), 479–80
high-resolution scanning transmission electron microscopy (HR-STEM), 446, 467
HL, *see* hydrogel layer
  hollow microneedles, 280, 284–87, 301, 309–10, 320
HR-STEM, see high-resolution scanning transmission electron microscopy
human umbilical vein endothelial cells (HUVECs), 21, 217–18, 220, 223–24, 230–31
Huntington's disease (HD), 190, 194
HUVECs, *see* human umbilical vein endothelial cells
hydrodynamic resistance, 31, 506, 510, 512
hydrodynamics, 437, 505–6, 510, 524
hydrogel layer (HL), 170, 174, 187, 218, 220
hydrogels, 102, 115, 118, 130, 162, 169–71, 174–75, 217–20, 229, 292–94, 312, 315, 479
  biological, 170
  drug-loaded, 129
  natural, 228
  patterned, 215
  polymerized, 217, 219, 239
  soft, 130
  swellable polymers/cross-linked, 293
  unpolymerized, 239
hydrophilicity, 232, 276, 318
hydrophobicity, 410, 440

IgG, *see* immunoglobulin G

immunoglobulin G (IgG), 317, 484–85, 487
indium tin oxide (ITO), 68–69, 78, 310, 360–62, 365
induced pluripotent stem cells (iPSCs), 230, 242
interfacial tension, 440, 456
interstitial flow, 221, 223, 226–27, 231, 238
in vitro vascularization, 220, 224–25
in vivo drug delivery, 124, 129
iPSCs, *see* induced pluripotent stem cells
irreversible electroporation, 35, 369, 371, 375–77, 379, 404
ITO, *see* indium tin oxide

Jakob disease, 194
Janus microparticle, 114
Janus micro-robots, 113–14, 124–25, 131–32, 139
  artificial, 95
  biocompatible, 136
  self-thermophoretic Au-PSS/PAH, 132
Joule heat, 406
Joule heating, 54–55, 348, 351, 361, 406, 414, 420, 505
Jurkat cells, 65

kinetic energy, 105, 459–60

lab-on-a-chip (LOC), 2, 38, 83, 195–97, 404, 413, 421–23, 431–32, 475, 491, 505
LaMer mechanism, 433
LaMer nucleation, 447
laminar flow, 19, 175, 435–36, 439–40, 447, 474, 491
laminar streamlines, 512
Laplace pressure difference, 116
laser ablation, 431, 485, 489
lasers, 16, 22, 24–25, 50, 54, 56, 58–59, 63–64, 78, 81–83, 114, 403, 476, 485, 488–89
  continuous, 22
  direct, 286, 515
  femtosecond, 485
  fiber, 62
  high-power, 403
  infrared, 490
  near-infrared, 24
  pulsed, 23
lattice Boltzmann system (LBM), 510–11
lattices, 218, 220, 255, 295, 434, 457, 510
layer-by-layer deposition, 124, 316
LBM, *see* lattice Boltzmann system
lift force, 79–80
LIGA, *see* lithography, electroplating, and molding
ligands, 157, 185, 450, 452, 473–74
Lipinski's rule, 301
lithography, electroplating, and molding (LIGA), 286, 288
LOC, *see* lab-on-a-chip
  disposable, 195
longitudinal plasmon resonance (LPR), 454
longitudinal surface plasmon resonance (LSPR), 452, 457, 464, 481
LPR, *see* longitudinal plasmon resonance
LSPR, *see* longitudinal surface plasmon resonance

MACS, *see* magnet-activated cell sorting
magnet-activated cell sorting (MACS), 504–5

magnetic fields, 3, 13, 99–105, 112–13, 137, 355–56, 363, 478, 489, 513
- earth's, 112
- external, 99, 101, 104, 112–13, 115, 129, 363
- external rotating, 119, 129
- high-strength, 137
- rotating, 103–4, 126
- rotational, 101, 104
- static, 137
- uniform, 104

magnetic forces, 99–103, 137, 173
magnetic micro-robots, 99, 104
magnetic properties, 102–3, 113, 429, 458–60, 475, 484
magnetic resonance imaging (MRI), 99, 101, 376, 476, 478–80
magnetoelectroporation (MEP), 356, 363–64
matrix-assisted laser desorption, 269
matrix metalloproteinases (MMPs), 126, 222, 239
Maxwell's equation, 347
MEA, *see* microelectrode array
- flat, 177
- standard, 167

mechanical properties, 64, 214, 228, 232, 234, 242, 286, 295, 297
mechanical strength, 283, 287–88, 290, 292, 294, 298, 300, 309
mechanoporation, 3, 17, 21–22, 39, 344
MeHA, *see* methacrylated hyaluronic acid
MEMS, *see* microelectromechanical systems
MEP, *see* magnetoelectroporation
mesenchymal stem cells (MSCs), 222, 230, 378
metallic nanomaterials, 432, 434, 484, 487
metallic NPs, 432, 434–35, 438, 446–48, 452, 461, 475, 477, 481, 484, 491
methicillin-resistant *Staphylococcus aureus* (MRSA), 515
methicillin-susceptible *Staphylococcus aureus* (MSSA), 515
methacrylated hyaluronic acid (MeHA), 307–8
μFDs, *see* microfluidic devices
- compartmentalized, 152
- four-chamber, 175
- high-throughput, 195
- hydrogel gradient, 152
- multidomain, 182
- three-compartment, 154
- traditional, 152

Mg-based micro-robot, 108, 124
microarray, 4, 9–10, 32, 39, 257, 268–69
microbeads, 503, 512, 514, 519–20
microbubbles, 130
microcannons, 130–31
microcavities, 371
microchambers, 34, 156, 168
microchannels, 28, 163–65, 168–70, 175–76, 216–19, 224, 258, 359, 362–63, 414, 416, 436–47, 462–63, 465–66, 471–72, 506, 514
- axon-guiding, 165
- central, 233
- circular, 465
- circular PDMS, 219
- curved, 505
- curved double spiral, 510
- diffuse, 505
- hollow, 239
- hydrogel-based, 217

laminin-filled, 176
long, 419
microneedle-induced, 312
outgrowth, 159
semicircular, 219
separate media, 163
tapered, 506–9, 512, 519
thin, 184
microconduits, 156, 180
microdevices, 416, 422–23
four-channel, 185–86
microdrillers, 130–31
rolled-up, 130
sharp-edged, 118
microdroplets, 4–7, 12, 39, 355
microelectrode array (MEA), 152, 156, 160, 166–69, 177–80, 196, 370–71, 377
cavity-based, 371
microelectrodes, 166, 196, 344, 370–71
coplanar, 371
microcavity, 371
virtual, 349
microelectromechanical systems (MEMS), 285, 404, 413, 421, 431, 435
microenvironments, 38, 149, 151, 154, 163, 169, 171, 173, 175, 183, 188–89, 196, 214, 223, 226
biochemical, 178
biological, 213
bone, 156
bone-mimicking, 238
chemical, 178
complex, 242
controlled, 178, 185
external, 230
hypoxia, 229, 237
hypoxic, 240
local figure-sized, 155
localized, 154, 163
muscle-mimicking, 238
native, 241
organ, 214
microfabricated chips, 404, 421
microfabrication, 2–4, 38, 96, 153, 215–16, 223, 243, 323, 401, 404, 413, 416, 504
microfluid, 169, 352, 370, 445, 511
microfluidic channels, 20–21, 162, 165, 216, 223, 225–26, 355, 357, 359–60, 369–71, 435, 470, 472, 511, 513
fixed, 359
long, 19
multiflow, 27
six-compartment-based, 485
straight, 21
tapered, 511
microfluidic chips, 14, 25–26, 28, 225, 370–71, 402, 413, 465, 481, 485, 489–91, 514, 516
array-based, 65
droplet, 355
hybrid, 514
sandwiched, 15
microfluidic devices (μFDs), 3–5, 17–19, 35–38, 149–55, 160, 163–68, 170, 173–85, 188–89, 193, 196–97, 221–27, 401–4, 412–20, 430–32, 435–36, 458–59, 461–62, 472, 474–76, 478–81
first compartmentalized, 154
glass-based, 357, 413
high-throughput, 3
hybrid, 230
hydrogel, 490
laminar-based, 439
miniature, 404
passive tapered, 521
on-chip EL, 414
single-step, 479
tapered, 503, 506, 508, 510, 515
three-chamber, 176
three-compartment, 155, 182

microfluidics, 24, 26, 149, 151–54, 169, 171, 197, 213, 215, 229, 431, 435–37, 439–40, 490–91, 519
compartmentalized, 154, 179
condensed, 523
droplet-based, 458, 485
easy-to-use, 242
multiphase, 437–38
tapered, 509, 515, 519, 522
microfluidic systems, 39, 78–79, 168, 173, 190–91, 193, 215, 226, 239, 431
advanced, 194
compartmentalized, 193
pulsed mixing, 445
tapered, 503
microfluidic wafers, 258, 267
droplet-type, 269
microgrippers, 130–31
microgrooves, 154, 156–57, 166, 170, 172–73, 176, 181, 193
axon-guiding, 155
high-resistance, 154
microholder, 128–29
microholes, 169, 353–54, 363, 414, 416
microinjection, 17, 21–22, 310
micromixer, 437, 463, 504
rotary, 463
static, 465
three-phase, 471
micromolding, 217–18, 220, 286, 307, 317
micromolds, 289–90
microneedle arrays, 279, 284–85, 299, 310, 313–15, 317–18, 320
dissolvable, 318
hollow, 285
hollow stainless-steel, 285
hydrogel-forming, 297
polymeric, 317
pyramidal-shaped, 307
microneedle patches, 290, 298, 302, 306, 308, 314, 319, 321–22
composite, 317
dissolvable starch/gelatin, 310
double-layered, 307
flexible, 285
multilayered polymeric, 294
polymeric, 314
silk fibroin, 294
transdermal, 316, 321
microneedles, 22, 132, 217–18, 275–324
microparticles, 54, 229, 277, 279, 290, 503, 505, 511–13, 515
dielectric, 63
encapsulated hydrogel, 292
encapsulated PNIPAAm, 292
Janus, 125
multisize, 510
micropillars, 36, 175
meniscus-pinning, 159
micropipes, 258, 267
cross-shaped, 258
micropores, 290, 414, 416–17
double-conical, 118
micropumps, 286, 444–45, 463, 465
piezoelectric, 444
microreactors, 355, 436, 441, 444, 460, 465, 467, 470
coaxial-flow, 471
continuous-laminar-flow, 441
distributed mixing, 441
interconnected, 477
spiral, 444, 472
untreated, 448
micro-robots, 96–116, 118–20, 122–26, 128–31, 133–34, 136–39
actuated/controlled, 115
bare, 125
bicomponent, 125
biocompatible, 138

biodegradabale, 125
biosensing, 133–34
controlled artificial, 95
double-helical, 126
flexible tail, 122, 137
helical tail, 137
hybrid, 108
magnetic-gradient-pulled, 101
mobile, 133
polyaniline/Pt bilayer, 118
receptor-modified, 133
rectangular, 104
rolled-up, 119
smallest artificial, 121
soft, 125, 127
surgical, 130
synthetic, 112, 138
templated, 119
two-component, 102
untethered, 97, 99, 132
water-powered, 108
microstructures, 214, 216, 221, 223, 234, 276, 290
hollow, 217–18
narrow microfluidic, 239
tortuous, 235
trapping-based, 372
microswimmers, 95–96, 98–99, 139
biological, 101, 109
living, 95
magnetized, 112
micro-total analysis systems (μTAS), 404, 413, 422–23, 504, 524
microvascular networks (μVNs), 215, 219, 221–22, 224–25, 229–232, 238
microvasculature, 213–16, 226, 228, 233
engineered, 213
perfused human, 240
microvessels, 213, 216, 218, 220–23, 234, 238, 240, 242
engineered, 215, 234, 241–42
lumenized, 222
perfusable, 223
vasculogenesis-derived, 236
microvortex manipulation (MVM), 505
microvortices, 4, 11–12, 39, 78, 82
symmetric, 12
microwells, 9–10, 15–16, 28, 32–34, 37–38, 65, 167–68
empty, 16
no-cell, 15
migration, 65, 165, 174, 191
electrophoretic, 380
intracellular, 380
ion, 54
lateral-laminar-flow, 514
millirobot, 97
MMPs, *see* matrix metalloproteinases
MNs, *see* motor neurons
motor neurons (MNs), 154, 186, 194
MRI, *see* magnetic resonance imaging
MRSA, *see* methicillin-resistant *Staphylococcus aureus*
MSCs, *see* mesenchymal stem cells
MSSA, *see* methicillin-susceptible *Staphylococcus aureus*
μTAS, *see* micro-total analysis systems
multiwall carbon nanotubes (MWCNTs), 351
mural cells, 218, 227, 230, 238, 242
MVM, *see* microvortex manipulation
μVNs, *see* microvascular networks
complex, 216
functional, 238
perfusable, 223, 238
stable, 230
vasculogenesis-derived, 222, 238

MWCNTs, *see* multiwall carbon nanotubes

nanobullets, 130–31
  high-speed, 131
nanochannel, 359–60, 363
nanochannel electroporation (NEP), 357–60, 378
nanocrystals, 434, 442, 455, 459, 467, 469
  hcp, 459
  monodisperse semiconductor, 438
  sharp-edged, 457
nanoelectroporation, 360, 377–78
  single-cell, 344
nanoelectroporation chip, 362
nanofountain probe (NFP), 359, 361
nanomaterials, 429, 431–32, 434, 436, 476, 479–80, 484, 489, 491
  carbon-based, 481
  composite, 477–78, 485
  HDL-derived, 479
  hybrid, 429
  iron oxide, 478, 489
  metal oxide, 477
  plasmonic, 488
nanoparticles (NPs), 81, 107, 125, 127, 315, 349–50, 430–37, 441–52, 454, 456, 458–67, 469, 472, 474–76, 479–81, 483, 489–91
  cross-linked hyaluronic acid, 479
  fluorescent, 478
  hollow gold, 464
  latex, 279
  lipoplex, 360
  metallic, 430, 436, 457
  monodispersed, 475
  PNIPAM@Au, 485
nanoplates, 450, 463, 468, 474
  flat, 463
  flat-surfaced, 450
  ultrathin gold, 463
nanopores, 358
  localized, 379
nanorods, 452–54, 460, 464, 467–68, 473–74, 476–77
  low-aspect-ratio, 474
nanostructures, 460, 464
  low-dimensional, 457
  periodic, 487
  undefined, 465
near infrared (NIR), 63, 78, 131–32, 368, 464, 476, 489
n-DEP, *see* negative dielectrophoresis
negative dielectrophoresis (n-DEP), 52–53, 75
negative pressure, 34–35, 353–54, 357, 359, 363
NEP, *see* nanochannel electroporation
nerve growth factor (NGF), 152, 187
nervous system, 150, 156–57, 164, 173, 194, 197
  central, 115, 128, 136, 185, 188, 232
  peripheral, 188
net force, 52–54, 56–59, 61, 99
net scattering force, 57, 61
neural stem cells (NSCs), 150–51, 172, 189–90
neuronal cells, 152, 154, 166–69, 176, 178, 190
neuronal networks, 157–58, 177–78, 188, 191
  arrayed, 159
  dissociated, 167
  space-controlled, 184
neurons, 149–50, 152–54, 156–62, 164–65, 167–68, 170–73, 175–90, 193–96, 377–78, 407

neuroscience, 149–54, 156, 158, 160–62, 164, 166, 168, 170, 172, 174, 176, 178, 180, 182, 194–97
Newtonian fluid, 98
NFP, *see* nanofountain probe
NGF, *see* nerve growth factor
NHLFs, *see* normal human lung fibroblasts
NIR, *see* near infrared
normal human lung fibroblasts (NHLFs), 221, 223–24, 235
NPs, *see* nanoparticles
  anisotropic, 435
  antigen-loaded, 318
  bimetallic, 485
  drug-loaded, 314
  embedded, 127
  encapsulated, 125
  glucose-responsive, 314
  iron oxide, 122–23, 479–80
  low-aspect-ratio, 447
  magnetite, 110, 112
  metal, 461, 475, 481, 490
  metal-based, 479
  metal oxide, 479
  nonspherical, 450
  pH-sensitive, 314
  PLGA, 318
  polydispersed, 447
  polymeric, 319
  pyramidal, 475
  spherical, 474–75
  structured, 460
  thin, 450
NSCs, *see* neural stem cells
nucleation, 407, 409–10, 431–33, 446–47, 449–50, 461, 470
  heterogeneous, 474
  rate-independent, 450

OCT, *see* optical coherence tomography
ODEP, *see* optically induced dielectrophoresis
OET, *see* organic electrochemical transistor
on-chip electrical lysis, 404, 412–15, 421–23
optical coherence tomography (OCT), 476–77
optically induced dielectrophoresis (ODEP), 76–77
optical manipulation, 50, 68, 83, 195
optical trapping (OT), 50–51, 56–61, 63, 65–67, 77, 83, 353–354, 359
optical traps, 56, 59–64
optical tweezers, 38, 50–51, 56, 63–68, 83, 353, 355
optoelectronic tweezers, 51, 67–77, 365
  negative, 68
  phototransistor-based, 71
  positive, 68
optoporation, 17, 22, 24, 344
  automated, 24
  high-throughput, 24
organic electrochemical transistor (OET), 50–51, 67, 83, 483
organ-on-a-chip, 184, 214, 242–43
Ostwald ripening, 433, 447
OT, *see* optical trapping

Parkinson's disease, 38, 155
patch-clamp technique, 177
PB, *see* PDMS barriers
PCR, *see* polymerase chain reaction
p-DEP, *see* positive dielectrophoresis
PDMS, *see* polydimethylsiloxane
PDMS barriers (PB), 218
  valve-enabled, 165
Péclet number, 227, 231
PEF, *see* pulse electric field

perfusion, 181, 232–33, 439
continuous, 216
dextran, 226
FITC-dextran, 225
heterogeneous, 239
permeability, 100, 227, 233–34, 237, 279, 281–83, 302, 406, 419, 423, 455
electrical, 405
endothelial, 232, 237
improved, 277
increasing membrane, 379
low, 232
lower solute, 234
paracellular, 163
selective, 17
vascular, 226
vessel, 232
vessel wall, 238
permeation, 280, 294, 300, 304, 321
enhanced, 324
increased, 303
percutaneous, 301
PFF, *see* pinching flow fractionation
photolithography, 3, 68, 130, 258, 443
photothermal therapy, 132, 139, 313, 487–89
pinching flow fractionation (PFF), 505
plasmid DNA, 21, 316, 380
plasmon resonance, 432
localized surface, 23
longitudinal, 454
longitudinal surface, 452
transverse surface, 464
POC, *see* point-of-care
point-of-care (POC), 38–39, 404, 476, 481, 490, 504, 524
Poisson ratio, 513
polarity, 52, 107, 152, 172–73, 197
polydimethylsiloxane (PDMS) ,25–26, 166–67, 181–82, 217, 219–20, 229–30, 242–43, 290, 353–54, 357–60, 373, 413, 463, 465–66, 514–16
polymerase chain reaction (PCR), 30–31, 36–38
quantitative, 31, 514
polymeric microneedles, 277, 287–88, 321
polymer microneedles, 287, 296, 316
hollow, 286
swellable, 318
polymers, 116, 122, 286–87, 289–91, 293–98, 309, 316, 413, 432, 438, 447, 463, 474, 478–79
biodegradable, 291, 316
deposited, 118
dissolvable, 288–89, 316
hyaluronic acid, 478
melted, 289
natural, 162
stiff, 130
superswelling, 294
synthetic, 138, 292
water-soluble, 283, 288
polystyrene (PS), 10, 12–13, 72, 111, 118, 122–23, 243, 280–81, 291, 297, 503–4, 512–13, 516–18, 520–21, 524
polyvinylpyrrolidone (PVP), 290, 297–98, 303, 313, 317, 448–51, 457, 465, 469, 471–73, 475, 484
pores, 19, 23–24, 118, 120, 228, 232, 234, 239, 279, 281, 283, 287, 353, 364, 405–12, 418
circular, 285
hydrophobic, 17, 345
intermediate, 407
micron-size, 281, 287
permeable, 344
unplugged, 279

positive dielectrophoresis (p-DEP), 52, 75
positron emission tomography, 376, 476
propulsion, 96–115, 119, 125–26, 128–29, 133, 137–38
  biological, 97, 100, 109
  bubble, 97, 100, 105–6, 137
  bubble-free directional, 138
  fuel-free, 99
  helical, 100–101, 104
  helical wave, 103
  high, 134
  high-speed, 108
  planar, 101
  planar wave, 100–101, 104
  self-phoretic, 114
  self-thermophoretic, 124
  water-powered, 136
PS, *see* polystyrene
pulse electric field (PEF), 412–14, 417, 420–21
PVP, *see* polyvinylpyrrolidone

qRT-PCR, *see* quantitative real-time polymerase chain reaction
quantitative real-time polymerase chain reaction (qRT-PCR), 37
quantum dots, 359, 479–80

radiation pressure, 77, 353, 368
  direct, 50
  direct optical, 83
radio frequency (RF), 353–54
Raleigh scattering, 57
Raman scattering, 484
Raman signals, 484–85
Raman signature, 484
rapid electrokinetic patterning (REP), 50–51, 67, 77–79, 81–83
Rayleigh instability, 122
RBC-Mg micro-robot, 136
RBCs, *see* red blood cells
reactive ion etching, 120, 279, 285, 306
red blood cells (RBCs), 65, 73, 76, 125, 136, 413–14, 419
reducing agents, 430, 447, 449–50, 457, 461, 470–71, 474–76, 491
reduction, 132, 188, 418, 435, 446, 449–50, 452, 454, 470, 485, 521
REP, *see* rapid electrokinetic patterning
resonance, 420, 454, 488
  cellular spin, 420
  optical, 467
  plasmonic, 489
reversible electroporation, 351, 371
Reynolds number, 11, 97–98, 103, 439
RF, *see* radio frequency
RNA, 2, 22, 35–36, 67, 154, 256, 272, 376
robots, 96–98, 105
  autonomous, 126
  conventional tethered, 97
  millimeter-sized, 102
  miniature, 96
  miniaturized, 96, 139
  small-scale, 96
  spiral, 103

scaffolds, 162, 169
  cellular, 162
  layered hydrogel, 187
scanning electron microscopy (SEM), 362, 446
scattering, 58–59, 484
  effective, 56
  isotropic, 56

scattering forces, 56–59, 62
SCEP, *see* single-cell electroporation
high-frequency continuous-wave, 369
micro-/nanochannel-based, 363
reversible, 369
SCTs, *see* single-cell technologies
sedimentation, 449, 503, 507, 519, 521–22, 524
gravity-induced particle, 81
rapid, 507, 522
segmented flow, 438–40, 455–56, 472, 474, 491
continuous, 472
slug-based, 441
self-diffusiophoresis, 114
self-electrophoresis, 114
self-thermophoresis, 97, 100, 113–14
SEM, *see* scanning electron microscopy
separation, 2, 18, 30, 75, 178, 267, 278, 293, 484, 503–8, 510, 512–14, 517–19, 521–24
hydrodynamic, 503, 505, 515, 522
hydrodynamical, 506
micro-object, 521
multiparticle, 504
nucleic acid, 514
positive, 524
size-based, 503
SERS, *see* surface-enhanced Raman spectroscopy
real-time, 485
shear forces, 4, 21–22, 219, 439
shear stress, 149, 165, 175, 196, 216, 219, 226–27, 231–32, 236, 238–39
single-cell analysis, 6, 28, 161, 404, 422–23, 431
microfluidic-based, 1
single-cell electroporation (SCEP), 345, 348–49, 351–53, 357–59, 363, 365, 367–71, 376–77, 381
parallelized, 361
precise, 343
single-cell micro-/nanoelectroporation, 19, 344
single-cell technologies (SCTs), 2
single-cell trapping, 3–4, 6, 9, 11, 15, 33, 196, 351
high-throughput, 9
siRNAs, 349, 378
skin permeability, 279, 301, 304, 309, 311, 319
SMCs, *see* smooth muscle cells
smooth muscle cells (SMCs), 217, 222, 230, 235
soft lithography, 3, 515, 524
regular, 515
template-based, 232
solid microneedles, 278–79, 283–84, 287, 296, 303–4, 307, 314, 320
solid-phase extraction (SPE), 514–15
SOP, *see* standard operating procedure
SPE, *see* solid-phase extraction
sprouting, 222, 224–25, 227–28, 230–31, 235, 237
endothelial, 227–28
irregular, 235
vessel, 218
sprouts, 223–24, 226–28
standard operating procedure (SOP), 380
Stokes equations, 98, 509, 512
Stokes law, 98
Stokes number, 519
strain, 111–12, 116–17, 119, 232, 234, 255
mechanical, 158, 233
physical, 188

SU-8, 115, 286–87, 297, 300, 365, 413, 443, 515
surface-enhanced Raman spectroscopy (SERS), 434, 472, 477, 484–85, 487–88
surface plasmon resonance, 432
swellable microneedles, 293–94
swelling, 36, 118, 130, 292–94, 322
swelling/deswelling, 130
synapses, 152, 157, 166, 169, 180–81, 183
  corticostriatal, 184
  functional, 181
  transcription-mediated, 183

tails, 103–5, 261–62, 265, 405
  flexible, 105, 122–23, 137
  nonpolar lipid, 403
  trilayer, 120
targeted drug delivery (TDD), 95, 128–29, 139, 305–6, 319
TDD, *see* targeted drug delivery
TEC, *see* thermal expansion coefficient
TEER, *see* transendothelial electrical resistance
TEM, *see* transmission electron microscopy
therapeutics, 1–2, 4, 6, 8, 10, 12, 14, 16, 18, 20, 22, 24, 26, 28, 139
  cellular, 2
  protein, 277, 312
  temperature-susceptible, 317
thermal expansion coefficient (TEC), 117
T-junction, 4–5, 30, 441–44, 491
TME, *see* tumor microenvironment
TMP, *see* transmembrane potential
toxicity, 18, 39, 343–44, 380, 429, 476, 479
  low, 349
transdermal drug delivery, 275–76, 293–94, 301, 309, 322, 324
transendothelial electrical resistance (TEER), 233–34
transfection, 18–19, 21–22, 24, 195, 343–45, 348–49, 358–60, 363, 381
  cellular, 17–18
  gene, 413
  nanolocalized, 345
  non-uniform, 343
  plasmid, 360
  safe, 363
  single-cell, 348
  uniform, 365
transmembrane potential (TMP), 345, 347, 371
transmission electron microscopy (TEM), 446
transverse surface plasmon resonance (TSPR), 464
trapping, 3–4, 6, 8–10, 12–15, 33–36, 57–58, 60, 63–64, 73, 78–79, 81, 158, 161, 353–54, 357–59
  direct, 158
  hydrodynamic, 9
  optoelectrokinetic, 81
  optoelectronic, 50
  single object, 60
  whole-cell, 57
traps, 9, 11, 15, 60–62, 65–66, 68, 158, 161, 168, 353, 355, 357, 359, 363, 373–75
  consecutive, 375
  extracellular, 33
  fiber-optic, 62
  hydrodynamic, 4
  mechanical, 353, 369
  non-Gaussian laser beam, 60
  passive hydrodynamic cell, 372
  p-optoelectronic tweezer, 77
  stable, 64
TSPR, *see* transverse surface plasmon resonance

tubular micro-robots, 106, 108, 115–20, 133–34
bubble-propelled, 116, 134
tumor angiogenesis, 227, 229, 235–37, 241
tumor cells, 28, 132, 134, 236–40, 423
circulating, 25, 76, 134, 481, 504
intravasated, 236
rare, 509
tumor intravasation, 236–38
tumor microenvironment (TME), 238–39, 242
two-photon polymerization, 286
laser-based direct, 286

ultrasound, 95, 108–9, 376, 476
ultraviolet (UV), 16, 109, 111, 288, 456, 484
ultraviolet-visible (UV-vis), 446, 453–54, 457, 466, 485
UV, *see* ultraviolet
UV-vis, *see* ultraviolet-visible

vaccination, 318, 321
vaccines, 276, 279, 286–87, 299, 302, 316–18, 321, 323–24
cellular, 22
transdermal, 321
vascular endothelial growth factor (VEGF), 221–23, 227–29, 237
vascularization, 214–15, 223, 231, 234, 241–42
vascularized microtissues, 215–16, 221
functional, 242
pathological, 213
vascularized tumor models, 239–43
microfluidics-based, 241
vascular networks, 164, 188, 190, 194, 219–23, 225, 227
vasculature, 216, 226, 239–40, 242
angiogenic, 235
dysfunctional, 237
vasculogenesis, 216, 221–25, 227, 229, 231, 241
VEGF, *see* vascular endothelial growth factor
virtual electrodes, 68–70, 350
viscosity, 53, 77, 122, 261, 298, 456, 506
dynamic, 98–99

wafer, 120, 256–59, 264–65, 267, 269
one-piece multifunction, 267
silicone, 515
silicon-on-insulator, 371
waveguide, 368–69
WBCs, *see* white blood cells
WEC, *see* wireless endoscope capsule
pill-sized, 136
white blood cells (WBCs), 73–74, 419, 505, 514
wireless endoscope capsule (WEC), 101–2, 134, 136

X-ray, 286, 288
X-ray powder diffraction, 460

Y-channel, 444
Y-channel microchip, 445
Y-junction, 441, 491
Young's modulus, 513
Y-shaped microchannel, 444–45
Y-shaped mixer, 465

Zyomyx's microarray protein wafers, 258

For Product Safety Concerns and Information please contact our EU representative GPSR@taylorandfrancis.com Taylor & Francis Verlag GmbH, Kaufingerstraße 24, 80331 München, Germany

Batch number: 10726507

Printed by Printforce, the Netherlands